D0142183

Physiology of the Bacterial Cell

PHYSIOLOGY
OF THE
BACTERIAL CELL
A Molecular Approach

Frederick C. Neidhardt
UNIVERSITY OF MICHIGAN
ANN ARBOR

John L. Ingraham
UNIVERSITY OF CALIFORNIA
DAVIS

Moselio Schaechter
TUFTS UNIVERSITY, BOSTON

Sinauer Associates, Inc. • Publishers
Sunderland, Massachusetts

THE COVER

Electron micrograph of negatively stained *Pseudomonas aeruginosa* (×19,000). Photograph by J. J. Cardamone, Jr., University of Pittsburgh School of Medicine/Biological Photo Service.

PHYSIOLOGY OF THE BACTERIAL CELL: A MOLECULAR APPROACH

 For information address Sinauer Associates, Inc., Sunderland, Massachusetts, 01375, U.S.A.

Library of Congress Cataloging-in-Publication Data

Neidhardt, Frederick C. (Frederick Carl), 1931–
Physiology of the bacterial cell : a molecular approach / Frederick C. Neidhardt, John L. Ingraham, Moselio Schaechter.
p. cm.
Includes bibliographical references (p.).
ISBN 0-87893-608-4 (alk. paper)
1. Bacteria—Physiology. 2. Escherichia coli—Physiology. I. Ingraham, John L. II. Schaechter, Moselio. III. Title.
QR84.N39 1990
589.9'01—dc20 90-9446
CIP

Printed on paper that meets the guidelines for permanence and durability of the Committee on Production Guidelines for Book Longevity of the Council on Library Resources.

Printed in U.S.A.

6 5 4 3 2 1

We wish to pay tribute to
Ole Maaløe,
Jacques Monod,
and
Roger Y. Stanier
for their unique contributions
to the study of the bacterial cell

Contents

4. Assembly and Polymerization: The Bacterial Envelope 102

5. Biosynthesis and Fueling 133

6. Quest for Food 174

7. Growth of Cells and Populations 197

8. The Effects of Temperature, Pressure, and pH 226

9. Genetic Adaptation: The Genome and Its Plasticity 247

10. Genetic Adaptation: Genetic Exchange and Recombination 275

11. Coordination of Metabolic Reactions 302

12. Regulation of Gene Expression: Individual Operons 320

13. Regulation of Gene Expression: Multigene Systems and Global Regulation 351

14. Cell Cycle 389

15. Growth Rate as a Variable 418

16. Cellular Differentiation 442

17. Physiological Ecology 463

Preface

Bacteria teach us life's potential at the single-cell level. Thanks to molecular devices of consummate elegance and precision, these minute organisms display an intriguing set of properties, including

- the highest rates of metabolism and reproduction known among living things
- an adaptability that permits them as a group and as individual cells to grow under a wide variety of conditions, including some that are marginal for the very existence of biological molecules
- an ability to cope with a feast-or-famine existence and to withstand a multitude of damaging agents
- an ability to colonize surfaces buffeted by liquid currents and protected by the elaborate antibacterial measures of higher organisms

This book is designed to introduce college and graduate students to these properties of bacteria and how they are studied. Our thesis is that bacterial physiology is best presented as a coherent story in which the leitmotifs are as important as, or more important than, the individual characters and incidents. This book, then, is far from exhaustive; it does not pretend to deal with each molecular device, or even with each major lifestyle encountered in the bacterial world. We have focused on the central ideas of bacterial physiology and on how those ideas have come about. Our aim has been to provide a meaningful overview for the one-time visitor to the microbial world and a framework for advanced inquiry by students committed to a deeper study of bacteria. For this reason facts about structure and function have been presented in the context of their biological significance for the bacterial lifestyle.

It has been tempting to parade the triumphs of bacterial studies, because so much of what we understand about the molecular biology of all cells derives from work on these organisms. We have worked toward a balance between presenting the wealth of information that we know and identifying the major gaps in our understanding of cell growth and survival.

Some readers will recognize the close kinship of this book to a previous one, *Growth of the Bacterial Cell* (Ingraham et al., 1983). Early in 1988 we began to prepare a second edition of that work, but as we went along the product resembled less and less a revision of the original book. Our eventual decision to produce a new book freed us from any constraints we might have felt to adhere strictly to the format and philosophy of the earlier work.

Two major conceptual differences separate the earlier book from our current work. First, growth as the core of bacterial physiology has been supplemented in this work by emphasis on processes of survival and colonization. Second, focus on *Escherichia coli* and *Salmonella typhimurium* as the paradigms of free-living cells has been replaced by a balanced presentation of the major topics of bacterial physiology throughout the microbial world. No text on the molecular aspects of bacterial physiology can avoid heavy treatment of the enteric bacteria—that emphasis comes about simply from the realities of modern bacterial research. But presentation of the molecular physiology of enteric bacteria has been placed in the perspective of the rich opportunities for exploration that lie in the broader world of bacterial physiology.

Fans of *Growth of the Bacterial Cell* have urged the retention of certain of its features in the present text: the original definitions of metabolic processes (fueling, biosynthetic, polymerizing, and assembly pathways), the tracing of pathways of metabolism in the order of assembly through fueling (rather than the more customary reverse order), the cost accounting of growth, the emphasis on global control systems, and the treatment of regulation at the whole-cell level. This we have done.

Which bring us to a final comment. The coauthor of *Growth of the Bacterial Cell,* our friend and colleague Ole Maaløe, died in the spring of 1988, shortly after work on the present text commenced. We three have each experienced the pleasure of a scientific sojourn in Copenhagen as a guest of Ole—and we have each enjoyed the scientific and personal companionship of this unusual person. His unique contribution to bacterial physiology has been chronicled in a recent Festschrift (Schaechter et al., 1985). To encapsulate his contributions to microbial physiology: Ole was the supreme synthesizer and integrator of information about the working of the bacterial cell. His talent at whole-cell analysis and interpretation sharpened the thinking of all post-World War II microbial physiologists.

Jacques Monod is unsurpassed for his penetrating analysis of bacterial growth and his brilliant conception of the operon. Roger Stanier was a wizard in the mastery of the whole sweep of microbial metabolism and physiology. Thus, each of the three individuals to whom we dedicate this book shared a love of the bacterial cell and brought to its study a unique and indelible perspective. Twentieth century bacterial physiology can in large measure be epitomized by their contributions.

ACKNOWLEDGMENTS

The authors were aided in their work by careful and insightful study of the manuscript by Leo W. Parks, William R. Sistrom, and Linc Sonenshein. These individuals corrected errors, offered suggestions, and provided encouragement at a critical stage in the creative process.

We wish to acknowledge our continued debt to H. Edwin Umbarger, who inspired much of our treatment of metabolism, and to Charles Yanofsky for his patient and critical assistance with the *trp* operon, both of which subjects are carried over from the precursor to the current text. Donald Nierlich provided a perceptive and detailed review of the precursor volume; his suggestions have been valuable.

The copy editor, Jodi Simpson, challenged our science as well as our English. The artist, John Woolsey, was a creative and patient executor of our ideas. Carol Wigg did a masterly job of shepherding the job from floppy disc to printed page. Andy Sinauer provided encouragement for the project and valuable guidance in countless matters.

F. C. N., J. L. I., and M. S.
January, 1990

Ingraham, J. L., O. Maaløe and F. C. Neidhardt. 1983. *Growth of the Bacterial Cell.* Sinauer Associates, Inc., Sunderland, Massachusetts.

Schaechter, M., F. C. Neidhardt, N. O. Kjeldgaard and J. L. Ingraham. 1985. *Molecular Biology of the Bacterial Cell,* Jones and Bartlett, Boston.

Physiology of the Bacterial Cell

1

Composition and Organization of the Bacterial Cell

INTRODUCTION

One of the many ways to observe the growth of bacteria is to spread some on the surface of a nutrient agar block mounted on the stage of a light microscope. Individual cells can be seen to increase their volume and mass by invisible means until a division produces two offspring, each of which proceeds directly to repeat the cycle.

In a pure culture the cells produced in this manner are uniformly able to divide, and do so in a reproducible and constant time. Direct microscopy shows that if a large number of cells from a liquid culture are spread on nutrient agar, very few fail to divide, and microcolonies of remarkably uniform size develop. In fact, for hundreds of testable characteristics, the cells (except for rare mutants) are homogeneous. Thus, bacterial growth and division must be organized in a manner that ensures genetic continuity and phenotypic homogeneity; and the overall process, in spite of the short time required (20–30 minutes for some bacteria), must be a display of the fundamental property of living systems—self-replication.

Bacteria are ideal subjects with which to study self-replication—or cell growth, as we shall call it—because as a group they emerge from over two billion years of evolution as masters of rapid growth. Their existence today is certainly related to their ability to grow rapidly when conditions permit and to survive when conditions prevent growth.

Rapid growth alone is not sufficient to explain the remarkable success of bacteria in colonizing this planet. As small, free-living organisms, bacteria face major challenges that stem from their limited ability to exert any control over their environment. Lack of control over their environment is, in fact, part of the price paid by bacteria for their smallness and their streamlining for rapid growth. Multicellular forms of life have evolved specialized structures that create for their cells a remarkably constant internal environment, and they have generated the myriad survival mechanisms we associate with plants, animals, and the protists (protozoa, algae, and fungi). In many organisms, these mechanisms are so highly developed that the generation time must be measured in years or decades rather than in minutes. Bacteria, on the other hand, have learned to cope unicellularly.

The problems created by this choice are enormous. Consider the soil as a habitat. Near the surface, there are abrupt fluctuations in the several physical and chemical characteristics that matter to living cells. The temperature, the amount of moisture, the availability and nature of organic nutrients and of inorganic ions needed for growth, the pH, the oxygen tension, and the presence of toxic chemicals and of damaging solar radiation all are subject to change. Yet hundreds, if not thousands of species of bacteria successfully occupy this inhospitable niche. Even those bacteria that are found chiefly in seemingly more constant environments, such as those that live in or on large animals, experience special challenges to survival and growth. To inhabit the mouth successfully, for example, a bacterium must avoid being washed away by saliva or by food and liquids ingested by the animal. Many species of bacteria solve such problems by elaborating sticky polysaccharides or even highly specialized adhesive proteins to enable them to colonize surfaces.

To account fully for the success of bacteria, we must understand how they

- survive a feast-or-famine existence
- occupy areas that are subject to liquid currents
- cope with damaging agents

In this book we discuss how bacteria can be used to study growth, and we assess where we are in understanding this process. In doing so, we are forced to consider how the legendary adaptability of bacteria is achieved.

We begin by asking what a bacterium is made of. The answer includes many details about molecules and structures and how they are organized. These details may appear overwhelming at first sight, but they are necessary to our story, and they can in fact be rather easily managed because of their direct relevance to the central process of growth. We omit many biochemical details that can be found in any of the several excellent texts (Stanier et al., 1986; Gottschalk, 1986) dealing with microbial chemistry; we choose rather to emphasize how these details relate to growth.

CHEMICAL COMPOSITION OF BACTERIAL PROTOPLASM

By centrifugation or filtration, one can collect the bacterial cells from a pure culture growing in liquid medium. Assume that filtration of a culture has yielded 100 mg of packed bacterial mass. This biomass can be analyzed by any of a number of methods to learn its chemical composition. An aliquot can be dried to determine its water content, and a portion of the dried material can be used to ascertain its elemental composition as though it were a single substance. Typically one finds that approximately 70% of the packed cell mass is water—significantly less than the 90% found in cells of higher organisms.

Elemental assay of the dry mass of *Escherichia coli* cells reveals a fairly typical composition of protoplasm (all values are approximate): 50% carbon, 20% oxygen, 14% nitrogen, 8% hydrogen, 3% phosphorus, 2% potassium, 1% sulfur, 0.05% each of calcium, magnesium, and chlorine, 0.2% iron, and a total of 0.3% trace elements including manganese, cobalt, copper, zinc, and molybdenum. From these data, one can anticipate that most of the bacterial mass consists of organic compounds—a correct guess—but otherwise the elemental analysis is too unsophisticated for our purpose, and we must turn to the analytical techniques of organic chemistry.

The second column of Table 1 shows the results of organic analysis of the dry mass of cells harvested by rapid filtration from a bacterial culture of a given strain in STEADY-STATE GROWTH (i.e., at an unchanging rate and with constant average cell size and composition) under specified conditions of culture medium and temperature.

Right at the start it should be noted that biochemical data are meaningful only if attention has been given to specify (1) the organism, (2) the growth environment, and (3) the state of growth. These parameters have a profound effect on biochemical results but often are not adequately documented in the reports of experiments.

The organism / Although all bacteria, as PROCARYOTES, share a basic cellular organization (to be examined later in this chapter), the more than 5,000 species of bacteria are distinguished by many individual characteristics, including chemical composition. An average composition could perhaps be roughly estimated (e.g., 50% protein, 25% RNA, 3% DNA, and 22% other constituents), but this would be too imprecise for our purposes and would, in fact, not necessarily be the composition of any real cell. We must specify our cell.

Specification is accomplished by naming the genus, species, and strain; for example, *Escherichia coli* strain B. Some clarification of these terms is in order. The concept of species is rooted in the sexual reproduction by which most organisms other than bacteria increase their numbers. A population of sexually reproducing organisms that are free to interbreed will share a gene pool and assort new mutations among its members. If one portion of the population becomes isolated and can no longer interbreed with the other, the two subpopulations are likely to evolve along different lines, a divergence leading eventually to their physiological incapacity to interbreed. At that point, speciation is said to have occurred, and the decision whether any two subpopulations are the same or are different species is made on the basis of this criterion. Of course, many other characteristics may distinguish one group from the other.

Table 1. Overall macromolecular composition of an average *E. coli* B/r cell[a]

Macromolecule	Percentage of total dry weight	Weight per cell (10^{15} × weight, grams)		Molecular weight	Number of molecules per cell	Different kinds of molecules
Protein	55.0	155.0		4.0×10^4	2,360,000	1,050
RNA	20.5	59.0				
23S rRNA			31.0	1.0×10^6	18,700	1
16S rRNA			16.0	5.0×10^5	18,700	1
5S rRNA			1.0	3.9×10^4	18,700	1
transfer			8.6	2.5×10^4	205,000	60
messenger			2.4	1.0×10^6	1,380	400
DNA	3.1	9.0		2.5×10^9	2.13	1
Lipid	9.1	26.0		705	22,000,000	4[b]
Lipopolysaccharide	3.4	10.0		4346	1,200,000	1
Murein	2.5	7.0		$(904)_n$	1	1
Glycogen	2.5	7.0		1.0×10^6	4,360	1
Total macromolecules	96.1	273.0				
Soluble pool	2.9	8.0				
building blocks			7.0			
metabolites, vitamins			1.0			
Inorganic ions	1.0	3.0				
Total dry weight	100.0	284.0				
Total dry weight/cell		2.8×10^{-13}g				
Water (at 70% of cell)		6.7×10^{-13}g				
Total weight of one cell		9.5×10^{-13}g				

[a]In balanced growth at 37°C in glucose minimal medium, mass doubling time, *g*, of 40 minutes. The data are assembled from Dennis and Bremer (1974), Maaløe (1979), F. C. Neidhardt (unpublished), Roberts et al. (1955), and Umbarger (1977).
[b]There are four classes of phospholipids, each of which exists in many varieties as a result of variable fatty acyl residues.

For bacteria, which are haploid and reproduce only asexually, the very concept of species seems inapplicable. Furthermore, their short generation time and the ease with which they can be found to evolve genetically in the laboratory does not lead one, a priori, to expect that any large, random population of bacteria coexisting in the same ecological niche would consist of identifiable clusters of closely similar biotypes. But, such indeed is the case. Bacteria can be ordered in taxonomic units, unfortunately called SPECIES, each of which consists of a large number of independently isolated STRAINS—also called ISOLATES—that share the vast majority of their observable characteristics. Furthermore, species can be arranged according to their degree of similarity (and presumed relatedness) into GENERA (singular: GENUS). Similar genera are clustered into FAMILIES, and for some kinds of bacteria, families into ORDERS (Holt and Krieg, 1984, 1986). The name "*Escherichia coli*," therefore, refers to a group of bacteria that are independently obtained from nature as separate strains, or isolates, and that share a large number of qualitative and quantitative characteristics (genus: *Escherichia*; species: *coli*). Not surprisingly, these characteristics can be readily determined in the laboratory by an experienced microbiologist.

One strain of *E. coli* is called K-12; others are called B, ML30, C—and so on. By appropriate biochemical tests, these strains can easily be recognized as *E. coli* and not strains of the related species *Escherichia freundii*. Strains (isolates) of the same species usually have 80%, or greater genetic homology. Of course, they can still differ in very many genes. Unfortunately, the term strain is also used to designate mutant variants of a give isolate. A histidine-requiring derivative of *E. coli* strain K-12 is usually given a "strain number." No term has come into common use to designate the tens of thousands of genetic variants of *E. coli* strain K-12 other than the word strain itself. The resulting ambiguity (strain for isolate and strain for variant) has fostered an unfortunate sloppiness in reporting genealogies of variants. At one point considerable effort had to be expended to clarify the confusing state of *E. coli* strain K-12 genealogies (Bachmann, 1972).

For many laboratory strains (original sense) of *E. coli*, long cultivation and storage under different conditions in different laboratories has led to the inadvertant selection of variants that have different growth rates and different chemical composition in a given medium. For these strains it is necessary to identify their laboratory source. This problem is greatest for *E. coli* strain K-12, less so for strain B and for bacteria of other genera and species. The problem of strain variation among laboratories can be virtually eliminated by the modern practices of lyophilization or low temperature (−80°C) storage of strains.

The medium / Bacteria have control devices that bring about adaptive adjustment of the levels of their major constituents in response to the chemical and physical environment. These adjustments are an important aspect of the cell's mastery of rapid growth and survival in nature. Bacteria have little ability to modify their environment to advantage, so they rely mostly on internal adjustment of their own growth machinery. As we shall see in Chapter 15, many aspects of cell composition vary with growth rate, including the total content of ribonucleic acid. In addition, at the same growth rate in different media, an organism will produce different sets of proteins and other specific components. Therefore, meaningful and reproducible chemical measurements on bacteria must involve careful definition of the growth conditions.

A simple but effective check to help assure that one has used the desired strain of bacterium and has constructed the desired medium in any experiment is to measure the (exponential) growth rate of the cells (Chapter 7). This parameter should be invariant from experiment to experiment and from laboratory to laboratory. Its value should always be included in the other data reported on experiments in bacterial physiology.

The state of growth / When bacterial cells of a single strain have been growing in a medium for sufficient time to complete their chemical adjustments to that growth condition, the exponential increase in population mass occurs at a constant and reproducible rate characteristic of the strain and growth condition. For a time—until the medium is changed as a result of the growth of the bacteria—a steady-state situation exists, one in which cell growth is said to be BALANCED because every cellular component increases by the same constant factor per unit time. It is this growth state that in practice is the *only* reproducible state of these cells, and every effort must be made to achieve it before commencing serious measurements. Why balanced growth is important and how it is achieved in practice is dealt with in detail in Chapter 7.

Now, with these three elements (strain, medium, state of growth) specified, we can examine the data of Table 1, column 2, that pertain to our overall analysis of bacterial growth:

- macromolecules constitute the preponderance (96%) of the cellular dry mass
- proteins constitute over half of this amount
- there is a higher content of RNA in bacterial cells than there is compared to that in most cells of higher organisms
- there are many familiar substances found in all organisms, but two major ones are unique to bacteria: murein and lipopolysaccharide (about which we have more to say later in this chapter)

DYNAMIC ASPECTS OF PROTOPLASMIC COMPONENTS DURING GROWTH

A simple experiment can tell us much about the dynamic relationships of these substances in growing cells. Adding glucose uniformly labeled with ^{14}C to a culture growing with this sugar as its sole organic source of carbon enables one to follow the flow of this element through the various classes of molecules. This experiment is done by sampling at intervals after adding labeled glucose to a culture already in a steady state of growth. The samples may then be treated with 5% trichloroacetic acid, a treatment separating molecules that are acid soluble from those that are acid insoluble—that is, roughly separating low-molecular-weight substances from macromolecules. The insoluble material can then be fractionated into general classes, such as proteins, nucleic acids, and fat-soluble molecules, or further into subclasses of these. The fractions can then be assayed for radioactivity.

Eventually every organic molecule in the cell will approach a specific level of ^{14}C activity equal to that in the glucose substrate, but the time course of labeling will be governed by the metabolic characteristics of each molecule. As a class, the small quantity of soluble, low-molecular-weight molecules attain values near their maximal specific activity within a small fraction of a generation time—a few minutes. This behavior, shown idealized as curve A in Figure 1, reveals that they are

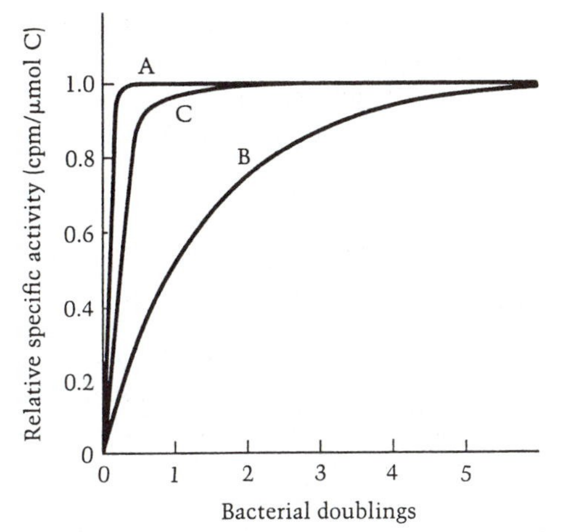

Figure 1

Idealized time course of labeling of organic molecules following addition of [^{14}C]glucose to a culture of bacteria in steady-state growth in glucose minimal medium. Curves A, B, and C are idealized data for different classes of macromolecules described in the text. cpm, counts per minute.

formed and consumed rapidly during growth; that is, there is a rapid flow through the pool of these substances in the cell. This behavior is typical of molecules that include the building blocks destined to be polymerized into macromolecules, metabolic intermediates on their way to becoming such building blocks, and molecules that will be excreted as waste products. Macromolecules, on the other hand, as a class become labeled *pari passu* with growth: after one doubling of bacterial mass, their specific radioactivity is roughly half that of the glucose substrate; after two doublings (a fourfold increase in mass), it is 75% that of the glucose; and so on. This labeling pattern, idealized as curve B in Figure 1, is characteristic of substances that are stable components of the cell. They include most proteins, DNA, some RNA species, and other large molecules.

Some compounds acquire label with a time course that is apparently a composite of curves A and B. Such behavior is usually seen for molecular species that label quickly because they are present in small quantities and are converted into other, stable products but also are formed by degradation of macromolecules. Pools of ribonucleotides exhibit this behavior as a result of mRNA degradation. Curve C in Figure 1 represents one of the resulting complex labeling patterns.

Of special interest are certain large molecules, such as messenger RNA (mRNA), that become labeled by time course C—a labeling pattern indicating metabolic lability—and certain small molecules, such as the coenzyme NADP, that become labeled by time course B—a labeling pattern indicating metabolic stability, which is a property of molecules with catalytic functions in cell growth.

It is clear that this simple experiment can reveal much about the general role of compounds in cellular growth. The complexity of labeling of the nucleotide pool, however, serves as a cautionary note to avoid both simple expectations and unwarranted conclusions from such experiments. Later, in Chapter 7, we return to this point when we discuss precise measurements of rates of synthesis of cellular components by radioactive incorporation.

SIZE AND CHEMICAL COMPOSITION OF AN AVERAGE BACTERIAL CELL

We have more to discuss concerning the chemical makeup of bacteria, and it is now appropriate to refine our analysis by considering, not a 100-mg lump of bacterial mass, but individual cells, the biological units that constitute the mass.

How many cells are in our sample? Several techniques can tell us. We can suspend a measured portion in a suitable solution and count the particles with an electronic particle counter (or count them more laboriously one by one under a microscope using a chamber of measured volume). Bacteriologists frequently use a simpler method that relies on the fact that normally greater than 99% of the cells taken from growing cultures are viable, that is, can grow and divide to form a clone of descendants. Spreading a measured volume of a suitably diluted suspension of cells over an agar plate results, after an incubation time of 10–18 hours (longer for the more slow-growing species), in macroscopic colonies that are easy to count. (A colony of 10^7 cells is easily seen with the unaided eye.) With the use of appropriate duplicate and control samples, an adequate tally can be obtained.

In the example we have been using (a culture of *E. coli* strain B/r grown in glucose minimal medium at 37°C), we would find that 100 mg (wet weight) of bacterial protoplasm consists, astonishingly, of roughly 100 billion cells (10^{11} cells). The precise value obtained as an average of many measurements of this sort is 1.05×10^{12} cells per gram of biomass in such a culture. The average cell in this population must, therefore, have a wet mass of 9.5×10^{-13} g, or 2.9×10^{-13} g dry mass.

At this point we encounter one of the central facts about bacteria—their extraordinary smallness. Although the variation of size among different kinds of bacteria is rather large (the smallest are almost the size of the largest viruses, the largest surpass the smallest cells of higher organisms), for the most part they are the primary occupiers of the 1-μm domain. The average *E. coli* cell, for example, is a rod-shaped body with diameter and length of approximately 0.5 and 2 micrometers (μm), respectively. A major consequence of small size is a very large surface: volume ratio, which makes possible the rapid transfer of nutrients and waste materials between the interior of the cell and its environment, and, consequently, a high metabolic rate and rapid growth.

What is the meaning of the term *average cell*, in the sense we have been using it? Is it a cell midway in age and size between those in the population just formed by cell division and those just about to divide? Consider that upon division (in *E. coli* and most other bacteria, this occurs by binary fission) one old cell becomes two young ones. In a random population proceeding regularly through the cell cycle, therefore, there are twice as many just-born cells as there are cells ready to give birth. The idealized frequency distribution of cell age is therefore given by

$$f(x) = 2^{1-x}$$

where x is the age of cells measured on the scale from 0 to 1 cell generation time (Powell, 1956). This relationship is illustrated in Figure 2. The cell described by dividing the total biomass, or any measured component of our randomized, steady-state population, by the total number of cells in that population is one that is approximately 44% along the cell cycle in age and, if individual cells increase in mass exponentially, is approximately 33% larger than when it was born. It is the average cell presented in Table 1.

The amount of each chemical component of the average cell presented in Table 1 has been carefully determined by both direct and indirect measurements. These values are used throughout our analysis of growth. They enable us to calculate how much of each component must be made for the cell to reproduce itself and what this biosynthesis costs (Chapters 3, 4, and 5). They provide a framework for describing the apparatus of macromolecular synthesis (Chapter 4) and how it must be regulated (Chapters 11, 12, 13, and 15). And they provide us with an inventory of molecules to be assigned to the various structures and compartments of the cell (Chapter 2). For these purposes, it will be necessary to complete two tasks: we must express the amount of each chemical class of molecule in terms of number of molecules per cell, and we must then ascertain how many different molecular species are represented in each of the major chemical classes.

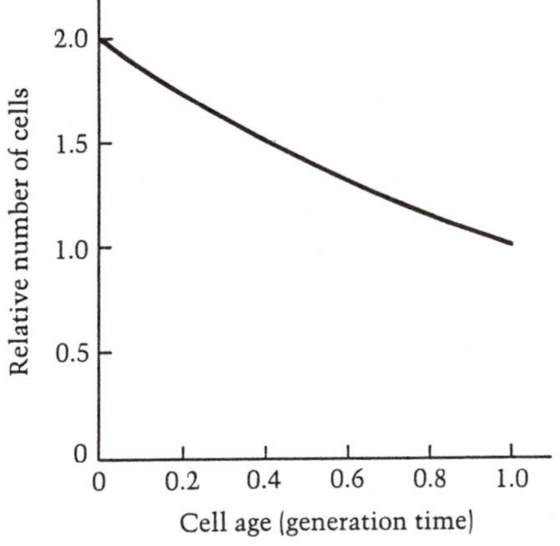

Figure 2

Idealized frequency distribution of cell age in a steady-state culture increasing by binary fission.

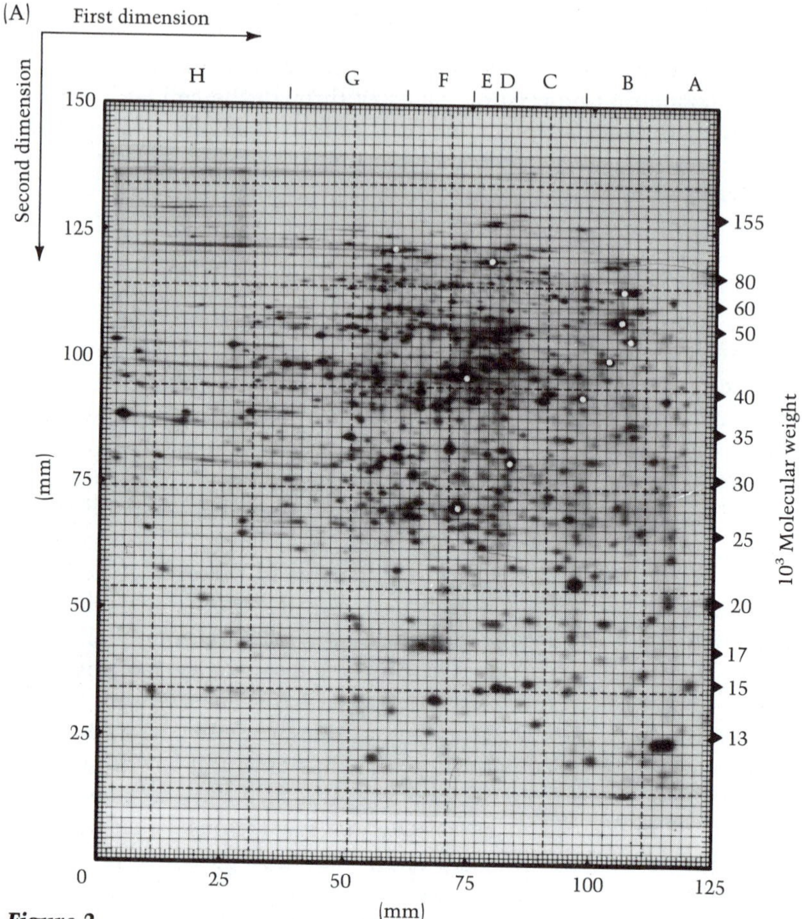

Figure 3

Autoradiograms of two-dimensional gels of extracts of *E. coli* strain B/r grown in glucose minimal medium at 37°C and labeled with $^{35}SO_4$. A grid overlay provides coordinates for individual spots. The letter designations A to I, from the acidic to the basic end of the first (horizontal) dimension, indicate zones each containing proteins with similar isoelectric points; the values of the isoelectric points increase from right to left. The approximate molecular weight scale from the bottom to the top of the second (vertical) dimension was determined by using the average migration distance on these two gels of proteins of known molecular weight. These parameters (designated in the order of the dimensions: first × second) are used to designate proteins; for example, a protein with coordinates 51 × 81 in panel A is in

VARIETY AND NUMBERS OF MOLECULES

Proteins

Proteins constitute 55% of the cell's dry mass, so our average cell contains 155 femtograms (1 femtogram = 10^{-15} g) of protein. How many molecules does this represent? How many kinds of proteins are there in the cell? And what is their range of molecular abundance.

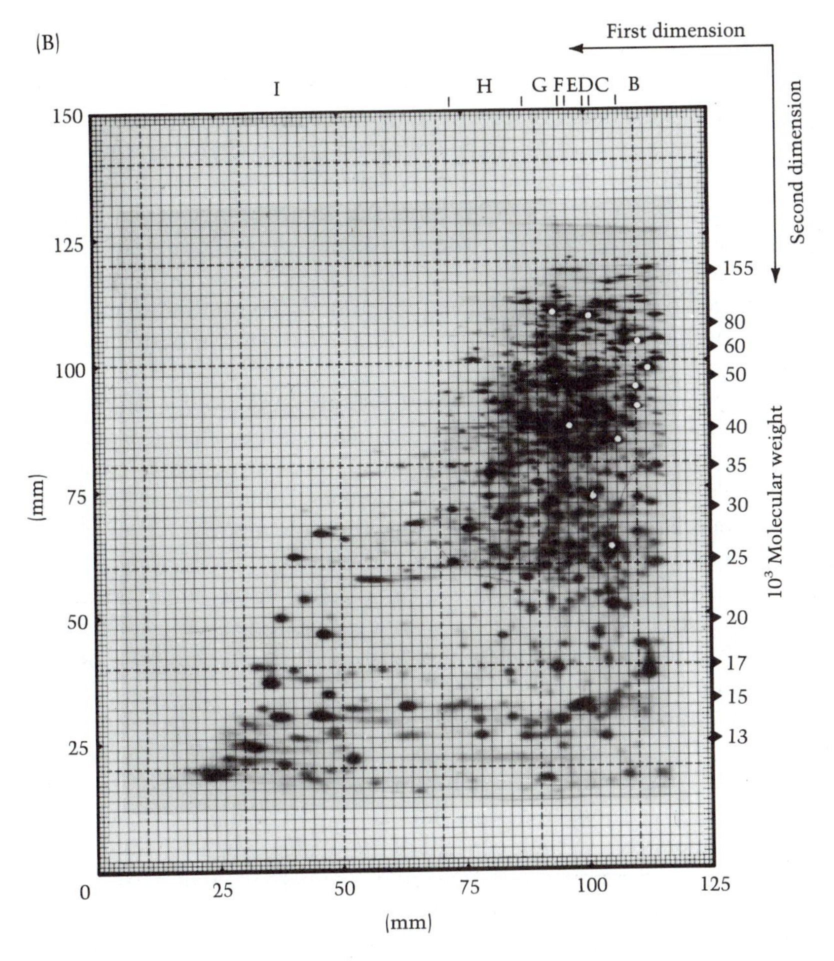

zone G and has an apparent molecular weight of 35,000; it is called protein G35.0. These two panels are the reference patterns of protein spots for B strains of *E. coli*. That is, the position of a protein spot from another extract run on a separate gel is transformed (by comparison with neighboring landmark spots) into the coordinates that the spot would have in the reference gel. (A) Isoelectric focusing to equilibrium. (B) Nonequilibrium electrophoresis. Very basic proteins do not enter the isoelectric focusing gel in the equilibrium method. In the nonequilibrium method, electrophoresis begins with the extract loaded at the acidic end, and electrophoresis is stopped before the basic proteins migrate out of the gel and when the proteins of zone A have just begun to enter the gel. (From Bloch et al., 1980.)

We can get reasonably accurate answers to each of these questions from analyses of two-dimensional gels (O'Farrell gels) that resolve total cell protein . In this technique, proteins are separated on the basis of their net charge by means of ISOELECTRIC FOCUSING in one dimension and on the basis of their apparent size by SODIUM DODECYLSULFATE (SDS) GEL ELECTROPHORESIS in the second dimension. Figure 3 depicts autoradiograms of dried gels prepared from a reference culture identical to the one used for the chemical analyses given in Table 1, except for the presence of [^{35}S]sulfate in the growth medium, which is used to label cell proteins in their methionine and cysteine residues. Figure 3A is an autoradiogram on which approximately 1,500 radioactive spots are visible on the original film. With few exceptions, each spot is a different polypeptide chain—the product of a different gene. But the pH range of the carrier ampholytes in the gel tube of the first dimension is not sufficiently alkaline to permit entry of the many basic proteins of the cell. Figure 3B is an autoradiogram of a gel prepared by a modified method to circumvent this problem; it reveals the approximately 300 major basic proteins of the cell.

There is some evidence that the 1,800 proteins (polypeptides) displayed by the O'Farrell technique is a close approximation of the number of proteins present in significant amounts (30 or more molecules) in an *E. coli* cell growing in this medium. (Consider what sort of evidence might support this contention.) Because a rough approximation of the molecular weight of each of these 1,800 polypeptides is given by its vertical position on the gel, the O'Farrell display permits an estimate of the average molecular weight of *E. coli* polypeptides—approximately 40,000. Our reference cell must therefore contain on the order of 2.4 million protein molecules.

For the third question, a glance at Figure 3 is sufficient to indicate an enormous range of abundance among these different protein species. Optical methods for measuring the intensity of individual spots, confirmed by measurement in a scintillation counter of their radioactivity, reveal that spots vary in their content of molecules over a 10^5-fold range. Evidence indicates that the relative amount of individual proteins in the spots on the gels reflects in most instances their cellular abundance. We therefore face the interesting fact that the cell contains over 100,000 molecules of some proteins and only a handful of others. As we shall see, most proteins are remarkably stable in *E. coli*, so we can anticipate powerful mechanisms for controlling the synthesis of individual proteins and thus for controlling the expression of individual genes.

Control mechanisms not only set the rate of synthesis of individual proteins at appropriate values with respect to one another, they adjust these rates in response to the environment. The two autoradiograms reproduced in Figure 4 illustrate the magnitude of these adjustments. They show amounts of individual proteins synthesized in one strain of *E. coli* during (A) steady-state growth in acetate minimal medium and (B) steady-state growth in glucose-rich medium. The adjustments of levels of individual proteins to these circumstances is readily apparent visually.

Approximately 350 of the protein spots on these gels have already been identified as known enzymes, factors, and structural proteins. This information forms the beginning of a cellular protein database for *E. coli* (Phillips et al., 1987).

RNA

Bacterial cells are rich in RNA. One-fifth of the dry mass of our reference cell growing in glucose minimal medium is RNA. The prepon-

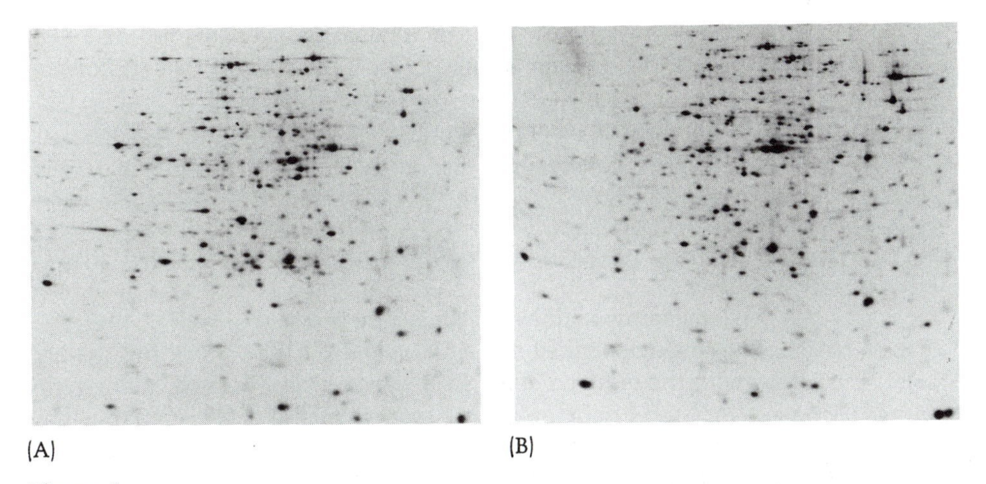

Figure 4

Comparison of proteins in cells of *E. coli* strain B/r in two different conditions. The two-dimensional gels (both of them produced by equilibrium isoelectric focusing) were prepared in the same manner as those shown in Figure 3A. (A) Steady-state growth in acetate minimal medium. (B) Steady-state growth in glucose rich medium. (F. C. Neidhardt, unpublished.)

derance (81%) of this is ribosomal RNA (rRNA), which is composed of three molecular species (23S, 16S, and 5S rRNA) that are present in equimolar amounts at 18,700 copies each per cell. Because one copy of each of these three kinds of RNA is present in each ribosome, this number also indicates the number of these particles present in the cell.

Transfer RNA (tRNA) constitutes most of the remainder (15%). There are approximately 8.6 femtograms of tRNA per cell. At a molecular weight of 25,000 each, there are somewhat over 200,000 tRNA molecules per cell—10 times the number of ribosomes. The complete nucleotide sequences of many tRNA species are known. How many species of tRNA are present in the *E. coli* cell is not yet known with certainty, but the number is estimated to be near 60. The cellular abundance of the different molecular species of tRNAs varies considerably.

Messenger RNA amounts to only 2.4 femtograms (4% of the total RNA). This number, however, indicates not the rate of synthesis of mRNA (which is approximately equal to the rate of making rRNA) but only the amount that is present at any time. The experiment described earlier, in which radioactive glucose was added to a culture in a steady state of growth, can be used to illustrate the extreme metabolic lability of mRNA. The pattern of acquisition of label corresponds generally to curve C in Figure 1. More sophisticated experiments reveal that the average half-life of mRNA in our reference cell is approximately 1.3 minutes. Not all molecular species of mRNA are equally labile (the range is from 0.5 to 10 minutes); the reason for this variation is not yet known. A further contrast with rRNA and tRNA is the structural heterogeneity of mRNA. The actual number of kinds of mRNA at any one time in the cell is not known with certainty, but one arrives at an estimate of

600–900 kinds by assuming that the typical genes coding for the 1,800 proteins in the cell are arranged in OPERONS (gene clusters that are cotranscribed into a single mRNA) of two to three genes each.

It is not by accident that bacteria like *E. coli* are rich in RNA and that their mRNA is metabolically labile; these two facts fit well with the bacterial property of rapid growth, as we shall see in Chapter 15.

DNA

As far as is known, all bacteria have their essential genetic material arranged as a single, naked molecule of DNA, called a CHROMOSOME (by analogy with the more complex structure in higher cells.) Although the GENOME (the sum of all the genes of an organism) of a bacterial cell is contained on a single chromosome, many bacteria have smaller, accessory molecules of DNA called PLASMIDS. The genes carried on plasmids constitute a variable, accessory portion of a cell's genome. Our reference cell lacks plasmids.

Chromosomal DNA constitutes approximately 9 femtograms of our reference cell. The complete, single chromosome of *E. coli* is known, with considerable precision, to be a circular, covalently closed, double-stranded molecule of 4,720,000 base pairs (4,720 kilobase pairs, or kbp for short). Therefore its molecular weight is about 2.5×10^9, and, stretched out, it is approximately 1 mm long. How such a structure is arranged within a cell that is only 1/500th this length is an interesting story told later in Chapter 2. Our reference cell, caught in the act of replicating its DNA, has slightly over two complete chromosomes that are duplicates of each other.

What can 4,720 kbp do for a cell? Recall that the average molecular weight of *E. coli* proteins is 40,000 and the average molecular weight of an amino acid residue in protein is 110. Therefore the average protein has 364 amino acids. Because it takes three nucleotide bases in DNA to specify each amino acid residue in a protein, the average gene size in *E. coli* is approximately 1.1 kbp. So, there cannot possibly be more than 4,300 protein-encoding genes. In fact, there must be considerably fewer, because some of the DNA encodes the stable RNA species, and another portion is taken up as intragenic control sequences. Reasonable corrections for the amount of DNA that does not encode protein lead to a "best guess" of 3,800 possible protein-encoding genes. We have already seen that 1,800 gene products are detected by the O'Farrell technique in cells under any single growth condition. The ability to know both these numbers for the same cell demonstrates the value of concentrating attention on the chemical and molecular parameters of a given organism as a cell paradigm.

The genome sizes of other bacteria are known less precisely. Cells of the genus *Mycoplasma* hold the record for smallest genome— approximately one-fifth the size of that of *E. coli*. There is considerable variation in one easily determined parameter of DNA, its guanine plus cytosine (GC) content. Species of *Mycoplasma* have a GC content of 23%, whereas—at the other extreme—*Geodermataphilus* has 75%. This range is greater than that found throughout the rest of the biological world.

Lipid

From the human perspective, one expects fats to serve as accumulated reserve material. Some bacteria do accumulate poly-β- hydroxybutyrate, but no bacteria accumulate neutral fats as reserve material; rather, lipids are a functional part of their membranes.

The lipids of *E. coli*, exclusive of lipid A of the lipopolysaccharide (see below), are entirely phospholipids; the major ones are phosphatidylethanolamine (75%), phosphatidylglycerol (18%), and cardiolipin (5%). There are only traces of phosphatidylserine (Figure 5). The major fatty acids found in these molecules are 43% palmitic (16:0), 33% palmitoleic (16:1), and 25% *cis*-vaccenic (18:0) acids (Figure 6). Together the membrane phospholipids amount to 26 femtograms in our reference cell.

Lipopolysaccharide

The structure of lipopolysaccharide (LPS) varies among species of bacteria, and even among strains, but all Gram-negative bacteria (ex-

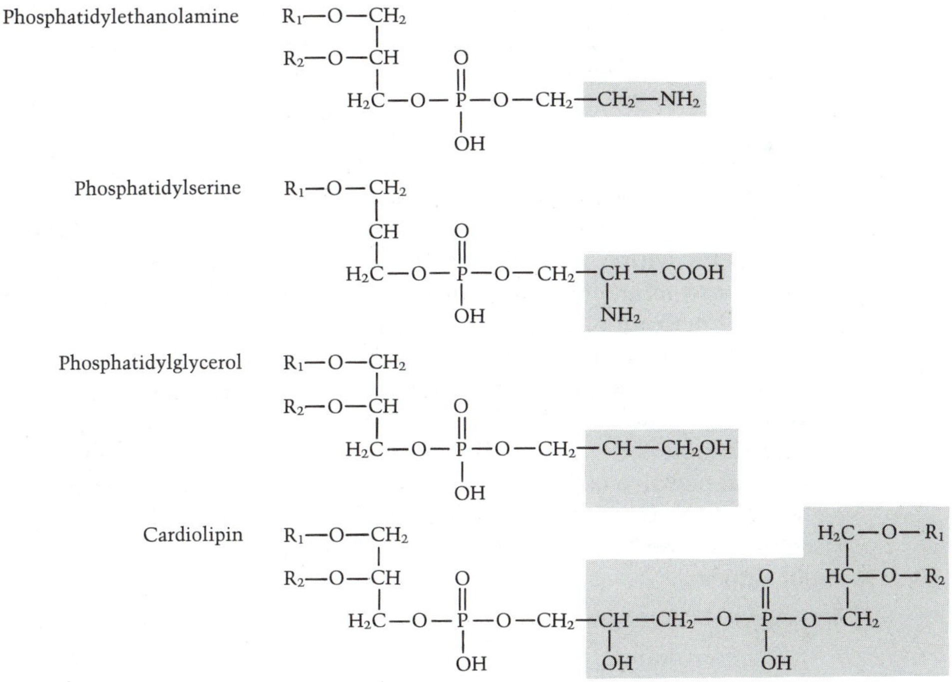

Phosphatidylethanolamine

$R_1—O—CH_2$
$R_2—O—CH$
$H_2C—O—P(=O)(OH)—O—CH_2—CH_2—NH_2$

Phosphatidylserine

$R_1—O—CH_2$
CH
$H_2C—O—P(=O)(OH)—O—CH_2—CH(NH_2)—COOH$

Phosphatidylglycerol

$R_1—O—CH_2$
$R_2—O—CH$
$H_2C—O—P(=O)(OH)—O—CH_2—CH—CH_2OH$

Cardiolipin

$R_1—O—CH_2$
$R_2—O—CH$
$H_2C—O—P(=O)(OH)—O—CH_2—CH(OH)—CH_2—O—P(=O)(OH)—O—CH_2$
$HC—O—R_2$
$H_2C—O—R_1$

Figure 5

Structures of common bacterial phospholipids. The four types shown are all found in *E. coli*. R_1 and R_2, fatty acyl residues.

$CH_3(CH_2)_{12}COOH$	Myristic acid
$CH_3(CH_2)_{10}CHOHCH_2COOH$	β-Hydroxymyristic acid
$CH_3(CH_2)_{14}COOH$	Palmitic acid
$CH_3(CH_2)_5CH{=}CH(CH_2)_7COOH$	Palmitoleic acid
$CH_3(CH_2)_5CH{-}CH(CH_2)_7COOH$ (with CH_2 bridging the two CH)	*cis*-9,10-Methylenehexadecanoic acid
$CH_3(CH_2)_5CH{=}CH(CH_2)_7COOH$	*cis*-Vaccenic acid

Figure 6

Structures of common fatty acids in bacteria. The six molecules shown are found in *E. coli*, but only the upper three are prevalent.

plained later in this chapter) possess it in some form or other. Lipopolysaccharide is responsible for the protection of the outer membrane of Gram-negatives (Chapter 2).

The LPS of *E. coli* and other enteric bacteria is an extremely complicated molecule, the elucidation of which is a triumph of structural chemistry and microbial genetics. A diagram of the general structure of LPS is shown in Figure 7. All 10.0 femtograms of LPS are found in the outer surface of the outer membrane of this cell.

Murein

Murein is a peptidoglycan found in all bacteria, Gram-positive and Gram-negative alike, except for the archaebacteria and the wall-less bacteria known as the mollicutes. In Gram-positive bacteria, as we shall see, murein forms the bulk of the cell wall. Although present in lesser amounts in Gram-negative cells such as *E. coli*, it nonetheless plays the major role in maintaining the structural integrity of the cell. How it does so can perhaps be intuitively understood from its structure (Figure 8); the lattice of murein forms a covalent shell (or better, hauberk) of great strength around the entire cell. In many Gram-positive cells (and never in Gram-negative cells), teichoic acids are associated with the murein. TEICHOIC ACIDS (Figure 9) are polymers of glycerol or ribitol joined through phosphodiester bonds and carrying one or more amino acid or sugar substituents.

Carbohydrate

During growth on glucose, a small proportion of the substrate molecules are polymerized to form the polysaccharide glycogen, with a reserve function believed to be analogous to that in higher organisms.

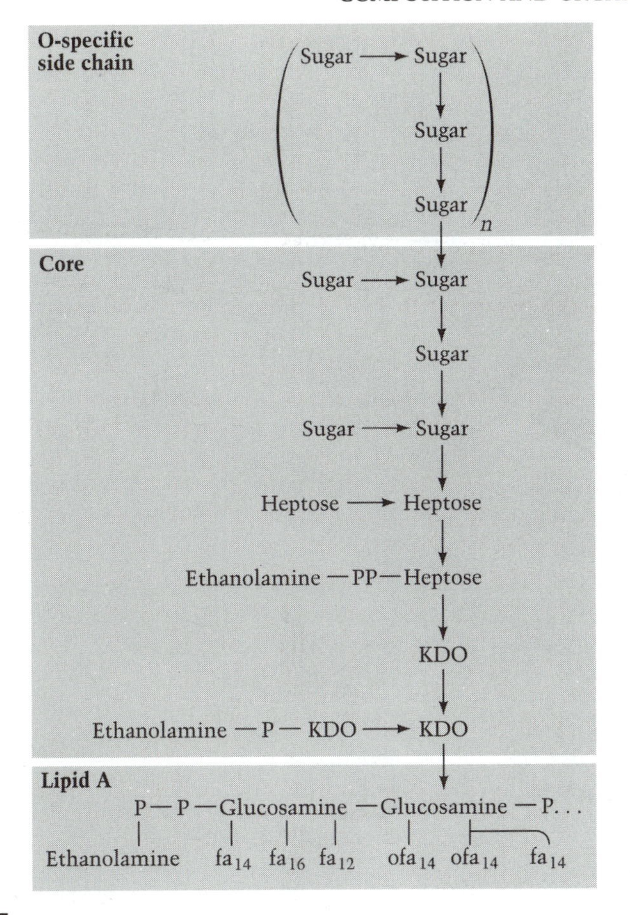

Figure 7

General diagram of a subunit of lipopolysaccharide. The structure shown is based on analyses of the envelope of enteric bacteria, particularly species of *Salmonella*. LIPID A is a phosphorylated glucosamine disaccharide esterified with fatty acids (fa), commonly including dodecanoic (fa_{12}), tetradecanoic (fa_{14}), 3-hydroxytetradecanoic (ofa_{14}), and hexadecanoic (fa_{16}) acids. The CORE is an oligosaccharide that commonly includes a heptose (L-glycero-D-mannoheptose) and KDO (commonly called keto-deoxy-octanoic acid, but properly, 3-deoxy-D-mannooctulosonic acid). The O-SPECIFIC SIDE CHAIN is usually longer than the core oligosaccharide and consists of many repeating tri-, tetra-, or pentasaccharide subunits, including (in different species) a variety of unusual sugars. Lipid A and the core are quite similar among all enteric species, but the O-specific side chain is highly strain specific. (Modified from Nikaido, 1973.)

Our reference *E. coli* cell contains approximately 7 femtograms of glycogen. This tiny amount is insufficient to support the carbon needs for growth for more than a minute or so in the event the medium becomes depleted of substrate. Nor would the energy made available by metabolizing this amount of glycogen serve the maintenance energy needs for

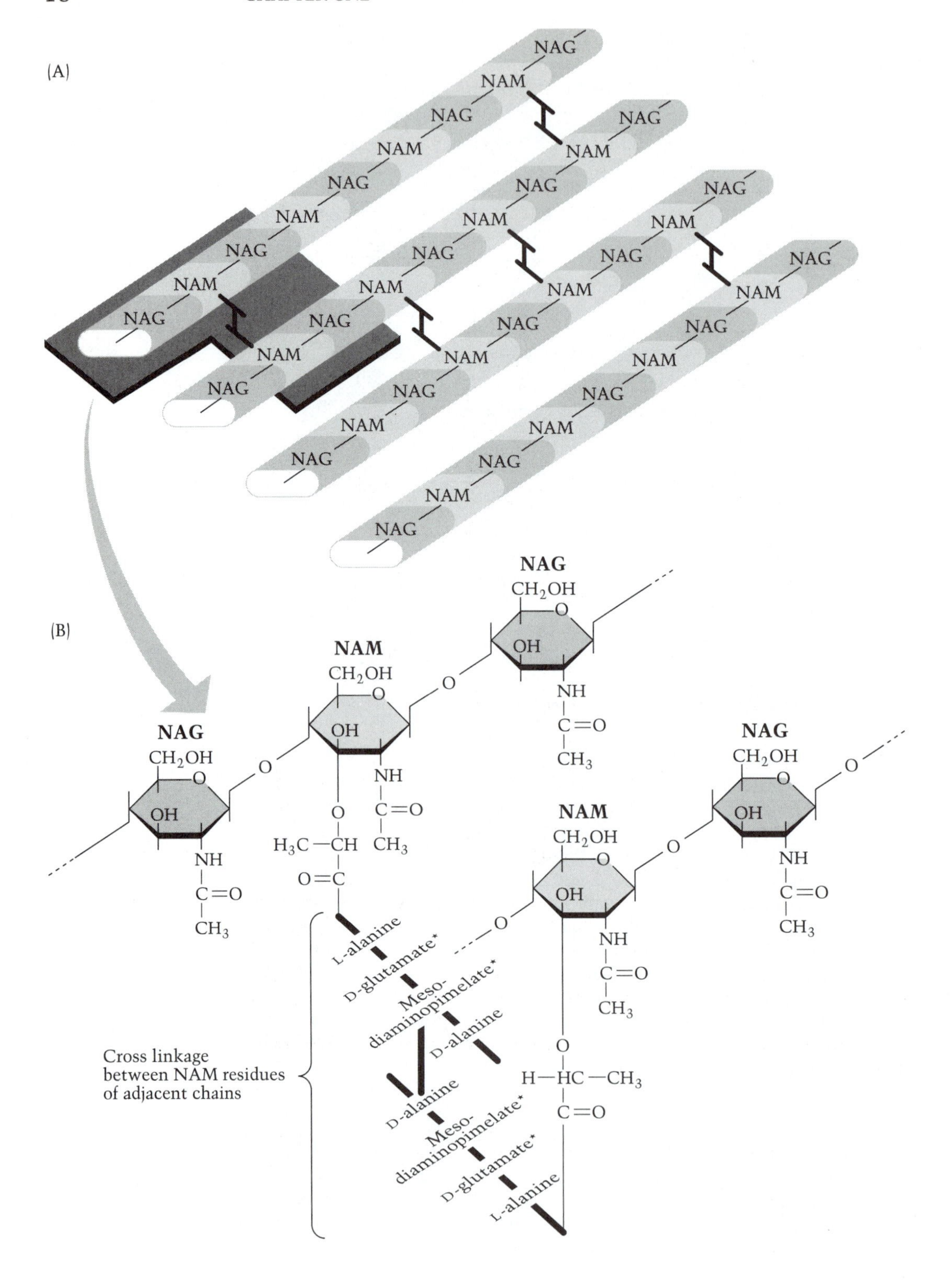
(A)
NAG
NAM
NAG
NAM
NAG
NAM
NAG
NAM
NAG
(B)
NAG
CH_2OH
OH
NH
C=O
CH_3
NAM
CH_2OH
OH
O
H_3C-CH
O=C
NH
C=O
CH_3
L-alanine
D-glutamate*
Meso-diaminopimelate*
D-alanine
D-alanine
Meso-diaminopimelate*
D-glutamate*
L-alanine
Cross linkage between NAM residues of adjacent chains
O
H—HC—CH_3
C=O

◄ *Figure 8*

Diagrammatic representation of the murein of *E. coli*. (A) The arrangement of *N*-acetylglucosamine (NAG) and *N*-acetylmuramic acid (NAM) residues linked by β-1,4 glycosidic bonds. (B) The shaded area of A is enlarged to show the cross-linkage between tetrapeptides attached to NAM residues of adjacent chains. The murein of other species may differ in the nature of the amino acids at positions 2 and 3 of the tetrapeptide and in the nature and frequency of the cross-link (noted by asterisks in the figure). (Modified from Ghuysen, 1968.)

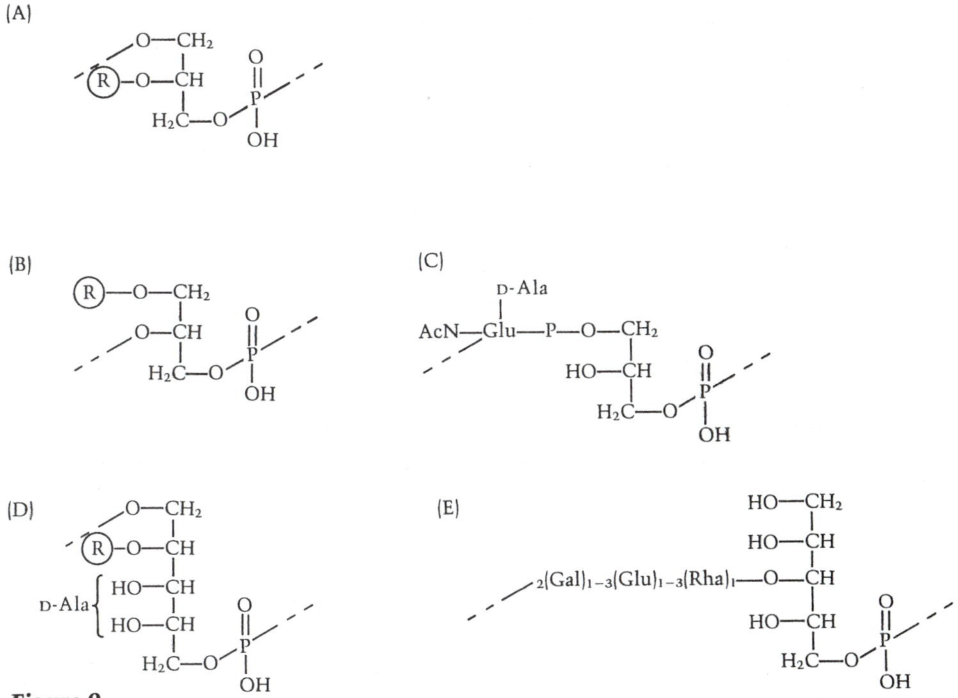

(A)

O—CH_2
(R)—O—CH
H_2C—O—P(=O)(OH)

(B)

(R)—O—CH_2
O—CH
H_2C—O—P(=O)(OH)

(C)

D-Ala
AcN—Glu—P—O—CH_2
HO—CH
H_2C—O—P(=O)(OH)

(D)

O—CH_2
(R)—O—CH
D-Ala { HO—CH
HO—CH }
H_2C—O—P(=O)(OH)

(E)

HO—CH_2
HO—CH
$_2$(Gal)$_{1-3}$(Glu)$_{1-3}$(Rha)$_1$—O—CH
HO—CH
H_2C—O—P(=O)(OH)

Figure 9

Structure of teichoic acids. The repeating units of five teichoic acids from different Gram-positive species are shown. (A) Glycerol teichoic acid of *Lactobacillus casei* strain 7469 (R = D-alanine). (B) Glycerol teichoic acid of *Actinomyces antibioticus* (R = D-alanine). (C) Glycerol teichoic acid of *Staphylococcus lactis*; D-alanine appears in the 6-position of *N*-acetylglucosamine. (D) Ribitol teichoic acids of *Bacillus subtilis* (R = glucose) and *Actinomyces streptomycini* (R = succinate); D-alanine is attached to position 3 or 4 of ribitol. (E) Ribitol teichoic acid of the Type 6 pneumococcal capsule. (From Stanier et al., 1986.)

long in the absence of growth (Chapter 8). The real significance of the glycogen reserve seems to be that it can become enormously increased when growth is restricted by other than the depletion of the carbon and energy sources. Under these conditions, the cell converts available substrate into large quantities of intracellular glycogen, sequestering it from competing organisms in a form readily useable when growth is again possible.

Although not all bacteria make glycogen in this manner, the property is widespread. Some bacteria that cannot make glycogen store excess supplies of carbon in the form of a polyester, poly-β-hydroxybutyrate. Other bacteria seem to lack the ability to store excess carbon in any form (Chapter 4).

Organic molecules of low molecular weight

The small portion of the cell that consists of molecules of low molecular weight—approximately 8.0 femtograms in our *E. coli* example—consists of many hundreds of different kinds of molecules. In general they are (1) precursors (building blocks) of macromolecules—amino acids, nucleotides, sugars, and fatty acids, (2) metabolic intermediates on the way to becoming precursors or to being excreted; (3) cofactors of enzymes; and (4) polyamines bound to DNA. Many—perhaps most of them—are known (Table 2).

Inorganic ions

Our inventory of small molecules is completed by listing the inorganic ions found in bacterial protoplasm. The major anions depend to some extent on the anionic composition of the medium, but they always include phosphate and usually sulfate and chloride. Potassium (K^+) is the major cation in most cases (many bacteria seem able to dispense with sodium) and it is involved in active adjustment of the internal osmotic

Table 2. Small molecules present in a bacterial cell growing in glucose minimal medium

Molecule	Approximate number of kinds
Amino acids, their precursors and derivatives	120
Nucleotides, their precursors and derivatives	100
Fatty acids and their precursors	50
Sugars, carbohydrates, and their precursors	250
Quinones, polyisoprenoids, porphyrins, vitamins, other coenzymes and prosthetic groups, and their precursors	300

Table 3. Inorganic ions present in the bacterial cell growing in glucose minimal medium

Ion	Function
K^+	Principal cation, cofactor for certain enzymes
NH_4^+	Principal form of inorganic N for assimilation
Mg^{2+}	Cofactor for a large number of enzymes
Ca^{2+}	Cofactor for certain enzymes
Fe^{2+}	Present in cytochromes and other enzymes
Mn^{2+}	Cofactor for several enzymes
Mo^{2+}	Present in several enzymes
Co^{2+}	Present in vitamin B_{12} and its coenzyme derivatives
Cu^{2+}	Present in several enzymes
Zn^{2+}	Present in several enzymes
Cl^-	Not required for many bacteria
SO_4^{2-}	Main source of S in most media
PO_4^{3-}	Participant in many metabolic reactions

pressure. Ammonia (NH_4^+), magnesium (Mg^{2+}), calcium (Ca^{2+}), and iron (Fe^{2+}, Fe^{3+}) are present in significant amounts; manganese (Mn^{2+}), molybdenum (Mb^{2+}), cobalt (Co^{2+}), copper (Cu^{2+}), and zinc (Zn^{2+}) are found in quite small amounts, bound to a few enzymes. Altogether the inorganic molecules of the cell amount to only 1% of the dry mass, but their functions (Table 3) are essential for growth.

Observations

One comes away from this overview of the chemical composition of the bacterial cell with a number of impressions. First, for all its smallness, the bacterial cell is chemically complex. This complexity is of a manageable order; techniques now exist that enable one to resolve and study complex mixtures of molecular species, and one can foresee the complete definition of the molecules present in a bacterial cell in the near future. Second, the composition of the bacterial cell is extremely orderly. A given strain growing in a given environment achieves a predictable steady-state composition and readjusts that composition in a predictable and precise manner when the environment changes. Finally, although there are some molecules unique to bacteria (lipopolysaccharide, murein, and teichoic acid), on the whole there is a high degree of commonality between the biochemistry of bacteria and of other organisms—a fundamental observation often referred to as the "unity of biochemistry."

We shall have frequent occasion to refer to these three aspects of bacterial cell composition: its *complexity, orderliness,* and *commonality*.

ORGANIZATION OF THE PROCARYOTIC CELL

It was only with the development in the 1950s of techniques for preparing very thin sections of bacteria (100 nm or less in thickness) and the availability of the electron microscope that the ultrastructure of bacterial cells was revealed. A properly prepared and stained section, such as the one through a cell of *E. coli* shown in Figure 10A, can show almost all of the important cellular structures and organelles. When such micrographs became available, it was quickly recognized that the internal structure of bacterial cells was strikingly different from that of other cells.

The PROCARYOTIC BODY PLAN of bacterial cells is a unique design that bears little similarity to the EUCARYOTIC plan shared by all cells of animals, plants, and protists. Procaryotic cells contain no organelles bounded by UNIT MEMBRANES (phospholipid bilayer membranes) inside the cell membrane, and their genetic material is never assembled into the complex structures (also called chromosomes) found in eucaryotic cells. Procaryotes contain no endoplasmic reticulum, Golgi apparatus, lysosomes, mitochondria, chloroplasts, microtubules, mitotic fibers, or membrane-bounded nucleus. Their overall intracellular appearance is that of densely packed cytosol surrounding an amorphous, rather fibrous-appearing, less electron-dense NUCLEAR REGION. Functional features also distinguish procaryotic cells from eucaryotic cells—some of these are listed in Table 4.

Being procaryotic means being small, lacking internal membranes and membrane-bounded compartments, having intimate contact between DNA and the abundant protein-forming machinery, lacking

Figure 10

(A) Transmission electron micrograph of a thin section of a dividing cell of *E. coli* strain B/r. The locations of some important subcellular structures are indicated. The thin murein layer that lies between the outer membrane of the wall and the cytoplasmic membrane is not revealed. The section is approximately 50 nm thick. ×23,500. (Courtesy of Jack Pangborn.) (B) Shadowed electron micrograph of *Vibrio alginolyticus* showing the pattern of insertion of flagella at the cell surface. This organism, like *E. coli*, produces numerous unsheathed peritrichously arranged flagella; unlike *E. coli*, it produces, in addition, a single, sheathed, polar flagellum (arrow). ×10,900. (From de Boer et al., 1975.) (C) Nucleoids in *E. coli*. Cells in exponential growth were rapidly deposited onto silver membrane filters under a constant supply of oxygen; they were then rapidly frozen. The ice (at −50°C) was substituted with acetone containing the fixative OsO_4. Samples were then embedded, thin-sliced, and stained with uranyl acetate. The lower copy of the micrograph has been labeled to delineate the two compact nucleoids (N) surrounded by the ribosome-rich cytosol (C) of this cell. Freeze-substitution leads to preparations with more compact nucleoids than those in specimens fixed by aldehyde, and reveal changes in nucleoid appearance with different physiological conditions. (Micrographs supplied by J. Hobot, W. Villiger, J. Escaig, and E. Kellenberger.)

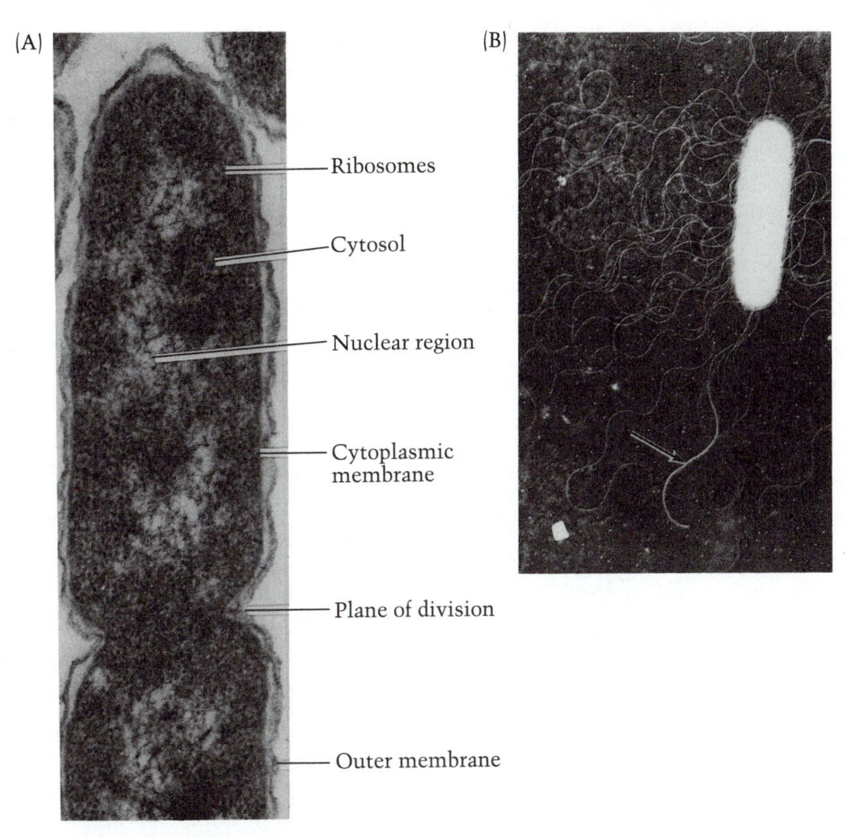
(A)
Ribosomes
Cytosol
Nuclear region
Cytoplasmic membrane
Plane of division
Outer membrane
(B)

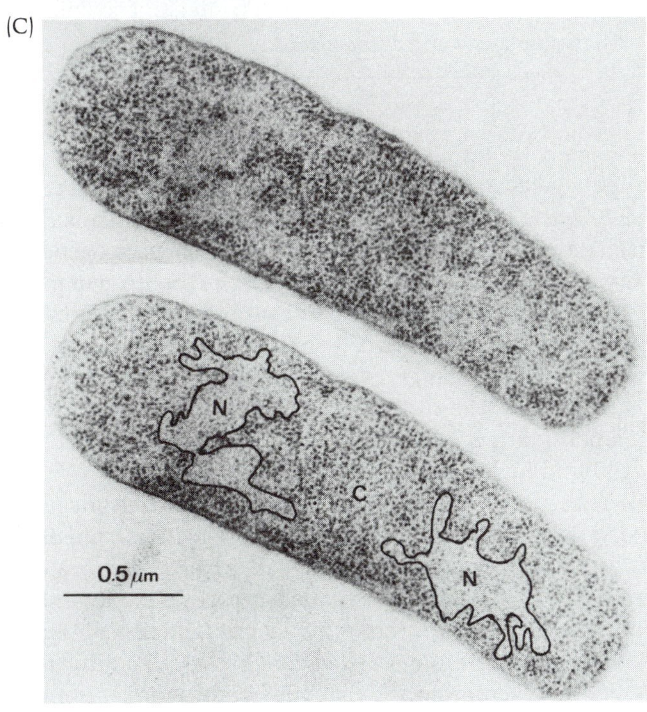
(C)
N
C
N
0.5 μm

Table 4. Comparison of procaryotic and eucaryotic cell organization[a]

Structural/functional feature	Procaryotes	Eucaryotes
Endoplasmic reticulum	Absent	Present
Number of chromosomes	1	> 1
Chromosome with histones	Absent	Present
Nucleolus	Absent	Present
Genetic exchange by conjugation	Plasmid-mediated, unidirectional	By gamete fusion
Nuclear membrane	Absent	Present
Golgi apparatus	Absent	Present
Lysosomes	Absent	Present
Mitochondria	Absent	Present
Chloroplasts	Absent	Present in plants
Glyoxosomes	Absent	Sometimes present
Microtubules	Absent	Present
Ribosomes	70S structure	80S structure
Phagocytosis	Absent	Sometimes present
Pinocytosis	Absent	Sometimes present
Cellular endosymbionts	Absent	Sometimes present
Ameboid motion	Absent	Sometimes present
Cytoplasmic streaming	Absent	Present
Site of electron transport	Cell membrane	Organelle
Cell wall with murein	Present, except in mycoplasma and archaebacteria	Absent

[a]Adapted from Stanier et al., 1986.

complex organelles, concentrating organized electron transport systems at the cell surface, having highly developed transport mechanisms that actively concentrate a variety of nutrient solutes, and possessing genetic material, a cell division apparatus, and organelles of locomotion of minimal structural complexity. The ensemble of these features creates an unmistakable picture: the procaryotic cell shown in Figure 10 is a cell that appears customized for a high metabolic rate leading to rapid growth and division. Indeed, the METABOLIC RATE (amount of substrate or oxygen consumed per hour per unit of cell mass) of procaryotic cells is typically 10- to 100-fold higher than that of eucaryotic cells.

A closer look at Figure 10A reveals the important macromolecular structures. The ENVELOPE that bounds the cell can be seen to be composed of an inner CELL MEMBRANE surrounded by a wall and, sometimes, an OUTER MEMBRANE. The CELL MEMBRANE (also called CYTOPLASMIC MEMBRANE) appears as a double layer, a pattern that is typical for the phospholipid bilayers or unit membranes common to all cells. The WALL is a thin layer of murein—the saclike macromolecule that confers structural strength and helps determine cellular shape. The outermost portion of the envelope, the outer membrane, resembles the cell membrane in appearance and in certain chemical and physical properties. In contrast to the murein layer, the outer membrane is a significant barrier to the passage of macromolecules and certain small molecules. Thus, the region between the cell membrane and the outer membrane has a unique solute composition differing significantly from that of the cytosol and that of the surrounding medium. This region, called the PERIPLASM, is found (as we shall shortly see) only in the envelope of Gram-negative bacteria.

Interior to the cell membrane, the dense CYTOSOL appears to be granular because it is densely packed with RIBOSOMES, which at this magnification can be individually resolved as discrete oval-shaped bodies. The NUCLEAR REGION, or NUCLEOID, appears to ramify through the central region of the cell almost as though its shape were determined by the surrounding cytosol (Figure 10D). Very little additional internal structure can be seen.

Typically, *E. coli* bears two types of surface appendages, FLAGELLA (singular: flagellum) and PILI (singular: pilus) or FIMBRIAE (singular: fimbria). These structures are found in many bacterial species. They almost never appear in sections like Figure 10A and D but are readily apparent in micrographs of whole cells (Figure 10B). Flagella and pili, although bearing gross structural resemblance, serve quite different functions. They both arise from the cell membrane and in the main are aggregates of a single or a small number of proteins. Flagella are helix-shaped organelles of locomotion. Pili are rod-shaped organelles of adhesion. Certain types of pili, termed SEX PILI, allow mating pairs of bacteria to stick together in the conjugal transfer of DNA from one cell to another. Others allow bacteria to stick to surfaces, a process of considerable selective advantage in many natural environments, and essential to some infectious processes. Sometimes such adhesive interactions are remarkably specific, as when certain bacteria adsorb only to the cells of a particular tissue of a higher animal.

Although not shown in Figure 10, many bacterial cells are surrounded by an extensive layer of a hydrophilic polymer, usually polysaccharide but in some instances other polymers. This CAPSULE may extend beyond the cell for many cell diameters and undoubtedly is essential to the cell's ability to compete in a variety of natural environments. It provides another mode of adhesion to surfaces, and it also protects many PATHOGENIC (disease-causing) bacteria against phagocytosis by macrophages and polymorphonuclear leukocytes of higher animals.

Bacteria can be divided into two broad classes—Gram-positive and Gram-negative—on the basis of their differential abilities to retain a crystal violet–iodine stain when treated with organic solvents (alcohol or acetone). In 1884 a Danish microbiologist, Christian Gram, discovered this staining technique, which has become known as the Gram stain. Those bacteria that retain the stain are termed GRAM-POSITIVE and those that lose it, GRAM-NEGATIVE. This staining property depends on the morphology and composition of the bacterial envelope, even though

Gram-positive and Gram-negative bacteria differ in a number of important aspects in addition to envelope structure. A few bacterial species possess the key structural and compositional features of Gram-positive cells but exhibit a weak or variable ability to stain Gram-positive, particularly in old cultures after growth has ceased; such bacteria are considered to be Gram-positive. A few other species (such as *Mycobacterium tuberculosis*) with the cytological characteristics of Gram-positive cells are coated with a waxy layer that prevents penetration of the crystal violet stain unless the cells are pretreated with heat or a detergent; these Gram-positive bacteria are said to be ACID-FAST, because, once a dye has penetrated, it cannot be removed with dilute mineral acids.

The ultrastructure of the *E. coli* outer envelope is characteristic of Gram-negative bacteria: a very thin murein layer, the periplasm, and the outer membrane-like layer. The structure of *Bacillus subtilis*, a typical Gram-positive bacterium (Figure 11), differs strikingly. Outside the cell membrane is a single, thickened wall layer, composed largely of murein, together with TEICHOIC ACID. The wall is not bounded by the outer membrane that can be seen in Gram-negative bacteria, and there is, therefore, no periplasm. The diagram in Figure 12 compares the architecture of the Gram-positive and Gram-negative cell envelopes.

The Gram-positive–Gram-negative distinction is a profound one. It reflects not only a fundamental difference in wall structure but other important biochemical, physiological, and genetic differences as well. The majority of, but not all, bacteria are members of these two groups. Those that fall outside these groups are of two classes. One group is closely similar to other bacteria, but completely lack walls—the MOLLICUTES. The other group, the ARCHAEBACTERIA, produce walls that do not contain a typical murein, and in many other respects differ from the typical procaryotic cell.

The archaebacteria are of great interest to students of evolution. Archaebacteria are clearly procaryotic, but on the basis of the sequence of bases in their ribosomal RNA, the chemical composition of their lipids, and their wall structure, they seem to be only distantly related to

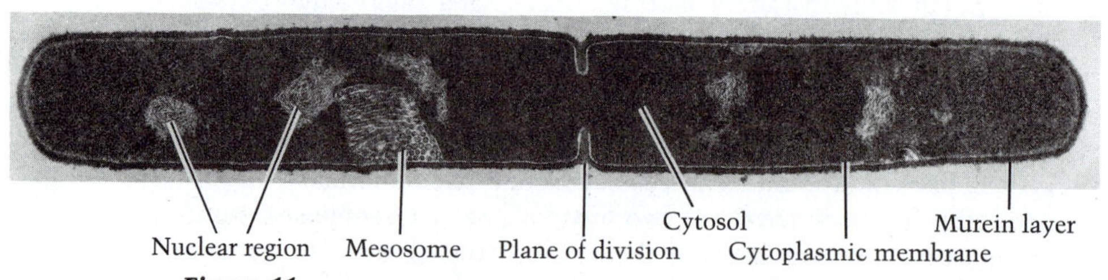

Figure 11

Electron micrograph of a thin section through a dividing cell of *Bacillus subtilis,* showing the significant intracellular structures. Mesosomes are tubular involutions of the cytoplasmic membrane; they are artifacts of fixation, but they emerge often at regions associated with cell division. (Courtesy of Jack Pangborn.)

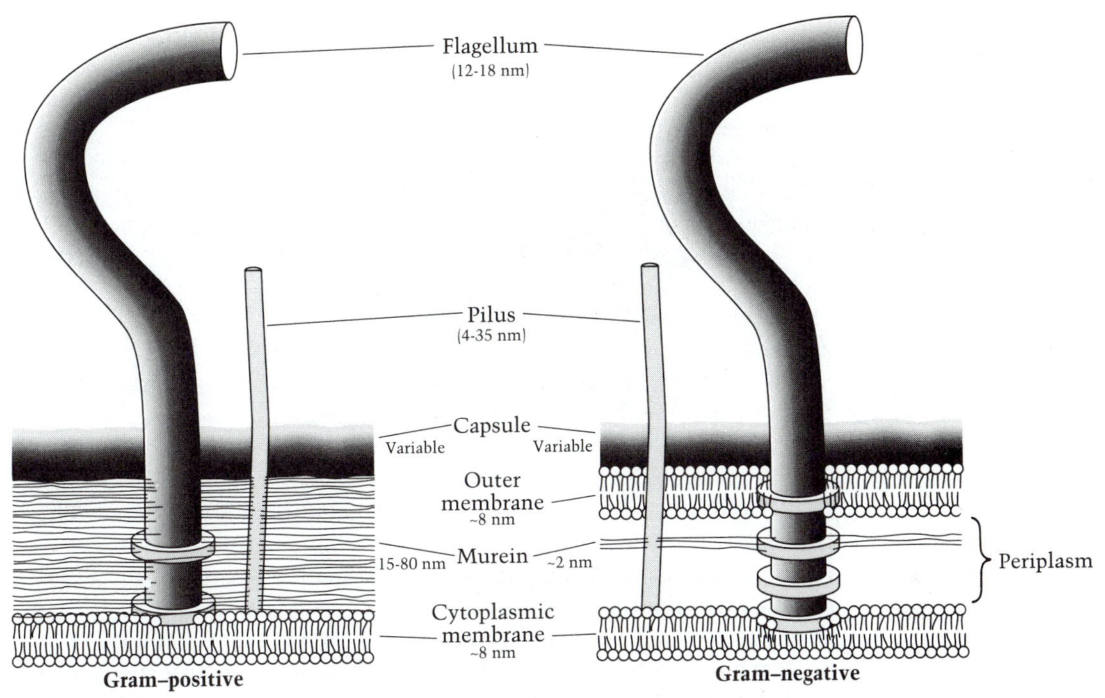

Figure 12

Comparison of the structure of the Gram-positive and Gram-negative cell envelopes, showing the major molecular components and their approximate dimensions. The region between the outer membrane and the cytoplasmic membrane of the Gram-negative envelope is called the periplasm.

other bacteria (EUBACTERIA). Indeed, it appears they are no more closely related to eubacteria than they are to plants or animals. Archaebacteria constitute a small group of procaryotes, largely found in specialized or extreme habitats, including hot acidic niches for the thermoacidophiles, nearly saturated salt solutions for the extreme halophiles, and highly reducing (low redox potential) conditions required for the growth of methane producers. Wall composition and ultrastructure is quite variable among archaebacteria. Some contain a heteropolymer, called PSEUDOMUREIN, that resembles the murein of bacteria but lacks two of its characteristic components, muramic acid and diaminopimelic acid. Other archaebacteria have walls composed of protein subunits, with or without traces of glucosamine; still others have heteropolysaccharide walls. Because of such chemical diversity, it is not possible to select one organism as having a typical archaebacterial ultrastructure. A thin section of the methanogen *Methanospirillum hungatii* (Figure 13) reveals a typical procaryotic appearance: the lack of internal membranes and the dense ribosomal packing of the cytosol. However, the wall, which is composed of protein and is surrounded by an external sheath, differs dramatically in appearance from that of either a Gram-positive or a Gram-negative cell.

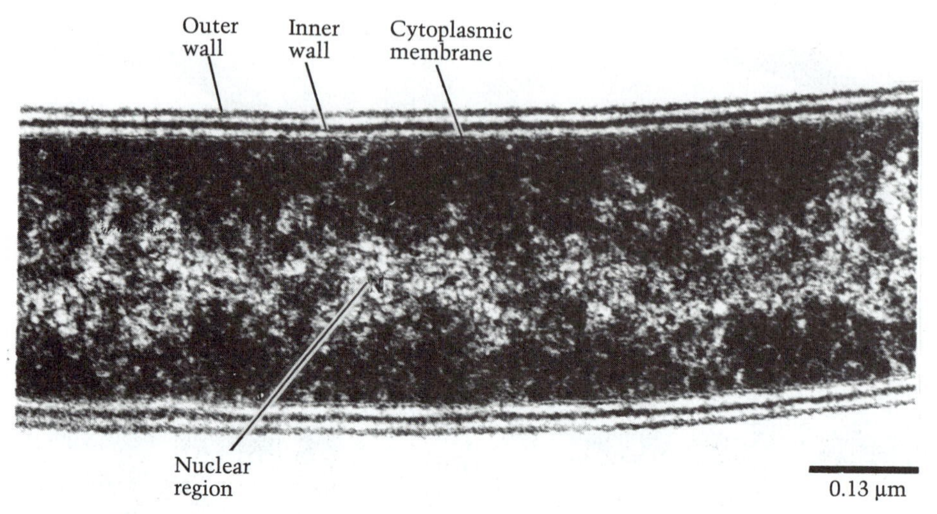

Figure 13

Transmission electron micrograph of a section of the archaebacterium *Methanospirillum hungatii*. Although the variation in ultrastructure among archaebacteria is considerable and no single one can be said to be representative of the group; most are quite different in structure from either Gram-positive or Gram-negative bacteria. A longitudinal section revealing the cytoplasmic membrane and the double-track appearance of the cell wall. The nuclear region is similar to that seen in both Gram-positive and Gram-negative bacteria. (Courtesy of J. G. Zeikus; Zeikus and Bowen, 1975.)

SUMMARY

1. Bacteria generally have been selected in nature for high metabolic rates and rapid growth and for a great capacity to adapt to environmental change.
2. Their ease of cultivation and the ready availability of powerful techniques for biochemical and genetic analyses make bacteria an excellent paradigm for the study of cell growth.
3. Bacteria are chemically complex. Over 95% of their mass is macromolecular. Most of their molecules resemble those of other cells (exceptions: lipopolysaccharide, murein, teichoic acid); the "unity of biochemistry" includes bacteria.
4. Bacteria are small and their genomes are the smallest known for any cell. The various proteins that can be made by a bacterium such as *E. coli* are few enough that they can be resolved and catalogued by modern methods.
5. Bacteria display biochemical orderliness. Their chemical composition is reproducible under any given growth condition, and the adjustments they make when growth conditions change are similarly predictable.

6. The bacterial cell body plan differs in fundamental ways from that of nonbacterial cells. The procaryotic organization of bacteria helps make their rapid growth possible.
7. The interior of the bacterial cell is plain; its envelope, fancy. Great biological individuality is expressed in the molecular variations of the enveloping layers of the cell. This individuality is a basis for grouping eubacteria into major classes (mollicutes, Gram-positive bacteria, Gram-negative bacteria).

STUDY QUESTIONS

1. A newly isolated bacterial species is found to have a single chromosome with a molecular weight of 1×10^9. Estimate the maximum number of protein-encoding genes it is likely to have if a two-dimensional (O'Farrell) gel reveals that the average molecular weight of its proteins (polypeptides) is 35,000. Assume that 10% of the genome does not code for protein and that 1% codes for stable RNA species.
2. If the average transcriptional unit (mRNA molecule) in the *E. coli* cell described in Table 1 contained the information for only a single polypeptide, how many molecules would there be per cell?
3. Predict and draw the time course of radioactive labeling (i.e., curve A, B, or C of Figure 1) of a protein species that has a half life of 30 seconds in a cell with a generation time of 2 hours.
4. Predict and draw the time course of radioactive labeling (i.e., curve A, B, or C of Figure 1) of a phospholipid species that is completely stable in a cell with a generation time of 2 hours.
5. For the reference culture of *E. coli* described in Table 1, calculate the number of ribosomes in a newborn cell.
6. For the reference culture of *E. coli* described in Table 1 calculate the LPS content of the oldest cell in the culture.

SUGGESTED READING

General microbial biochemistry

Gottschalk, G. 1986. *Bacterial Metabolism*, 2nd Edition. Springer-Verlag, New York.
Stanier, R. Y., J. L. Ingraham, M. L. Wheelis and P. R. Painter. 1986. *The Microbial World*, 5th Edition. Prentice-Hall, Inc., Englewood Cliffs, New Jersey.
Neidhardt, F. C., J. L. Ingraham, K. B. Low, B. Magasanik, M. Schaechter and H. E. Umbarger (eds.). 1987. *Escherichia coli and Salmonella typhimurium: Cellular and Molecular Biology*. American Society for Microbiology. Washington, D.C.

Microbial taxonomy

Bergey's Manual of Systematic Bacteriology. 1984 (Vol. 1), 1986 (Vol. 2). J. G. Holt (ed. in chief) and N. R. Krieg (ed.). Williams and Wilkins, Baltimore.
Starr, M. P., H. Stolp, H. G. Truper, A. Balows and H. G. Schlegel (eds.). 1981. *The Prokaryotes* (2 vols.). Springer-Verlag, Berlin, Heidelberg, and New York.

2

Structure and Function of Bacterial Cell Parts

INTRODUCTION

Bacteria have evolved a unique repertoire of structural strategies that allow them to compete for scarce food, survive in hostile environments, and occupy distinct spatial and chemical niches in the environment. They have done this in a way that is species specific, but, in the aggregate, typically bacterial. The fundamental gulf between procaryotes and eucaryotes is nowhere to be seen more clearly than at the level of structural organization. In fact, it was structural organization that led to the original definition of procaryotes, thereby separating them from eucaryotes. Thus, a consideration of bacterial anatomy immediately serves to tell what is typical of these forms of life.

BACTERIAL ENVELOPE

Like all cells, bacteria first perceive the characteristics of their environment at their surface. It follows, then, that structural and functional properties of the bacterial envelope directly reflect the adaptive strategies of bacteria. The components of the bacterial envelope are adapted to take up nutrients, often from highly dilute solutions, and to exclude certain toxic compounds. In addition, surface components participate directly in the adherence of bacteria to surfaces and interfaces and in the transfer of genetic information. Bacteria invest heavily in these endeavors and much of their genetic potential (upward of one-quarter of the total) is devoted to the synthesis, regulation, and maintenance of the components of the cell envelope. Individual bacterial species vary greatly in this regard: hardly any other of their intrinsic properties is more typical of a given species than are their cell membranes, wall, and surface appendages. Indeed, many of the traits used to distinguish bacteria taxonomically rely on differences of their surface components.

CELL MEMBRANE

All bacteria possess a cell membrane of the usual bilayer, or "unit-membrane" structure. Like the membranes of most eubacteria, that of *E. coli* consists of phospholipids (Figure 5 in Chapter 1) and upward of 200 different kinds of proteins. Approximately 70% of the mass of the membrane is attributable to proteins, a considerably higher proportion than in mammalian cell membranes. Sterols are absent, except in certain mycoplasmas. On the other hand, a class of polycyclic compounds called the HOPANES (derivatives of pentacyclic triterpenoids; Figure 1) are widely distributed among bacteria. Hopanes have structural features similar to those of steroids and may well have analogous condensing effects on membranes.

Basically, the cell membrane is an osmotic barrier modified by the presence of specific transport systems. Because most of the macromolecular syntheses and metabolic reactions occur in the cytoplasm, the cell membrane is the true boundary between the cell interior and the outside. The high protein content of the bacterial cell membrane suggests that this structure is involved in multiple and unique functions. Indeed, the cell membrane is the site where many complex metabolic processes take place to an extent unknown in the rest of the living world (Table 1).

A model for the structure of the bacterial cell membrane is shown in Figure 2. At or near the temperatures that permit growth, membrane lipids are mostly in a semifluid state. Individual lipid molecules are generally free to exchange places with one another within each leaflet of the membrane, thus the whole resembles a two-dimensional fluid. Nevertheless, there must be regions of considerable order within the membrane because some lipid molecules are not free but are closely bound to specific membrane proteins. Little is yet known about the intimate architecture of the cell membrane. Suggestions for the existence of differentiated membrane areas have caused some confusion in the past. When bacteria are fixed for preparing ultrathin sections, the cell membrane undergoes multiple convolutions visible in the electron micro-

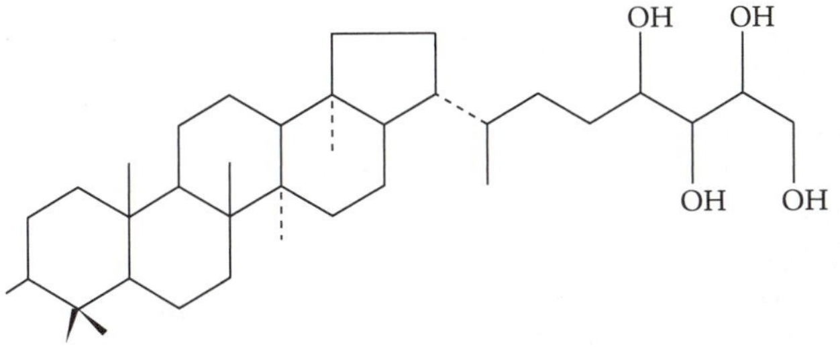

Figure 1

Structure of a hopane, a polyterpenoid found in many species of bacteria. Other compounds of this class are present and may play a similar role in maintaining membrane rigidity.

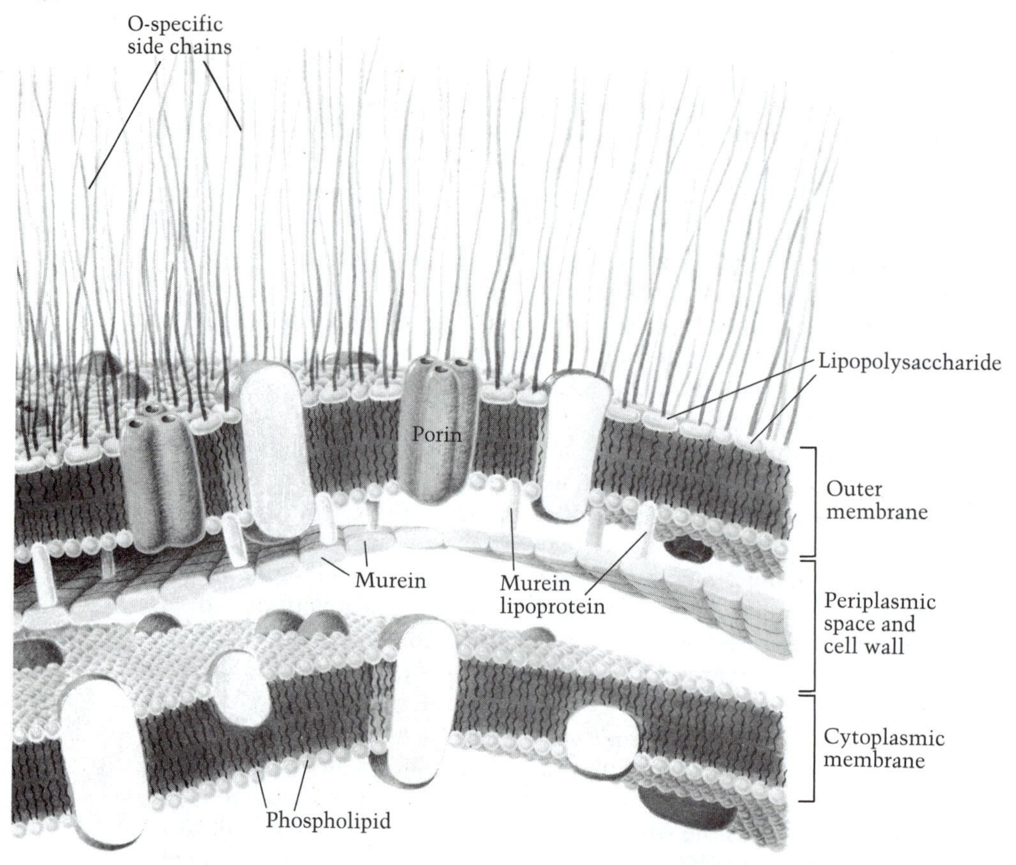

Figure 2

Model of the *E. coli* cell envelope, showing its various layers and their approximate thicknesses.

Table 1. Functions of the bacterial cell membrane

Osmotic barrier
Transport of specific solutes (nutrients and ions)
Synthesis of membrane lipids (including lipopolysaccharide in Gram-negative cells)
Synthesis of wall murein
Assembly and secretion of extracytoplasmic proteins (membrane periplasmic, outer membrane, extracellular)
Respiratory electron transport
Chromosome segregation (probably)
Chemotaxis (both motility *per se* and sensing function)

scope. Considerable importance has been attributed to the resulting onion-like structures, known as MESOSOMES. Although mesosomes have turned out to be artifacts of fixation, they tend to be formed at specific locations on the membrane, a finding that suggests that the membrane is indeed differentiated into specific regions.

CELL WALL

As cells go, bacteria are particularly tough; that is, they are hard to break by mechanical means, and they retain their shape even under harsh conditions. Much of the toughness of bacteria is related to their cell wall, which provides rigid mechanical support and prevents pressure from bursting the cell (osmotic lysis). In addition, the cell wall represents a chemical and physical defense against noxious chemicals that may harm the cell membrane. Gram-positive and Gram-negative bacteria differ considerably in the structure of their cell walls and in the role of this structure in coping with environmental changes (Figure 12 in Chapter 1). Any portion of the wall of Gram-positive cells looks like a thick blanket, that of most Gram-negative cells like a flimsy sheet. In reality, even the thin Gram-negative wall has considerable tensile strength.

Gram-positive cell wall

The Gram-positive cell wall consists of a thick multimolecular coat (Figure 3) of MUREIN (a type of PEPTIDOGLYCAN) with lesser amounts of other polymers, notably TEICHOIC ACIDS, interspersed (Figure 9 in Chapter 1). This polymeric fabric consists of many layers wrapped around the length and width of the cell, thereby forming a sac that determines the size and shape of the organism. The murein can be isolated intact as a structure called the MUREIN SACCULUS (Figure 4). The shape of a bacterium depends on the shape of the sacculus: most bacteria look like rods (BACILLI), spheres (COCCI), or helixes (SPIRILLA, SPIROCHETES); a few look like spindles, starfish, or flattened polyhedra; still others assume more complex, differentiated shapes (Chapter 16).

The structure of murein is shown in Figure 8 in Chapter 1. All the *N*-acetylmuramic acid residues of the glycan backbone are substituted with a tetrapeptide, approximately 25% of which in *E. coli* are cross-linked by a peptide bond between the free amino group of the diaminopimelic acid of one tetrapeptide and the carboxyl group of the terminal alanine of another. There is considerable diversity among species of bacteria in the frequency of cross-links and in the nature of the amino acid at positions 2 or 3 of the tetrapeptide. A common alternative is a cross-link between an exclusively bacterial amino acid, DIAMINOPIMELIC ACID, and the terminal alanine. In some species, the two strands are linked, not directly, but via a peptide chain—in *Staphylococcus aureus*, a pentaglycine. Some of the cross-links extend above and below the plane of the murein sheet to make multiple layers of murein. The

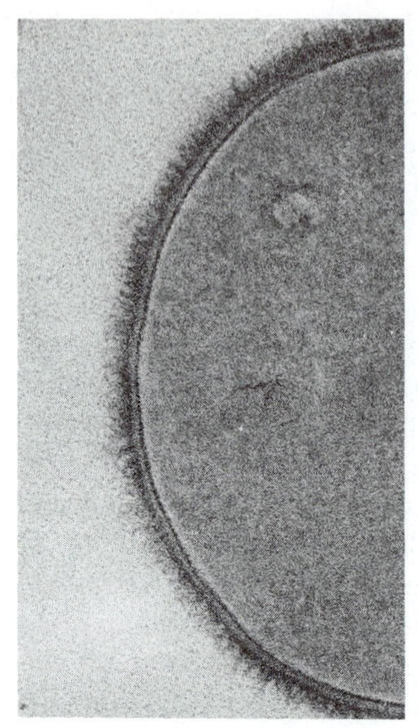

Figure 3

Ultrathin section of *Bacillus subtilis* prepared by freeze-substitution. The cytoplasmic membrane is covered by a thick cell wall. (Courtesy of T. J. Beveridge.)

Figure 4

Purified murein sacculi of *Bacillus megaterium*. The white spheres are particles of latex 0.25 μm in diameter, and are included to show the scale. (From Stanier et al., 1954.)

mechanical strength of murein is derived partly from the fact that it is a single molecule and partly from the helical arrangement of its glycan strands.

The other major components of the Gram-positive wall are teichoic acids (Figure 9 in Chapter 1) and teichuronic acids, which may account for up to half the wall mass in some species. These compounds are highly antigenic and are found in great variety; that is, they vary in the nature of the sugar backbone and the kind and location of various substituents. Because a bacterial species produces a consistent type of teichoic acid, these molecules are useful as taxonomic markers (e.g., the *Staphylococcus aureus* "polysaccharide A" or the *Enterococcus faecalis* "group D carbohydrate" are distinct teichoic acids). The function of teichoic acids is still a matter of speculation. Some of them have lipid substituents (such as LIPOTEICHOIC ACID, Figure 5) and appear to be functionally distinct from the wall teichoic acids—they may serve to anchor the wall to the underlying membrane. In *Streptococcus pyogenes*, lipoteichoic acid is associated with a protein called M-PROTEIN that protrudes from the cell membrane to the exterior of the cell. The long molecules of M-protein are decorated with lipoteichoic acid; together, these two compounds make microfibrils that give the surface of the organism a fuzzy

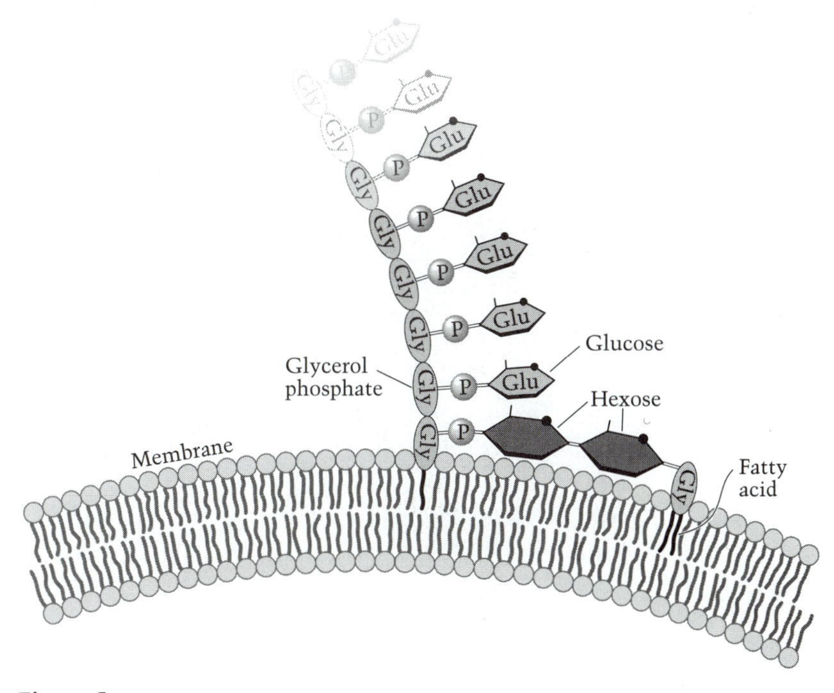

Figure 5

Tentative structure of lipoteichoic acid. The molecule is anchored to the membrane by its fatty acids. Gly, glycerol; glu, glucose.

look (Figure 6). These microfibrils facilitate the attachment of *S. pyogenes* to animal cells.

How does the cell wall contribute to the protection of the cytoplasmic membrane from chemical agents? The multilayered Gram-positive wall impedes the passage of hydrophobic compounds, because the amino sugars and charged amino acids of murein and the phosphates of the teichoic acids make the wall highly polar. Thus, the cells are surrounded with an ample hydrophilic layer. Consequently, Gram-positive cells can often withstand noxious hydrophobic compounds, such as the bile salts in the intestine of vertebrates.

The thick hydrophilic coat may explain why certain bacteria are Gram-positive. We deal here with exclusion in reverse: the hydrophobic dye–iodine complex, once formed within the cells, apparently cannot escape through the thick murein layer of these organisms. The murein meshwork is permeable to hydrophilic compounds, and nutrients such as sugars or amino acids readily penetrate it to reach the cell membrane. The murein layer does not appear to present a particular obstacle to DNA, and many Gram-positive cells can be genetically transformed with DNA from other cells (Chapter 10).

The rigid murein meshwork allows bacteria to survive in media that are commonly hypotonic—having a lower osmotic pressure than that of their cytoplasm. In the absence of a rigid corset-like structure to push against, the membrane would burst and the cells lyse. This outcome can be demonstrated experimentally by removing the murein by the action

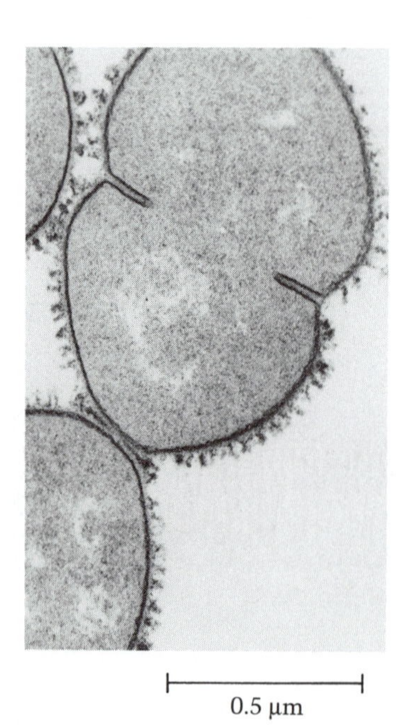

Figure 6

Thin section of *Streptococcus pyogenes*, showing external fuzz of M protein. (Courtesy of R. M. Cole.)

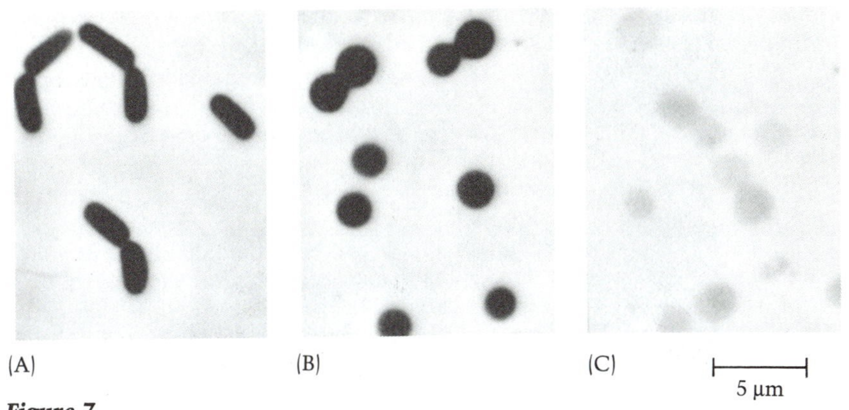

Figure 7

Protoplasts of *Bacillus megaterium*. (A) Intact cells. (B) Protoplasts formed by the action of lysozyme in the presence of 0.5 *M* sucrose. (C) Ghosts (i.e., empty membranes) formed after osmotic rupture of the cytoplasmic membrane in a hypotonic medium. (Micrograph by C. Weibull.)

of the hydrolytic enzyme, LYSOZYME, which is present in many human and animal tissue fluids. Treatment with lysozyme causes bacteria to lyse in an environment with low osmotic pressure. If lysozyme-treated bacteria are kept in an isoosmotic medium, such as 0.5 *M* sucrose, they do not lyse but do become spherical. Such structures are called PROTOPLASTS (Figure 7). When lysozyme is removed, the protoplasts, if handled gently, may revert to their original rod shape.

Gram-negative cell wall and the outer membrane

Gram-negative bacteria have evolved a radically different solution to the problem of protecting the cytoplasmic membrane. Their murein layer is much thinner than that of Gram-positive bacteria, and they make a completely different structure—an OUTER MEMBRANE, which is built up outside a thin murein layer (Figures 8, 9, and 10). The outer membrane is chemically distinct from the usual biological membranes and has the ability to resist damaging chemicals. It is a bilayered structure, and its inner leaflet resembles in composition that of the cytoplasmic membrane. Its outer leaflet, on the other hand, has a unique constituent in the place of phospholipids. This component is the bacterial LIPOPOLYSACCHARIDE, or LPS, a complex molecule not found elsewhere in nature. As a result, the leaflets of this membrane are extremely asymmetrical, and the properties of this unusual bilayer differ considerably from those of a regular biological membrane. The ability to exclude hydrophobic compounds is unusual among biological membranes.

Lipopolysaccharides consist of three portions (Figure 7 in Chapter 1):

- A lipid, called LIPID A, anchors the LPS in the outer leaflet of the membrane. Lipid A is an unusual glycolipid, because the fatty acids attached to its disaccharides are shorter than usual (14 carbons long

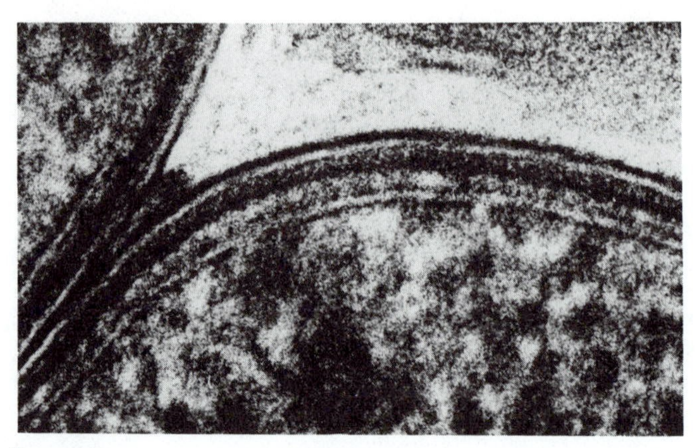

Figure 8

Ultrathin section of *E. coli* strain K-12 prepared by freeze-substitution. The cytoplasmic membrane and the outer membrane appear as double-tracked structures separated by a heavy layer, presumably of murein, and the periplasmic space. (Courtesy of T. J. Beveridge.)

instead of the more common 16 or 18); phosphate groups are also attached to the disaccharide. All of these fatty acids are saturated and have hydroxyl group substituents.

- A short series of sugars constitutes the CORE. Its structure is relatively constant among Gram-negative bacteria and includes two characteristic sugars, KETO-DEOXYOCTONOIC ACID (KDO in Figure 7 in Chapter 1) and a HEPTOSE.
- A long carbohydrate chain, up to 40 sugars in length, makes up the O ANTIGEN. The hydrophilic carbohydrate chains of the O antigen cover the bacterial surface. Although this is a loose structure, at least compared with the murein layer of Gram-positive bacteria, it is highly effective in excluding hydrophobic compounds. Mutants that make no O antigen become sensitive to compounds like bile salts and certain antibiotics to which the wild type is resistant.

Exclusion of hydrophobic compounds in Gram-negative bacteria, as in Gram-positive bacteria, is accomplished by surrounding the cells with hydrophilic polysaccharides, but these differ in structure and organization in the two groups. The outer membrane, however, presents an apparent impasse: because of its lipid nature, it can be expected to exclude hydrophilic compounds as well. In that case, no compounds, hydrophobic or hydrophilic, could cross the outer membrane. While solving the problem of protection of the cytoplasmic membrane, Gram-negative bacteria appear to have created a new one. How do these organisms transport their nutrients? Do they copy in the outer membrane the active transport devices of the cytoplasmic membrane? This strategy not only would be a wasteful investment, but also could make the outer membrane just as sensitive to environmental challenges as the cytoplas-

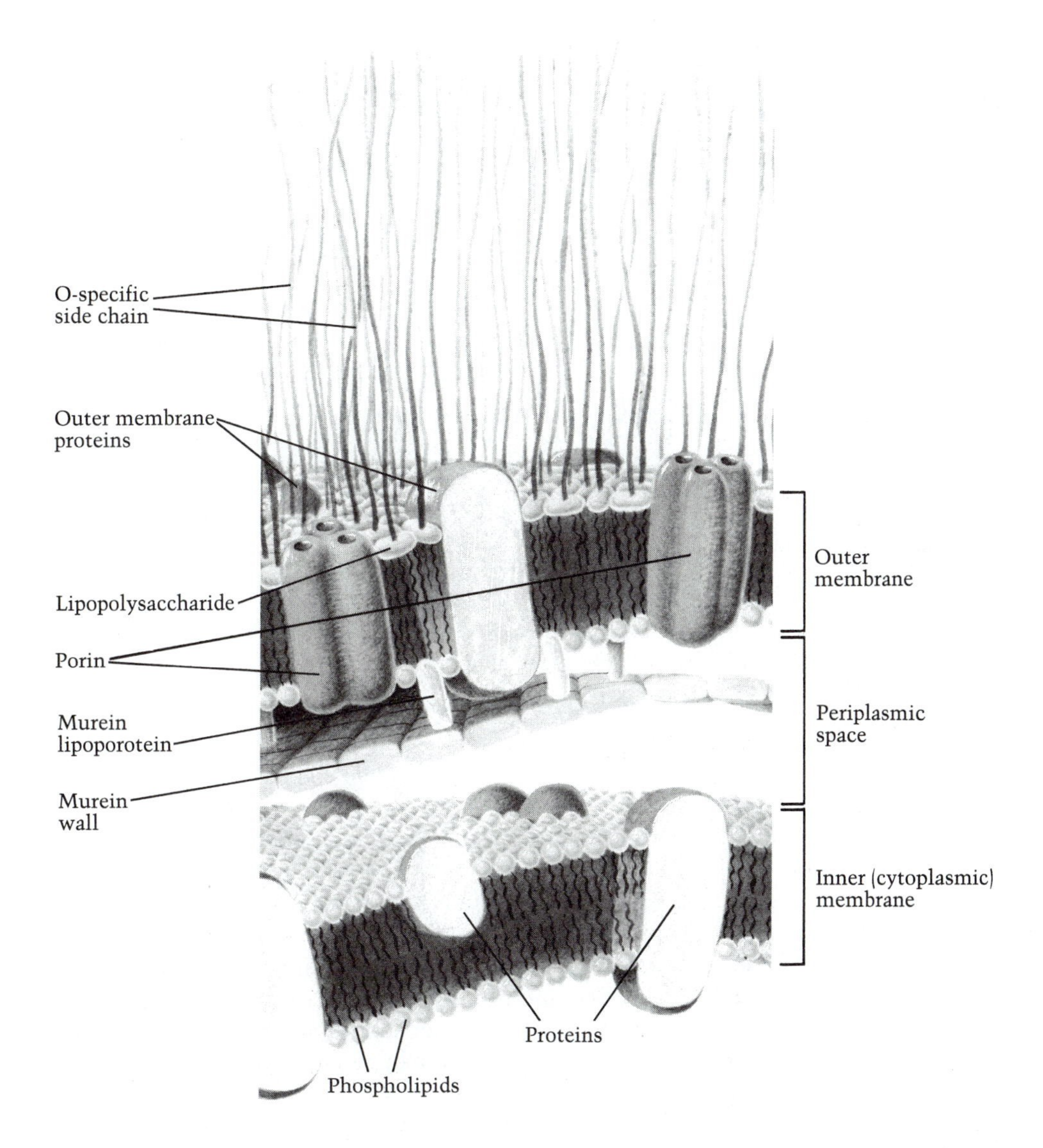

Figure 9

Model of the Gram-negative cell envelope.

mic membrane is. Once again, bacteria have found an interesting solution: the outer membrane has special channels that permit the passive diffusion of hydrophilic compounds like sugars, amino acids, and certain ions. These channels consist of protein molecules, aptly called PORINS. Porin channels are narrow, just the right size to permit the entry of compounds smaller than 600–700 daltons (Figure 9). The channels are small

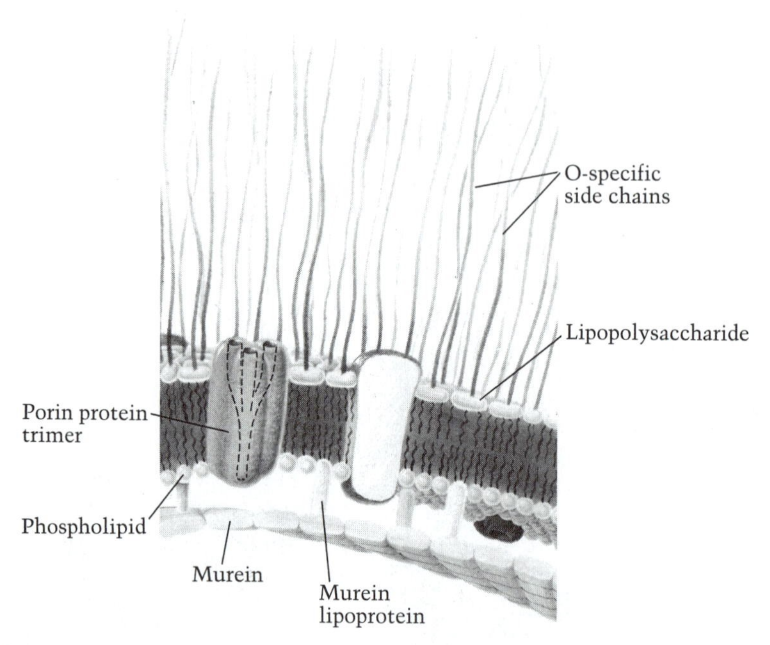

Figure 10

Model of the outer membrane of Gram-negative bacteria. (After Nikaido and Nakae, 1979.)

enough that hydrophobic compounds would come in contact with the polar "wall" of the channel and thereby be excluded. To some extent, *E. coli* can select porins with different channel sizes in response to the osmotic properties of the medium, a point we shall discuss in Chapter 13.

Certain hydrophilic compounds necessary for the cell's survival are larger than the exclusion limit of porins. These larger molecules include vitamin B_{12}, sugars bigger than trisaccharides, and iron chelates. Such compounds cross the outer membrane by separate, specific permeation mechanisms, which utilize proteins especially designed to translocate each of these compounds. Thus, the outer membrane allows the passage of small hydrophilic compounds, excludes hydrophobic compounds, large or small, and allows the entry of some larger hydrophilic molecules by especially dedicated mechanisms.

The outer membrane barrier constitutes both an advantage and a hazard to Gram-negative bacteria. For example, some bacteriophages use proteins in the outer membrane as attachment sites on their host bacteria. On the other hand, the outer membrane confers considerable resistance to many antibiotics. Broadly speaking, Gram-negative bacteria are more resistant than Gram-positive species to antibiotics, especially penicillin.

The peculiarly Gram-negative solution to the problems of protecting the cytoplasmic membrane has unexpected biological consequences. The LPS of the outer membrane is highly reactive when introduced into animals. The lipid A component has a large number of biological activities, depending on its concentration. In small doses it elicits fever and activates a series of immunological and biochemical events that lead to the mobilization of host defense mechanisms. As Lewis Thomas has pointed out, LPS advertises the presence of bacteria in tissues. In large doses, this compound, also known as ENDOTOXIN, can cause shock and even death. The O antigen portion, as part of its name denotes, is antigenic. The O antigens come in many varieties and are used by microbiologists to define species and subspecies of Gram-negative bacteria.

The outer membrane is not an entirely separate structure; it is connected to both the murein layer and the cytoplasmic membrane. The connection with the murein layer is mediated by two types of interactions, the most important being that involving an outer membrane lipoprotein. This protein is present in some 700,000 copies per cell, which makes it the most abundant protein (numerically, not by weight) in *E. coli.* Its role is still not clear, and it is apparently nonessential, because mutants defective in the structural gene for this protein are fully viable. About one-third of the lipoprotein molecules are covalently linked to murein and help hold the two structures together. The other type of interaction is the tight (but probably not covalent) association of some of the outer membrane porins with murein. Together, lipoproteins and porins provide in each cell over 400,000 strutlike contacts between the outer membrane and the cell wall.

The connection between the outer membrane and the cytoplasmic membrane is less well understood. These two structures can be mechanically separated by plasmolysis, that is, by placing bacteria in a hypertonic medium. Under the electron microscope, however, sections of plasmolyzed cells show that the two layers appear to remain united at ZONES OF ADHESION. These junctions are also known as Bayer's junctions, after their discoverer. They number about 200 per *E. coli* cell. It is not known whether these structures are permanent or transient, periodically undone and remade. Considerable effort has gone into studying the zones of adhesion, because they have been implicated in diverse phenomena such as the passage of proteins to the outer membrane and the exterior of the cell, the deposition of new LPS molecules, and the attachment of certain bacteriophages. Unfortunately, it is difficult to define a priori the physical properties of isolated zones of adhesion and, consequently, to study them by cell fractionation.

In recent years, Rothfield and colleagues (MacAlester et al., 1983) discovered that there are special zones of adhesion girding the whole circumference of *E. coli.* These annular rings are formed on both sides of the cell division septum (Figure 11). Therefore, they have become known as PERISEPTAL ANNULI. The role of these structures is not yet known, but they may define separate domains of the *E. coli* envelope and perhaps participate in the segregation of chromosomes (Chapter 14).

Can one convert Gram-negative bacteria into protoplasts? Given the thinness of the Gram-negative murein layer, one assumes that this conversion should be easy. However, the Gram-negative cell's murein is protected by the outer membrane, which is impervious to lysozyme.

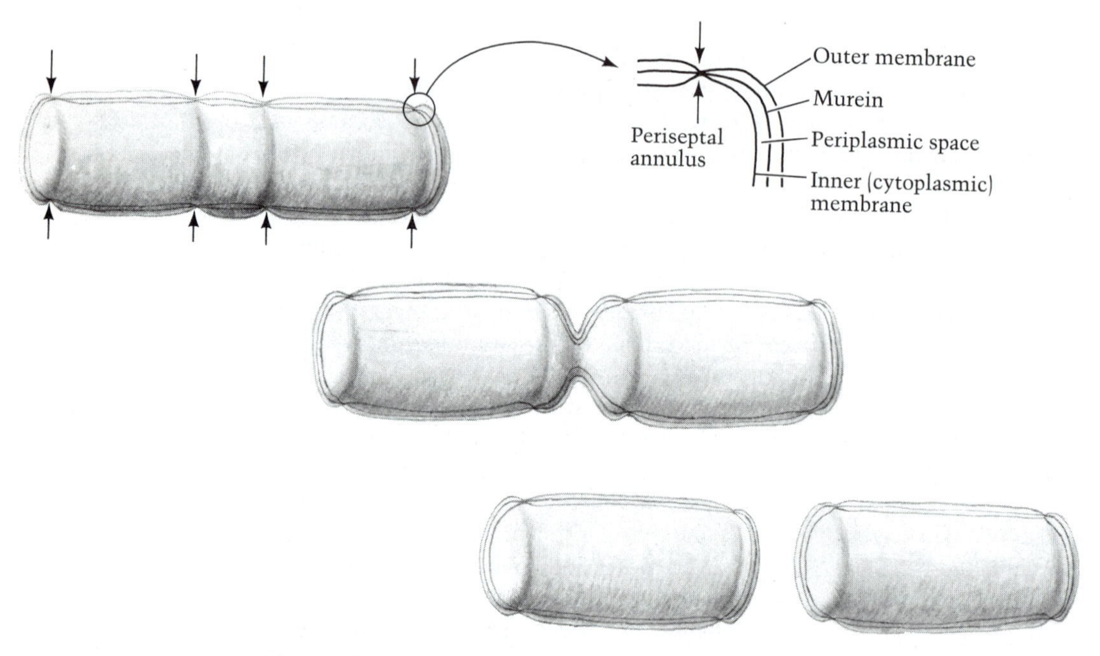

Figure 11

Periseptal annulus of *Salmonella typhimurium*. The murein–outer membrane attachment at the periseptal annulus is shown in a cell plasmolysed with 10% sucrose. (Redrawn from MacAlester et al., 1986.)

Two general techniques can be used to remove the murein layer of these organisms and, in an isoosmotic medium, generate osmotically sensitive protoplast-like forms.

- The outer membrane can be made permeable to lysozyme by the use of the bivalent ion chelator EDTA, which loosens the structure of LPS. Freezing certain strains of *E. coli* also achieves the same thing.
- The synthesis of new murein can be prevented by the addition of penicillin. Because in the presence of this drug cell growth is not inhibited, bacteria continue to synthesize their constituents, which, when accumulated in sufficient amounts, make a bulge through the murein layer. Eventually, the whole cell is extruded and becomes spherical.

These procedures do not lead to the complete removal of the outer membrane. For this reason, it has become customary to reserve the term *protoplasts* for Gram-positive cells and to use the noncommittal word SPHEROPLASTS for these Gram-negative forms. Spheroplasts are usually "dirty protoplasts"; they are defined as spherical forms that are osmotically sensitive but still contain some adherent outer membrane and entrapped murein. Like the Gram-positive protoplasts, spheroplasts may revert to the original cell shape once lysozyme or penicillin is removed from the medium.

Periplasm

The dual membrane system of Gram-negative bacteria creates a compartment called the PERIPLASMIC SPACE, or PERIPLASM, on the outside of the cytoplasmic membrane (Oliver, 1987). This compartment is approximately 20–40% of the cell volume, which makes it far from insignificant. The periplasm contains the murein layer and a gel-like solution of proteins that facilitates nutrition and inactivates toxic compounds. These proteins are released from cells by osmotic shock or by making spheroplasts. As shown in Table 2, the periplasmic proteins include

- Binding proteins that help soak up sugars, amino acids, vitamins, and minerals from the medium
- Degradative enzymes (phosphatases, nucleases, proteases, etc.) that break down large and impermeable molecules to digestible size
- Detoxifying enzymes that inactivate certain antibiotics, like the β-lactamases that destroy penicillins and cephalosporins

Gram-positive bacteria have no periplasm because they have no outer membrane. However, the cell wall is not freely permeable to very large molecules. Thus, it can be expected that proteins that are secreted across the cell membrane will be retained for some time between the membrane and the thick murein wall. Might this region be a primitive equivalent of the more sophisticated Gram-negative periplasm?

Does the periplasm have the same osmotic pressure as the external medium? This condition is what would be expected if the outer membrane were totally "leaky." Measurements of growing Gram-negative bacteria have shown that the periplasm has the same osmolarity as the interior, thus an osmotic gradient exists between the periplasm and the cell exterior. The osmolarity of the periplasm appears to be regulated at least in part by a high concentration of unusual compounds, the so-called MEMBRANE-DERIVED OLIGOSACCHARIDES. These periplasmic com-

Table 2. Representative periplasmic proteins from *E. coli*

Binding proteins
For amino acids (e.g., arginine, leucine)
For sugars (e.g., galactose/glucose, arabinose)
For vitamins (e.g., thiamine, vitamin B_{12})
For ions (e.g., phosphate, sulfate)
Degradative enzymes
Various phosphatases
Proteases
Endonuclease I
Detoxifying enzymes
β-lactamases (penicillin and cephalosporin inactivating enzymes)
Aminoglycoside-phosphorylating enzyme

pounds contain 8–10 glucose units in branched arrangements, multiply substituted with glycerophosphates. The synthesis of these compounds is sensitive to the osmotic conditions of growth and is almost completely suppressed in media of high osmolarity. These compounds may play a role in the osmoregulation of the periplasm because they are charged and have a molecular weight (approximately 2,400) that does not allow them to cross the outer membrane. However, mutants impaired in the synthesis of membrane-derived oligosaccharides grow normally even in dilute media, a result indicating that much remains to be learned about the physiological function of these compounds.

Acid-fast cell wall

A few Gram-positive bacterial types, notably the tubercle and leprosy bacilli, have developed yet another solution to the problems of environmental challenge to the cytoplasmic membrane. Their cell walls contain large amounts of WAXES, which are complex, long-chain hydrocarbons substituted with sugars and other modifying groups. Some of these waxes are known as MYCOLIC ACIDS. Having such a protective cover makes these organisms nearly impervious to many harsh chemicals, including acids. If a dye is introduced into these cells, for instance by brief heating, it cannot be removed by dilute hydrochloric acid, as would be the case in all other bacteria. These organisms are therefore called ACID-FAST or acid-resistant.

In addition to the waxy coat, the acid-fast wall contains murein, polysaccharides, and lipids. The wax enables the organisms to resist not only the action of many noxious chemicals but also killing by white blood cells. All of this is at a cost; these organisms grow very slowly, possibly because the rate of uptake of nutrients is slow—perhaps limited by their waxy covering. Some acid-fast bacteria, such as the human tubercle bacillus, divide only once every 24 hours, even under optimal conditions.

Crystalline surface layer

A crystalline surface structure represents yet another variation on the theme of the organization of the bacterial envelope. This structure consists of an outer layer of protein subunits arranged in a crystalline array that may be square, hexagonal, or oblique (Figure 12). It resembles in appearance the chain mail of medieval armor. This structure is sometimes known as the S-LAYER or RS-LAYER, and is located outermost on the cell surface. An S-layer is present on a large number of species that embraces all the major groups of bacteria. In the case of some archaebacteria, it is the only layer external to the cytoplasmic membrane. In Gram-positive cells, the S-layer is external to the murein wall; in Gram-negative cells, it is external to the outer membrane. In both Gram-positive and Gram-negative organisms, this structure is sometimes a multilayered sheet several molecules thick.

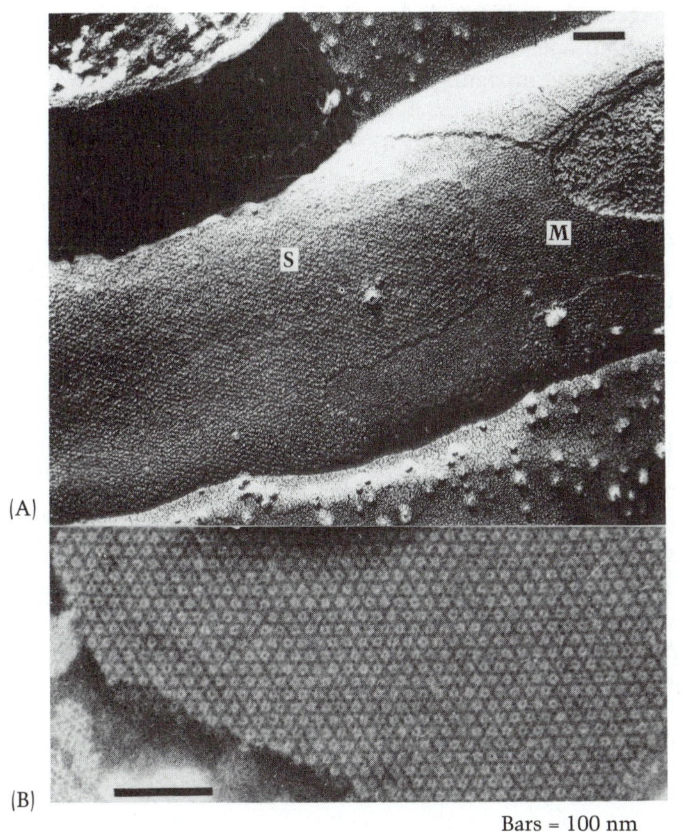

Figure 12

Crystalline layers of *Aquaspirillum serpens.* (A) Replica of a freeze-cleaved and etched preparation showing a regular hexagonal array (S) on the external surface of the cell wall. Part of this layer has been stripped away to reveal the cleaved surface of the outer membrane (M). (Reproduced with permission from F. L. A. Buckmire and R. G. E. Murray, 1970. (B) A negatively stained fragment of the S-layer, showing the arrangement of units. Each unit is a hexamer of a 140-kDa polypeptide. (Courtesy of R. G. E. Murray.)

An S-layer is usually made up of a single kind of protein molecule, sometimes with carbohydrates attached. The isolated molecules are capable of self-assembly, that is, of making sheets similar or identical to those present on the cells. It is possible to think of the S-layers as a primitive lipid-less membrane, consisting exclusively of proteins or glycoproteins. Not surprisingly, S-layer proteins are quite resistant to proteolytic enzymes and protein-denaturing agents.

The function of the S-layer is not entirely understood; but, as may be expected from its appearance, it probably plays a protective role. In some organisms (e.g., *Aeromonas salmonicida,* a fish pathogen), it helps protect from phagocytosis. It may serve both to guard bacteria from infection by phages and, by functioning as receptors, to promote infection by other phages. The S-layer also participates in the adherence of certain bacteria to surfaces. It is fair to say, however, that much needs to be learned about this intriguing structure.

In conclusion

The bacterial envelopes are unusually varied in structure and chemical composition, a diversity reflecting the enormous number of environmental challenges that are encountered by microorganisms. This diversity appears to be the price paid for unicellular free life. An overview of the variety of uses to which bacterial envelope components are put is presented in Table 3.

CAPSULE

Many bacteria, Gram-positive and Gram-negative alike, secrete a slime of amorphous structure. In most cases, this slime is a high-molecular-weight polysaccharide, either heteropolymeric (i.e., containing more than one kind of monosaccharide) or homopolymeric. The synthesis of this material is conditional; that is, it depends on the growth medium and is not essential at all times. Growth occurs readily without capsule formation, at least under laboratory conditions. However, the capsule is frequently a major determinant of a bacterial cell's ability to colonize a certain niche (e.g., *Streptococcus mutans* on the teeth). The capsule is also a major line of bacterial defense against phagocytes,

Table 3. Bacterial strategies utilizing envelope components

Some bacteria beset by	Have developed strategies involving
High internal osmotic pressure and life in dilute environments	Wall strength and integrity
Noxious enzymes	Barrier network of small pore size
Organic poisons and antibiotics	Destructive specific enzymes or a permeability barrier or mutation of target system
Toxic metal ions in the face of need for trace metals	Appropriate ligands for tapping and selective chelators for retaining essential ions
Predatory animalcules (viruses, protozoa, phagocytes, *Bdellovibrio* sp.; see Chapter 17)	Making or alteration or receptors; barrier layers (including S layers); physicochemically unpalatable surfaces, capsules, etc.
Strange and unsettling DNA	Appropriate nucleases
Elusive substrates or environments	Attachment mechanisms; swimming and gliding; chemotaxis; phototaxis
Variations in the physicochemical environment	Gas vesicles; effective and selective diffusion barriers; more appropriate membrane lipids

Source: From Koval and Murray (1986).

Table 4. Bacteria often possessing large capsules

Species	Composition	Common habitat[a]
Streptococcus salivarius	Levans (fructose polymers)	Human mouth
Streptococcus pneumoniae	Complex polysaccharides	Human respiratory tract
Haemophilus influenzae	Poly-ribitol phosphate	Human respiratory tract
Bacillus anthracis	Poly-D-glutamate	Soil
Azotobacter vinelandii	Polyuronides (uronic acids polymers)	Soil
Leuconostoc mesenteroides	Dextrans (glucose polymers)	Soil, vegetable matter

[a]Capsule formation may not take place in the most common habitats.

because its slipperiness impedes uptake of the bacteria by phagocytic cells. It is not surprising that many bacteria that must travel through the blood to reach a target organ are encapsulated; examples of such pathogens are the agents of bacterial meningitis—the meningococci and *Haemophilus influenzae.* A list of famous capsule makers is shown in Table 4.

FLAGELLA

The FLAGELLUM (plural: flagella) is the organ of bacterial locomotion (Macnab, 1987a). It is a helical filament that is driven by a motor at its base and rotates relative to the bacterial surface, thereby propelling the cell through the medium. Bacterial flagella impart motion by rotation only, not by bending, as is the case with eucaryotic flagella (thus, for bacteria, the term *flagellum,* which means "whip," is misleading). Flagella are a rare example of rotating shafts in biology, their presence allowing one to come to the conclusion that biological wheels do indeed exist.

Not all bacteria are motile, and some motile bacteria do not possess flagella. Other forms of motility exist, as seen in the so-called gliding bacteria (Chapters 6 and 16). Flagellated bacteria can be differentiated by the number and position of these organelles (Figure 13). Some species, such as members of the genus *Pseudomonas,* have only a single, polar flagellum. *Escherichia coli* has approximately 10 flagella; its relative *Proteus,* several hundreds. In these species, flagella are inserted all over the surface; such cells are called PERITRICHOUS ("hairy all over").

Each flagellum is composed of three parts, which have different molecular complexities (Table 5). Outermost is the long helical FILAMENT, which extends 5 to 10 μm into the medium—several times the length of the cell (Figures 14 and 15). The filament is connected via a HOOK to the BASAL BODY, a complex structure that anchors the flagellum to the cell envelope and serves as the motor that turns the flagellum.

The filament is composed of a protein, FLAGELLIN; there are several thousand copies of this protein in each flagellum. In a few species (e.g., *Caulobacter*), flagella are composed of two types of flagellin, but in most

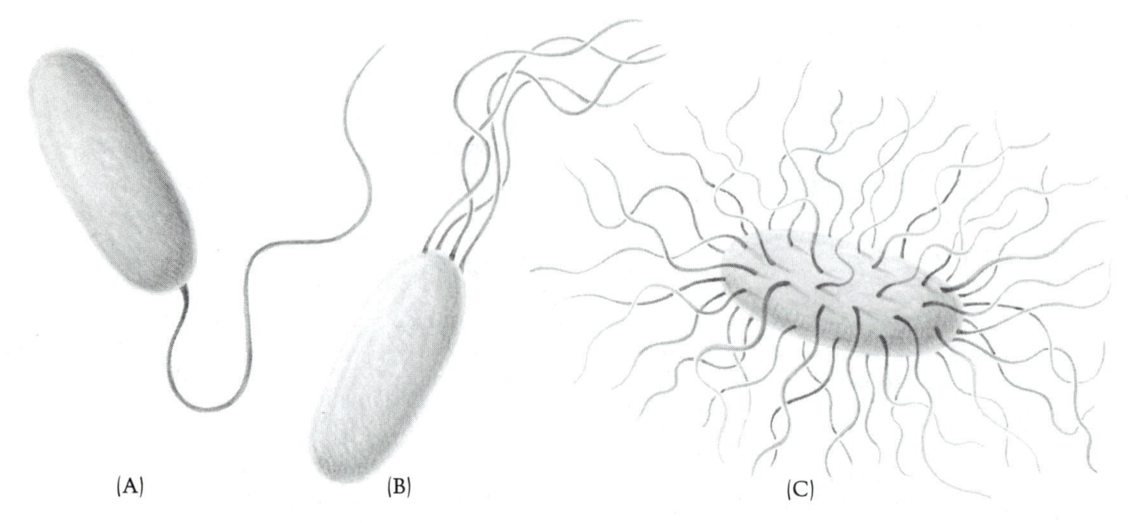

Figure 13

Arrangements of flagella in different bacteria. (A) Polar, single flagellum. (B) Polar, multiple flagella. (C) Peritrichous flagella. (Modified from Schaechter et al., 1989.)

Table 5. Some of the components of the *E. coli* flagella

Components[a] (approx mol wt)	Location
14,000	Basal body
27,000	Basal body
38,000	Basal body
65,000	Basal body
42,000	Hook
30,000	Hook
35,000	Hook-filament junction
60,000	Hook-filament junction
17,000	Cytoplasmic membrane
22,000	Cytoplasmic membrane
32,000	Cytoplasmic membrane
34,000	Cytoplasmic membrane
14,000	Cytoplasm
21,000	Cytoplasm
53,000	Filament
55,000	Filament

Source: From Macnab (1987a).

[a]Several other proteins are needed for flagellar assembly and function, but their location or site of activity are not necessarily known.

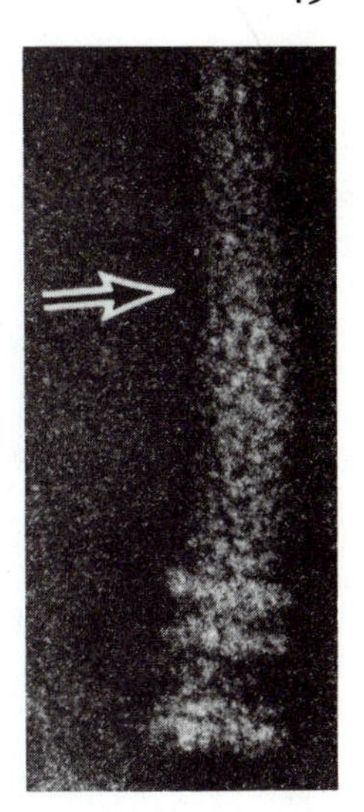

Figure 14

An isolated flagellum from *E. coli.* The filament, the hook (arrow), and the four rings of the basal body are revealed by negative staining. (From DePamphilis and Adler, 1971.)

only a single type is found. Flagellins are highly antigenic, and some of the immune responses to infections are directed against these proteins. Mechanically speaking, flagellins are strikingly rigid, about 100 times more so than the F-actin of muscle. The diameter of a flagellum is about 20 nm in *E. coli* and is constant throughout its length. The wavelength of the helical turns varies among species but is typically 2 to 2.5 μm per turn.

Flagellin molecules aggregate spontaneously to form the characteristic structure of the flagellar filament. Isolated filaments can be dissociated into a solution of flagellin in vitro; if the ionic environment is properly adjusted in the presence of a primer, the flagellin molecules spontaneously reaggregate to form filaments indistinguishable from the natural product. This event is a fine example of biological morphogenesis taking place by the self-assembly of molecules. In vitro, the rate of elongation is constant, and "growth" of the filament is from the end that would be distal to the cell.

The hook is a short curved structure that connects the flagellar filament to the cell. It appears to act as the universal joint between the motor in the basal structure and the filament. The molecular structure of the hook is also quite simple. Like the filament, it is an aggregation of a single type of protein molecule called hook protein. The hook is slightly larger in diameter than the filament and has a constant length—approximately 80 nm in *E. coli.*

The flagellar hook is connected to the basal body, which is a small but complex structure embedded in the cell surface. The basal body is composed of 15 or more proteins that aggregate to form a rod to which four rings are attached (at least in Gram-negative cells; Figures 14 and 15). The rings appear to act as bushings or as "stators," on the one hand anchoring the structure in the various layers of the cell envelope, on the other hand allowing the rod (the "rotor") to rotate. It is not yet known how the rod portion of the basal body is physically retained on the cell surface. As shown in Figure 14, each ring corresponds to a specific layer of the Gram-negative cell envelope.

As might be expected, the variation in envelope structure between Gram-positive and Gram-negative bacteria is reflected in a difference in the basal bodies of their flagella. Gram-positive cells have only two rings—one embedded in the cell membrane and another associated with the teichoic acid component of the wall.

What makes flagella turn? No one really knows the details but the energy source for the rotation of the basal body and its attached hook and filament is known to be the protonmotive force, or membrane potential. This energy is generated by the electron transport chain during respira-

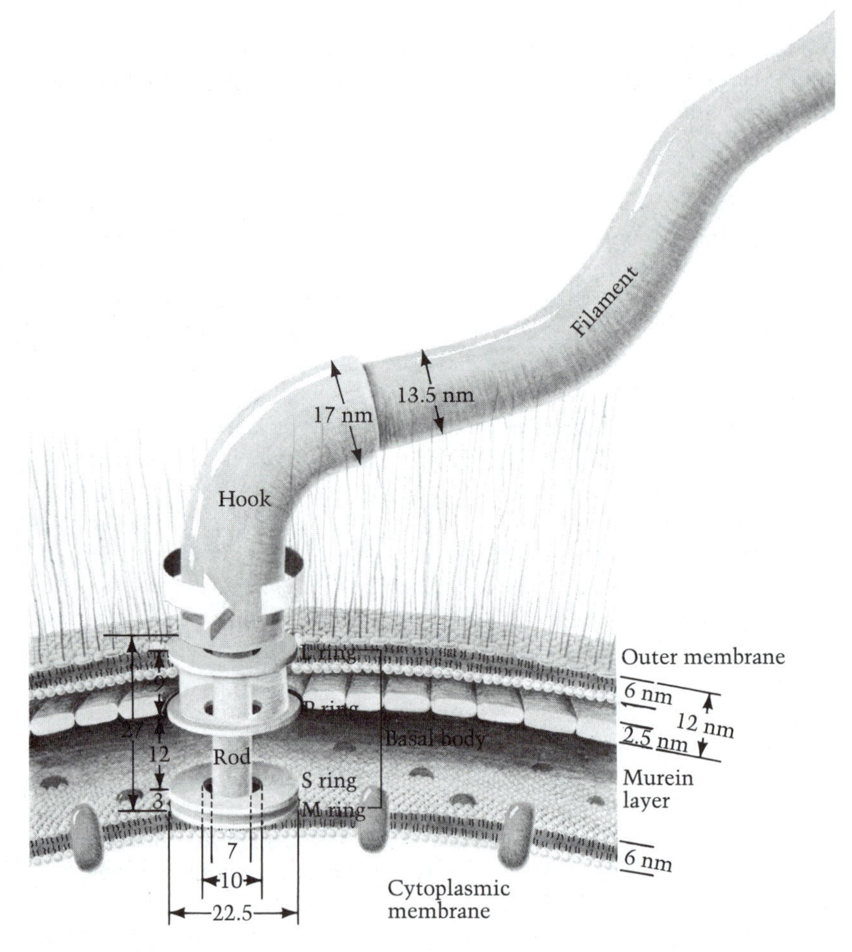

Figure 15

Basal body structure of the *E. coli* flagellum. The relationship of the four rings of the basal body and the envelope of *E. coli* are shown. (After DePamphilis and Adler, 1971.)

tion or, anaerobically, by hydrolysis of ATP. The flagellar motor is a highly efficient device and requires the passage of only about 1,000 protons per turn (Chapter 6).

Flagellar motors are part of a complex behavioral system that enables motile bacteria to move toward environments favorable for growth and away from hostile environments (Macnab, 1987b). How is this done? Flagellar rotation can take place in either direction, clockwise or counterclockwise. The choice of direction makes a lot of difference in the outcome: counterclockwise rotation propels the bacterium smoothly forward, a process called SWIMMING; clockwise rotation leads to TUMBLING. The reason for this difference is that the flagellar filaments are normally left-handed helixes. Thus, rotation in the counterclockwise direction exerts a pushing motion. You may see this for yourself by constructing a left-handed helix with a paper clip and turning it in either direction. Counterclockwise rotation allows the flagellar filaments to sweep around the cell and to make a common bundle that can operate in concert. Clockwise rotation causes the bundle to disperse.

Swimming is normally interrupted by episodes of tumbling, and the length of time for each episode is determined by compounds in the environment. The flagellar motor responds to chemical stimuli called ATTRACTANTS or REPELLANTS, or, more accurately, to gradients of concentration of such compounds. The concentration of an attractant or repellant determines the length of time flagella will turn counterclockwise and clockwise. The net result of this behavior is CHEMOTAXIS, the ability of motile bacteria to swim toward attractants and away from repellants. Chemotaxis is described in more detail in Chapter 6.

PILI

Pili (singular: pilus, or "hair") are a second type of protein structure projecting beyond the bacterial surface (Eisenstein, 1987). They are also known as fimbriae. Pili are organelles of attachment to surfaces and exhibit remarkable specificity. A typical *E. coli* cell has 100 to 300 of these organelles. Pili originate from the cell membrane and extend 0.2 to 2 μm into the medium (Figure 16). They are composed of structural proteins termed PILINS. Some pili contain a single type of pilin, others contain more than one. Minor proteins, sometimes located at the tip of pili, are responsible for the attachment properties. Pilin molecules are arranged helically to form a straight cylinder. In some respects, the structure of a pilus is similar to that of a flagellum; however, a pilus is a straight rod rather than a helix, it does not rotate, and it lacks a complex basal body.

Certain pili, termed SEX PILI, play a vital role in bacterial conjugation. They form the initial attachment between mating pairs. Other pili, termed TYPE 1 PILI or COMMON PILI, are involved in the attachment of bacteria to other interfaces, notably the surface of eucaryotic cells. Thus, the gonococcus (*Neisseria gonorrhoeae*), *E. coli*, and other pathogens attach to membranes of the urinary or genital tracts by means of their pili. The attachment results from specific binding between pili and specific receptors—probably glycoproteins—of the host cell surface. The term ADHESINS has been given to the minor proteins in pili that play this role in host–parasite interactions.

Pili of different bacteria are antigenically distinct and elicit the formation of antibodies of different specificities. Thus, antibodies against the pili of one bacterial species will not impede the attachment of another species. Even within one species, bacteria such as the gonococcus have evolved the ability to make pili of different antigenic types. By switching the type of pili, a gonococcal cell can still adhere to cells after having elicited a powerful antibody response to its original type of pili.

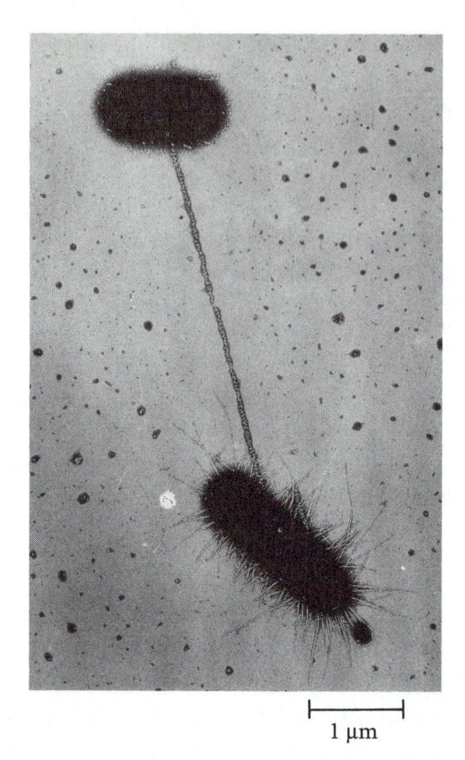

Figure 16

Pili of *E. coli*. The cell, covered with numerous hair-like appendages, is a genetic donor connected to a recipient cell (without appendages) through an F1 pilus. The other appendages are called Type 1 pili and play no role in conjugation. (Courtesy of C. Brinton Jr.)

POLYSOMES

The ribosomes that appear to fill the cytosol (Figure 10A in Chapter 1) are complex organelles that catalyze the process of translation of the information encoded in messenger RNA (mRNA) into protein (Noller and Nomura, 1987). Several ribosomes traverse the same molecule of mRNA simultaneously, thereby forming structures called POLYSOMES. At any moment, 80–90% of the ribosomes of a growing bacterium are found in polysomes and are actively engaged in protein synthesis. Some remarkable electron micrographs by O. Miller enable one to visualize the role of polysomes in protein synthesis and gene expression. The one shown in Figure 17 illustrates the expression of a single gene. A set of polysomes of increasing length is seen attached to a segment of DNA. The increasing lengths of the polysomes (from left to right in the picture) indicate that transcription of DNA into mRNA was proceeding in that direction. Indeed, RNA polymerase can be discerned as a small object at each DNA–RNA junction. The picture illustrates the proposition that polysomes are forming while mRNA is still being synthesized. Because it is known that ribosomes attach near the end of the mRNA molecule, the picture establishes the fact that ribosomes move down the

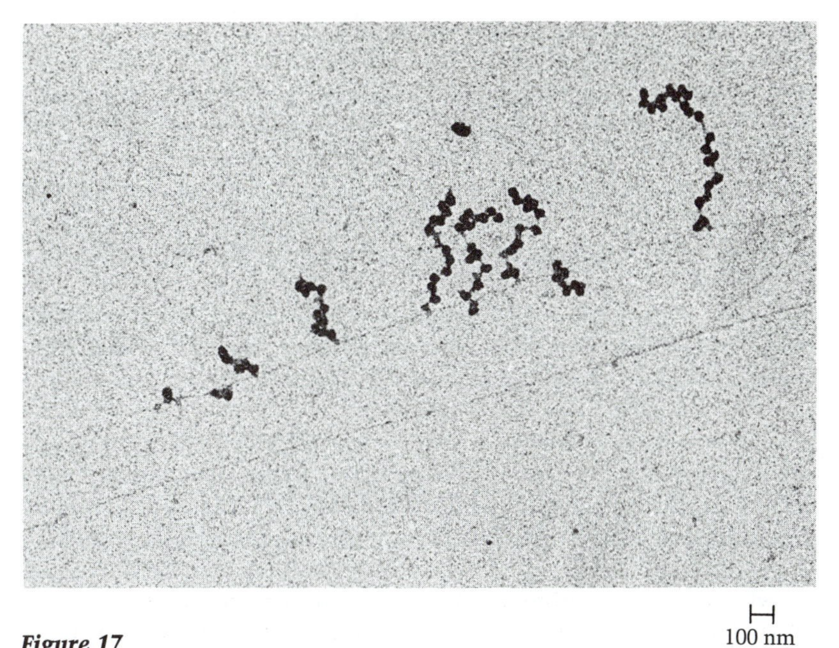

Figure 17

Polysomes attached to transcripts of a single gene. The two thin, nearly horizontal lines are strands of DNA. Polysomes are attached to the upper strand. Their length increases from left to right, an observation indicating that transcription occurred in the same direction. Molecules of RNA polymerase appear as small objects at the intersection of the polysomes and the DNA. Nascent protein molecules, presumably attached to the ribosomes, cannot be resolved. (From Miller et al., 1970. Copyright 1970 by the A.A.A.S.)

mRNA molecule, synthesizing protein as they go. The nascent protein molecules attached to the individual ribosomes are not visible in this micrograph.

The procaryotic body plan permits translation (protein synthesis) and transcription (mRNA synthesis) to occur simultaneously. In eucaryotic cells, transcription takes place within the membrane-bounded nucleus, whereas translation occurs largely in the cytoplasm on the endoplasmic reticulum. This difference is a fundamental distinction between the two forms of life, one that has important physiological consequences. We shall see later how the procaryotic arrangement permits speedier adjustment of gene expression.

Ribosomes are complex ribonucleoprotein particles, about the size of the smallest viruses (Figure 18). They are roughly heart shaped, approximately 25 nm in the largest dimension. The molecular architecture of ribosomes is unique in the biological world; they are made up of over 50 different proteins and 3 kinds of RNA. In *E. coli,* ribosomes are composed of 38% protein and 62% RNA (called rRNA for ribosomal RNA). Thus, ribosomes differ from viruses and other structures in that RNA is their major structural element by weight. A notion that is currently under serious consideration is that ribosomal RNA plays not only a structural but also a functional role (although the function still awaits discovery). This notion is made more plausible by the finding of RIBOZYMES—RNA molecules that can function as enzymes.

Procaryotic ribosomes sediment in a centrifugal field at a velocity of 70 svedberg units (S) and on this basis are called 70S ribosomes. Eucaryotic ribosomes are 80S, and those of mitochondria are highly variable. By reducing the concentration of Mg^{2+}, bacterial ribosomes can be dissociated into two subunits—a 50S subunit and a 30S subunit (svedberg units are not linearly additive). This dissociation takes place in vivo each time the synthesis of a protein molecule is completed. The smaller subunit contains a single RNA molecule (16S) and 21 different proteins, which are designated S (for "small" subunit) 1 through S21. The larger subunit contains two RNA molecules (5S and 23S) and 32 different proteins, designated L (for "large" subunit) 1 through L32. All ribosomal proteins are encoded by different genes and thus differ in primary structure, with the exception of L7 and L12, which differ only by the presence of an acetyl group on the amino terminus of L7. Most are small, the majority ranging in molecular weight from 7,000 to 25,000. They are quite basic, as would be expected of proteins that make stable complexes with RNA. A number of these proteins are modified posttranscriptionally by acetylation or methylation; the significance of this fact is not understood. With the exception of L7/L12 (which is present in four copies per ribosome), all the ribosomal proteins and rRNA molecules are present as a single copy per ribosome. This finding implies that all ribosomes in a

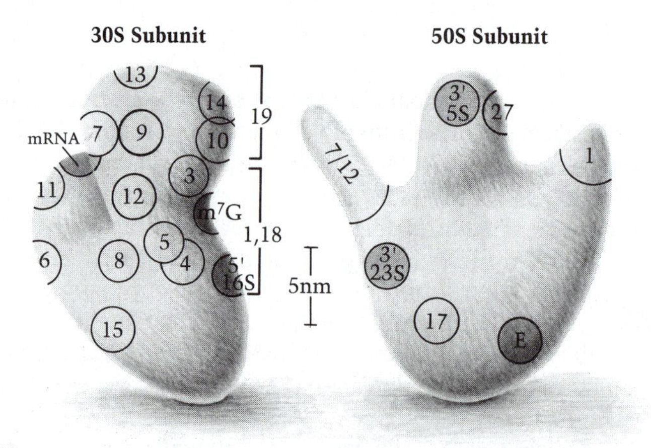

Figure 18

The shape of *Escherichia coli* ribosomal subunits. The open circles indicate the known positions of some of the ribosomal proteins. Shaded circles show the known positions of some of the ends of the RNAs (3′5S, 5′16S, 3′23S), the site of entry and exit of mRNA (mRNA), the exit site of the nascent polypeptide (E), and the site of a methylated guanine on the 16S RNA (m^7G). Only the exterior face of the subunits is shown.

given bacterium are alike and that "specialized ribosomes" dedicated to the synthesis of distinct proteins do not exist.

Both the external and the internal architectures of ribosomes are beginning to be elucidated, thanks to a combination of powerful physical and biochemical techniques. These techniques include electron microscopy, neutron and X-ray diffraction, chemical cross-linking with bifunctional reagents that bind to proteins and rRNAs, fluorescent probes, and immunoelectron microscopy. In addition, the constituents of ribosomes can be isolated and, when mixed together under appropriate in vitro conditions, reassemble into active ribosomes. The order of addition of the components is crucial for proper reconstitution, a condition suggesting that specific interactions must take place between individual molecules (see Chapter 3 for details). Moreover, the complete nucleotide sequence of the 5S, 16S, and 23S rRNA molecules is now known, knowledge that enables one to speculate about the degree of secondary structure (e.g., double-strandedness) in the molecules. Certain general facts are clear: rRNA molecules contain significant double-stranded regions; they serve as an extended core to which specific proteins are bound at specific regions; the relative position of the proteins is definite and fixed. Thus, ribosomes are complex and highly asymmetric structures, which makes the study of their detailed architecture an awesome task.

Although most studies on the bacterial ribosomes have been on those from *E. coli,* sufficient information from other species is now available to reveal a remarkable similarity among all procaryotic ribosomes. It turns out that both their rRNAs and ribosomal proteins have been highly conserved through evolution. This similarity is dramatically illustrated by the finding that proteins taken from 30S ribosomes of *E. coli* will bind specifically to the 16S rRNA of other enteric bacteria, of species of *Bacillus,* and of diverse bacteria such as *Anacystis nidulans* (a cyanobacterium), *Clostridium pasteurianum* (a strict anaerobe), *Chromatium vinosum* (a phototrophic purple-sulfur bacterium), and even certain archaebacteria. The implication from these findings is that gross variations on the central theme of the procaryotic ribosome are not possible, or, at least, that evolution has not figured out how to make drastically different functional ribosomes. The high degree of conservation in 16S rRNA has been used extensively to study relationships among organisms that span wide taxonomic distances (Woese, 1988, Chapter 9).

NUCLEOID

The DNA of the bacterial cell is contained in a region called the NUCLEOID (Drlica, 1987). This term (meaning "nucleus-like") emphasizes a fundamental structural distinction between the organization of the genome in procaryotes and the membrane-bounded true nucleus of eucaryotes. In ultrathin sections of a bacterium examined under the electron microscope (Figure 13 in Chapter 1), the nucleoid appears to be an amorphous body located roughly in the center of the cell, with irregular lobules radiating outward. This appearance is not an artifact of fixation and embedding, because similar structures can be seen in living cells. Visualizing the living nucleoid requires special manipulations, because normally it is optically obscured by the ribosome-rich cytoplasm. Thus, if cells are suspended in a medium with a refractive index matching that of the cytoplasm, the contrast between the nucleoid and its surroundings is enhanced and its structure becomes visible. Under these conditions, it is possible to follow nucleoids of growing bacteria under the phase contrast microscope (Figure 19). As can be seen by time-lapse photography, the nucleoids appear to divide without the morphological complexities seen in eucaryotic mitosis. Rather, nucleoids just seem to pull apart.

Bacteria contain from one to several nucleoids, depending on their rate of growth (Chapter 15). On the basis of the cellular DNA content, it can be calculated that each nucleoid contains a complete chromosome of the organisms. In other words, bacteria have one CHROMOSOME, and the terms *chromosome, DNA content,* and *nucleoid* refer to the same unit, although they are used in different contexts. Chromosomal DNA of bacteria is a single, covalently linked, ring-shaped molecule of staggering dimensions. It is about 1,000 times longer than the bacterium that contains it, or over 1 mm long. For such a long molecule to be accomodated, the DNA must be exceedingly thin. The DNA of *E. coli* is, in fact, only about 30 Å (3 nm) in thickness, and its length-to-width ratio is over 300,000 to 1. If DNA were magnified to the size of regular spaghetti, like that served in the United States, the length of the *E. coli* DNA would be the equivalent of 200 regular platefuls with each spaghetti linked end to end to another, and the last one on one plate linked to the first one of the

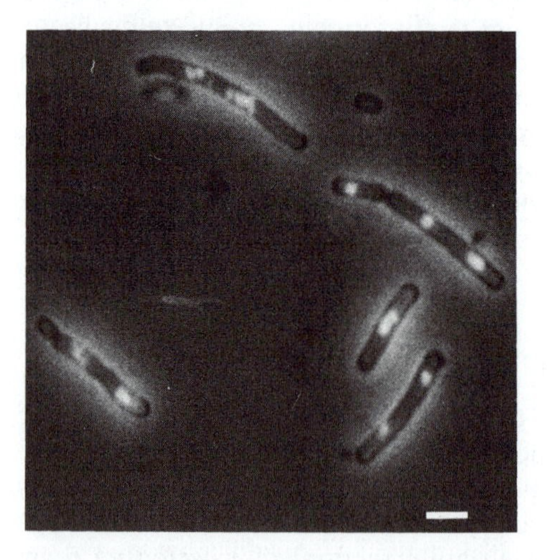

Figure 19

Nucleoids of *E. coli.* Cells were treated with ethidium bromide, a fluorescent dye that intercalates into DNA. Cells were observed under a fluorescence microscope. The bar represents 1 μm. (From Drlica, 1987.)

next. Thus, DNA is a remarkable molecule, not only for its informational content, but for its physical properties as well.

Nucleoids constitute perhaps 10% of the cells' volume. Why is the DNA not dispersed throughout the bacterial cell instead of being confined to a small region? Within the nucleoid, the local concentration of DNA reaches 10–20%, which is several orders of magnitude higher than what can usually be achieved in solution with DNA of such immense molecular weight. A 0.1% solution of DNA in a test tube is a gel! DNA in the living cell must then be in a special physical state. What little is known about the ability of DNA to fold into such a compact structure comes from in vitro studies. When bacteria are broken by mechanical means using common buffers, the lysate makes the medium highly viscous, because the DNA molecules "explode" into long strands. However, if the ionic strength of counterions such as Mg^{2+} is high (e.g., 0.1 *M*), the nucleoids remain condensed and the lysate does not increase in viscosity. Thus, in living cells the ionic environment within and around the nucleoids must induce a special state of condensation of the DNA. Note that DNA packaged within viral particles may achieve an even greater state of condensation.

Studies with isolated nucleoids have revealed that DNA condensation requires supercoiling of the molecules. Measurements were made to determine how many single-strand breaks (nicks) had to be introduced to relax all the supercoiling of the DNA of isolated nucleoids. The number of such nicks defines regions of topological constraint on the rotation of the double helix. From the number of nicks required for complete relaxation, it has been estimated that nucleoids contain 50 loops of supercoiled DNA. The supercoiled state of DNA has also been determined in vivo, without removing nucleoids from the cell. Sinden and co-workers (1980) made use of the fact that the compound psoralen binds to double-stranded DNA at a rate proportional to its degree of supercoiling, or torsional strength. Nicks can be introduced by gamma irradiation of intact cells; thus, the amount of psoralen bound as a function of the number of nicks produced allows one to estimate the number of supercoil domains in vivo. The number obtained was reasonably close to that observed in vitro.

How does a molecule as physically complex as the bacterial DNA replicate, segregate, and become transcribed into messenger RNA? Clearly, it is not easy to visualize how such a enormous "plate of spaghetti" can keep from becoming hopelessly tangled as it fulfills its many functions. The implications of nucleoid structure and supercoiling for DNA replication, segregation, and transcription are discussed in Chapters 3 and 14.

STORAGE GRANULES AND OTHER INCLUSION BODIES

Frequently, granular inclusion bodies are seen in sections of procaryotes (cf. Shively, 1974). In the case of *E. coli,* these structures are electron-dense bodies of uneven appearance, 20–100 nm in diameter, and consisting of glycogen. Thus, they are storage deposits that accumulate in response to excess carbon in the medium when the source of nitrogen, sulfur, or phosphorus is limiting or when the pH is low. Upon carbon starvation, these granules disappear as the glycogen is used up.

A number of bacteria, including many enteric bacteria, the spore-formers, and the cyanobacteria, store carbon as glycogen. Others, including pseudomonads and rhizobia, store carbon in poly-β-hydroxyalkane granules (Figure 20). Still others, including certain photosynthetic bacteria, have the capacity to store carbon in either of these forms. Some bacteria (e.g., *Acinetobacter*) appear to be incapable of storing carbon in any insoluble form.

Carbon-rich compounds are not the only nutrients stored by bacteria. Certain species store phosphate as granules of polyphosphate and sulfur as elemental sulfur.

In some bacteria, inclusion bodies exist not to store energy-rich compounds but to fulfill highly specialized roles. Thus, some bacteria are magnetotactic, that is, they respond to magnetic fields (Chapter 17) by virtue of possessing small intracellular iron grains.

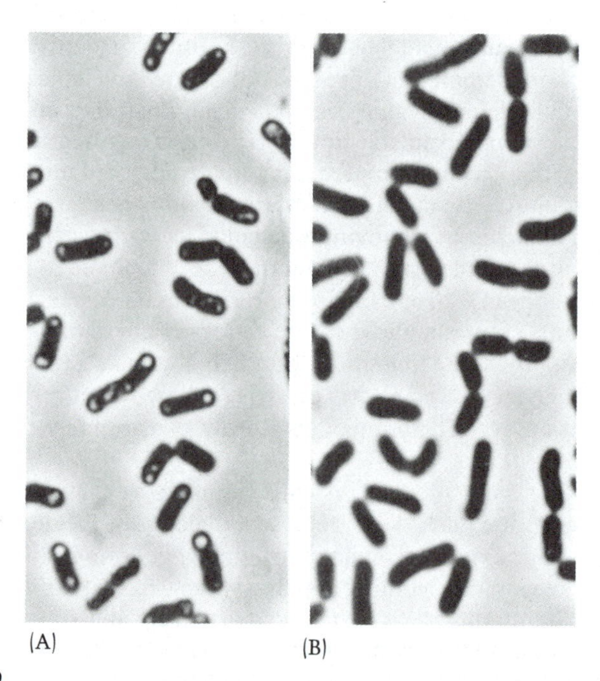

Figure 20

The formation and utilization of poly-β-hydroxyalkane in *Bacillus megaterium* (phase contrast, ×2,200). (A) Cells grown in a high concentration of glucose and acetate. All cells contain one or more granules of poly-β-hydroxyalkane (light areas). (B) Cells from the same culture after incubation for 24 hours in the absence of a carbon source. (Micrographs by J. F. Wilkinson.)

VARIATIONS ON THE ORGANIZATIONAL THEME

In addition to the profound differences in the organization of bacterial cells that distinguish the three major classes—Gram-positive bacteria, Gram-negative bacteria, and archaebacteria—a number of other important, but less fundamental, variations on the main organizational theme occur among bacteria. Their study constitutes the richly diversified field of general microbiology—not the topic of this book, but one extensively considered in others (e.g., Stanier et al., 1986). Some of the more obvious variations in procaryotic cell organization are discussed throughout our book. In particular, you should refer to bacterial variations in the generation of ATP (Chapter 5), in the means of locomotion (Chapter 6), in the response to physical and chemical agents (Chapter 8), in differentiation (Chapter 16), and in the "social life" and ecology of bacteria (Chapter 17).

SUMMARY

1. The cytoplasmic membrane of bacteria not only is the site of transport of metabolites but also is involved in a large number of other activities, including synthesis of lipids, respiratory metabolism, and chromosome segregation. Accordingly, in addition to phospholipids, it contains a large number of proteins.
2. In most bacteria, the cytoplasmic membrane is surrounded by a cell wall responsible for the shape of the cells. Usually the wall is composed of murein, a lattice-like structure made up of glycan backbones cross-linked through amino acid side chains. Removing the cell wall results in the formation of spherical, osmotically sensitive forms called protoplasts or spheroplasts.
3. In Gram-positive bacteria, the cell wall is thick, because it is made up of many layers of murein. In some, the wall also contains teichoic and teichuronic acids. Molecules of a modified type of teichoic acids, the lipoteichoic acids, are anchored in the membrane. The thick hydrophilic wall excludes hydrophobic compounds.
4. In Gram-negative bacteria, the wall is one or a few murein layers thick. The cytoplasmic membrane is protected by an outer membrane which contains lipopolysaccharide in its outer leaflet. Polysaccharide chains, located on the outer membrane and facing the outside of the cell, impede the passage of hydrophobic compounds.
5. Small-molecular-weight compounds traverse the Gram-negative outer membrane through pores made up of proteins called porins. In addition, specific mechanisms carry out the transport of certain larger compounds.
6. In Gram-negative bacteria, the inner and outer membranes are joined at many zones of adhesion. One special zone of adhesion—the periseptal annulus—goes around the circumference of the cell.

7. The acid-fast bacteria are surrounded by a layer of waxes that protects the cells from drying and caustic chemicals. These bacteria are called acid fast because, once stained, they are resistant to decolorization with acid solutions.
8. Many bacteria are surrounded by a layer of viscous polymers, usually polysaccharides, known as the capsule. Capsules play important ecological roles, for example, in adhesion of the bacteria to surfaces or in their protection from phagocytes.
9. Some bacteria move by the rotating action of flagella, which are long, helical, protein filaments embedded in the envelope layers by means of a series of rings. Flagellar motion is responsible for chemotaxis—the movement toward attractant and away from repellants. Gliding bacteria move without flagella.
10. Pili, which are thin protein rods, are present on the surface of some bacteria and permit the organisms to adhere to surfaces. Some are involved in chromosomal transfer during bacteria conjugation.
11. In growing bacteria, most mRNA molecules are being translated by more than one ribosome at a time, thereby forming polysomes. In turn, many polysomes are attached to DNA by nascent chains of mRNA. Transcription and translation are thus spatially and temporally linked in bacteria. Bacterial ribosomes have smaller subunits and shorter rRNA chains than those found in eucaryotic cells.
12. The bacterial DNA is about 1,000 times longer than the cell; thus, it must be repeatedly folded to form a compact structure. This folding is due in part to supercoiling along discrete regions of the molecule. Counterions, such as magnesium, also play a role in maintaining the shape of this structure.
13. Many bacteria can store energy-rich compounds in so-called inclusions or storage granules. These granules may be composed of polysaccharides, polyphosphates, lipids, or elemental sulfur. The contents of the granules are used up when the cell faces starvation.

STUDY QUESTIONS

1. The envelope layers of *Escherichia coli* are thought to have the following approximate thickness: outer membrane, 8 nm; periplasm, 10 nm; cell membrane, 8 nm. Assuming that an *E. coli* cell is close to 1.2 μm in length and 0.7 μm in width, calculate the proportion of the cell volume occupied by each of the envelope layers.
2. Discuss the major strategies used by different groups of bacteria to protect their cytoplasmic membrane.
3. What are protoplasts; spheroplasts? How are they made? For what are they used in research?
4. What would be the physiological consequences of different mutations with impairments in the biosynthesis of the main components of lipopolysaccharide?
5. How do Gram-negative bacteria allow the entry of compounds with a molecular weight of 700 daltons and greater?

6. Contrast the structure of the basal body of flagella in Gram-positive and Gram-negative bacteria.
7. What roles do pili and capsules play in the ecology of certain bacteria? How might this problem be studied experimentally?
8. Review the structure of bacterial ribosomes. What makes these structures unique?
9. What keeps the DNA in a condensed form within bacteria?
10. Discuss how environmental factors may affect the main subcellular constituents and storage devices of bacteria.

SUGGESTED READING

Topics in this chapter are covered in the following chapters from *Escherichia coli and Salmonella typhimurium: Cellular and Molecular Biology*, F. C. Neidhardt, J. L. Ingraham, K. B. Low, B. Magasanik, M. Schaechter and H. E. Umbarger (eds.). 1987. American Society for Microbiology, Washington, D.C.

Drlica, K. The nucleoid. (Volume 1; Chapter 9)
Eisenstein, B. Fimbriae. (Volume 1; Chapter 8)
Macnab, R. M. Motility and chemotaxis. (Volume 1; Chapter 49)
Macnab, R. M. Flagella. (Volume 1; Chapter 7)
Nikaido, H. and Vaara, M. Outer membrane. (Volume 1; Chapter 3)
Oliver, D. B. Periplasm and protein secretion. (Volume 1; Chapter 6)
Park, J. T. The murein sacculus. (Volume 1; Chapter 4)

Other suggested readings include:

Shively, J. M. 1974. The inclusion bodies of prokaryotes. Annu. Rev. Microbiol. 28:167.
Sleytr, U. B. and Messner, P. 1983. Crystalline surface layers of bacteria. Annu. Rev. Microbiol. 37:311.
Smit, J. 1987. Protein surface layers of bacteria. In *Bacterial Outer Membranes as Model Systems*. M. Inouye (ed.). Wiley, New York, p. 343.

3

Assembly and Polymerization: The Bacterial Interior

INTRODUCTION

We return now to the observation with which we began our study of the bacterial cell in Chapter 1—through the light microscope we watched cells make faithful copies of themselves from ingredients in the medium. In both Chapters 1 and 2 we considered the chemical and structural end products of the reproductive process, now we turn our attention to the process itself. In this chapter and continuing into Chapters 4 and 5, we consider the chemistry of synthesis of a bacterial cell. We begin by considering how metabolically skillful chemoheterotrophs like *E. coli* make a new bacterial cell from glucose and some salts. Then we turn our attention to how other kinds of bacteria reproduce themselves using different sources of their constituent elements and other means of obtaining metabolic energy. In these respects, the variations among bacteria are considerable: for example, different species can probably use all naturally occurring sources of carbon and can exploit, as a source of energy, a vast variety of naturally occurring exergonic reactions. But in spite of this dramatic diversity, we find an overriding theme of unity: most bacteria contain about the same set of macromolecules and they put them together in about the same way.

OVERVIEW OF METABOLISM

Approximately 2,000 distinct metabolic reactions participate in the synthesis of an ordinary bacterial cell. On the basis of their primary function in growth, these metabolic reactions can be catagorized as assembly reactions, polymerization reactions, biosynthetic reactions, and fueling reactions (Figure 1).

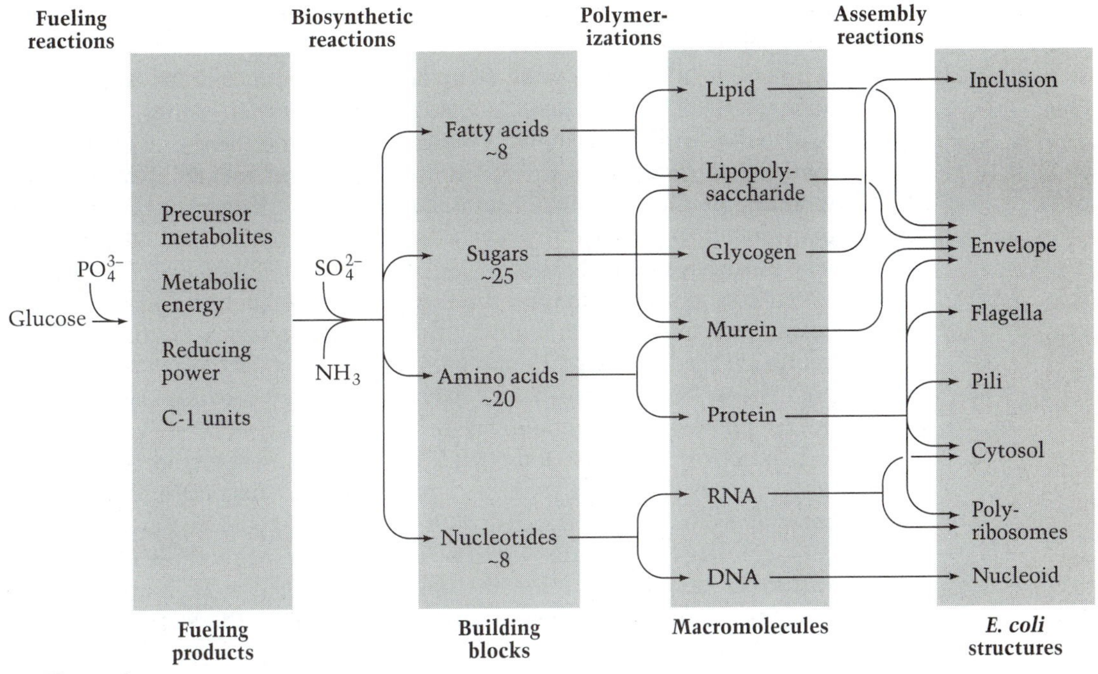

Figure 1

Overview of metabolism leading to the chemical synthesis from glucose of a chemoheterotroph like *E. coli*.

ASSEMBLY REACTIONS involve the chemical modification of macromolecules, their transport to prespecified locations in the cell, and their association to form cellular structures: envelope, appendages, nucleoid, polysomes, inclusions, and enzyme complexes. In some cases, the structures form by spontaneous association of macromolecules—SELF ASSEMBLY. In other cases, other macromolecules must aid the process—DIRECTED ASSEMBLY.

POLYMERIZATION REACTIONS consist of the directed, sequential linkage of activated molecules into long (sometimes branched) chains. All the macromolecules are formed from BUILDING BLOCKS that include 20 amino acids, 8 nucleotides, numerous sugars, and fatty acids. Polymerization of building blocks into proteins, RNA, DNA, and glycogen occur inside the cell, whereas the final steps of their polymerization into lipopolysaccharide, capsule, and murein occur outside the cell membrane.

BIOSYNTHETIC REACTIONS produce the building blocks of polymerization reactions; they also produce cofactors and related compounds including signaling molecules called ALARMONES. The hundreds of biosynthetic reactions are grouped into functional units called BIOSYNTHETIC PATHWAYS, each consisting of from one to a dozen sequential reactions that produce one or more building blocks. Biosynthetic pathways may be linear, branched, or in some cases, interconnected; each pathway is controlled en bloc and in some cases part or all of the pathway consists of enzymes made from a single mRNA molecule transcribed from a set of contiguous genes, an OPERON. All cellular biosynthetic pathways begin with one or another of a small group of elite compounds called PRECURSOR METABOLITES. There are just 12 of these precursor metabolites, yet they lead to the synthesis of about 75 building blocks and coenzyme products. With a few exceptions, the routes from precursor metabolites, to end products are the same in all bacteria, but species differ in the number of pathways they possess. To grow, a species that lacks a particular pathway must have its end product available in the medium.

FUELING REACTIONS produce the 12 precursor metabolites and reducing power (as reduced pyridine nucleotides); they also conserve metabolic energy (in the form of protonmotive force, ATP, or some other source of high-energy bonds) needed for biosynthesis. They produce the energy needed for polymerizations and all other endergonic processes of the cell (e.g., transport, motility, and repair). Fueling reactions include all those pathways referred to as CATABOLIC (pathways that degrade and oxidize substrates) and AMPHIBOLIC (pathways that produce both energy and precursor metabolites). Certain of these reactions—notably those that constitute the CENTRAL PATHWAYS (including glycolysis, the TCA cycle, and the pentose pathway)—are well-nigh universal, but otherwise there is a rich diversity of fueling reactions among various bacterial species. AUTOTROPHS, for example, have two separate sets of fueling reactions: one that conserves energy and reducing power, another that makes precursor metabolites from CO_2. Even within a single heterotrophic species, the reactions used when presented with glucose are somewhat different from those used when presented with a different substrate, for example, succinate or acetate. Fueling reactions differ also in facultative anaerobes, depending on whether they are growing aerobically or anaerobically. In some cases, evolution has lead to alternative pathways for handling the same substrate (e.g., glucose by either the glycolytic pathway or the Entner-Doudoroff pathway). The sole (but significant) biochemical unity in these various fueling reactions is their metabolic purpose: forming precursor metabolites and conserving energy and reducing power.

The assignment of certain other cellular reactions to one or another of these four metabolic reaction groups is somewhat arbitrary. REPAIR REACTIONS (which chiefly involve DNA) can be included with polymerizations. TRANSPORT REACTIONS (which enable uptake and retention of nutrients) can be included with the fueling reactions. ASSIMILATION REACTIONS (incorporation of elements into cell components) are divided between two groups: assimilations of nitrogen and sulfur are biosynthetic reactions; assimilation of phosphorus is an integral part of the fueling reactions. METABOLIC TURNOVER REACTIONS (the continuous breakdown and resynthesis of macromolecules) are divided between fueling and polymerization reactions. Finally, the reactions of motility and chemotaxis, which are valuable for locating nutrients and favorable ecological niches and for avoiding toxic material, are behavioral; they do not fit comfortably into any of the four reaction groups.

It has been customary to trace the metabolism of glucose (or other substrates) through the central pathways, into biosyntheses, polymerizations, and then cellular structures. The reverse order makes more sense, because macromolecules and cellular structures define the nature and magnitude of the needs for building blocks and energy. Then, the required energy, reducing power, and precursor metabolites must be met by fueling reactions.

ASSEMBLY

Although assembly is the only aspect of metabolism that can be observed under the microscope, it is the least understood step of the entire process. Complete analysis of how structures become assembled must wait until their complex molecular anatomy is understood in detail. Fortunately, the physical and genetic tools appropriate for these investigations are rapidly becoming available.

We know that subcellular structures form by the condensation of macromolecules, but how does this happen? Do these structures simply form spontaneously by aggregation of their macromolecular parts, or does a special construction machinery guide and energize the process? Are macromolecules made where they are assembled into larger structures, or are they transported to these sites? These questions are, of course, interrelated and the answers depend on the structure being assembled.

Assembly involves a combination of self-assembly and guided assembly, on-site construction, and delivery of prefabricated units. Self-assembly, in which the participating macromolecules themselves provide the necessary information and energy, is a major element in the morphogenesis of most subcellular structures. But many assembly processes are guided and/or energized by preexisting structures, by special scaffolding, or by other morphogenetic aids. In certain cases, large cellular structures are constructed directly from precursors too small to be called macromolecules; in other cases, channels and vectorial reactions carry ready-made macromolecules from their site of synthesis to the assembly site.

In this chapter we consider the assembly of cellular structures—the nucleoid, polysomes, and the cytosol—that lie inside the cytoplasmic membrane. The particularly challenging problem of assembly outside the cell membrane is considered in Chapter 4. The macromolecules needed for each of the cell's components were described in Chapter 1; those needed to form intracellular structures are summarized here (Figure 2).

Assembly of the nucleoid

Synthesis of the chromosome (see later in this chapter) and the finished structure of the procaryotic nucleoid (Chapter 2) are now understood in some detail, but the link between them—the assembly of the nucleoid—remains almost a complete mystery. The action of topoiso-

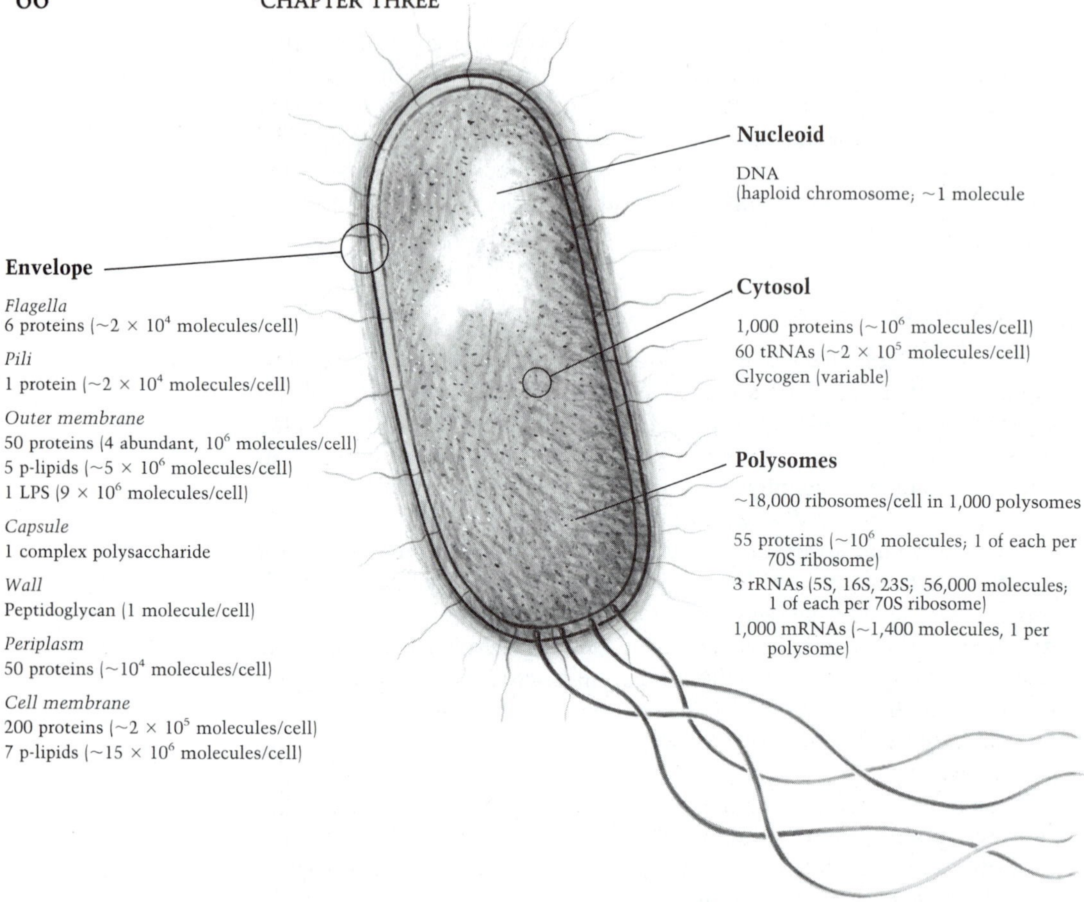

Nucleoid

DNA
(haploid chromosome; ~1 molecule

Envelope

Flagella
6 proteins (~2 × 10^4 molecules/cell)

Pili
1 protein (~2 × 10^4 molecules/cell)

Outer membrane
50 proteins (4 abundant, 10^6 molecules/cell)
5 p-lipids (~5 × 10^6 molecules/cell)
1 LPS (9 × 10^6 molecules/cell)

Capsule
1 complex polysaccharide

Wall
Peptidoglycan (1 molecule/cell)

Periplasm
50 proteins (~10^4 molecules/cell)

Cell membrane
200 proteins (~2 × 10^5 molecules/cell)
7 p-lipids (~15 × 10^6 molecules/cell)

Cytosol

1,000 proteins (~10^6 molecules/cell)
60 tRNAs (~2 × 10^5 molecules/cell)
Glycogen (variable)

Polysomes

~18,000 ribosomes/cell in 1,000 polysomes

55 proteins (~10^6 molecules; 1 of each per 70S ribosome)
3 rRNAs (5S, 16S, 23S; 56,000 molecules; 1 of each per 70S ribosome)
1,000 mRNAs (~1,400 molecules, 1 per polysome)

Figure 2

Macromolecular composition of the bacterial interior. This figure is based on the data presented in Table 1, Chapter 1.

merases introduces superhelicity into the 40 to 200 constrained domains of the nucleoid; but how the borders of these domains become gathered into the central core and how the entire structure becomes attached to the cytoplasmic membrane is unknown. Mutant strains blocked at various steps in the assembly process would be extremely useful, but none has been isolated.

Assembly of polysomes

Polysomes are the most abundant subcellular structures in a growing bacterial cell. The reference bacterial cell discussed in Chapter 1 (Table 1) has approximately 1,000 polysomes, each consisting of certain associated molecules (accessory factors, enzymes, and tRNA molecules) and about 20 ribosomes attached to an RNA molecule. The associated molecules cycle between the soluble phase and the polysomes. And, of course, each polysome has a spectrum of polypeptide chains, averaging half the size of whatever protein the ribosomes are making.

Polysomes are assembled from ribosomal subunits, a molecule of mRNA, and a molecule of *N*-formyl-methionyl-tRNA (fMet-tRNA) in a process aided by protein initiation factors. This assembly is actually the first step in TRANSLATION (protein synthesis) and is discussed in the section dealing with the polymerization of amino acids into proteins. Here we deal with assembly of the major components of the polysome, the 30S and 50S ribosomal subunits.

The growing bacterial cell contains no significant pools of rRNA and ribosomal proteins (r-proteins), a condition implying that the assembly process is rapid and efficient. In fact, the entire complex process of ribosomal assembly in *E. coli* occurs in only 2–5 minutes.

Processing of rRNA

The genes encoding the three rRNA molecules (5S, 16S, and 23S) contained in each ribosome are clustered and transcribed as a single unit—that is, they are arranged in operons. In *E. coli* there are seven such operons (*rrnA* to *rrnE, rrnG,* and *rrnH*), the locations of which are shown in Figure 3. The general structure of each of these operons is

16S gene—spacer tRNA—23S gene—5S gene—distal tRNA

Although the rRNA genes—with some slight sequential heterogeneities (one *rrn* operon contains two 5S genes)—are the same in all operons, the tRNA genes differ. Table 1 shows which tRNA genes are associated with the various *rrn* operons. The redundancy of rRNA-encoding genes (a highly unusual property for procaryotes, which are not known for having repetitive DNA) probably reflects the extraordinarily high demand for the products of these genes: under certain conditions, more than half of the cell's transcripts come from *rrn* operons. But redundancy alone cannot account for this high rate of transcription, because, all together, the *rrn* operons make up only 0.5% of the *E. coli* genome. Multiple highly active promoters are largely responsible. All *rrn* operons are transcribed from two tandem promoters separated by about 100 bp. One operon, *rrnB* has two additional promoters. It is not clear why some tRNA genes are included in *rrn* operons, but those present are the ones needed in the highest quantities.

Simple transcription of any of these operons would yield a useless giant molecule of RNA. But this does not occur. As the transcript is being made, a group of ribonucleases (RNases) cut and trim it into usable ribosomal and tRNA precursors. About 20% of the transcript is processed away (Figure 4). Four different enzymes perform PRIMARY PROCESSING, cutting the transcript into precursors of rRNA (p16S RNA, p23S RNA, and p5S RNA) and tRNA (ptRNA). Some r-proteins bind to the precursor forms of the rRNA molecules, forming ribonucleoprotein particles containing the p16S RNA and the p23S RNA plus p5S RNA. At this point, SECONDARY PROCESSING begins, utilizing at least three addi-

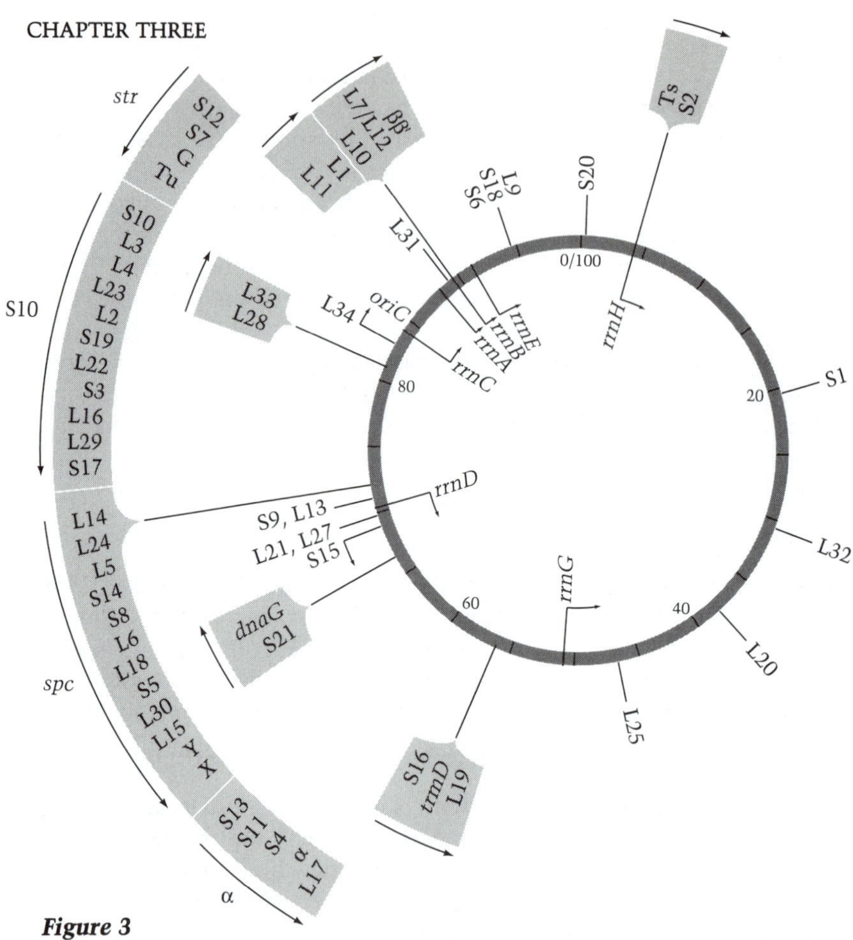

Figure 3

Genetic map of *E. coli,* showing the location of genes encoding ribosomal components. Ribosomal components are encoded in the seven operons, *rrnA* to *rrnG;* the r-protein components are indicated by the names of their products. L, large; S, small. Arrows indicate direction of transcription. (After Jinks-Robertson and Nomura, 1987.)

Table 1. Ribosomal transcription units of *E. coli* K-12: Chromosomal locations and associated tRNA genes

rRNA operon	Map position (minutes)	Spacer tRNA genes	Distal tRNA genes
rrnA	86	$tRNA^{Ile_1}$, $tRNA^{Ala_{1B}}$	None
rrnB	89	$tRNA^{Glu_2}$	None
rrnC	84	$tRNA^{Glu_2}$	$tRNA^{Asp_1}$, $tRNA^{Trp}$
rrnD	72	$tRNA^{Ile_1}$, $tRNA^{Ala_{1B}}$	$tRNA^{Thr_1}$
rrnE	90	$tRNA^{Glu_2}$	None
rrnG	56	$tRNA^{Glu_2}$	None
rrnH	5	$tRNA^{Ile_1}$, $tRNA^{Ala_{1B}}$	$tRNA^{Asp_1}$

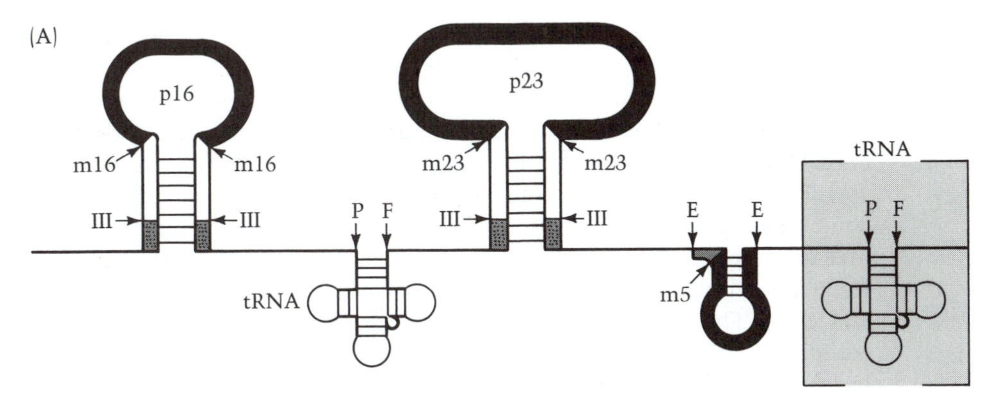

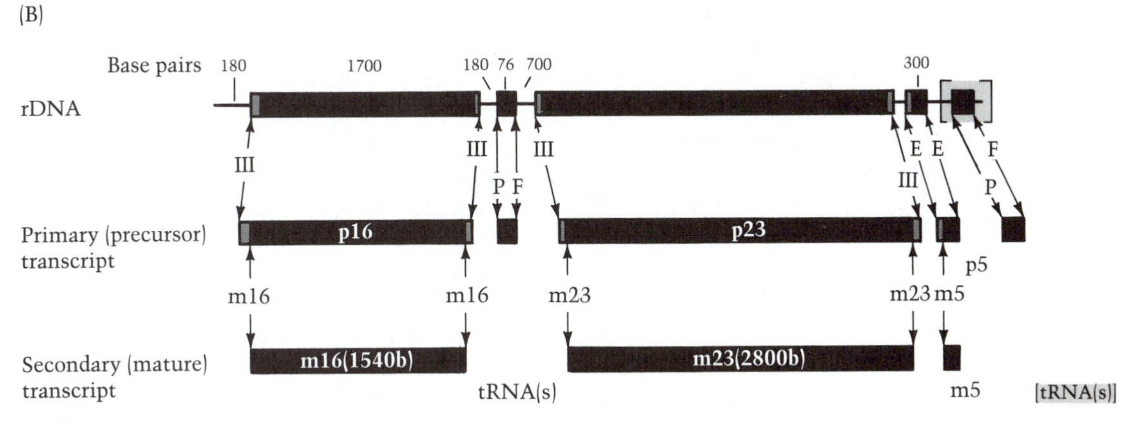

Figure 4

Processing of rRNA. (A) The structure of a hypothetical unprocessed transcript from an *rrn* gene. As the transcript is polymerized, cuts are made at the cleavage sites indicated by arrows: cuts labeled III are made by RNase III, those labeled P by RNase P, F by RNase F, E by RNase E, m16 by RNase m16, m23 by RNase m23, and m5 by RNase m5. (B) The sequence of processing steps is shown without indicating when r-proteins are bound. Numerals on the rDNA are the number of base pairs in the various segments of the operon; p and m indicate precursor and mature respectively. (Modified from Apirion and Gegenheimer, 1981.)

tional RNases: RNase m16, RNase m23, and RNase m5 for the final trimming to size of the 16S, 23S, and 5S rRNA precursors. In addition, specific enzymes catalyze the transfer of methyl groups from *S*-adenosylmethionine to certain bases in each of the rRNA molecules.

Assembly of ribosomal subunits

The complicated task of coordinating the expression of 52 r-proteins arranged in 21 transcriptional units (Chapter 2) is considered in Chapter 15, so for the moment we need only assume an efficient, coordinated, and non-rate-limiting synthesis of these proteins. Each subunit assembles by a sequence of reactions in which various r-proteins bind in a certain definite order to the rRNA core of the particle. Ribosomal subunits assemble spontaneously in vitro from mature rRNA and r-protein molecules by a series of reactions that probably resemble the natural ones. However, assembly of either 30S or 50S subunits in vitro, which is usually performed at 0°C, stops when the subunits are only partially completed. Assembly continues only if the partially assembled ribosomal intermediates, called RI_{30} and RI_{50}, are heated (to 37° and 44°C, respectively), thereby converting them to activated forms (called RI^*_{30} and RI^*_{50}). The intermediate RI^*_{30} accepts six more proteins to form the completed 30S subunit. In contrast, RI^*_{50} accepts eight more proteins but does not become a mature 50S subunit until it is heated to 50°C. The several heat-activation steps are mysterious and obviously nonphysiological. In the cell, there may be a protein that obviates heat activation, or maybe activation is completely unnecessary if the reaction sequence begins with p30 and p50 RNA, as it does in vivo.

By sequentially adding purified r-proteins, it has been possible to construct a map showing the sequence of in vitro ribosomal assembly (Figure 5). But it should be kept in mind that the assembly of ribosomes in the cell may well follow different, although similar paths.

Assembly of the cytosol

If we could remove the envelope, the nucleoid, and the polysomes from a bacterial cell, what would remain is the CYTOSOL. Mostly it is protein (about 80% by dry weight), but it also contains tRNA and a variable amount of storage polymer (glycogen in *E. coli*), plus small quantities of a large number of metabolites, building blocks, vitamins, cofactors, ATP, and inorganic ions. All but a small fraction of the fueling and biosynthetic reactions of metabolism occur here. Their products can be supplied directly to the macromolecule-producing machinery of the cell, because no membranes separate the cytosol from the polysomes (where protein is made), the nucleoid (where DNA and RNA is made), or the inner surface of the cell membrane (where wall and membrane subunits are collected). This barrier-free design may well be the principal reason for the high rate of metabolism and growth of bacterial cells. It has implications for regulation of gene expression as well.

Figure 5

Assembly maps of ribosomal subunits. Arrows indicate the facilitating effect on the binding of one protein by another. Thick arrows indicate major relationships, thin arrows weaker ones. Arrows from RNA to protein indicate binding of that protein directly to RNA in the absence of other proteins. (A) The 30S subunit. Proteins above the dotted line are those that form the RI_{30} particle. (Modified from Held et al., 1974, with information from Laughrea and Moore, 1978.) (B) Assembly of the 50S subunit. (From Nierhaus, 1982.)

(A)

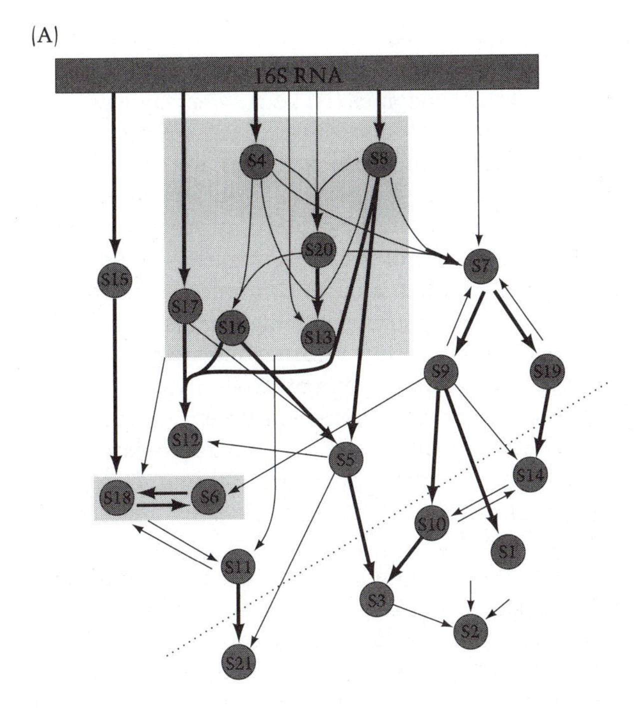
16S RNA
S4
S8
S20
S7
S15
S17
S16
S13
S9
S19
S12
S5
S14
S18
S6
S10
S1
S11
S3
S2
S21

(B)

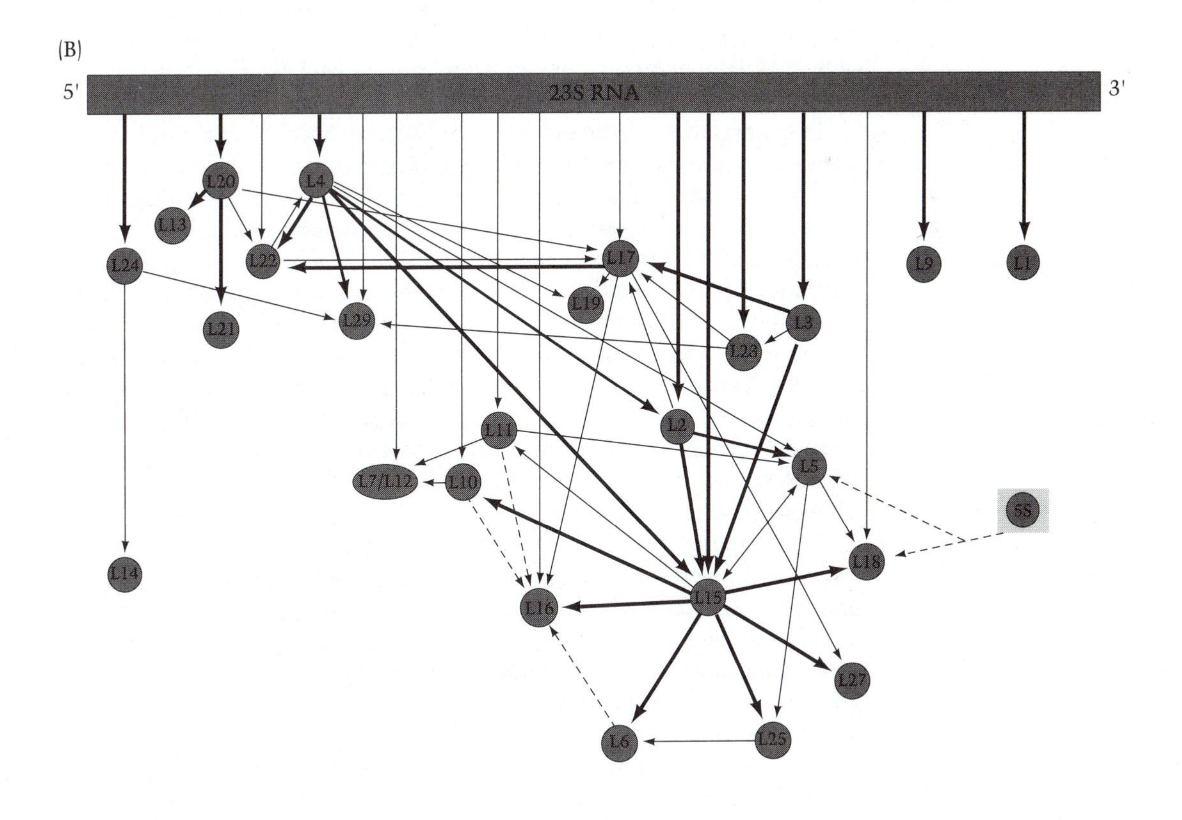
5'
23S RNA
3'
L20
L4
L13
L24
L22
L17
L9
L1
L19
L21
L29
L3
L23
L11
L2
L7/L12
L10
L5
5S
L14
L18
L16
L15
L27
L6
L25

It is a matter of some debate whether the cytosol is highly organized or is a disordered solution—Is it assembled in an ordered way that facilitates metabolism or is it a "bag of enzymes"? The answer is not known, but we have hints. First, the cytosol is a concentrated solution. (One limitation to in vitro studies of biochemical reactions and processes is that cell-free extracts rarely approach the protein concentrations found in the cytosol.) As a result, it is quite possible that protein–protein interactions can provide a degree of molecular organization in vivo that is totally lost when the cell is broken and its contents are diluted. It seems likely that enzymes with cooperative or sequential functions (for example, members of a biosynthetic pathway) are in physical contact in the cytosol. Direct and indirect evidence supports this view. Certain enzymes of the arginine and other biosynthetic pathways aggregate even in relatively dilute solution. The indirect evidence is more extensive: concentrations of most metabolic intermediates in biosynthetic pathways are vanishingly small, as though these molecules are handed baton-like from one enzyme to the next.

This situation, however, is far from universal—and for good reason. The products formed by many enzymes, particularly products of the fueling reactions, must be supplied to the many pathways that branch out from these central reactions. In these cases, metabolic intermediates are found in readily measurable amounts. The rate of solute diffusion over the very small intracellular distances probably does not appreciably diminish the rate of flow through these systems (with little structural order). Whatever order does exist in the cytosol is probably brought about by self-assembly of protein aggregates.

In general, there is no evidence for the existence of diffusion barriers within the cytosol, but there are some interesting exceptions. The cytosol of certain bacteria contain organelles bounded by protein (rather than phospholipid) membranes. Many phototrophic and chemolithotrophic aquatic bacteria produce GAS VESICLES that are believed to regulate the buoyancy of the cell and thereby enable it to float at an appropriate depth in the water. Cells of one group of photosynthetic green bacteria enclose all their photosynthetic pigments in cigar-shaped bodies called CHLOROBIUM VESICLES; even more elaborate is the stack structure, called a THYLAKOID, in which the photosynthetic apparatus of cyanobacteria is arranged. A large number of autotrophs contain POLYHEDRAL BODIES or CARBOXYSOMES; these vesicles contain ribulosebisphosphate carboxylase, the key enzyme of CO_2 fixation.

The remaining assembly reactions of the cytosol are the macromolecule-processing reactions. The tRNA molecules, whether originating as a joint transcript with rRNA or separately, are modified by a moderately large number of reactions carried out by specific enzymes. These modifications occur throughout the synthesis and folding of the primary transcript. In general, procaryotic tRNA is less decorated than eucaryotic tRNA, but nevertheless a wide variety of specific methylations, thiolations, rearrangements, and dehydrations of the four bases occur along with the elaborate substitutions of certain bases at particular positions. Because there is at least one tRNA-modifying enzyme for each kind of modified nucleoside at each specific position, there must be over 40 such enzymes.

Many proteins must be processed after translation, sometimes by authentic assembly reactions. Two examples of protein modifying reactions are the cleavage of the signal peptide from secreted proteins and the addition of diglyceride to the lipoprotein of the outer membrane. In the assembly category we also include the attachment of prosthetic groups and the binding of cofactors to enzymes. But many other posttranslational reactions modify the activity of a protein rather than determine its assembly. In the modification category we include reactions that methylate, acetylate, adenylate, or phosphorylate proteins. Usually these reactions can be reversed; they are used primarily to activate or inactivate enzymes or signal proteins. Table 2 lists a few examples. Bacteria probably have fewer proteins that are subject to these covalent modifications than slower growing eucaryotes. Induction and repression of enzymes in bacteria are fast and effective devices by which rapidly growing cells can alter enzymic activity and, of course, they are more economical than building unused protein (Chapter 11).

The granular inclusions found in many bacteria during certain growth conditions (e.g., limiting nitrogen but ample carbon) consist of reserve material: either glycogen, poly-β-hydroxyalkanes, or both. (No procaryote is known to store neutral fat.) Glycogen is deposited fairly evenly throughout the cytoplasm; no special assembly reaction is involved. The granules of poly-β-hydroxyalkanes are larger and readily seen as refractile bodies surrounded by a protein membrane. Similarly, granules of polyphosphate and inorganic sulfur are bounded by a protein membrane.

Table 2. Examples of bacterial protein modifications

Modification type	Protein
Chiefly assembly reactions	
Signal peptide cleavage	Secreted proteins
Formation of S–S bonds	Many proteins
Addition of lipid moiety	Murein lipoprotein
Addition of sugar moiety	Membrane glycoprotein
Attachment of prosthetic group	Many enzymes
Chiefly modulation reactions	
Phosphorylation	Ribosomal protein S6; isocitric dehydrogenase; many proteins with regulatory function
Methylation	Chemotactic signal transducers
Acetylation	Ribosomal protein L7
Adenylation	Glutamine synthetase

POLYMERIZATION: INTRODUCTION

The assembly processes we have just considered form cellular structures at prodigious rates. They use macromolecules that for the most part are chemically similar to those in eucaryotic cells but that must be supplied more that 10-fold faster than in any eucaryotic cell. To learn something about how these rates are achieved, we examine the polymerizations that occur inside the procaryotic cell. The polymerizations and associated assembly reactions that make the cell envelope are considered separately in Chapter 4; they have special features because they occur essentially outside the cell.

To provide a quantitative framework for our discussion of the chemical synthesis of a bacterial cell, we follow a specific metabolic task—the formation of 1 g (dry weight) of the bacterium described in Table 1 in Chapter 1 (*E. coli* B/r growing in glucose minimal medium). Altogether, the cellular assembly reactions (considered above and in Chapter 4) needed to accomplish this task require 550 mg of protein, 205 mg of RNA, 31 mg of DNA, 91 mg of phospholipids, 34 mg of lipopolysaccharide, 25 mg of murein, and 25 mg of glycogen; when assembled, about 1.05×10^{12} cells are produced (Figure 6). In this chapter and the next, we examine how each of the macromolecules is made, and we determine the quantity of building blocks and energy needed for polymerization. In Chapter 5 we examine how the building blocks are made by the biosynthetic reactions and how the need of biosynthetic reactions (precursor metabolites, energy, and reducing power) are supplied by the fueling reactions. As we progress we keep a running tally of the costs of each of these processes, finishing with a complete ledger of costs for making 1 g of our reference cell.

With respect to most biological polymerization reactions, the meaning of the term BUILDING BLOCK can be confusing because it differs from that applied to ordinary chemical polymerizations. Nylon, for example, is a class of heteropolymers of a dicarboxylic acid and a diamine. These two building blocks are both reactants in polymerization and constituents of the polymer. But in the case of most biological polymerizations, the reactants and polymer constituents are different. For example, in the case of DNA, deoxynucleoside triphosphates are the reactants, deoxynucleoside monophosphates are the polymer constituents. We follow the usual convention of calling the polymer constituents building blocks; the reactants are activated forms of these building blocks. The synthesis of activated forms of the building blocks for DNA and RNA are considered along with biosynthesis in Chapter 5; those for protein and glycogen are considered here.

POLYMERIZATION OF DNA

The essence of the mechanism by which DNA is replicated was included in the paper by J. D. Watson and F. H. Crick that announced the double helix structure of the molecule in 1953. They stated that "each

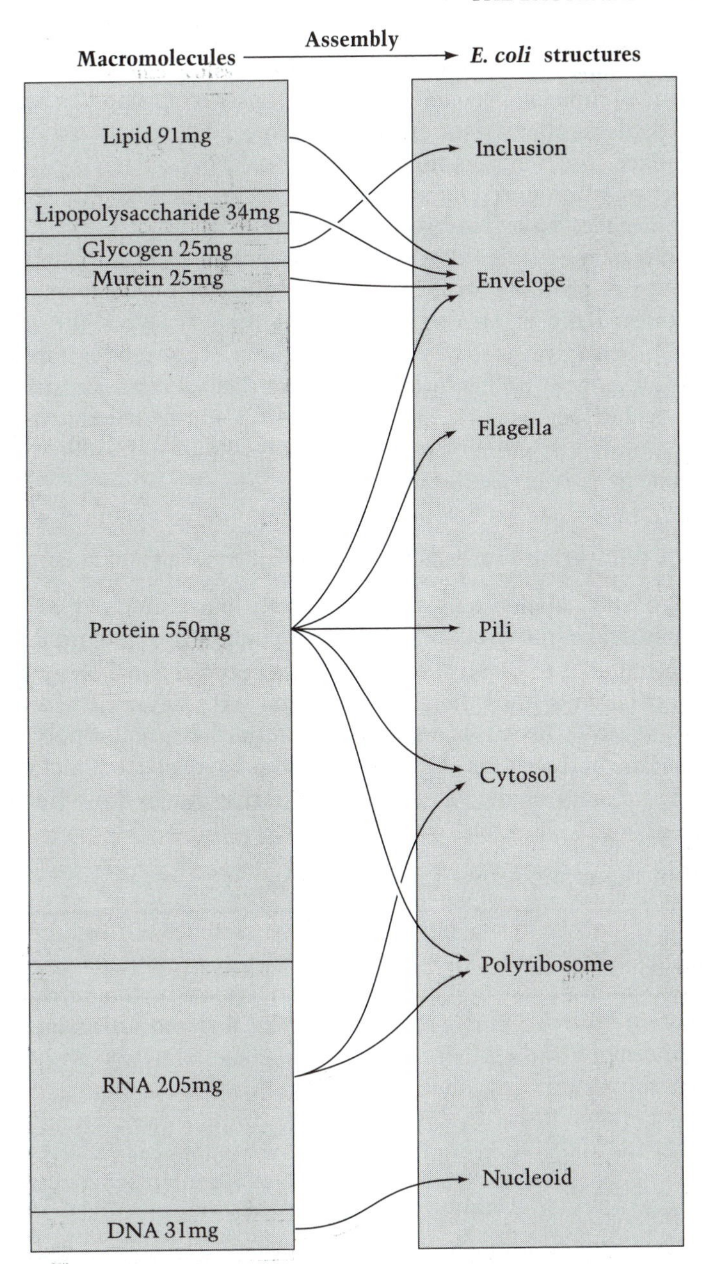

Figure 6

Macromolecules needed to produce 1 g (dry weight) of *E. coli* B/r. The chemical composition of this cell (Table 1 in Chapter 1) was used to derive the amounts of each macromolecule needed for the assembly processes. The capsule has been omitted.

chain acts as a template for formation on itself of a new companion molecule." This model suggests that at the point of replication the double-stranded molecule separates and nucleoside triphosphates pair with their complementary bases on the two exposed single strands and are polymerized. Such replication is SEMICONSERVATIVE: each new double strand consists of an old strand and a newly synthesized one. The succeeding decades have proved the correctness of this elegantly simple model but have revealed that it is biochemically quite intricate.

Groups of proteins form in a loose multienzyme complex—sometimes termed the REPLICATION APPARATUS—that catalyzes the reaction (Table 3). In addition, it has been determined that replication of any chromosome, be it procaryotic or eucaryotic, does not start at random locations, but begins at a specific initiation site called the ORIGIN OF REPLICATION. A TERMINUS of replication is located about 180° from the origin on the *E. coli* circular map.

DNA replication proceeds with remarkable speed and accuracy

Around 800 nucleotides are polymerized per second (at 37°C) and about 1 mistake (incorrect pairing) is made per 10^{10} nucleotide copies. Once initiated at the origin, replication proceeds at a *uniform pace* and in *both directions* until the two replication forks have reached the terminus (Figure 7). Because in growing bacteria the speed of polymerization is almost constant, DNA replication is regulated, not by the intrinsic velocity of the replication apparatus, but by how frequently

Table 3. Proteins involved in the replication of *E. coli* DNA

Protein	Molecules per cell	Gene	Function
DnaB	20	*dnaB*	DNA helicase (unwinding); priming
SSB	500	*ssb*	Binding to ssDNA
Primase	75	*dnaG*	RNA primer synthesis
DNA polymerase III	20		Replicative DNA polymerase
α subunit		*polC (dnaE)*	DNA polymerase
ε subunit		*dnaQ*	3′ to 5′ proofreading exonuclease
θ subunit		Unknown	Unknown
β subunit		*dnaN*	DNA polymerase accessory protein
τ subunit		*dnaZX*	DNA polymerase accessory protein
γ subunit		*dnaZX*	DNA polymerase accessory protein
δ subunit		Unknown	DNA polymerase accessory protein
DNA polymerase I	300	*polA*	Primer removal, gap filling
DNA ligase	300	*lig*	Sealing DNA nicks
DNA gyrase			DNA supercoiling
α subunit	250	*gyrA*	Nicking-closing
β subunit	150	*gyrB*	ATPase

Source: After McMacken et al. (1987).

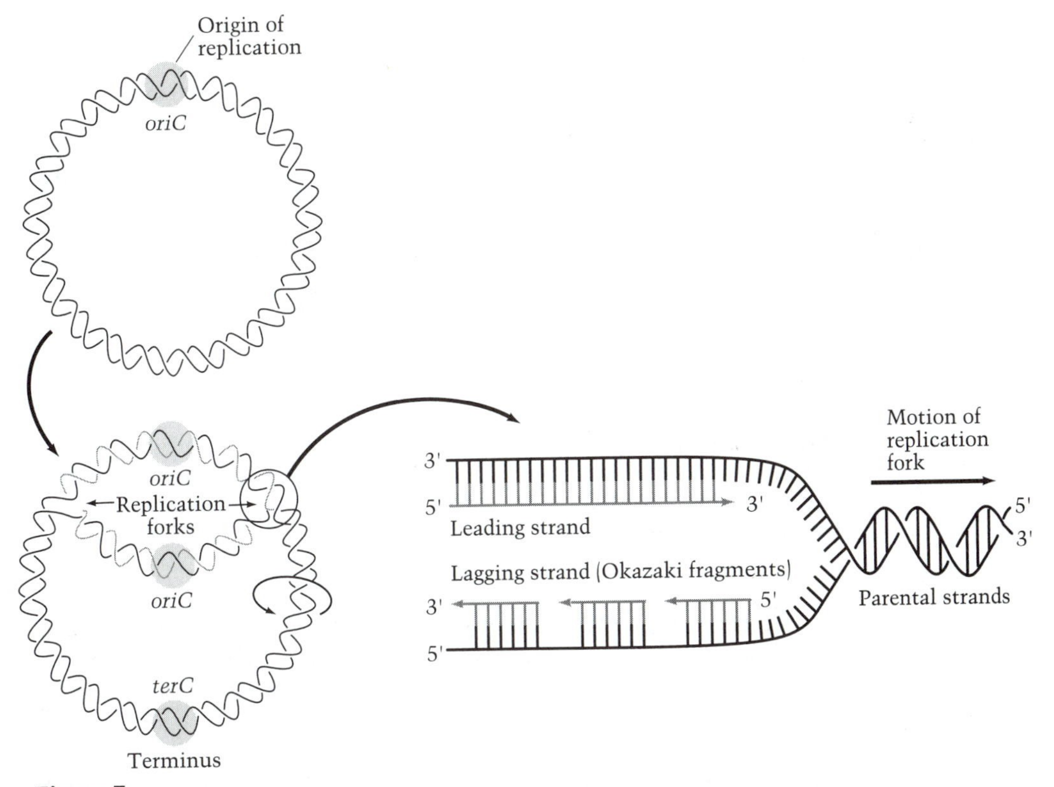

Figure 7

The replication of the circular *E. coli* chromosome. Replication is initiated at the origin, *oriC*, and proceeds bidirectionally toward the terminus, *terC*. The semiconservative nature of DNA synthesis at the advancing replication fork is depicted in the insert. (Modified from McMacken et al., 1987.)

the process is initiated at the origin. New DNA synthesis takes place in the 5′ to 3′ direction (relative to the phosphate linkages between adjacent deoxyribose residues). Consequently, the synthesis of each strand must proceed by somewhat different mechanisms because the two template strands of the parent double helix are antiparallel. The strand synthesized in the same direction as the overall direction of replication is conventionally called the LEADING STRAND, the other the LAGGING STRAND.

Replication apparatus

A scheme for how nucleotides are polymerized at the replication fork is shown in Figure 8. The first step in the process is separation of the two strands of the double chain. This separation requires the participation of a HELICASE, a strand-separating protein called DnaB. ATP is used, because energy is required to break hydrogen bonds between complementary bases and to unwind the strands. The helicase moves along the parent DNA in a 5′ to 3′ direction, although it remains attached to the lagging

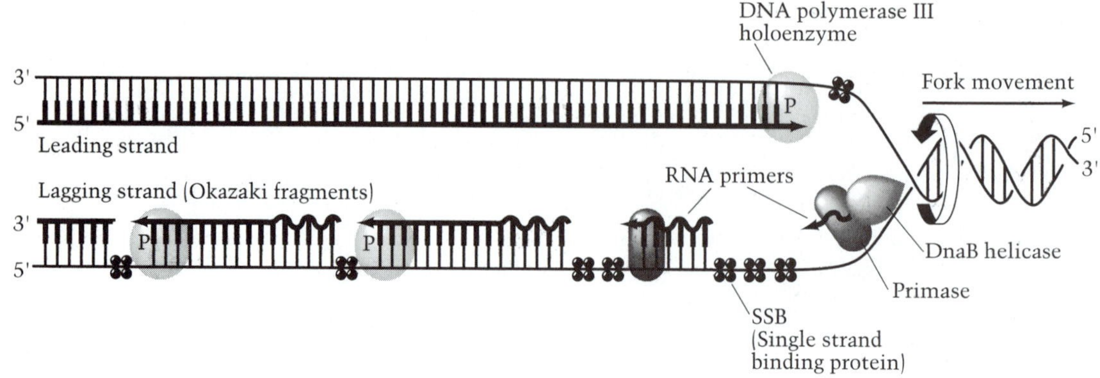

Figure 8

Hypothetical scheme for the mechanism of action of *E. coli* replication proteins at the replication fork. The action of DNA polymerase I in removing the primer and of DNA ligase in sealing nicks are not shown; see text. (After McMacken et al., 1987.)

strand. The energy requirement is diminished by the action of a tetrameric SINGLE STRAND-BINDING PROTEIN (called SSB), which binds tightly to exposed single-stranded regions of DNA, thereby destabilizing adjacent regions of duplex DNA.

Once the strands are separated, 5′-nucleoside triphosphates pair with their complementary bases exposed on the single-stranded region of the TEMPLATE strand (the parent strand). Polymerization is then carried out by action of a DNA POLYMERASE, which in *E. coli* is one of three enzymes with such activity. The one actually involved in the elongation of the DNA molecule is DNA polymerase III. This enzyme catalyzes the formation of a phosphodiester bond between the α-phosphate group of the nucleoside triphosphate and the terminal 3′ OH group of the growing strand. The acceptor molecule is called a PRIMER, and, as we see below, it is not the same for the two strands of DNA. In the process, pyrophosphate (P_i–P_i) is released from each nucleoside triphosphate. Thus, to synthesize DNA, DNA polymerase requires

- dATP, dGTP, dCTP, and dTTP as substrates
- a single-stranded region of DNA that serves as a template
- nucleic acid with a free 3′-OH end that serves as a primer of the reaction

How do the syntheses of the leading and lagging strands differ from each other? Replication of the leading strand is the simpler of the two; it is a continuous process, because DNA polymerase proceeds along the template molecule behind the helicase and the single strand-binding protein. Replication of the lagging strand, on the other hand, takes place by a "backing up" mechanism. Short stretches are made discontinuously in a direction opposite that of overall replication. They are then joined to make a continuous daughter strand. Replication of the two strands share central common features, but lagging strand synthesis requires the activity of several extra enzymes.

Direct evidence for discontinuous synthesis on the lagging strand was obtained by R. Okazaki, who exposed growing cells to [^{3}H]thymidine for brief periods (a technique called PULSE LABELING) and found that some of the newly synthesized DNA could be isolated as short, single-stranded fragments (500 to 2,000 nucleotides in length). These segments, waiting to be joined to the lagging strand, are called OKAZAKI FRAGMENTS. As stated earlier, DNA polymerases require a primer molecule with a free 3′-OH end, which is missing in the lagging strand. Thus, synthesis of each Okazaki fragment has the additional requirement for a primer. This function is served by a special RNA polymerase called PRIMASE, which does not itself require a primer. A short length of RNA is synthesized and serves as the primer for DNA polymerase, which then replicates the lagging strand. As shown in Figure 8, synthesis of an Okazaki fragment stops when the DNA polymerase "bumps" into the 5′ RNA end of the previous fragment. At this point, DNA polymerase I takes over and simultaneously hydrolyzes the RNA primer and replaces it with a DNA strand. Finally, the gap between DNA segments is sealed by the action of an enzyme called DNA LIGASE, which generates a continuous strand. Mutants deficient in DNA ligase accumulate Okazaki fragments in large amounts.

One may ask why a DNA polymerase did not evolve as a primer-independent enzyme, thus obviating the need for an RNA primer for each Okazaki fragment. The answer seems to lie in the necessity for proofreading. During replication, DNA polymerases make intrinsic mistakes as often as 1 incorrect pairing per 10^5 nucleotides, a level of inaccuracy far too great to preserve the cell's required level of genetic stability. PROOFREADING is accomplished in part by a built-in 3′-to-5′ exonuclease activity of DNA polymerase III, which removes its own mismatches by moving backward when needed. This enzyme cannot in fact catalyze replication in the forward direction unless there is a properly matched nucleotide pair behind it. Thus, a self-correcting polymerase cannot start chains de novo. Notice that the two-headed nature of DNA polymerase III is a unique piece of enzymological sophistication that permits both precision and efficiency in DNA synthesis.

The synthesis of the leading and lagging strands apparently take place in unison. How are these two different processes coordinated? The answer is not entirely clear for bacteria. There is evidence that a DNA polymerase involved in DNA synthesis of a bacteriophage is a dimer molecule that carries out the synthesis of both strands simultaneously.

Initiation apparatus

The initiation step of replication of the circular bacterial chromosome requires the origin of replication and at least one protein that recognizes its unique DNA sequence. We discuss the properties of each of these components of the initiation apparatus and how they were discovered.

The origin of replication has been defined in *E. coli* by a series of genetic and physiological experiments. Perhaps the most convincing one consisted of fragmenting *E. coli* DNA with restriction enzymes (Chapter 9) and determining which fragment was capable of autonomous replication after transformation into a recipient cell. It was found that a 245-bp sequence, known as *oriC*, was the only one that could replicate independently as a plasmid. Such plasmids, containing *oriC* and some adjacent genes, are known as MINICHROMOSOMES. Furthermore, when *oriC* is spliced into a foreign piece of DNA that cannot replicate by itself, it permits the replication of the foreign DNA as a plasmid. The identity of the origin region has been confirmed by a physiological experiment, using synchronously replicating cells (see Chapter 14 for details). When such cells are labeled with [^{3}H]thymidine, the first portion of the chromosome to be labeled is *oriC*.

The origin of replication possesses a number of unique features, including repeated sequences and an abnormally high number of sites for methylation of adenines. It is likely that these features allow the recognition and binding of specific initiation proteins. Some of these structural features are clearly important and are retained among several species of enteric bacteria. As expected, mutations in these highly conserved sequences lead to impairment of initiation.

The first protein involved in initiation is called DnaA. Its requirement for replication can be shown by the following in vitro studies. Workers in Kornberg's laboratory have developed a test-tube system that allows initiation from *oriC*. Under these conditions, the process is entirely dependent on the addition of DnaA. But DnaA is not required for the *synthesis* of DNA, as revealed by the phenotype of mutants that are temperature sensitive in the gene for this protein. When transferred to a nonpermissive temperature, such mutants continue to synthesize DNA until the replication forks reach the terminus. The process then stops, because initiation is not possible in the absence of active DnaA protein. When DnaA is present, initiation can be described as a four-step process.

- The first step is the formation of a complex between *oriC* and several molecules of DnaA in a chromatin-like, three-dimensional nucleoprotein structure. This unique complex permits the execution of several subsequent stages, including strand separation by the helicase DnaB.
- A number of other proteins attach to the partially unwound *oriC* DNA to form what is known as the PREPRIMING COMPLEX. The helicase DnaB proceeds along the DNA to separate the strands of downstream DNA. Meanwhile, the DNA helix becomes unwound by the action of an enzyme called TOPOISOMERASE.
- RNA primers are synthesized by the action of PRIMASE. This process takes place on both sides of the molecule, thus allowing replication in both directions.
- The DNA chains elongate by the mechanisms discussed earlier under replication. Bidirectional replication has been demonstrated by an ingenious visual experiment (Prescott and Kuempel, 1972). Cultures of *E. coli* were labeled in two stages: first with relatively low specific activity [^{3}H]thymidine for 15–20 minutes, followed immediately by

labeling for a few minutes at a greatly increased specific activity. The DNA was then gently stretched out on a glass surface and autoradiograms such as that shown in Figure 9 were obtained. Two short, dense tracks corresponding to the regions of high specific activity are joined by less dense tracks representing the lightly labeled DNA. (What would have been the result had DNA replication proceeded in one direction only?)

Termination

Chromosome replication ends when the two replication forks meet "head on" diametrically opposite the origin, at a region called *terC*. This terminus region is not simply the site 180° from the origin; it has special properties of its own. Its uniqueness was demonstrated by splicing *oriC* into other regions of the chromosome. In such mutants, replication still ends at *terC* and not 180° across from the new origin site.

Termination of replication and separation of the daughter chromosomes presents some complex topological problems, which bacteria have solved somehow. We can only imagine the confusion that would be created if termination did not take place in an orderly fashion. For example, without special intervention, the two daughter chromosomes would remain CATENATED, that is, connected like two links in a chain. Decatenation probably takes place by the action of DNA GYRASE, an enzyme that introduces breaks into catenated or supercoiled DNA and reseals their ends to make untwined duplex molecules. We still do not know

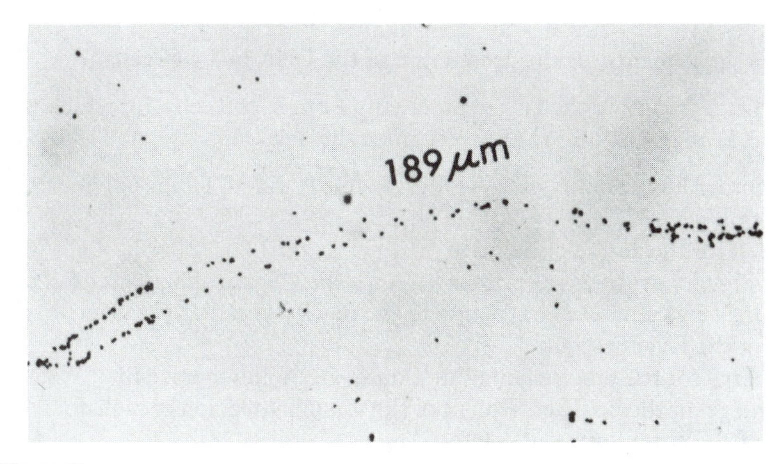

Figure 9

Autoradiograms of replicating *E. coli* DNA labeled successively at low and high specific activity of [^{3}H]thymidine. The picture shows a portion of the chromosome after labeling for 19 minutes at low specific activity and for 2.5 minutes at high specific activity. The two most recently labeled regions are separated by a 189-μm region of previously replicated DNA. Presumably, *oriC* is located halfway through this stretch. (From Prescott and Kuempel, 1972.)

how the replication apparatus comes off the finished DNA molecules and what happens to it later. At present, much remains to be elucidated about what happens during termination of chromosome replication.

The speed of replication

The intrinsic speed of the bacterial replication apparatus is about 800 nucleotides per second in the reference culture of *E. coli* described in Chapter 1. This rate is greater than that of eucaryotic cells and is 10 times faster than that of polymerization by RNA polymerase. Such a replication rate is all the more remarkable considering the great need for accuracy and the complexity of the reaction. It seems likely, therefore, that the proofreading function of bacterial DNA polymerase evolved hand-in-hand with the development of the rapid strand-separating and polymerizing activities of the replication apparatus.

The synthesis and proofreading reactions occur as replication is proceeding, and they keep pace with it. Completed molecules are further modified by methylation of certain bases, a mechanism widely used to distinguish the cell's own DNA from invading foreign DNA. Methylation lags behind replication by a few minutes.

Within 20 minutes of its completion, a new DNA molecule is segregated into a new daughter cell. Furthermore, as discussed in Chapter 14, a new round of replication may be started before the ongoing round is completed. This pattern of initiation allows the cell cycle to be completed in less time than is required to replicate the complement of DNA in a single chromosome.

Requirements for the production of the DNA in 1 g of cells

Let us return to the task of producing 1 g of *E. coli* cells, for which we need 31 mg of DNA. We now recognize the need for

- appropriate amounts of dATP, dCTP, dGTP, and dTTP for polymerization per se
- ATP for the action of helicase
- energy for synthesis of primer RNA for the Okazaki fragments and for ligating completed fragments to the newly synthesized complement of the lagging strand
- energy for the proofreading function of DNA polymerase III
- energy for the final adjustment of the torsional tension of each domain of the chromosome (Chapter 2)
- additional energy and methyl groups for strain-specific modifications of the newly synthesized DNA (Chapter 2)

The first of these—the provision of deoxyribonucleoside triphosphates as activated building blocks—is the only one that can be accurately specified. The other five requirements are of more or less uncertain magnitude, but collectively they are significant. Table 4 contains estimates of these needs.

Table 4. Requirements for the production of the DNA in 1 g of cells[a]

Factor required	Amount of building block required (μmol)	Amount of energy required (μmol ~P)
Activated building block[b]		
dATP	24.7	
dTTP	24.7	
dGTP	25.4	
dCTP	25.4	
Total nucleotides	100.2	
Energy for unwinding the helix[c]		100.0
Energy for discontinuous synthesis[d]		0.6
Energy for proofreading[e]		35.0
Energy for negative supercoiling		0.5
Energy for methylation		0.1
Total energy		136.8

[a]1 g of cells contains 31 mg of DNA or 100 μmoles of nucleotides; 1 μmol nucleotides = 309 μg nucleotides.
[b]Based on mole percent G+C = 51; the costs of making these are charged to biosynthesis (see Chapter 5).
[c]Based on 2 ATP to 2 ADP per nucleotide base pair separated.
[d]Based on synthesis and hydrolysis of a pentaribonucleotide from nucleoside triphosphates for every 1,000 bases in the lagging strand (10 ~P) plus 2 ~P to regenerate NAD used for ligation.
[e]Based on in vitro evidence indicating that, to achieve the fidelity observed in vivo, proofreading by DNA polymerase III contributes a factor of 10 to 200 to specificity (depending on the natures of the mispair) and costs 10–13% of the dATP and dTTP and 6% of the dGTP and dCTP being hydrolyzed to the respective nucleoside monophosphates. For the quantity of DNA made for the cells, 8.72 μmole of deoxyribonucleoside triphosphates would be hydrolyzed to monophosphates, or 8.72 × 2~P = 17.44 μmol ~P. We assume here that postreplication repair contributes an equal increment of accuracy—and at approximately equal cost—although this is not known.

Summing up the energy costs for synthesizing macromolecules, like any other form of bookkeeping, is somewhat arbitrary. We have chosen to include only the costs of polymerizing macromolecules from their building blocks, deoxynucleoside triphosphates, ribonucleoside triphosphates, and amino acids for the major macromolecules, DNA, RNA, and protein, respectively. The cost of synthesis of building blocks is tallied separately in Chapter 5. This system has the virtue of consistency, but it does not emphasize the difference in costs of making various building blocks. *Making nucleoside triphosphates (for nucleic acids) is much more costly than making amino acids (for proteins), and two high-energy phosphate bonds are lost in the form of pyrophosphate (P_i-P_i) for each nucleotide polymerized into a nucleic acid.* These comparisons are made in Chapter 4, where the costs of polymerization are summarized.

POLYMERIZATION OF PROTEIN

Owing to their rapid growth rate and high protein content, bacteria have evolved as highly efficient synthesizers of protein: the cell discussed in Chapter 1 can add approximately 300,000 amino acids each second onto growing peptide chains. But the challenge of protein synthesis is more than quantitative: a cell must synthesize over 1,000 different kinds of protein, all in the correct proportions within a somewhat restricted range of tolerance, if the organism is to compete successfully in nature. Moreover, as we shall see in Chapters 11, 12, and 13, the proportions of the many types of proteins can be modulated quantitatively and qualitatively in response to changes in the chemical or physical environment. As we shall see in Chapter 9, each type of protein is encoded by a separate gene (or in just a few cases, duplicate copies of the same gene), and the overwhelming majority of genes encode proteins. Only a few genes (22 for rRNA and up to 54 for tRNA) encode RNA as final products. Thus, protein synthesis and gene expression are essentially a single topic.

All bacterial proteins are synthesized according to the same two-step scheme:

- Using one of the strands of DNA as a template, a molecule of RNA polymerase synthesizes a complementary molecule of RNA (one to several genes in length). This process is termed TRANSCRIPTION, because it involves "rewriting" the gene's information in the form of a different informational molecule. The product of transcription is called messenger RNA, mRNA, or simply a TRANSCRIPT.
- Transfer RNA (tRNA) molecules with amino acids attached (aminoacylated) bind to three sequential bases (a CODON) on mRNA (Figure 10); and, through the intervention of a ribosome, the amino acids are polymerized into a protein in the order indicated by the sequence of bases in the mRNA. This process is called TRANSLATION, because the informational content of mRNA is converted into an amino acid sequence.

Some peptide bonds are made by processes in which ribosomes do not participate. Each amino acid is added to polypeptide antibiotics and to the short peptides contained in murein by a distinct enzyme-catalyzed reaction.

Transcription apparatus

Transcription can be viewed at two levels—biochemical and topological. Biochemically, it is quite simple: it is catalyzed by a single enzyme, DNA-DEPENDENT RNA POLYMERASE (RNA polymerase, RNA-P), which uses any ribonucleoside triphosphate (ATP, CTP, GTP, or UTP) as substrate and one DNA strand as template; it does not require a primer. A single step of the reaction can be written as

$$\text{RNA} + \text{ATP} \xrightarrow{\text{RNA-P}} \text{RNA-AMP} + P_i\text{-}P_i$$

Second letter

First letter	U		C		A		G		Third letter
U	UUU	Phe	UCU	Ser	UAU	Tyr	UGU	Cys	U
	UUC		UCC		UAC		UGC		C
	UUA	Leu	UCA		UAA	Ochre	UGA	Opal	A
	UUG		UCG		UAG	Amber	UGG	Trp	G
C	CUU	Leu	CCU	Pro	CAU	His	CGU	Arg	U
	CUC		CCC		CAC		CGC		C
	CUA		CCA		CAA	Gln	CGA		A
	CUG		CCG		CAG		CGG		G
A	AUU	Ile	ACU	Thr	AAU	Asn	AGU	Ser	U
	AUC		ACC		AAC		AGC		C
	AUA		ACA		AAA	Lys	AGA	Arg	A
	AUG	Met	ACG		AAG		AGG		G
G	GUU	Val	GCU	Ala	GAU	Asp	GGU	Gly	U
	GUC		GCC		GAC		GGC		C
	GUA		GCA		GAA	Glu	GGA		A
	GUG		GCG		GAG		GGG		G

Figure 10

The genetic code. The possible triplet codons of mRNA are listed with the amino acids they encode. A, C, G, and U represent the bases adenine, cytosine, guanine, and uracil, respectively. Amino acids are indicated by abbreviations: Ala, alanine; Arg, arginine; Asn, asparagine; Asp, aspartate; Cys, cysteine; Gln, glutamine; Glu, glutamate; Gly, glycine; His, histidine; Ile, isoleucine; Leu, leucine; Lys, lysine; Met, methionine; Phe, phenylalanine; Pro, proline; Ser, serine; Thr, threonine; Trp, tryptophan; Tyr, tyrosine; and Val, valine. Nonsense codons (darker), which cause termination of translation, are indicated by the terms amber, ochre, and opal.

By this step, the RNA molecule is lengthened by a single adenyl group (or C, G, or U residues, which are added by the same reaction), and one molecule of pyrophosphate (P_i-P_i) is released. The reaction is a nucleotidyl transfer from pyrophosphate to the 3′-OH of an RNA molecule.

But in spite of its biochemical simplicity, transcription is topologically complex. RNA-P must

- locate the proper place on the DNA helix to start the process
- separate the DNA strands in order to gain access to the base-pairing sites on the template DNA strand

- progress helically down the template strand, releasing the transcript behind it at a proper distance to prevent entanglement
- terminate transcription at the end of the transcriptional unit (operon or singly transcribed gene)

To summarize, the process of transcription that RNA-P catalyzes is a composite of three distinct phases: initiation, elongation, and termination. Not surprisingly, in view of their multifaceted metabolic assignment, RNA polymerases are rather complicated molecules. Those from eubacteria probably all share the same subunit structure, designated $\alpha_2\beta\beta'\sigma$ because it is composed of four different subunits, one of which (α) is present in two copies. The corresponding enzymes from archaebacteria are even more complex: they are composed of two large subunits and eight different small ones, resembling in this respect RNA-P from yeasts more than those from eubacteria.

Initiation of transcription occurs when the σ subunit of RNA-P locates specific DNA sequences, called PROMOTERS, which precede all transcriptional units. The promoter not only signals where transcription begins, it also determines which DNA strand will be transcribed, because the sequence of bases on the template strand determines how RNA-P binds to it. After RNA-P binds to the promoter region, it breaks the hydrogen bonds between the two DNA strands. The result is the formation of an "open complex," or "bubble," of single-stranded DNA (on the average, about 12 bp long) that extends from 9 bp behind (−9) to 2 bp in front (+3) of the pair (+1) encoding the first transcribed base. Then polymerization begins: RNA-P moves down the template DNA strand; the bubble moves with it (base pairs of DNA being opened in front and closed behind); and the nascent RNA transcript trails behind it, paired to the template DNA strand. When the nascent transcript is about eight nucleotides long, changes occur that allow the process to come up to maximum speed:

- The sigma factor (σ subunit), having fulfilled its function of promoter recognition, is released.
- The bubble grows to its mature size of about 18 bp in length. When the RNA–DNA hybrid has grown to a length of about 12 bp—just about long enough to complete one turn of the double helix—the hybrid begins to separate at its distal end. From this point on, RNA-P is followed by a 12-bp hybrid and then by a free strand of RNA, which is available for modification, processing, or translation into protein. Transcription continues at a constant rate, regardless of template, until a DNA sequence is encountered that signals termination. Then, RNA-P is released along with the completed transcript and the emptied DNA bubble closes.

Initiation signals

Initiation of transcription is a particularly critical step of procaryotic metabolism, because the frequency with which it occurs determines how many transcripts will be made from a gene—and this frequency can vary by four orders of magnitude. Some genes are transcribed less than once per cell generation; other genes are transcribed once every second. Frequency of initiation is set by the sequence of bases in the promoter region, but this basic frequency can be modulated over a wide range by a variety of regulatory mechanisms. Thus, initiation is appropriately considered to be an aspect of regulation of gene expression. Both promoter sequence and initiation-acting regulatory mechanisms are discussed in Chapters 12 and 13. Suffice it to say here that promoters vary in three ways:

- whether the promoter is strong (one at which initiation occurs at a high frequency) or weak (initiation occurs at a low frequency)
- which particular sigma species participates in the process (*E. coli* produces at least five different kinds of sigma molecules; other organisms—for example, *Bacillus subtilis*—produce more)
- which particular bacterium we are considering (promoters that are strong in one species can become weak or even nonfunctional when transferred to another species, because the interaction between the σ subunit and the promoter determines activity)

Termination signals

Termination is an obligate step that occurs at the end of transcripts, but in some operons it fulfills a regulatory function as well. Regulatory termination, called ATTENUATION, occurs before the operon's structural genes are transcribed and is discussed in Chapter 12. Here we consider the kinds of termination, SIMPLE or COMPLEX, that occur at the end of completed transcripts. Simple termination is mediated solely by a sequence of bases in the template DNA. Complex termination is also signaled by base sequences, but it requires the participation of an accessory protein, called rho. Complex termination is often called RHO-DEPENDENT TERMINATION.

Simple termination occurs after RNA-P has transcribed a G–C-rich region of DNA arranged in a pattern (an INVERTED REPEAT) that enables it to form a STEM-AND-LOOP STRUCTURE (Figure 11), followed by a region that contains a series of A residues. The stem-and-loop structure is thought to form within the transcription bubble, thereby causing the process to stall. The subsequent series of A-U pairs between transcript and template make weak connections that are easily broken, thereby releasing the transcript.

Rho-dependent termination occurs whenever RNA-P in its progress down the template strand pauses for at least 10 seconds. In vivo, pausing that results in rho-dependent termination is usually caused by formation of a stem-and-loop structure in the mRNA being transcribed. When this happens, rho interacts with the stalled RNA-P complex, one or more high-energy phosphate bonds are hydrolyzed, and the transcript is released. Any of a variety of artificial treatments (e.g., incorporation of a 3′-methoxy nucleotide or addition of the chelating agent, EDTA) that cause RNA-P to stall cause termination if rho and nucleoside triphosphates are present.

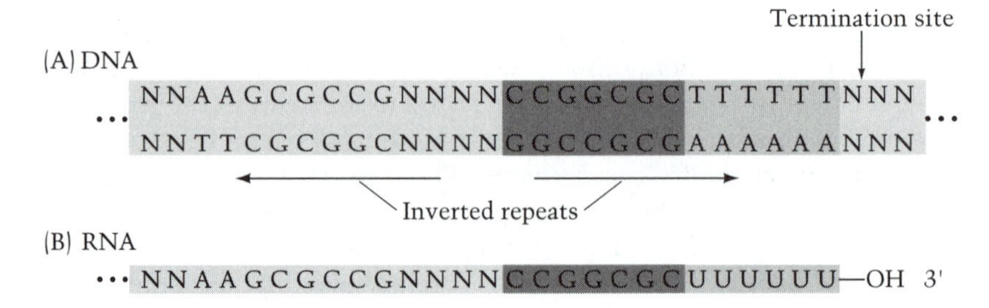

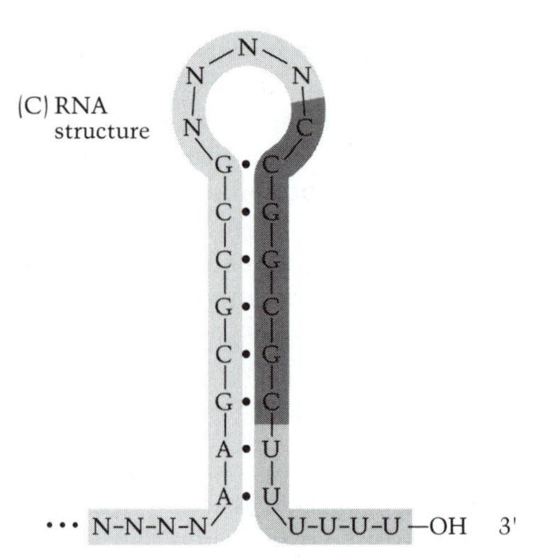

Figure 11

A hypothetical base sequence of a simple terminator. (A) A set of A·T pairs (light shading) is preceded by a set of G·C pairs (dark shading) and by an inverted repeat of some of these (horizontal arrows). Termination occurs at the site of the vertical arrow. Noncritical nucleotides are designated N. (B) The terminal portion of the messenger transcribed from the gene in (A). (C) The stem-and-loop structure that can form in the transcript as a consequence of hydrogen bonding (·) between regions encoded by the inverted repeat. This hypothetical terminator is derived from a number of simple terminator sequences in *E. coli* and its phages. (After Pribnow, 1979.)

Translation

As mentioned earlier, use of the terms *transcription* and *translation* to describe the steps of protein synthesis draws a useful analogy between this set of biochemical reactions and the processing of language. Translation from one language to another is dependent on someone's knowing both languages. In the molecular analogy, tRNA is the molecule that knows both languages: one portion of the molecule, the ANTICODON, pairs (according to base-pairing rules) with codons on mRNA; another portion carries a molecule of the amino acid that corresponds to that codon. One could imagine that tRNA must have been the first component of the translation system to evolve. The other components—ribosomes, various accessory proteins, and aminoacyl-tRNA synthetases—would then have evolved to increase the rate and accuracy of the process.

Amino acid attachment to tRNA (formation of the activated form of the protein building block), which is catalyzed by the aminoacyl-tRNA synthetase, takes place in two stages: first the amino acid reacts with ATP to form an enzyme-bound molecule of aminoacyl-adenylate; second, by a transfer reaction, aminoacyl-tRNA is formed:

(1) ATP + amino acid → aminoacyl-AMP + P_i-P_i

(2) aminoacyl-AMP + tRNA → aminoacyl-tRNA + AMP

It is convenient to divide the mRNA-directed polymerization of amino acids in the same way we divided transcription—into initiation, elongation, and termination. Synthesis of all bacterial proteins initiates with a formyl-substituted methionine residue (the formyl group, always, and the methionyl group, sometimes, is removed posttranslationally). Formyl-methionine (fMet) is usually encoded by the methionine codon, but less frequently (10% of the time in *E. coli*) it is encoded by the valine codon, GUG, and more rarely by the leucine codon, UUG. Of course, these codons occur frequently within genes, not necessarily at the beginning of a message. What, then, designates certain of them as START CODONS, the site where translation begins? Two Australian biochemists, Shine and Dalgarno, solved this puzzle. They found that the 3′ terminus of 16S rRNA contains a sequence that is complementary to the base sequences that precede the start codon by approximately 10 nucleotides on most mRNAs (Figure 12). It is generally believed that this complementarity helps designate the start codon; hydrogen bonding between the 16S rRNA and mRNA positions the 30S ribosomal subunit at the proper site to form an INITIATION COMPLEX.

The sequential steps leading to the formation of the initiation com-

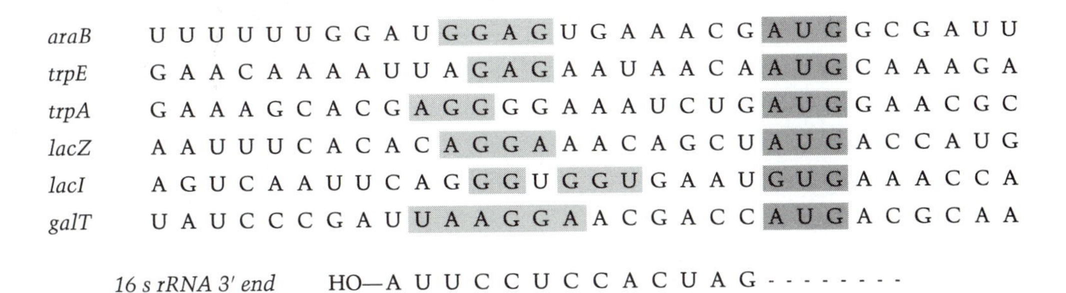

Figure 12

Sequence of the start regions of six mRNAs and the 3′ terminus of 16S rRNA from *E. coli*. Start codons are indicated by dark shading. Regions (Shine–Dalgarno sequences) complementary to 16S rRNA are indicated by light shading. (After Steitz, 1979.)

plex are shown in Figure 13. The 30S and 50S subunits of the 70S ribosome, released from a previously transcribed message, are dissociated by the intervention of GTP and three accessory protein initiation factors (IF1, IF2, and IF3), which bind to the 30S ribosomal subunit. These components also participate in forming the initiation complex. The IF2 protein facilitates the binding of fMet-tRNA to 30S ribosomal subunits; it also mediates a GTPase activity that is required to stabilize the initiation complex. The IF3 protein promotes binding of the 30S ribosomal subunit to mRNA and prevents association of ribosomal subunits in the absence of mRNA. The role of IF1 is less well defined; it enhances the activities of the other initiation factors and speeds dissociation of the 70S ribosome after a molecule of protein has been synthesized.

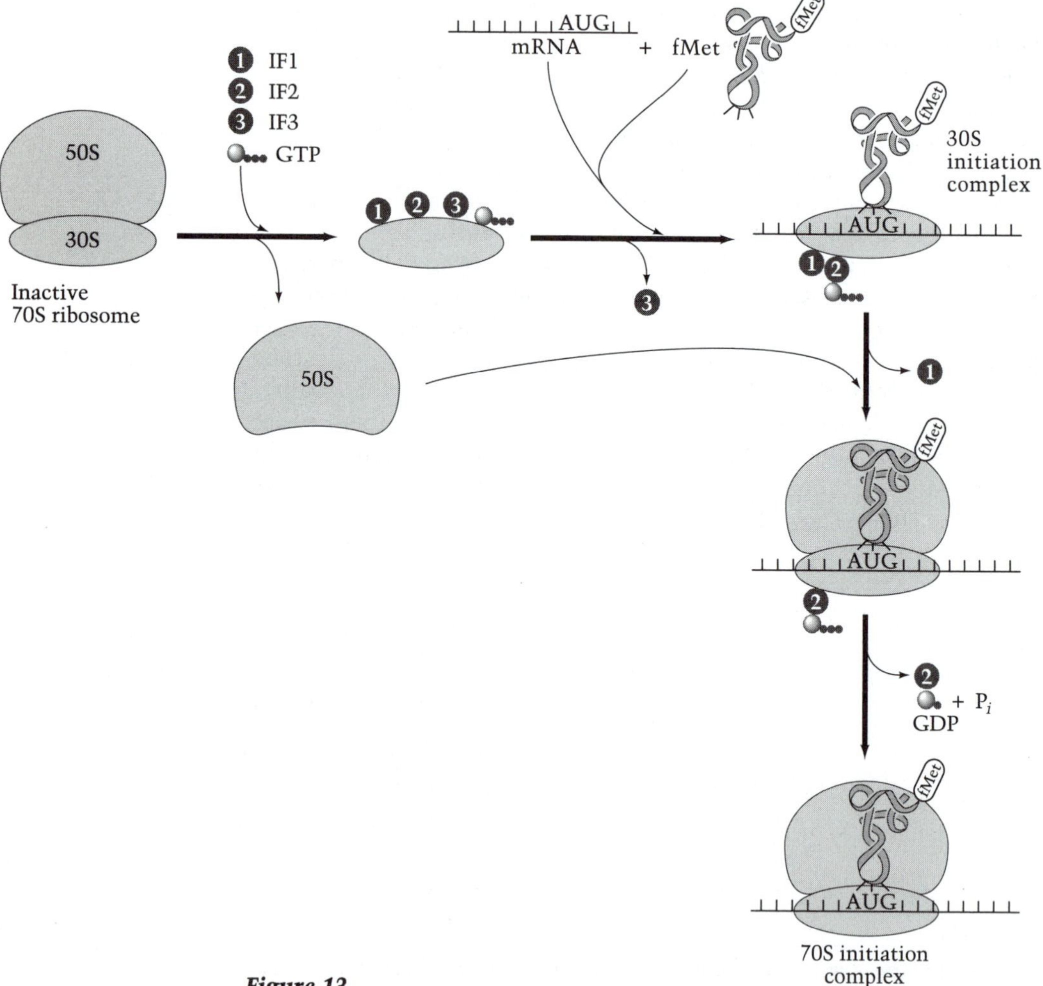

Figure 13

The sequence of reactions leading to the formation of the initiation complex. (After Hershey, 1987.)

Once the initiation complex has formed, a nascent protein begins to grow by operation of the ELONGATION CYCLE (Figure 14): each round of the cycle results in the formation of one peptide bond and the addition of one amino acid residue. The cycle requires the participation of three protein elongation factors (EF-Tu, EF-Ts, and EF-G) and the expenditure of two high-energy bonds (supplied by two molecules of GTP). To visualize the beginning of the cycle, think of a 70S ribosome bearing a partially com-

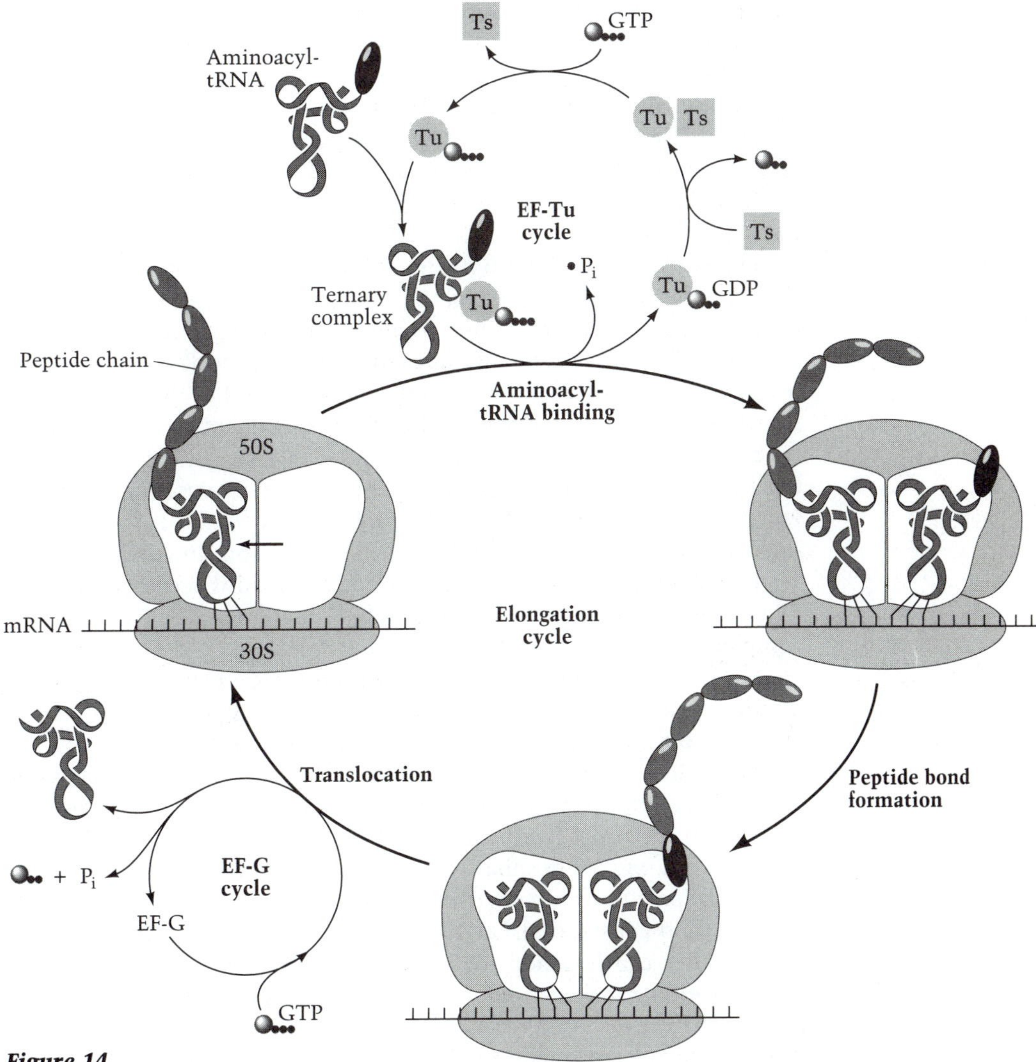

Figure 14

The elongation cycle. The steps of the cyclic process of elongation are shown with heavy arrows; the steps involving the recycling of EF-Tu through the action of EF-Ts are shown with light arrows. (After Hershey, 1987.)

pleted peptide chain, attached at a certain point to an mRNA. A 70S ribosome has three sites—each with differing properties—that bind tRNA. One of these, designated the **A SITE**, accepts the molecule of aminoacyl-tRNA. At this time, the **P SITE** is occupied by a molecule of tRNA to which a partially completed peptide chain is attached. The **E SITE** is occupied by a tRNA about to exit the ribosome. The anticodons of the tRNA molecules at all sites are positioned to pair with codons of the mRNA.

In the first reaction of the elongation cycle—AMINOACYL-tRNA BINDING—an aminoacylated tRNA complementary to the exposed anticodon enters the A site. This reaction requires the participation of two accessory proteins (EF-Tu and EF-Ts) and expenditure of energy in the form of hydrolysis of the γ phosphate of GTP. Aminoacyl-tRNA does not enter the A site alone, rather it enters as a ternary complex, aminoacyl-tRNA–EF-Tu–GTP. The high energy compound GTP is then hydrolyzed and EF-Tu–GDP and inorganic phosphate (P_i) are released. The protein EF-Ts removes GDP from the complex, so EF-Tu is free to form a new ternary complex.

The tRNAs in the A and P sites are properly positioned in the ribosome so that the amino group of the amino acid attached to the tRNA in the A site lies next to the terminal acyl group of the peptide in the P site. Breakage of the acyl bond and formation of the peptide bond results in transfer of the peptide (now one amino acid longer) to the tRNA in the A site—a reaction catalyzed by the 50S ribosomal subunit and requiring no expenditure of energy. The uncharged tRNA is then moved from the P site to the E site in preparation for TRANSLOCATION, a process by which the peptide-bearing tRNA is transferred to the P site and the ribosome is moved one codon in the 3′ direction down the mRNA. Translocation requires the participation of the third accessary protein, EF-G, and hydrolysis of the second γ phosphate of GTP. Translocation completes the elongation cycle.

The number of rounds of the elongation cycle, corresponding to the number of amino acids in the protein, results in the synthesis of a complete protein. TERMINATION, the hydrolysis of the peptidyl-tRNA and release of the completed peptide, occurs when the ribosome reaches one of the three termination (also called nonsense) codons, UAA, UAG, or UGA. Two accessory factors, RF (release factor)-1 or RF-2 and RF-3 participate, and one γ phosphate is hydrolyzed from GTP. (Interestingly, GTP is the sole source of energy used to polymerize amino acids; ATP is used for their activation.) The RF-1 protein is specific for UAG, RF-2 for UGA; either can function for UAA. Not everything has been learned about protein synthesis, not even in so well studied a bacterium as *E. coli.* There are proteins and RNA factors that participate in addition to those we have mentioned, and their precise roles are unclear. One essential player, for example is a small RNA molecule (about the size of a tRNA) called 4.5S RNA, which is required for protein synthesis—probably in the elongation cycle, but in a manner that is not yet understood.

Proofreading

As in the case of DNA synthesis, the overall process of protein synthesis is far more accurate than some of its individual steps. The fidelity is enhanced by a set of reversible reactions that correct errors that would otherwise result in an incorrect polypeptide sequence. Collectively, the process is called PROOFREADING.

The aminoacylation of tRNA by synthetases provides an example of how proofreading reactions work. A synthetase must recognize the correct tRNA and the correct amino acid. Misrecognition of either one is a potential error. The tRNA is misrecognized only rarely, because tRNA molecules are large, structurally complex, and differ significantly from one another. In contrast, amino acids are small and structurally similar; they are frequently misrecognized, because the slight structural differences between pairs of similar amino acids (e.g., isoleucine and valine, which differ by a single methylene group) contribute only a factor of 100 to 200 to the specificity of binding. How then does a synthetase distinguish these amino acids with a discrimination factor close to 10,000? The answer lies in selective reversal of the reaction. If in the first step of the reaction the correct AMP–amino acid complex is formed, the reaction will most probably proceed (the amino acid will be transferred to the cognate tRNA molecule). If, however, the incorrect AMP–amino acid complex is formed, it will most probably be hydrolyzed by the synthetase. The result of such selective reversal reactions is to correct mistakes and, thereby, increase fidelity. Sequential reactions that are selectively reversible increase overall fidelity by an factor that is the product of the increased fidelity contributed by the individual reactions. Not all the proofreading steps in translation are known, but their existence is a certainty.

The energy cost of proofreading in this instance of isoleucine and valine has been estimated at 0.01 ATP per isoleucyl-tRNA eventually passed on to the ribosome. For more dissimilar amino acids, the cost of accuracy will be lower.

The speed of protein synthesis

Although protein synthesis is fundamentally similar in all cells (they all use essentially the same genetic code and the same kinds of machinery), there are some significant differences between procaryotic and eucaryotic protein syntheses. These differences can best be interpreted in light of the 10-fold higher rate of protein synthesis in bacteria.

To sustain the rate of protein synthesis needed for doubling times that may be as short as 12 minutes, the bacterial cytosol must be rich in the machinery of protein synthesis. The PROTEIN-SYNTHESIZING SYSTEM (PSS)—which consists of ribosomes, mRNA, tRNA, the enzymes that make and modify these molecules, aminoacyl-tRNA synthetases, and the protein factors required for the initiation, elongation, and termination reactions—constitutes over one-half the total mass of rapidly growing bacterial cells. (How the size of the system is adjusted will be discussed in Chapter 15.) Natural selection has resulted in leaner, more streamlined PSS components than those found in eucaryotic cells. For example,

- bacteria use only 3 initiation factors; eukaryotic cells use 10
- bacterial ribosomes are smaller and have only one half the protein content of eucaryotic ribosomes
- bacterial mRNA is in many cases polycistronic (the transcript of an operon rather than a single gene)

In addition, the simpler organization in procaryotic cells avoids the translational complications of eucaryotic cells. Eucaryotic mRNA must be transported from the membrane-enclosed nucleus to the ribosomes in the cytoplasm. As a result, it must be protected against degradation during transport; this protection is achieved by CAPPING (attachment of a 7-methylguanosine triphosphate) at its 5′ terminus, by methylation at internal residues, and by POLYADENYLYLATION at the 3′ terminus. The central feature of the procaryotic body plan—a DNA molecule that is directly accessible to the cytosol—enables translation and transcription of genes to go on simultaneously and, thereby, eliminates the need for all of the devices involved in protecting and transporting mRNA.

The structural streamlining of the PSS in bacteria has several functional consequences, one of which is a difference in the initiation reaction. The use of base sequence complementarity between mRNA and 16S rRNA to help establish correct initiation of protein synthesis at the appropriate start codon turns out to be an exclusively bacterial device. Eucaryotic mRNA has been found in most instances not to contain a Shine–Dalgarno sequence at its 5′ terminus. It appears that cells of higher organisms can afford an initiation process (presumably a slower one) in which the 40S ribosomal subunit contains a methionyl-tRNA—GTP—initiation factor-2 complex that binds near the 5′ end of an mRNA and then scans the message for a start codon; after this step, numerous other initiation factors come into play. The eucaryotic initiation process presumably provides more opportunity for regulation of protein synthesis by restricting translation initiation of some mRNAs under certain conditions—a device not widely used to control expression of bacterial genes.

Requirement for the production of the protein in 1 g of cells

The production of 550 mg of protein requires

- appropriate amounts of each of the 20 amino acids found in *E. coli* protein
- sufficient energy (as ~P equivalents) for activation of each amino acid molecule and incorporation into protein
- energy for mRNA synthesis

- energy for proofreading at the synthetase and ribosome level
- energy for assembly and modification reactions

The requirement for amino acid building blocks is obtained by an amino acid analysis of total *E. coli* protein. The cost in high-energy phosphate bonds spent in the form of one ATP to AMP (i.e., 2 ~P) for activation, one GTP to GDP (i.e., 1 ~P) for the EF-Tu-mediated transfer of the aminoacyl-tRNA to the ribosomal A site, and one GTP to GDP (i.e., 1 ~P) for the EF-G-mediated translocation of peptidyl-tRNA to the P site comes to a total of 4 ~P per aminoacyl residue incorporated. There are two independent ways to calculate the energy requirement for mRNA synthesis. One, given in Chapter 15, is based on structural information about polysomes; the other is based on measurements of the rates of incorporation of nucleosides into RNA and amino acids into proteins by appropriate pulse-labeling of growing cells. Both methods yield an estimate of an average of 25–36 protein copies made from a single mRNA transcript. If we take 30 copies per transcript as our estimate and recall that each amino acid incorporated must be specified by a codon of three bases in mRNA, we arrive at a cost of 0.1 nucleotide (3/30) polymerized into mRNA per amino acid incorporated into protein; so, because each nucleotide polymerized into mRNA costs 2 ~P, the ~P cost per amino acid residue is 0.2 (0.1 × 2). The energy of proofreading is very uncertain, but estimates indicate it might be 0.1 ~P per amino acid incorporated. Finally, the cost for assembly and modification is also unknown. We shall estimate it by assuming that 10% of the cell's protein molecules are made with a signal sequence of 20 aminoacyl residues, which are cleaved off and hydrolyzed during translocation through the membrane. Altogether, synthesis of that quantity of oligopeptide is equivalent to approximately 0.6% of the total protein of the cell (10% of the cell's protein made with a signal peptide averaging 6% of the length of an average protein—20 is approximately 6% of 350 amino acids). The cost of assembly has therefore been estimated at an additional 0.6% of the cost of making protein. The values are summarized in Table 5.

POLYMERIZATION OF STABLE RNA

The polymerization of deoxynucleoside triphosphates into DNA requires the participation of a dozen or so proteins; the polymerization of amino acids into protein requires more than 60. In striking contrast, a single form of RNA-P (operating with different sigma factors) seems capable of making all the cell's RNA. For many transcripts, special factors such as rho are needed to assure termination, and perhaps most transcription is modulated by regulatory proteins that affect initiation. Nevertheless, these proteins all appear to be auxiliary factors not required in principle for accurate synthesis of RNA.

Table 5. Requirements for the production of protein in 1 g of cells[a]

Factor required	Amount of building block required (μmol)	Amount of energy required (μmol ~P)
Building block[b]		
Alanine	488	
Arginine	281	
Aspartate, Asparagine	458	
Cysteine	87	
Glutamate, Glutamine	500	
Glycine	582	
Histidine	90	
Isoleucine	276	
Leucine	428	
Lysine	326	
Methionine	146	
Phenylalanine	176	
Proline	210	
Serine	205	
Threonine	241	
Tryptophan	54	
Tyrosine	131	
Valine	402	
Total amino acid	5,081	
Energy for activation and incorporation[c]		20,324
Energy for mRNA synthesis[d]		1,016
Energy for proofreading[e]		508
Energy for assembly and modification[f]		122
Total energy		21,970

[a]1 g of cells contains 550 mg protein = 5,081 μmol amino acids; 1 μmol amino acids = 108 mg amino acids.

[b]Based on amino acid analysis of total *E. coli* B/r protein (murein amino acids excluded) by Teresa Philips; except the cysteine and tryptophan values are from Roberts et al. (1955).

[c]Energy is calculated by multiplying 4 ~P (sum of 2 ~P for ATP → ADP and 2 ~P for GTP → GDP) times 5,081 μmol amino acids.

[d]Assume 30 translations per mRNA transcript on the basis of life history of polysomes and independently from the information that 6.04×10^5 nucleotides are polymerized into mRNA per second by one genome-equivalent of cell mass while 6.46×10^6 amino acids are simultaneously incorporated into protein (Dennis and Bremer, 1974). Because the coding ratio is three bases per amino acid, the average mRNA sequence must be translated $[(6.46 \times 10^6)/(6.04 \times 10^5)] \times 3 = 32$ times. Each amino acid is incorporated at a cost of 0.1 nucleotide polymerized into mRNA, the cost of which is 0.1 X 2 (~P per nucleotide polymerized) = 0.2 ~P per amino acid residue. Cost = 0.2 ~P × 5,081 residues.

[e]The estimate of 0.1 ~P per amino acid for proofreading is extremely soft—direct measurements of excess ~P dissipation for proofreading *in vivo* is not technically feasible. We have chosen this value because it has been estimated as the cost of reducing the valine/isoleucine error (at tRNA charging) by 100-fold, and this error reduction is what is achieved for most tRNA-codon matches on the ribosome. Cost = 0.1 ~P × 5,081 residues.

[f]No estimate has been made for modification reactions; the cost of assembly is estimated as the cost of making the signal sequences of translocated proteins as described in the text. Cost = 0.6% × 20,324 μmol ~P.

The speed of stable RNA synthesis

Eucaryotic cells use different DNA-dependent RNA polymerases to make tRNA, rRNA, and mRNA. Those bacterial species that have been studied use only one enzyme for all three functions. We have already seen, while discussing synthesis of mRNA, how RNA-P recognizes promoters and initiates transcription. These same principles apply as well to the synthesis of STABLE RNA (ribosomal and transfer RNAs). All stable RNA molecules are made from precursor RNA molecules that must be processed by nucleases and then extensively modified to produce the mature product.

The tandem, extremely strong promoters for rRNA, along with their sevenfold redundancy (eightfold in the case of 5S RNA), enable synthesis of the large number of rRNA molecules needed during rapid growth. Also the clustering of these seven operons in the half of the operon proximal to the origin of chromosomal replication contributes to the gene dosage effect at fast growth rates, because genes near the origin of replication are present in more copies per cell than genes near the terminus of replication. The fact that all these operons are transcribed in the same direction as replication (Figure 3) is thought to be a necessary corollary of the high rate of their transcription to avoid collision between RNA and DNA polymerases (Chapter 9).

Requirements for the production of the stable RNA in 1 g of cells

Production of 197 mg of stable RNA (167 mg of rRNA and 30 mg of tRNA) in 1 g of *E. coli* cells requires

- the appropriate amounts of ATP, GTP, UTP, and CTP as building blocks
- an amount of energy (as ~P) equivalent to the segments of primary transcript that are removed and hydrolyzed
- the energy, methyl groups, and assorted minor constituents involved in the modification of these RNA species

Because the cost of mRNA turnover has been charged to protein synthesis (Table 5), the cellular content of mRNA can be treated here as though it were stable in calculating the building block requirement. These requirements are estimated in Table 6.

POLYMERIZATION OF GLYCOGEN

As discussed in Chapter 2, bacteria under certain conditions store carbon sources and other substances by producing large amounts of storage polymers (e.g., poly-β-hydroxyalkanes, glycogen, polyphosphate). Our reference cell accumulates glycogen—a small amount unless growth is restricted (e.g., by depletion of the nitrogen source) while glucose is plentiful.

The activated form of the building block of glycogen is the nucleoside diphosphate derivative (ADP-glucose) of the building block (glu-

Table 6. Requirements for RNA production in 1 g of cells[a]

Factor required	Amount of building block required (μmol)	Amount of energy required (μmol ~P)
Building blocks[b]		
ATP	165	
GTP	203	
CTP	126	
UTP	136	
Total nucleotides	630	
Energy for discarded segments of primary transcripts[c]		242
Energy for modification[d]		14
Total energy		256

[a]1 g of cells contains 197 mg stable RNA and 8.3 mg mRNA; (197 mg stable RNA = 167 mg rRNA + 30 mg tRNA)
[b]The average nucleotide residue was taken to have a molecular weight of 325.
[c]Approximately 20% of the primary transcript of *rrn* genes is discarded and a similar value is reasonable for tRNA gene transcripts. The energy cost is therefore 0.02 × 605 μmol of nucleotides in stable RNA, or 121 μmol times 2 ~P per nucleotide polymerized = 242 μmold ~P.
[d]Approximately 15% of the bases in tRNA are modified. We have estimated a cost of 1 ~P for each of the 92 × 0.15 = 14 μmol of modified tRNA bases, but this is undoubtedly an underestimate, given the great variety of complex decorations involved.

cose). Through the action of glycogen synthase, large polysaccharide chains of glucose joined in α-1,4 linkages are synthesized. A primer of at least four residues in length is required. In a second reaction—transglucosylation—small oligosaccharides are cleaved from the end of the chain by a branching enzyme, which then attaches them in α-1,6 linkages to another point (Figure 15). The ADP-glucose requirements for producing the low level of glycogen in a reference cells is given in Table 7.

SUMMARY

1. The bacterial cell synthesizes itself with as many as 1,000 to 2,000 chemical reactions. These can be usefully categorized as (1) assembly, (2) polymerization, (3) biosynthetic, and (4) fueling reactions on the basis of their primary function in growth. Assembly reactions involve the association of macromolecules to form cellular structures. Polymerization reactions consist of directed, sequential linking of activated forms of molecules called building blocks (amino acids, nucleotides, sugars, and fatty acids) into large polymers. Biosynthetic reactions produce the building blocks needed for polymerization, as well as coenzymes and related metabolic factors; the various biosynthetic pathways begin with one or more of 12 precur-

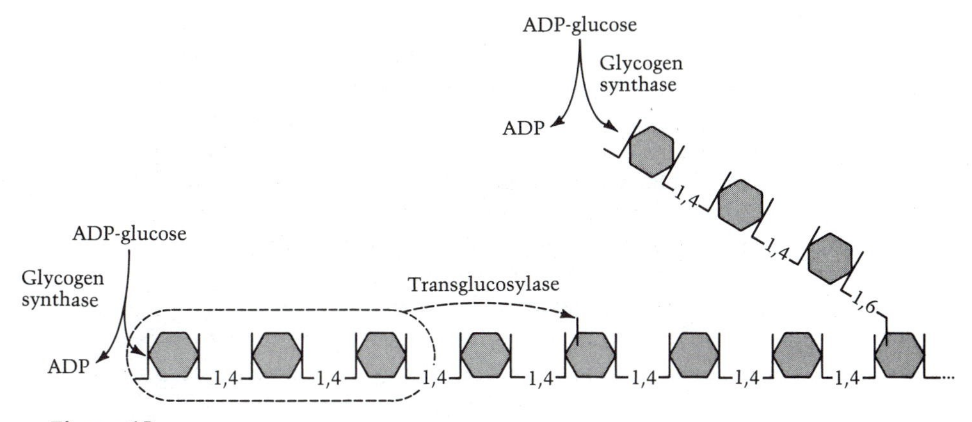

Figure 15

Synthesis of glycogen by bacteria. A small portion of a growing glycogen molecule is shown to illustrate the addition of glucose residues from ADP-glucose to the nonreducing ends of chains, a reaction catalyzed by glycogen synthase, and the formation of a branch by transfer of an oligosaccharide to an internal residue, a reaction catalyzed by a transglucosylase.

sor metabolites. Fueling reactions produce these precursor metabolites and the metabolic energy and reducing power needed for biosynthesis.

2. Assembly of cell structures occurs both spontaneously (self-assembly) and by specific mechanisms. Some macromolecules are made at the sites of assembly and others must be transported to them.
3. The nucleoid is formed by processes that involve methylation, supercoiling, and folding of newly replicated DNA.
4. Polysomes are formed by the factor-promoted association of ribosomal subunits to special sequences on mRNA. The ribosomal particles themselves are the product of elaborate assembly reactions that involve the processing of 5S, 16S, and 23S rRNA and the sequential attachment of over 50 different proteins to those molecules to gener-

Table 7. Requirements for the production of glycogen in 1 g of cells[a]

Factor required	Amount of building block required (μmol)	Amount of energy required (μmol ~P)
Building block: ADP-glucose	154	
Energy of polymerization		None

[a]1 g of cells contains 25 mg glycogen; 25 mg glycogen = 154 μmol glucosyl residues; 1 μmol residues = 162 μg residues.

ate the 30S and 50S ribosomal subunits. In vitro reconstruction of these structures from purified components has been achieved, but in vivo it might be guided by still unknown factors.

5. The cytosol is a fluid structure but probably has a loose organization brought about by affinities among proteins with cooperative or sequential functions. Some polypeptides must be processed in a variety of ways to become functional proteins. Certain bacterial species contain organelles within the cytosol, including vesicles of various sorts bounded by protein membranes.
6. Polymerization of DNA involves about a dozen proteins acting at a small number of sites (replication forks), where DNA is synthesized from activated building blocks at a nearly constant rate. Proofreading reduces the error frequency in DNA synthesis from 10^{-5} to 10^{-10}, at a cost of energy. Additional energy is required for strand unwinding, for introducing torsional tension (negative supercoiling), and for methylation of DNA.
7. Proteins are made at the direction of structural genes by the processes of transcription and translation; in bacteria, these processes are closely coupled. Transcription (mRNA synthesis) is biochemically simple but topologically complex. A single enzyme, RNA polymerase (functioning with different σ subunits), is responsible for all mRNA synthesis from activated building blocks. Translation of mRNA involves many proteins plus tRNA and ribosomes in a process that, as a result of sequential proofreading steps, is accurate to within 1 error per 10^4 aminoacyl residues. The protein-synthesizing system (PSS) constitutes over one-half the dry weight of a rapidly growing bacterial cell. Additional components of the PSS are still being discovered. Compared with those of eucaryotic cells, components of the bacterial PSS are streamlined for efficient and rapid operation: (1) initiation of translation requires fewer proteins, and ribosomes are smaller and simpler; (2) bacterial mRNA is mostly polycistronic; and (3) translation of mRNA occurs as it is being synthesized and, therefore, mRNA need not be processed and transported between cellular compartments.
8. Stable RNA is polymerized from activated building blocks by the same polymerase that makes mRNA. All stable RNA molecules are made from precursor molecules that must be processed by nucleases and then extensively modified to produce the mature product. There is sevenfold redundancy of rRNA operons, they are preceded by very strong tandem promoters and are clustered in the half of the chromosome proximal to the origin of replication; these factors help to generate the high rate of RNA synthesis needed to produce cells rich in PSS.
9. Long strands of α-1,4-linked glucose residues are polymerized from ADP-glucose by the action of glycogen synthase. Then a branching enzyme cleaves pieces off the ends of these strands and joins them at other points by α-1,6 linkages. The branched carbohydrate product is glycogen, a storage compound made by some bacteria, including our reference cell.

STUDY QUESTIONS

1. We neglected the contribution of cost of initiation of protein synthesis because it is quite small. How could you estimate it?
2. What aspects of procaryotic cell structure contribute to the efficiency and speed of transcription?
3. What is the physiological advantage to the cell of having redundant rRNA operons when so few other procaryotic genes are?
4. With respect to the three principal intracellular macromolecules, distinguish between their building blocks and the activated forms of them.
5. Why is proofreading necessary? What complications does the need for DNA proofreading contribute to DNA synthesis?
6. What purpose do Okazaki fragments serve?
7. Why does a cell need ribosomes?
8. Compare the cost of making protein and RNA from building blocks and from activated building blocks.
9. What metabolic function do fueling reactions serve?
10. What approach would you take to studying the process of assembly of the nucleoid?

SUGGESTED READING

Topics in this chapter are covered in the following chapters from *Escherichia coli and Salmonella typhimurium: Cellular and Molecular Biology*, F. C. Neidhardt, J. L. Ingraham, K. B. Low, B. Magasanik, M. Schaechter and H. E. Umbarger (eds.). 1987. American Society for Microbiology, Washington, D.C.

Drlica, K. The nucleoid. (Volume 1; Chapter 9)
Hershey, J. W. B. Protein synthesis. (Volume 1; Chapter 40)
Hoopes, B. C. and W. R. McClure. Regulation of transcriptional initiation. (Volume 2; Chapter 75)
Jinks-Robertson, S. and M. Nomura. Ribosomes and tRNA. (Volume 2; Chapter 85)
King, T. C. and D. Schlessinger. 1987. Processing of RNA transcripts. (Volume 1; Chapter 47)
McMacken, R., L. Silver and C. Georgopoulos. DNA replication. (Volume 1; Chapter 39)
Noller, H. F. and M. Nomura. Ribosomes. (Volume 1; Chapter 10)
Yager, T. D. and P. H. von Hippel. Transcript elongation and termination in *Escherichia coli*. (Volume 2; Chapter 76)

4

Assembly and Polymerization: The Bacterial Envelope

INTRODUCTION

The bacterial envelope carries out a large number of distinct functions (Figure 1 in Chapter 2); and, consequently, it is extremely complex. Its assembly has a typically bacterial flavor because its various constituents have unique chemical and architectural features. This area of research is active and exciting and has yielded important insights into cellular morphogenesis and its regulation.

Topologically, all the layers of the envelope are closed surfaces and must be physically continuous for cell integrity and viability to be maintained. Yet they must be able to expand during cell growth by the addition of new material. All the constituents of the envelope must grow coordinately and with special regard for their location: proteins and lipids must be correctly incorporated into their proper final destination. In the Gram-negative bacteria, these locations may be the cell membrane, the periplasm, the outer membrane, or the external environment. During each cell cycle, the newly assembled envelope must in some way facilitate the division of the cell into two viable offspring. In addition, environmental changes are first felt by the membrane, an interaction frequently leading to changes in some of the membrane's constituents.

How are the envelope components assembled with speed, accuracy, and adaptive flexibility? It is possible to construct only a general picture of the various aspects of this process. The overall process is similar in Gram-positive and Gram-negative bacteria but differs in detail because of the complexities introduced by the outer membrane and the periplasm of Gram-negative cells.

CELL MEMBRANE

Insertion of lipids

Assembling a cell membrane requires the construction of a lipid bilayer that contains several kinds of phospholipids and approximately 200 proteins in characteristic proportions and locations. For the most part, the enzymes responsible for phospholipid synthesis are located within the cell membrane itself. The building block and energy requirements for phospholipid production are shown in Table 1. Phospholipids have polar and hydrophobic regions and, in an aqueous medium, mutually attract and coalesce to form micelles (spheres of single molecular layers with a hydrophobic interior and a hydrophilic exterior) and, under some conditions, the familiar double-leaflet membrane. The polar groups become exposed at each surface, and the hydrophobic hydrocarbon chains are buried in the interior of the bilayer. Self-assembly obviously plays a large but, as discussed below, insufficient role in this process.

How does a bacterium assemble the outer leaflet of the cell membrane, which is exterior to its cytoplasm? It can be shown that newly formed phospholipid molecules appear first in the inner leaflet of the cell membrane and that some of them are then translocated to become part

Table 1. Requirements for phospholipid production[a]

Factor required	Amount of building block required (μmol)	Amount of energy required (μmol ~P)
Building block[b]		
Glycerol 3-phosphate	129	
Serine	129	
Fatty acyl-ACP ($C_{16:0}$, $C_{16:1}$, $C_{18:1}$)	258	
Energy for synthesis of phosphatidylethanolamine[c]		258

[a]1 g of cells contains 91 mg phospholipid other than lipid A; 91 mg phospholipid = 129 μmol phosphatidylethanolamine (PE); 1 μmol PE = 750 μg PE.
[b]Phosphatidylethanolamine, the major (ca. 75%) phospholipid in the cytoplasmic membrane is formed from glycerol 3-phosphate, serine, and fatty acids in the ratio 1:1:2. The next most prevalent phospholipid, phosphatidylglycerol, is formed from glycerol 3-phosphate and fatty acids in equal molar ratios. The difference (one serine replaced by one glycerol 3-phosphate) is small and, for simplicity, the total biosynthetic demand was calculated as though 100% were phosphatidylethanolamine. Also the three major fatty acids are combined as a single requirement. For details, consult a biochemistry textbook.
[c]Two ~P's are released in the formation of an intermediate, CDP-diglyceride. The fatty acids are provided as already activated building blocks—conjugated to the so-called acyl carrier protein. The cost of activation is therefore included later as a biosynthetic cost.

of the external leaflet. If this process were to occur by a simple "flip-flop" across the lipid layer, it would take considerable energy because it requires the passage of the polar group of the molecules through the hydrophobic interior of the leaflets. In the instance where it has been studied, this process appears not to require energy, because it takes place in the presence of several inhibitors that block the generation of metabolic energy (including cyanide and fluoride). This experiment was carried out with *Bacillus megaterium*, which, being a Gram-positive species, could be studied without the complication of having an outer membrane (Langley and Kennedy, 1979). How do lipids reach the outer leaflet of membranes, other than by a direct flip-flop? It has been proposed that the two leaflets are fused together around certain transmembrane proteins in a "hairpin bend," which joins the leaflets into a continuous layer (Figure 1). Such a structure would allow energy-independent diffusion of lipids between the two leaflets. In fact, topologically, the inner and outer leaflet would then become a single surface, without impediments for lateral diffusion. In this scheme, energy would be required for an earlier step, namely, the insertion of the proteins into the membrane and the creation of the hairpin bends.

Assembly of the lipid components of bacterial membranes is, however, not random but selective and regulated. Thus, the two leaflets of the cell membrane contain individual phospholipids in different proportions. In *B. megaterium*, for example, phosphatidylethanolamine is more abundant in the outer than in the inner leaflet. The distribution of phospholipids can be estimated by treating intact cells with nonpenetrating reagents that bind to molecules located in the outside but not in the inside leaflet. Note that asymmetry in the lipid bilayers does not happen spontaneously: synthetic membranes (liposomes) do not have asymmetric lipid distribution. The most striking example of asymmetric distribution is seen in the outer membrane of Gram-negative bacteria, where the outer leaflet contains all the cells' lipopolysaccharide and

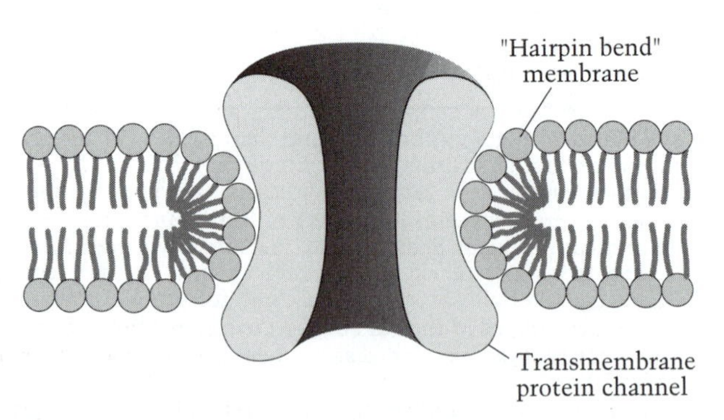

Figure 1

Hypothetical mechanism for movement of phospholipid across "hairpin bend" regions of membrane bilayers. Hairpin bends are created in the vicinity of transmembrane proteins, such as those forming the channels illustrated here. (From Langley and Kennedy, 1979.)

no phospholipid molecules. These examples strongly suggest that the incorporation of different classes of lipids into the two leaflets takes place not by diffusion alone but additionally requires specific lipid–lipid and lipid–protein interactions.

Does the synthesis of lipids play an active role in the insertion of proteins in the membranes? Is the membrane saturated with regard to proteins? The answer, within limits, is no, because when lipid synthesis is inhibited but protein synthesis is not, *E. coli* can increase the protein content of its membrane by 60% or more without losing viability. The converse is also true: when protein synthesis is inhibited, lipids continue to be inserted into the membrane. From these experiments, one concludes that the syntheses of membrane proteins and lipids are not coupled to each other. There are, however, limits to this uncoupling, and, given the chance, bacteria tend to adjust membranes to the proper lipid:protein ratio. Thus, it has been shown that when certain membrane proteins (fumarate reductase or a membrane ATPase) are grossly overproduced in *E. coli* strains harboring recombinant plasmids with genes for these enzymes, the amount of phospholipid also increases and the overall lipid:protein ratio remains constant. Making excess membrane material of this sort exceeds the capacity of the murein sacculus and the new material tends to form new tubular or vesicular structures that protrude into the cytoplasm (Figure 2).

Assembly of proteins into the cell membrane

Membrane proteins possess hydrophobic domains embedded in the hydrocarbon core of the lipid bilayer and hydrophilic domains exposed to the aqueous phases. Roughly speaking, transmembrane proteins are rivet-shaped, having two hydrophilic domains, one exposed to the interior, the other to the exterior of the cell. Some protein molecules exist individually within the lipid bilayer, some are associated with other proteins, often as elaborate structures (e.g., the membrane ATPase responsible for proton translocation that is coupled to the oxidative phosphorylation of ATP; see Chapter 5).

The force of attraction between hydrophobic molecules in aqueous media is considerable. Membrane proteins are notoriously insoluble in water and, unless a detergent is added to coat their hydrophobic surfaces, they rapidly aggregate into large complexes, stick to glass and other surfaces, and even denature. Hydrophobicity is sufficient to provide much of the energy for insertion of proteins into lipid bilayers, and, grossly, to distinguish between those proteins destined for the membrane and those that will remain in the cytosol. The primary structure (amino acid sequence) therefore ordains the cellular location of a protein by determining the extent and location of its hydrophobic surfaces when folded. It has been found, however, that for insertion into the membrane, extra energy must be supplied: protein translocation into and across membranes is inhibited by the addition of compounds (such as cyanide or dinitrophenol) that reduce the protonmotive force by collapsing the

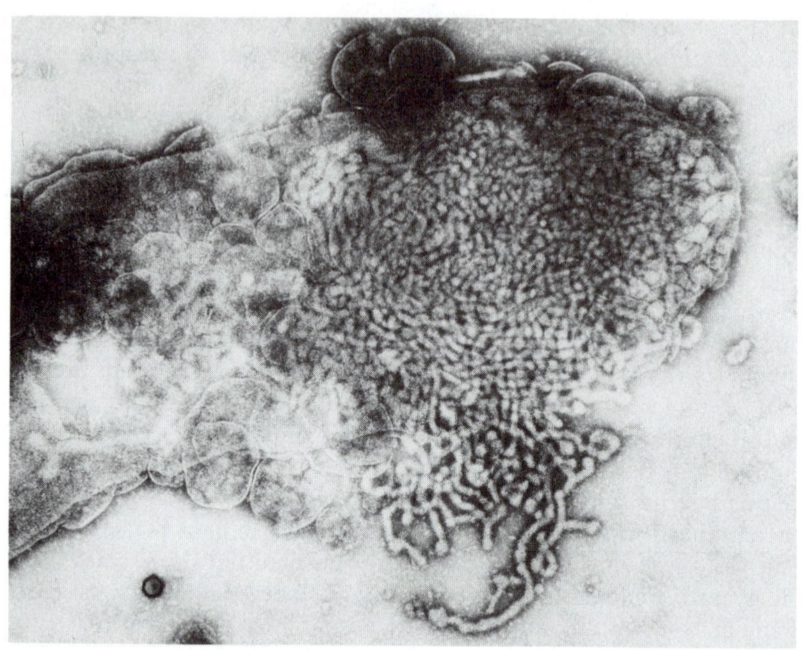

Figure 2

Hyperproduction of membrane-like material by an *E.coli* mutant that makes a 30-fold excess of fumarate reductase, a membrane protein. A negatively stained electron micrograph of a partially lysed cell shows a large amount of tubular membranous material extruding from the cytoplasm. (From Weiner et al., 1984.)

membrane potential. Additionally, ATP is required, possibly to effect the proper unfolding of the proteins.

However, the preceding dicussion can only be part of the story. Whereas many membrane proteins are hydrophobic, others are predominantly hydrophilic, with only small hydrophobic segments anchoring them to the lipid bilayer. The cell must cope with the problem of exporting these large hydrophilic domains across the membrane. Similarly, proteins that reside in the periplasm are hydrophilic and the same problem pertains to them. Clearly, special assembly devices are at work to guide the process of protein export and the orderliness of membrane structure. These mechanisms have been studied intensively since 1975. Many of the ideas have also come from research on protein secretion and membrane assembly in eucaryotic cells. This extensive work can only be summarized here, and the reader is encouraged to consult the critical discussions by Wu (1986) or Randall et al. (1987).

There are several aspects of protein assembly into membranes (or their secretion into the medium) that clearly go beyond the primary force of hydrophobicity. Thus, a large number of secreted proteins are made with an extra amino acid sequence at their amino terminus. These LEADER SEQUENCES (also known as SIGNAL PEPTIDES) differ from protein to protein but share certain characteristics (Figure 3), including

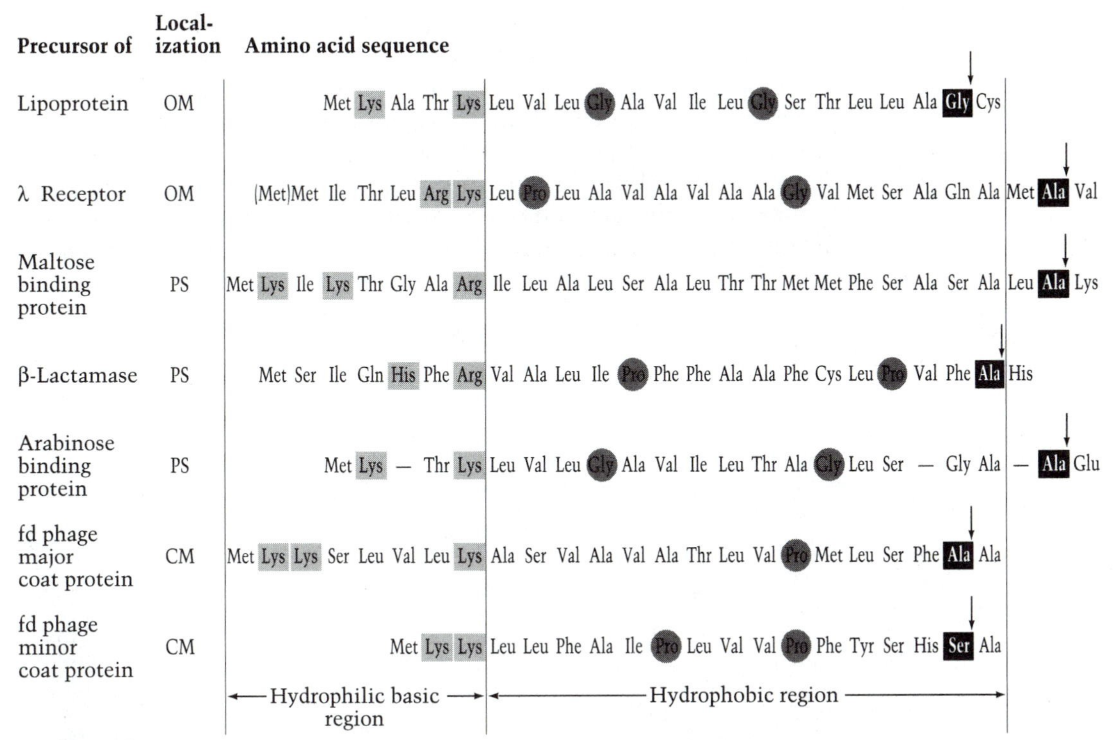

Figure 3

Leader sequences of proteins that enter the *E. coli* envelope. Basic amino acids are in gray squares, glycine and proline residues are in gray circles, and the residue at the cleavage site (arrowhead) is shown in black. The sequences are lined up at the last basic amino acid of the hydrophilic region. OM, outer membrane; PS, periplasmic space; CM, cell membrane. (After Osborn and Wu, 1980.)

- length of 15–40 amino acids, (most commonly about 20)
- a charged segment at the amino terminus, with one or two basic amino acids (e.g., lysine) followed by a stretch of hydrophobic amino acids, which usually includes two glycines or prolines
- a hydrophobic sequence followed by a stretch of about six amino acids that is thought to make a reverse turn in the chain

The synthesis of a membrane protein or a secretory protein begins in the usual way—on polysomes that are free in the cytosol. A new peptide, finished or unfinished, binds via the leader sequence to the cell membrane, where a complex export machinery takes charge and ensures that the peptide is threaded through the membrane. The importance of the leader sequence is demonstrated by showing that transport can be impaired when a single amino acid in the leader sequence is changed by mutation. In addition, proteins that normally are not exported are secreted when genetically fused to leader sequences. With such manipulations it has been shown that leader sequences are necessary, although not always sufficient, to direct export of proteins.

How does a protein molecule destined for export interact with the cell membrane? A hypothesis that unites many of the known facts, derived from studies on both procaryotic and eucaryotic systems, was proposed by Randall et al. (1987). The notion is that for a protein to be "competent" for translocation across to the membrane, it must be *unfolded and lack tertiary structure.* Folding is prevented or delayed by complexing with a cytosolic element, which in eucaryotes is known as the SIGNAL RECOGNITION PARTICLE (SRP). Bacteria also contain a cytosolic

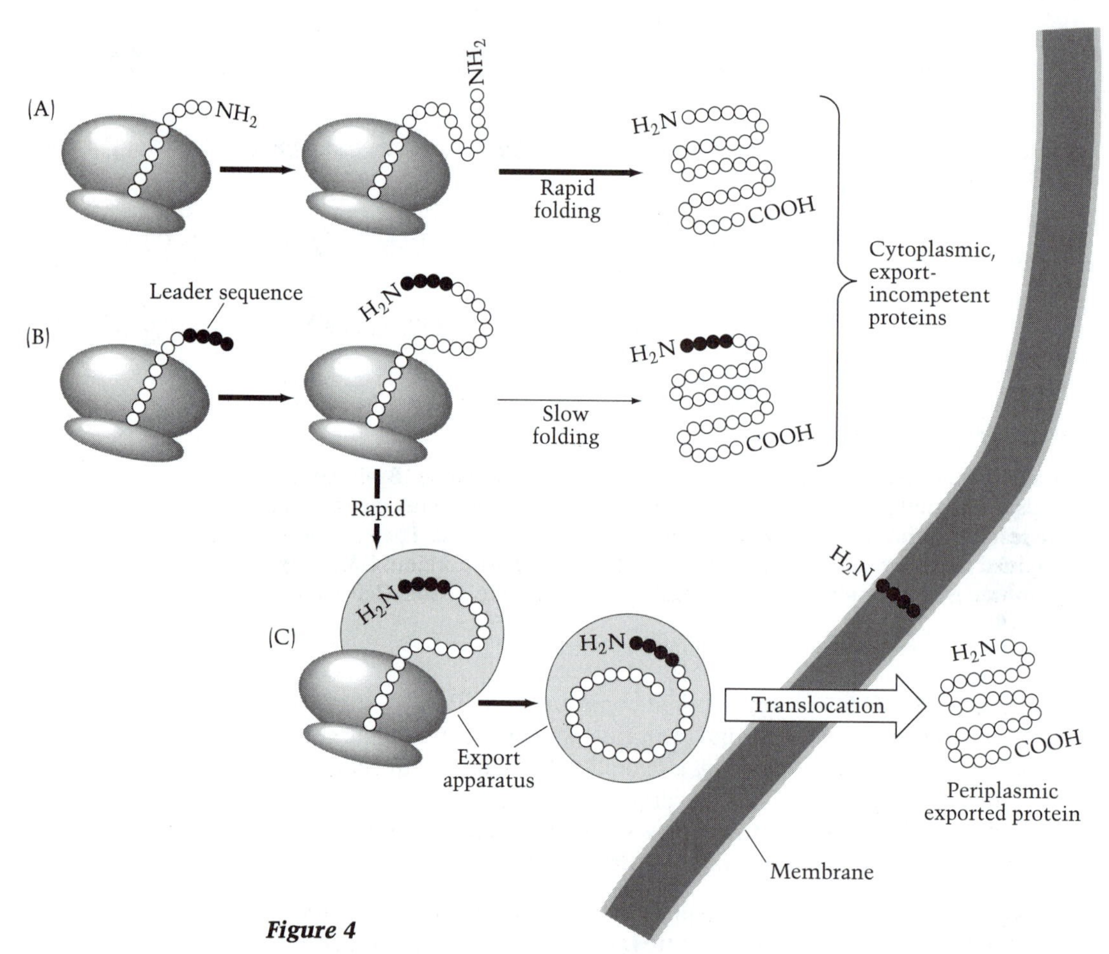

Figure 4

A model for how newly made proteins make the choice between an export and a nonexport pathway. The leader sequence bends over to interact transiently with a downstream portion of the nascent peptide (B). This interaction creates or exposes an element that is recognized by a component of the export apparatus (C). Binding by the export apparatus blocks further folding and maintains the peptide in an export-competent state. Peptides that escape this interaction (right side of B) or are synthesized without a leader (A) rapidly fold into a stable structure that cannot be exported. (After Randall and Hardy, 1989.)

protein that is equivalent to the signal recognition particle. Together, the leader sequence and the signal recognition particle are thought to bind to a receptor on the membrane (the eucaryotic DOCKING PROTEIN). These events ensure the delivery of the newly synthesized protein to the export machinery on the cell membrane. At this site, the leader sequence may enter directly into the lipid bilayer, or it may thread through a proteinaceous pore made up of the components of the export machinery. This model is represented schematically in Figure 4.

What is known about the membrane protein components of the bacterial export machinery? Investigators have isolated several mutants that are impaired in the secretion of membrane and extracellular proteins. Some of these mutants have general export defects, but others are specific for certain classes of proteins; that is, they are impaired in the secretion of outer membrane proteins but not of periplasmic proteins. It is not known at present whether these defects are in the recognition of leader sequences by a class-specific docking protein or in some subsequent step in the export reaction.

Many proteins destined to be incorporated into the membrane or to pass through it are translocated before they are fully synthesized (Figure 5). Polysomes that make these proteins are attached to the inner

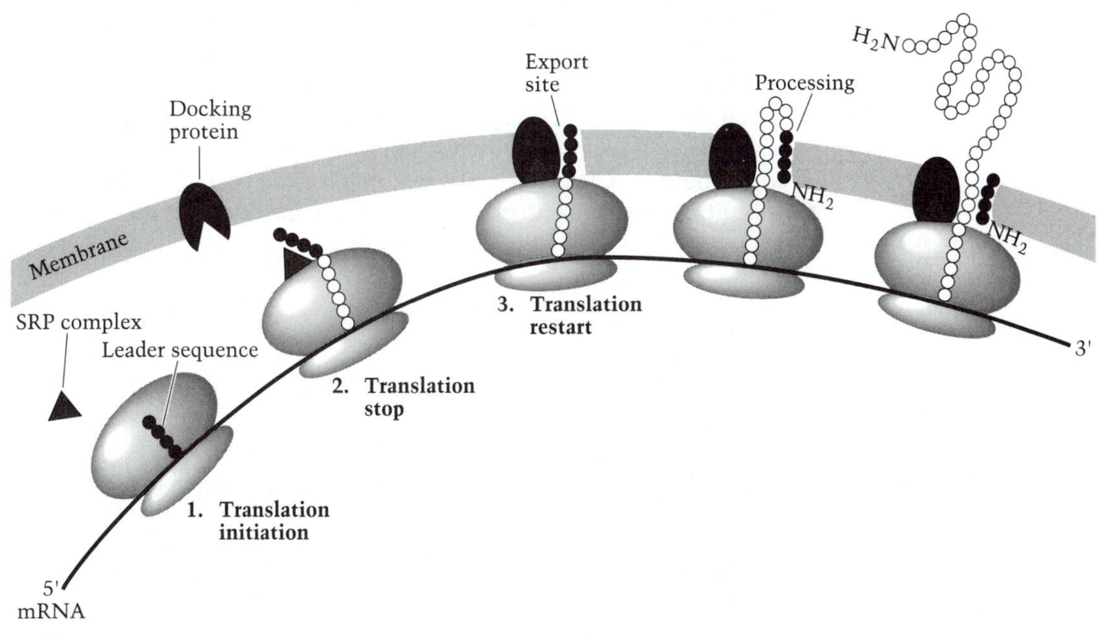

Figure 5

Cotranslational export of proteins. The process begins at the left of the illustration (1). The leader sequence is translated from the 5′ end of mRNA and is recognized by the signal recognition particle (SRP). (2) It is thought that this complex inhibits further translation. (3) This block is relieved when the complex (ribosome, SRP, and nascent peptide chain) interact with the docking protein on the membrane at the export site. (Modified from Silhavy et al., 1983.)

surface of the cell membrane by their NASCENT PEPTIDE (a peptide in the process of being synthesized). Nascent peptide chains protruding through the cell membrane can be labeled with nonpenetrating reagents that react with certain amino acid residues sticking through the membrane. It was found that such reagents can react with the amino terminus of *incomplete* peptide chains of, for example, the periplasmic protein alkaline phosphatase. These polysomes are attached to the membrane solely by the peptides they are synthesizing, as shown by the fact that they are freed from the membrane by the addition of puromycin (an antibiotic that dissociates nascent peptides from their ribosomes). This result makes it unlikely that the peptide chains are *pushed* through the membrane by the force of chain elongation.

Regardless of the nature of the events preceding translocation, the leader sequence is thought to participate directly in the threading of proteins across the membrane. One view of how this may occur is shown in Figure 6. According to this BENT LOOP MODEL, the leader sequence remains stuck in the membrane by virtue of hydrophobic attraction and, after a portion or all the rest of the protein has passed the membrane, is cleaved by a LEADER PEPTIDASE. Only then is a secreted protein fully mature. The leader peptidase is bound to the outer surface of the cell membrane. Although in *E. coli* a single leader peptidase splits most of the leader sequences, there is at least one other such enzyme, one that is specific for the lipoprotein of the outer membrane. The lipoprotein leader peptidase has been identified by mutation and by the fortuitous finding that the drug globomycin specifically inhibits its activity.

It is not entirely obvious why bacteria go to the trouble of cleaving the leader sequence. Cleavage is not required for the translocation step itself. Inhibiting cleavage either with drugs or by mutations in the leader

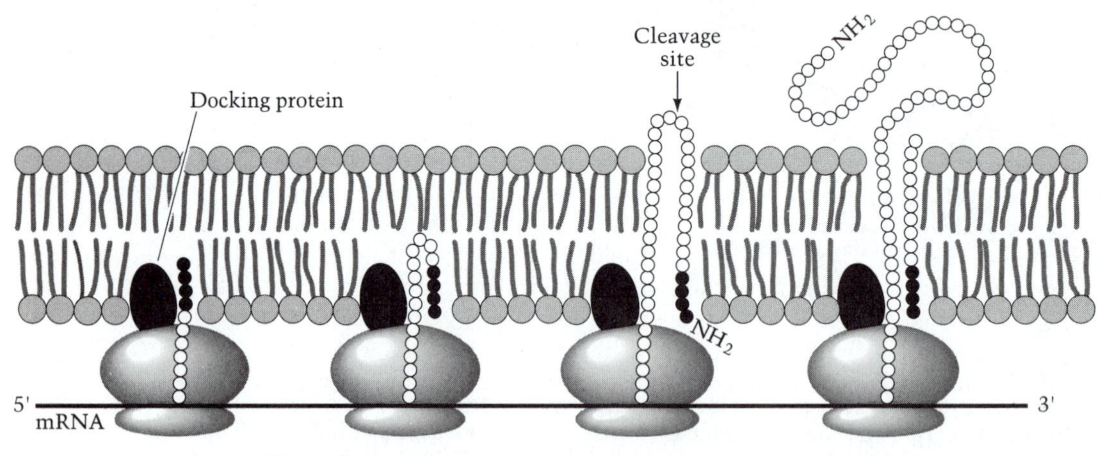

Figure 6

Bent loop model for the translocation of proteins that are secreted across the membrane. In this illustration, ribosomes are bound to the membrane by a hypothetical docking protein. (Modified from Inouye and Halegoua, 1980.)

sequence does not necessarily affect translocation. Furthermore, a still-attached leader sequence does not usually influence the enzyme activity of a protein. The best guess at this time is that removing the leader sequence allows rapid folding of the mature protein, an action that protects it from proteolysis. The cleaved leader sequence is not wasted but is recycled after further degradation into its component amino acids.

No important differences in the general aspects of cell membrane assembly have been found between Gram-positive and Gram-negative bacteria. Of course, features of the membrane in Gram-negative cells that relate specifically to the formation of the periplasm and the outer membrane are absent in Gram-positive cells. Conversely, features that deal specifically with the assembly and turnover of murein in the thick Gram-positive wall can be expected to be different in Gram-negative cells. The assembly of teichoic acids is also a purely Gram-positive subject, since these compounds are absent from Gram-negative bacteria.

Growth of the cell membrane

How does the membrane grow? Is new material inserted at a few specific zones or is it intercalated at many locations found all over the surface? Interest in the topology of membrane assembly has been high, in part because of the postulated role of the bacterial envelope in the segregation of the chromosome into daughter cells. The attachment of the chromosome to the envelope provides a plausible model for segregation, especially if we postulate that insertion of newly made membrane material gradually forces the two daughter chromosomes apart. This point is discussed in detail in Chapter 14.

The pattern of membrane growth has been studied by so-called segregation experiments. Such experiments provide indirect evidence for the mode of membrane growth by determining how membrane components from an original cell are distributed among its progeny. A culture labeled with a suitable radioactive precursor (for lipids this may be glycerol or fatty acids) is "chased" by growth in unlabeled medium, and autoradiograms are performed with the cells . If membrane synthesis takes place at only one site—say, the cell's middle—then each daughter cell will be labeled with half the "old" membrane after one generation, only 50% of the cells will be labeled after two generations, 25% after three generations, and so on (Figure 7). If, on the other hand, insertion occurs at many sites, all progeny cells will receive some of the "old" membrane material (all will be labeled, albeit with an ever-decreasing amount of label per cell after subsequent divisions). Notice that these experiments can only be interpreted if membrane components remain fixed at their site of insertion. In reality, this situation is not the case, because many, perhaps most, membrane lipids readily diffuse along each leaflet of the bilayer. Thus, if lipids are inserted at localized regions only, then they may well move away and appear to be inserted at many sites. What result is actually obtained? Several investigators have reported that the progeny of cells fed lipid precursors were all labeled, evidence leaving us with seri-

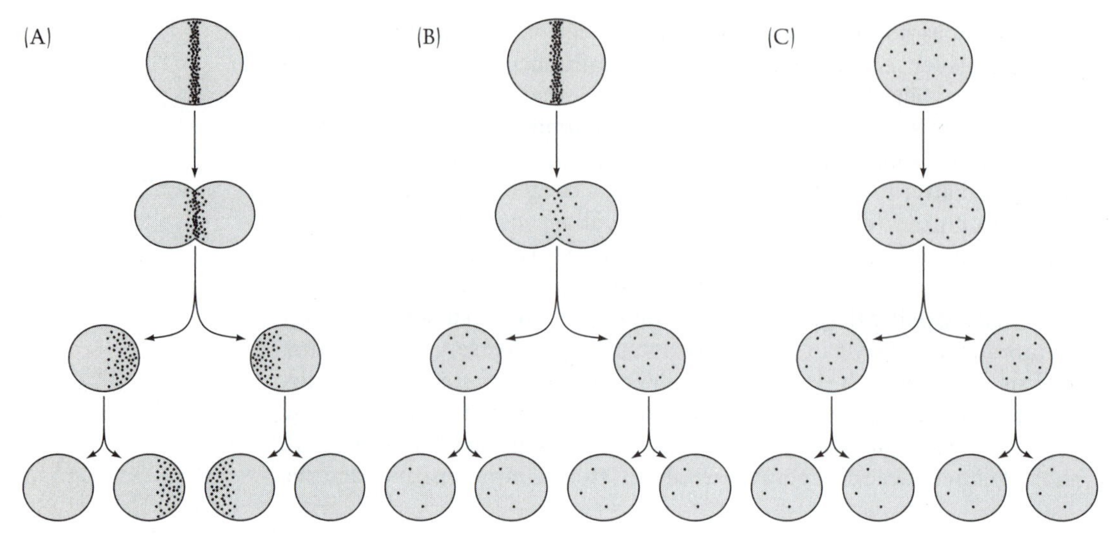

Figure 7

Three possibilities for membrane growth and the distribution during subsequent cell divisions. (A) Zonal growth with new material remaining in place. This pattern results in the partition of old membrane among some of the daughter cells only. (B) Zonal growth followed by distribution of newly made membranes. This pattern results in the partition of old membrane among all of the daughters. (C) Membrane growth at many sites. This pattern results in the same distribution of old membranes as seen in B.

ous uncertainty regarding the conclusions we can derive from such experiments.

However lipids are inserted, there is evidence that they are distributed among progeny cells as large complexes and not as individual molecules. When a segregation experiment was extended over many divisions, with both Gram-negative and Gram-positive cells, it was found that eventually some of the progeny cells did not have appreciable amounts of the original lipids (Green and Schaechter, 1972). This fact was determined by starting with a parent culture whose lipids were labeled with glycerol, washing away the excess labeled glycerol, and carrying out autoradiograms at different times of growth. After eight generations, an increasing proportion of cells had no radioactivity (as shown by an absence of photographic grains on these cells). The result was interpreted to mean that the membrane is divided into about 2^8, or 256, conserved subunits within which lipids are nonexchangeable. Nothing is known about the nature of these subunits (are they complexes of specific proteins and lipids?), and this finding awaits further study.

When it comes to segregation experiments with membrane proteins, the evidence is also equivocal. On the one hand, it was found that after induction of lactose permease and subsequent growth in noninducing medium, the enzyme is present only in some of the progeny cells, a result suggesting localized membrane growth (Kepes and Autissier,

1972). This experiment has not been confirmed, and other findings have led to the opposite conclusion. Thus, labeled heme molecules—presumably linked to cytochromes and other membrane proteins—are distributed among many progeny cells. So are flagella, which, because of their mode of insertion, also serve as a membrane marker. Furthermore, if membrane proteins were inserted at a few sites only, we would expect to find polysomes bunched up at such sites; this pattern has not been found. Thus, the preponderance of the evidence is that membrane proteins are inserted at many sites on the membrane.

PERIPLASM

The periplasm is the space between the two membranes of Gram-negative bacteria and contains at least 50 different kinds of proteins. These proteins are involved in the nutrition of the cell, either by reducing macromolecules to suitable size or by concentrating nutrients from dilute solutions. The periplasm and its proteins afford an especially rich opportunity to study the passage of proteins through the cell membrane. The making of spheroplasts—by treating *E. coli* with lysozyme and EDTA—releases the periplasmic proteins from the cells; these proteins can then be studied without the complication of the outer membrane.

The factors that determine the eventual cellular localization of secreted proteins are not well known; a general scheme is shown in Figure 8. The significance of the leader sequence in the secretion of periplasmic proteins has been studied by elegant experiments involving gene fusion. For example, a protein containing the leader sequence and most of the maltose-binding protein (a periplasmic protein) was fused to β-galactosidase (a typical cytosolic enzyme). The fused protein was found to enter the cell membrane but not the periplasm. Apparently, a stretch of amino acids in β-galactosidase stopped the passage across the membrane. Interestingly, the stuck hybrid molecules interfered with the translocation of normal outer membrane proteins, which accumulated in the cytosol with their leader sequence intact. This roadblock suggests that peptides destined for the periplasm and those destined for the outer membrane have, at least in part, a common pathway for translocation across the cell membrane. Mutations in the leader sequence prevent the hybrid protein from entering the cell membrane and from interfering with translocation.

WALL

The wall of a typical Gram-negative cell consists of a single, gigantic, sac-like molecule of murein, one or a few layers thick (Figure 9). In contrast, the wall of a Gram-positive cell (e.g., *Bacillus subtilis* or staphylococci) is made up of many layers of murein plus teichoic acids (Figure 10). The assembly of the wall components begins with the synthesis of precursors in the cytoplasm, their transport across the cell membrane, and their final polymerization. Eventually, a battery of

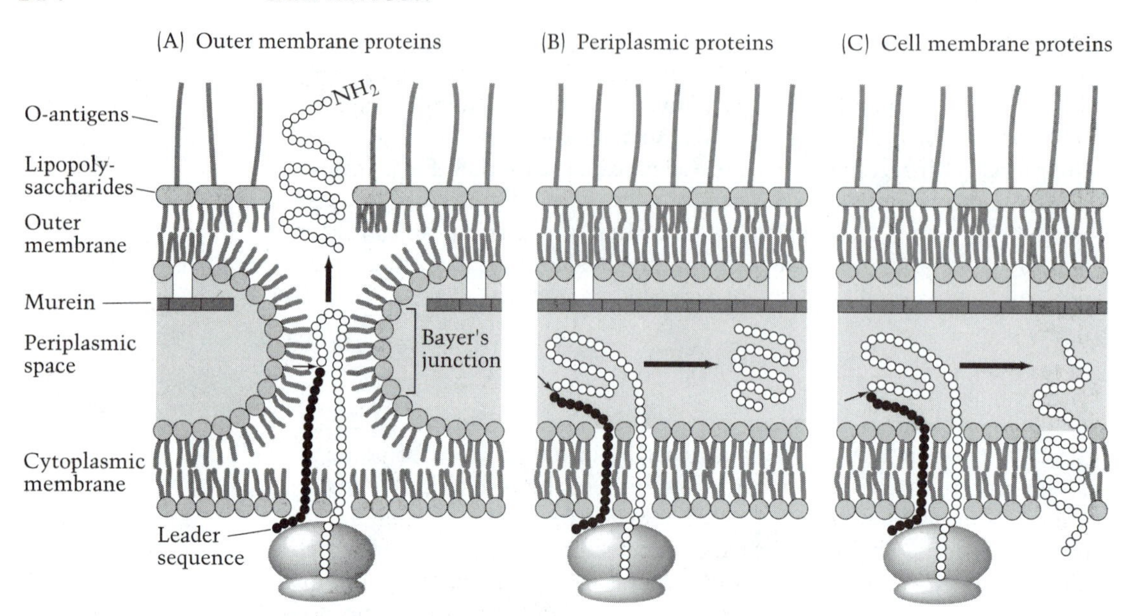

Figure 8

Possible mechanisms of secretion and final localization of envelope proteins in a Gram-negative bacterium. (A) Outer membrane proteins. It is assumed that when the nascent peptide is cleaved, the amino terminus interacts with LPS or proteins at the zones of adhesion between the cell membrane and the outer membrane. This process would lead the protein to the outer membrane. (B) Periplasmic proteins. Folding begins in the periplasm. Because of the hydrophilic properties of such proteins, they remain soluble in the periplasm. (C) Cell membrane proteins. Folding takes place as these peptides, which have considerable hydrophobic sequences, integrate into the membrane. (After Inouye and Halegoua, 1980.)

enzymes catalyzes covalent reactions that result in the extension, cross-linking between glycan strands, morphogenesis, and eventual septation of the murein sacculus. These proteins also have the unique ability to bind penicillin and other β-lactam antibiotics covalently and are known as PENICILLIN-BINDING PROTEINS, or PBPs. Not surprisingly, the β-lactam antibiotics—penicillins and cephalosporins (both of which come in many natural and semisynthetic varieties)—inhibit the action of these proteins. Indeed, the antibacterial action of these antibiotics is due to their ability to inhibit murein biosynthesis by binding to PBPs. Some of the properties and possible functions of PBPs are listed in Table 2. The number of PBPs varies among species.

Murein assembly takes place in distinct stages. For simplicity's sake, we shall limit them to the following steps (as known for *E. coli*):

- In the cytosol, the murein precursor UDP-*N*-acetylmuramic acid-pentapeptide is synthesized in several steps and then attached to a cell membrane carrier lipid called UNDECAPRENYLPHOSPHATE, or BACTOPRENOL. In the cell membrane, *N*-acetylglucosamine is added to form a complete repeating unit that is still attached to the lipid carrier.

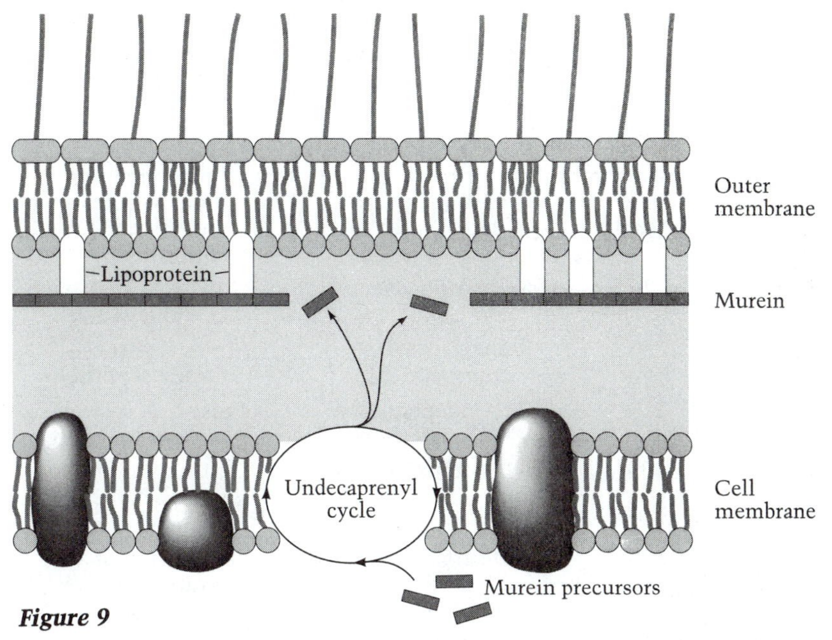

Figure 9

Wall assembly in Gram-negative bacteria. The diagram shows a region where murein precursors are shuttled through the cell membrane on undecaprenylphosphate; in the periplasmic space they are polymerized into glycan chains, which become covalently attached to the existing murein sacculus.

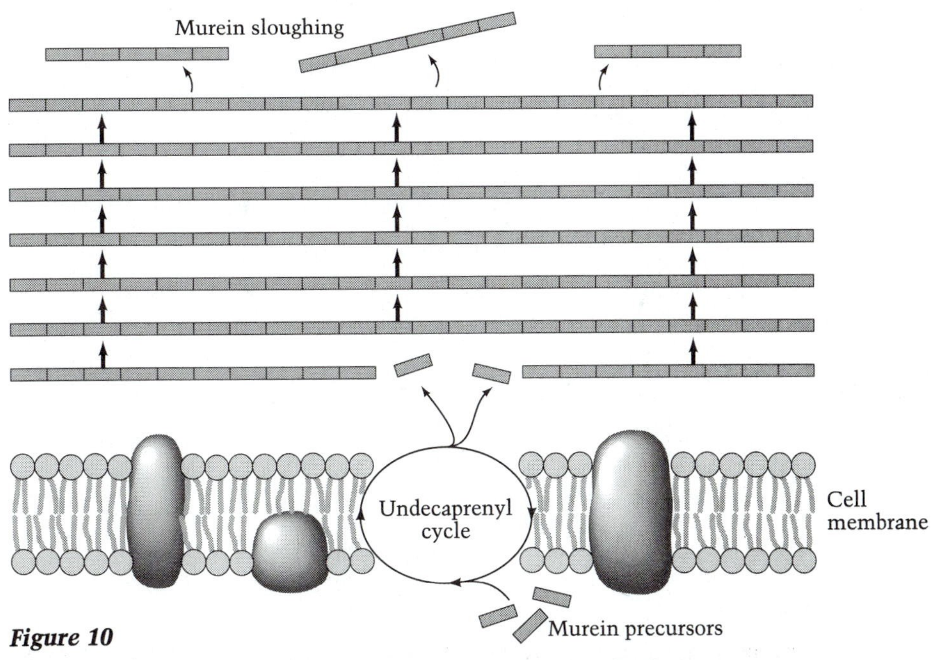

Figure 10

Wall assembly and turnover in Gram-positive bacteria. The dynamic nature of the wall is indicated by arrows, which show the progression of newly formed murein from its site of synthesis at the cell membrane to the periphery of the cell, where it is eventually sloughed off.

Table 2. Properties of penicillin-binding proteins of *E. coli*

PBP	Molecules per cell	Known enzyme activities	Possible function
1A or 1B	100 (each)	Transglycosylase–transpeptidase	Murein synthesis during cell elongation
2	20	Transpeptidase	Growth in rod shape; cell elongation
3	50	Transglycosylase–transpeptidase	Murein synthesis during septation
4	110	DD-endopeptidase, DD-carboxypeptidase	Cross-link hydrolysis in cell elongation
5	1,800	DD-carboxypeptidase	Destruction of unutilized pentapeptide
6	600	DD-carboxypeptidase	Destruction of unutilized pentapeptide
1C,7,8	Not known	Not known	Not known

Source: Adapted from Park (1987b).

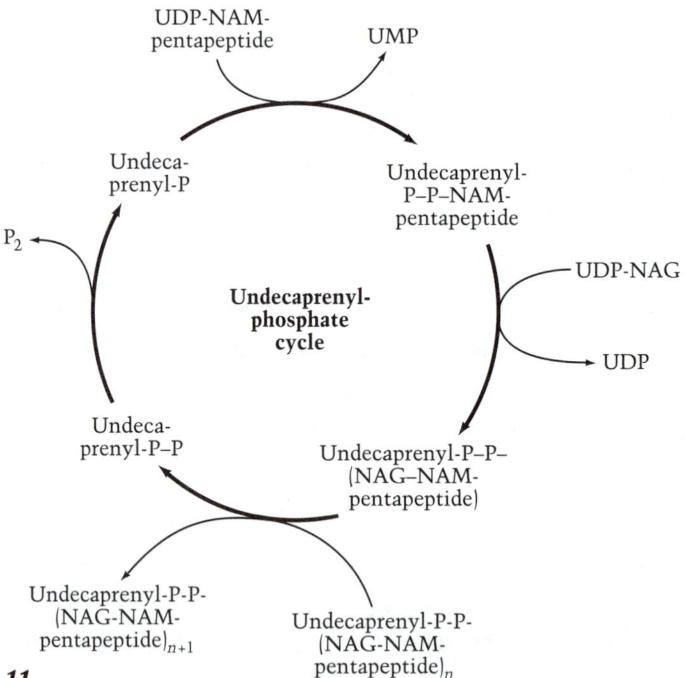

Figure 11

Undecaprenylphosphate cycle: assembly of murein precursors and their polymerization into glycans. P, phosphate; NAM, *N*-acetylmuramic acid; NAG, *N*-acetylglucosamine. (After Park, 1987b.)

- Murein repeating units are polymerized and simultaneously cross-linked by the activity of PBP 1A and 1B, each of which, interestingly, is a bifunctional enzyme that carries out both transglycosylation and transpeptidation. The precursor chains are bound to the lipid carrier, undecaprenylphosphate, which is later recycled (Figure 11).
- The precursor chains are inserted into the the cylindrical portion of the sacculus, thanks to the cutting of preexisting murein by PBP 2 and PBP 4. The lipid carrier is released to function cyclically in the synthesis and transport of precursors. The insertion of new precursor chains appears to take place sequentially; in other words, murein elongates unidirectionally around the circumference of the cell (Figure 12). It has been estimated that there are about 200 sites of murein insertion and that at 37°C the synthetic complex at each site takes about 8 minutes to go around the circumference of an *E. coli* cell. The evidence is discussed below.
- When the cell is ready to divide, the murein destined for the cell

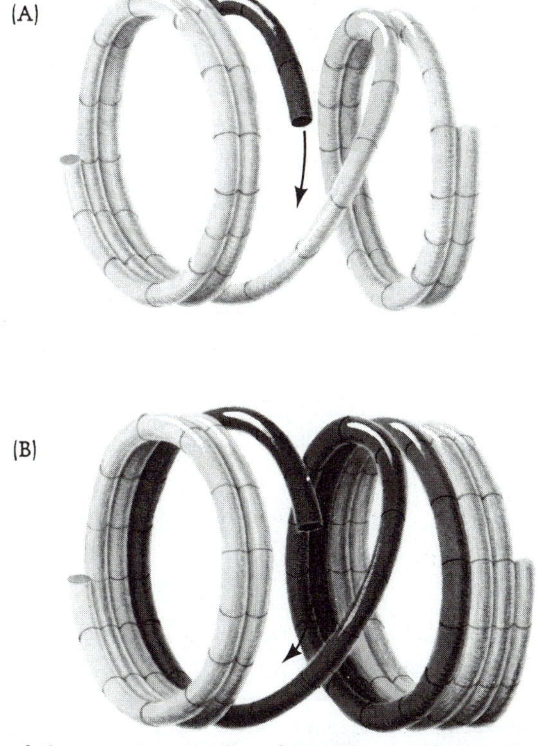

Figure 12

Elongation of the murein sacculus of *E. coli* according to the model of Burman and Park. The thick lines represent a pair of new murein strands inserted between preexisting strands. (A) Events after a few minutes of labeling. (B) Events 8 minutes later, when new strands have been inserted around the circumference of the cell. At present, it is not certain whether insertion is done by pairs of strands or by a single strand. (After Park, 1987a.)

septum is synthesized by another bifunctional enzyme, PBP 3. Nothing is known at present about how PBP 3 or the other transglycosylases–transpeptidases can distinguish between the synthesis of the cylindrical portion of the wall and the division septum.

It is not yet known why cells make several PBPs with similar activity. Is this a backup system that has evolved in case one or another of these essential enzymes becomes inactivated by mutation or environmental challenges?

The building block and energy requirements for murein production are shown in Table 3. Many of the fundamental reactions for assembling the murein sacculus in *E. coli* have their counterpart in Gram-positive cells, and indeed some were discovered there first. Nevertheless, wall growth in Gram-positive cells displays interesting differences from that of Gram-negative cells, differences due in part to the thicker wall structure of Gram-positive cells. Newly synthesized murein precursors are continually added to the existing wall at the innermost surface. Meanwhile, at the outermost surface, murein is being hydrolyzed and lost in the medium (Figure 10). The Gram-positive wall is, therefore, a dynamic structure in which murein components move outward as the result of continuous growth and turnover. Because the wall retains a constant thickness under normal conditions, there must be mechanisms that keep wall assembly and turnover in balance. It has been proposed that the wall is maintained intact by incorporating new units before cleaving existing stress-bearing sites. Koch (1988) has called this the "make-before-break" strategy.

Table 3. Requirement for the production of murein in 1 g of cells[a]

Factor required	Amount of building block required (μmol)	Amount of energy required (μmol ~P)
Building blocks[b]		
UDP-NAG	27.6	
UDP-NAM	27.6	
L-Alanine	27.6	
D-Alanine	27.6	
m-Diaminopimelic acid	27.6	
D-Glutamic acid	27.6	
Energy for forming the pentapeptide[c]		138.0

[a] 1 g of cells contains 25 mg murein; 25 mg murein = 27 μmol disaccharide subunits; 1 μmol disaccharide = 904 μg disaccharide.

[b] We consider the sugar building blocks to be their UDP derivatives, because NAM (*N*-acetylmuramic acid) is made from the UDP-derivative of NAG (*N*-acetylglucosamine). The amino acid building blocks will differ in different bacterial species (Chapter 1); and, in particular, there will in many species be an oligopeptide bridge between adjacent chains. In *Staphylococcus aureus*, for example, it is a pentaglycine bridge. For details, consult a biochemistry textbook.

[c] These ~Ps represent the five ATP → ADP hydrolyses that accompany construction of the pentapeptide.

Teichoic acids—anionic polymers that are found in the wall of Gram-positive bacteria—become linked to the murein glycan strands during assembly, although some may be secreted through the wall without becoming attached to murein. Assembly of teichoic acids appears to take place on a membrane carrier, probably undecaprenylphosphate. A membrane-bound form of teichoic acid, called LIPOTEICHOIC ACID, is attached to a glycolipid of the membrane rather than to the wall murein.

How does the cell wall replicate and segregate? For *Streptococcus pyogenes,* there is clear evidence that the cell wall is synthesized in a zonal way, starting at the midline and growing outward. Thus, each cell is composed of a hemispherical "old" and "new" cap (Figure 13). Zonal synthesis was demonstrated by treating cells with anti-wall fluorescent antibodies and allowing the cells to grow. In contrast, the major portion of the wall of *E. coli*—its cylindrical portion—is synthesized at many sites. This fact was shown by pulse-labeling cells with radioactive murein precursors for a short time and making autoradiograms. To obtain sufficient resolution, cells and their overlying photographic grains were examined under the electron microscope. The grains were found to

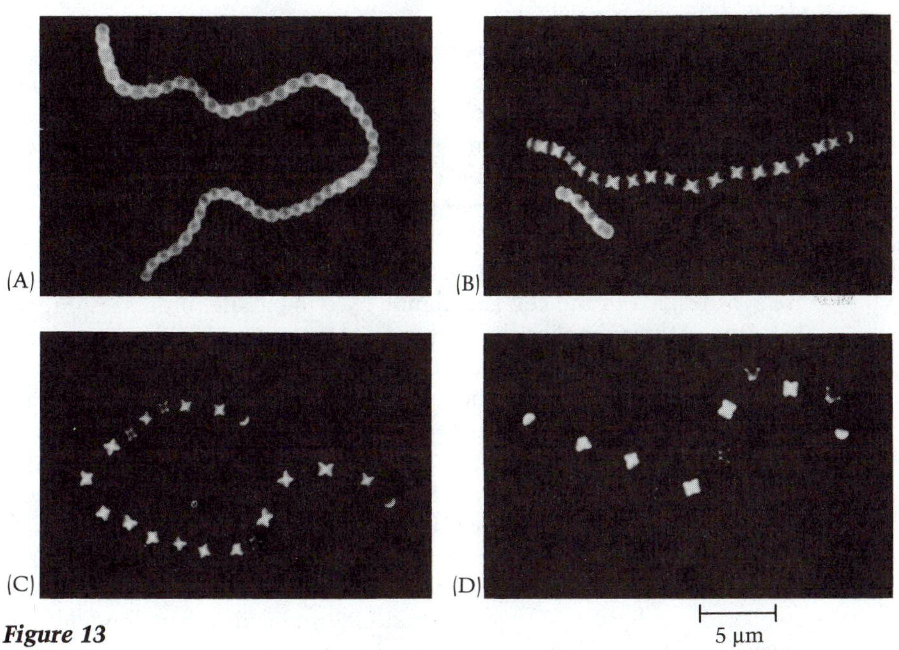

Figure 13

Distribution of the cell wall of *Streptococcus pyogenes* among daughter cells. The cell walls were labeled with fluorescent antibodies, excess antibody removed, and the cells examined under a fluorescence microscope. (A) Immediately after antibody treatment, all cells are labeled. (B) After 15 minutes of growth in the absence of antibody, new nonfluorescent material has been inserted around the equator of each cell. The previously labeled polar caps remain fluorescent. (C) and (D) Appearance of cells after 30 and 60 minutes of growth, respectively. (From Cole and Hahn, 1962. Copyright 1962 by the A.A.A.S.)

be evenly distributed along the length of the cells, a result suggesting that new and old wall became interspersed. As mentioned above, the estimate from biochemical experiments is that there are some 200 sites of murein insertion in the wall. Other data suggest that insertion does not take place at the pole wall, because, in *B. subtilis* and perhaps in *E. coli,* the murein of the poles is conserved as a single unit. It appears, then, that in rod-shaped bacteria the cylindrical portion of the wall, but not the poles, grows by extension at many sites. This conclusion invites the notion that cocci (at least some of them) in this respect are "all pole," the equivalent of the hemispherical poles of rod-shaped bacteria.

OUTER MEMBRANE

The outer membrane of Gram-negative bacteria deserves special attention: not only is it a specialized device designed to protect the cell membrane from injurious chemicals, but it is also involved in several important aspects of bacterial pathogenesis. Its most characteristic component—LIPOPOLYSACCHARIDE or ENDOTOXIN—is a powerful immunostimulant that provokes many of the reactions to infection, such as fever, inflammation, and, if present in high amounts, shock, coagulation of the blood, and even death.

The outer membrane consists of an inner leaflet of phospholipids, an outer leaflet of lipopolysaccharide, and a moderate number of different proteins embedded in each leaflet or spanning both. Each of these constituents is synthesized independently rather than by forming intermediate complexes containing all of them. These separate assembly reactions proceed in concert rather than in sequence (Figure 14).

Assembly of lipopolysaccharide / Lipopolysaccharide (LPS) subunits are synthesized in the cell membrane by two parallel processes (Figure 15). One pathway makes the repeating polysaccharide side chain, built up on the lipid carrier undecaprenylphosphate, and the other makes the core polysaccharide, built up on lipid A (which functions both as a primer and the carrier that transports the core across the cell membrane). The two sets of precursors are made at the inner surface of the cell membrane, then translocated by their respective carriers to the outer surface. Translocation is thought to be driven by the protonmotive force operating across the cell membrane. At the outer surface of the cell membrane, a transfer enzyme attaches the completed side chain to the core polysaccharide on lipid A. The individual LPS molecules then condense by mutual attraction into the two-dimensional array of the outer leaflet of the outer membrane. The requirements for LPS production are shown in Table 4.

Assembly of phospholipids / Translocation of phospholipids from their site of synthesis on the cell membrane to the outer membrane takes place by a mechanism that is also driven by the electrochemical potential across the membrane (protonmotive force) rather than by ATP.

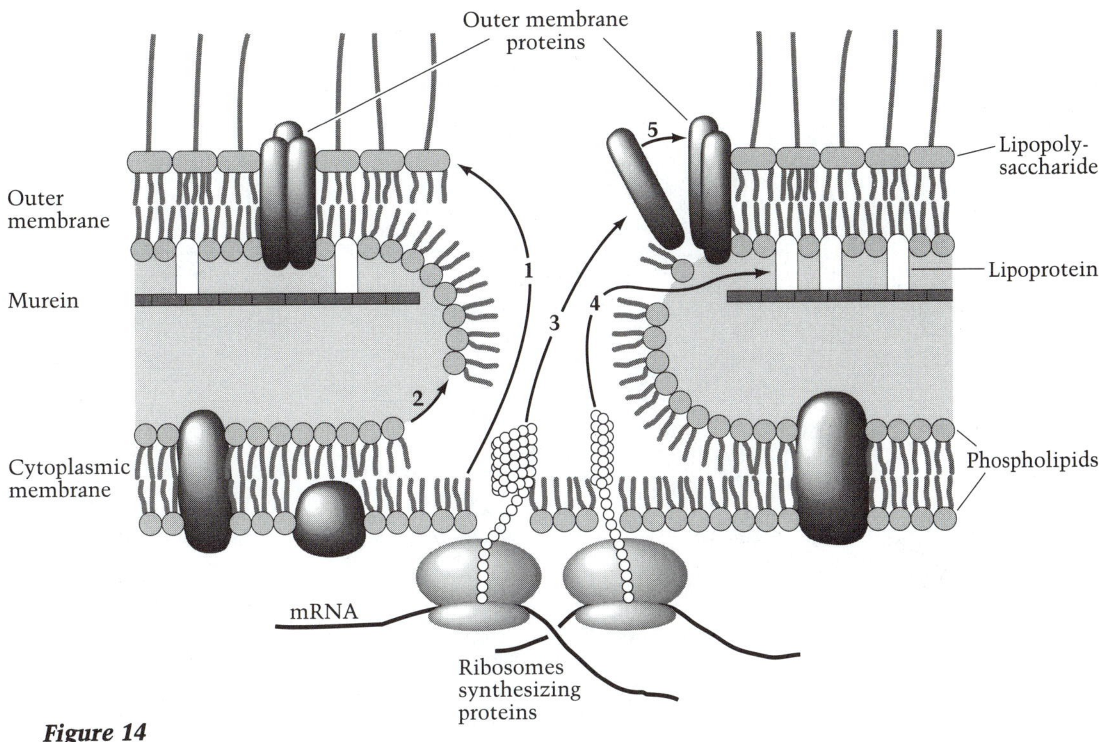

Figure 14

Assembly reactions of the outer membrane. (1) Lipopolysaccharide; (2) phospholipids; (3) outer membrane proteins; (4) lipoprotein; (5) self-assembly of the outer membrane.

Thus, phospholipid translocation requires the membrane to be energized and is suppressed by the addition of inhibitors that collapse the membrane potential (such as dinitrophenol).

Assembly of outer membrane proteins / As described in Chapter 2, the outer membrane contains three or four major proteins and as many as 50 less abundant ones. Like periplasmic and cytoplasmic membrane proteins, the outer membrane proteins are synthesized with leader sequences and utilize the cellular export machinery to cross the cell membrane. Are outer membrane proteins first exported from inside the cell to the periplasm and then inserted into the outer membrane? Apparently not. It is thought that outer membrane proteins are translocated via the zones of adhesion (Bayer's junction), thus permitting their direct interaction with LPS in the outer membrane (Figure 14). Association of newly made outer membrane proteins with these regions of the envelope was demonstrated by labeling ultrathin sections with gold-conjugated antibodies and examining them in the electron microscope.

Is LPS involved in determining which proteins are destined for the outer membrane? In some instances, LPS interacts with portions of the amino-terminal end of mature peptides, after the leader sequence has

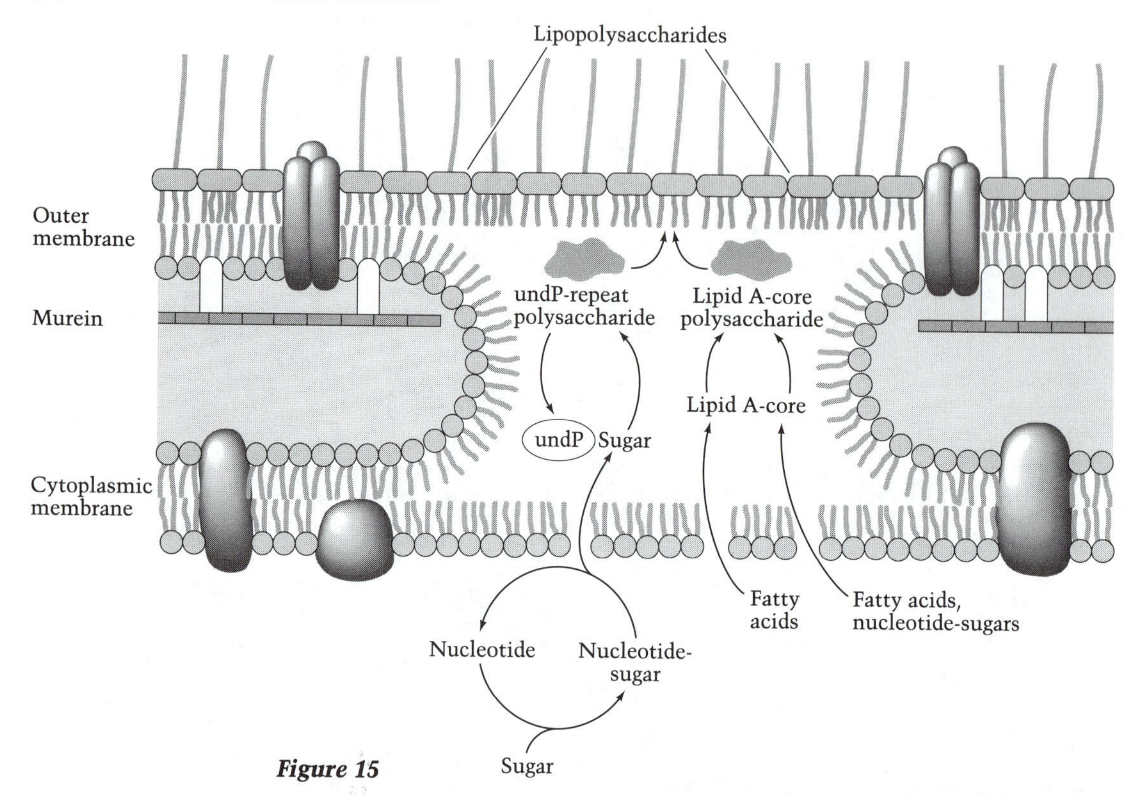

Figure 15

Assembly of the lipopolysaccharide. See text for details. undP, undecaprenylphosphate.

been removed. This interaction perhaps promotes the folding of the nascent peptide chains and guides their translocation and insertion into the outer membrane. This notion is supported by the finding that most major outer membrane proteins interact with LPS and that mutants defective in LPS do not incorporate normal amounts of outer membrane proteins.

Assembly of lipoprotein / The outer membrane contains a small protein that is the most abundant one (numerically, not by weight) in *E. coli* cells grown at 37°C. This protein is made with a leader sequence of 20 amino acids at its amino terminus (Figure 3). Shortly after cleavage of this leader, the amino-terminal cysteine of the mature peptide becomes modified by the addition of glycerol in a thioether linkage (Figure 16). This modification is followed by acylation with fatty acids at three sites, namely, the two remaining hydroxyls on the glycerol and the amino group of the cysteine. Presumably, the lipid end of the mature protein interacts with LPS and other lipids of the outer membrane, thereby facilitating its insertion into the membrane. The carboxy-terminal lysine of the mature lipoprotein projects inward into the periplasm. At

Table 4. Requirements for the production of the LPS in 1 g of cells[a]

Factor required	Amount of building block required[b] (μmol)	Amount of energy required (μmol ~P)
Lipid A building blocks		
TDP-glucosamine	15.7	
Fatty acid-ACP (β-OH-myristic, others)	47.0	
CDP-ethanolamine	7.8	
Core building blocks		
CMP-KDO	23.5	
ADP-heptose	23.5	
UDP-glucose	15.7	
CDP-ethanolamine	15.7	
Energy[c]		0

[a]1 g of cells contains 34 mg lipopolysaccharide; 34 mg LPS = 7.8 μmol LPS units; 1μmol LPS = 4350 μg LPS.
[b]These amounts are based on the truncated version of LPS found in some *E. coli* B strains. The mechanism of adding ethanolamine residues is assumed to be via CDP-ethanolamine rather than through serine and decarboxylation. In the abbreviations of the nucleotides (TDP, CMP, ADP, UDP, and CDP), T stands for thymine, C for cytosine, A for adenine, and U for uracil. ACP stands for acyl carrier protein. For details, consult a biochemistry textbook.
[c]The seeming lack of an energy requirement is because all the building blocks are used as activated derivatives, so the energy cost is charged to their biosynthesis.

some point in the process, this lysine reacts with a diaminopimelic acid residue on the side chain of the murein, thereby anchoring the lipoprotein and the whole outer membrane to the underlying wall (Figure 16). Only about one in three of the lipoprotein molecules (i.e., about 250,000 per cell) become cross-linked to murein, but this number appears to be sufficient for structural purposes.

Self-assembly of the outer membrane / Lipopolysaccharide not only helps translocate the outer membrane proteins, it is also involved in the overall morphogenesis of the structure. An organized hexagonal lattice strongly resembling the outer membrane can be produced spontaneously by simply mixing LPS and one or more major outer membrane proteins, such as the porins OmpF or OmpC. In this reconstructed structure, the porins are correctly condensed as trimers that form pores extruding through the LPS matrix. Furthermore, if LPS and outer membrane proteins are mixed in the presence of murein containing covalently attached lipoprotein, the membrane that is formed is bound to the lipoprotein and covers the surface of the murein (Yamada and Mizushima, 1978). From these results it can be concluded that the outer membrane is put together largely by a self-assembly process, guided by specific molecular interactions between proteins and LPS. The murein sacculus appears to serve as a scaffold for the assembly process.

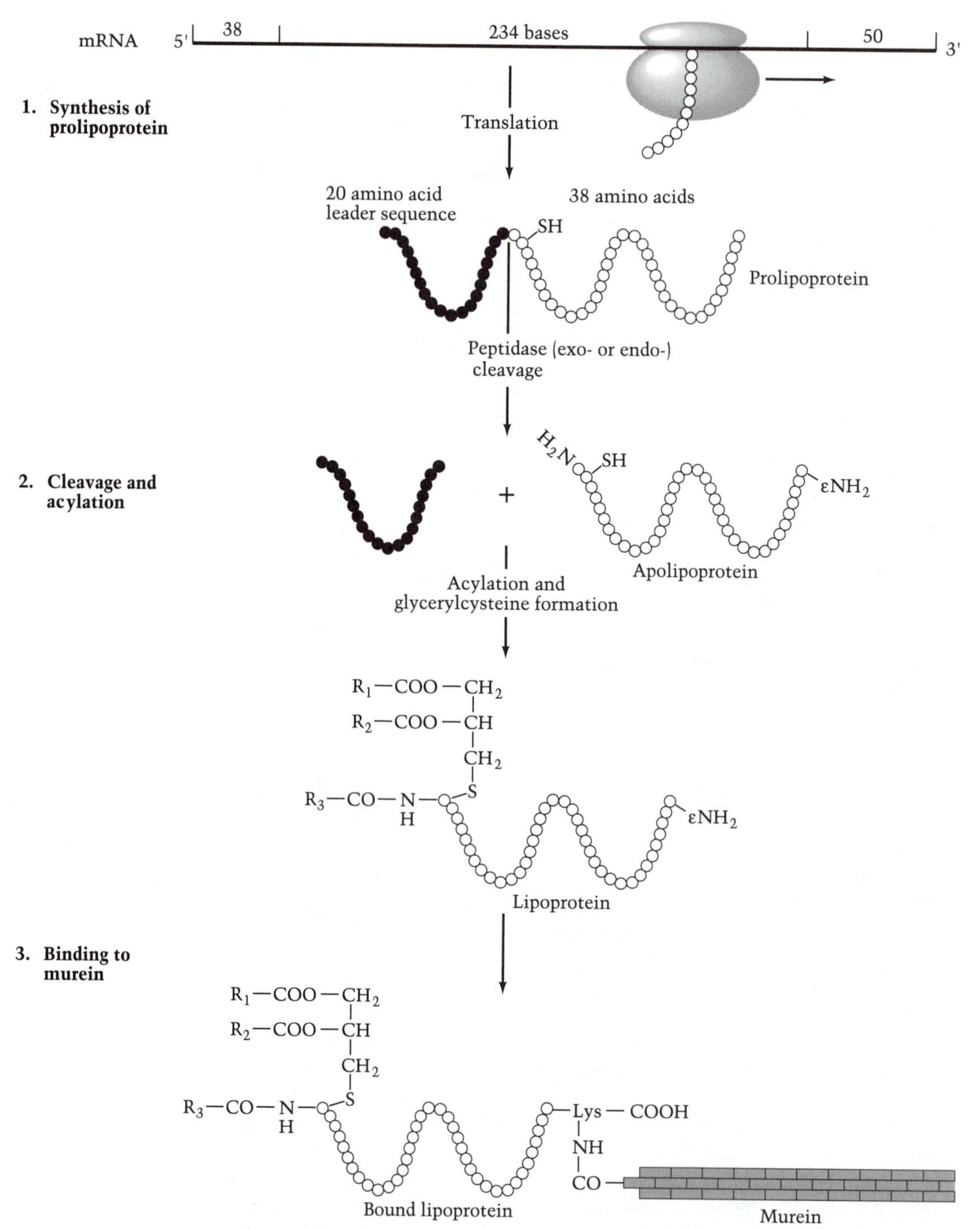

Figure 16

Posttranslational modification of lipoprotein. In step 1, prolipoprotein is cleaved by a specific leader sequence peptidase to yield apolipoprotein, the peptide of lipoprotein. In step 2, apolipoprotein is acylated by the addition of fatty acids, two on the aminoterminal glycerol, one on cysteine. In step 3, about one-third of the lipoprotein molecules are covalently bound at their carboxy termini to murein. (After Inouye, 1979.)

Perhaps the key to the assembly of the outer membrane is the existence of zones of adhesion between it and the cell membrane. There is convincing evidence, largely from electron microscopic studies, that outer membrane proteins, LPS, and even secreted proteins and capsular polysaccharides are inserted into or go through the outer membrane at the zones of the adhesions. However, the details of the mechanisms involved are speculative at present. It is likely that future work will add to our understanding of the role of these zones of adhesion, a fascinating structural quirk of Gram-negative bacteria.

FLAGELLA

Flagella are complex structures composed of a basal body, a hook, and a long filament. They are assembled by a series of intricate steps. At least 40 genes are required for flagellar assembly and function. The membrane plays a role in this process, as indicated by the finding that certain mutants in outer membrane proteins are nonflagellated. In *E. coli,* flagellar assembly appears to take place continuously throughout the cell cycle, whereas in specialized bacteria such as *Caulobacter,* the process has a temporal relationship with growth and the cell cycle. This difference is in keeping with the fact that *E. coli* has flagella located at random over its surface, whereas *Caulobacter* has polar flagella only.

Examining a large number of mutants that possess partial flagellar structures under the electron microscope has revealed an elaborate choreography (Figure 17):

- The process proceeds sequentially from the proximal to the distal end. The basal body is assembled first, then the hook, and lastly, the filament.
- The first recognizable structure is composed of the innermost rings of the basal body, incorporated into the cell membrane.
- The remainder of the basal body is then assembled and the hook is added to it, probably by extrusion through a central channel. The main protein of the filament—flagellin—plays no role in the assembly of the basal body or the hook. Nothing is known about how the length of the hook is precisely measured, other than the fact that mutants in certain genes lead to the formation of "superhooks." Thus, certain proteins probably function as "yardsticks" to determine the proper length of the hook.
- The filament is now made by extrusion of flagellin molecules through a central hollow core (Figure 18). Upon reaching the tip each molecule spontaneously condenses with its predecessors and thus elongates the filament. This phenomenon may well depend entirely on the properties of flagellin, because isolated flagellin molecules can, in the presence of a primer, self-assemble in vitro into structures indistinguishable from flagellar filaments. In vivo, the process can go on for a long time, although filament assembly slows down with increasing filament length. This slowing, plus mechanical breakage, may explain why flagella do not reach extraordinary lengths.

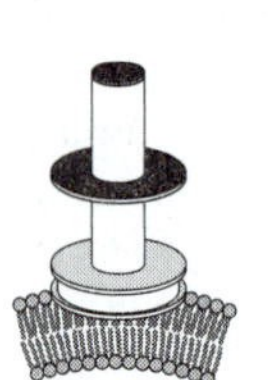

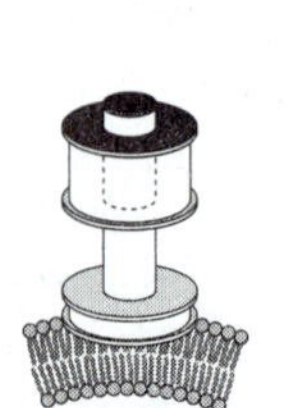

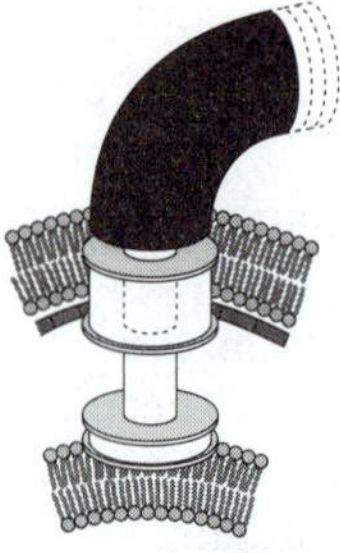

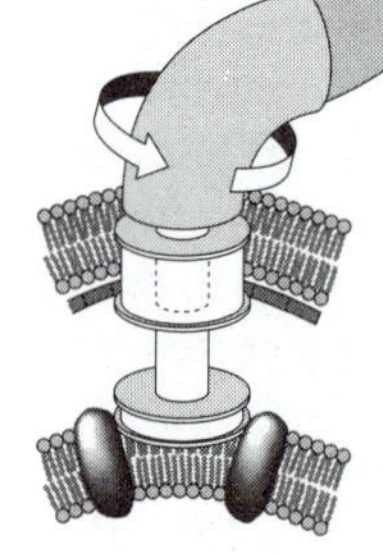

Figure 17

Flagellar assembly in *Salmonella typhimurium*. (1, 2) The M ring and the S ring of the basal body insert into the cell membrane. (3) The rod is added and the distal end of the rod is capped. (4) The P ring is added. (5) The basal body is completed by addition of the L ring. (6) The hook is made and finished (7) by addition of other proteins. (8) The flagellar filament is made. (9) Motility-enabling proteins are added to the cell membrane to finish flagellar assembly. (After Macnab, 1987.)

- It appears that only after the flagellum is assembled are several membrane proteins inserted near the basal body to render the structure functional. Thus, flagella cannot "turn on" until they are fully assembled.

As far as is known, assembly of flagella occurs in a similar fashion in all bacteria. Obviously, the details of basal body formation must be somewhat different in Gram-positive and Gram-negative cells. In addition, in the spirochetes, flagellar assembly takes place entirely within the periplasm. Whereas much remains to be learned about the details of flagellar assembly, it is a particularly illuminating example of macromolecular self-assembly.

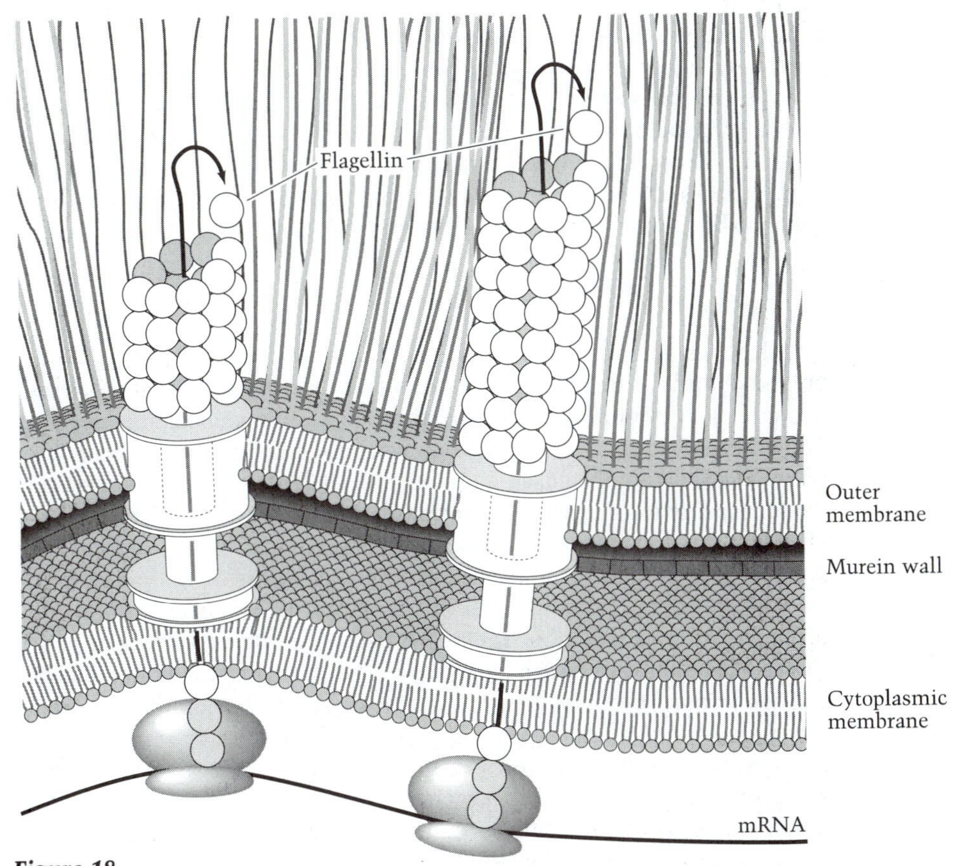

Figure 18

Growth of the flagellar filament by extrusion of flagellin subunits through the flagellar core. (Drawn partially from the studies of DePamphilis and Adler, 1971.)

PILI

Most Gram-negative and some Gram-positive bacteria make pili—straight protein rods that extrude from the surface and help the bacteria adhere to surfaces. The importance of pili in bacterial adhesion makes them occupy center stage in studies of pathogenesis and microbial ecology (see Chapter 17 for further discussion of this point). In addition, a specialized type of pili—the sex pilus—is specifically involved in bacterial conjugation.

At present, less is known about the assembly of pili than of flagella. Pilins, the major proteins of these structures, are capable of forming filaments in vitro by self-aggregation. These proteins are synthesized in the cytoplasm and cotranslationally translocated across the membrane. The best studied of these structures—the *E. coli* sex pili—appear to have an assembly process different from that of flagella; that is, pili are assembled by the addition of new molecules to the base. This difference is predictable, because, although pili contain a central channel, it is too narrow to allow the passage of pilin molecules. The mechanism that operates in the assembly of the more common, adherence-type pili is unknown. And at present, nothing is known about how the minor pili proteins (adhesins) that are involved in adhesion are placed at their specific sites, often at the tip of the structure. Are they extruded first, then pushed along by the rest of the pilin molecules?

CAPSULE

Although perhaps not a structure in the formal sense, the amorphous polysaccharide outer layer is of enormous ecological importance to some bacteria. For example, in pneumococci or meningococci the capsule allows the organisms to escape phagocytosis by polymorphonuclear white blood cells. The capsule presents an assembly problem similar in some respects to that of LPS. For both capsules and LPS, a large hydrophilic polymer must somehow be exported outside the cell membrane. And in both cases the solution is to build subunits on a lipid carrier, translocate them across the membrane (presumably through the zones of adhesion) and assemble the subunits on the outer surface. Again, undecaprenylphosphate is believed to be the carrier for capsular polysaccharide transport in *E. coli.* At the inner surface of the cell membrane, a repeating unit of three sugars is built up on the carrier by transfer from nucleoside diphosphate-sugar precursors. At the outer surface, the trisaccharides are linked together by head addition, that is, by the transfer of the growing polymer from its carrier to the next trisaccharide–carrier complex. It is known that *E. coli* capsular polysaccharides contain a phospholipid molecule in which one of the fatty acids is substituted by the polysaccharide chain. In this case, the capsular strands may be associated with or inserted into the outer membrane, thereby becoming firmly anchored. How common this is among other bacteria is not yet known.

One important group of capsular polysaccharides—the dextrans and levans—are assembled outside the cell by extracellular enzymes. They are made without the expenditure of ATP energy, using for transglycosylation the energy in glycosidic bonds provided by the precursor molecules—usually disaccharides. Note, however, that energy must have been expended in transporting the enzymes outside the cell.

Formation of capsular material proceeds more or less continuously, and the completed product is eventually sloughed off into the medium. The thickness of the capsule is, therefore, not fixed but varies with the nature of the growth medium and temperature. In general, bacteria growing on agar will be surrounded by a much thicker capsule than those growing in liquid medium. At times, the capsule of bacteria growing on solid media is truly enormous—dozens of cell lengths in radius. Such bacteria make colonies that look like drops of pure glue—translucent, wet, and very large (Figure 19).

With a few exceptions, such as *Bacillus* species that produce a capsule of poly-D-glutamate, bacterial exopolymers are polysaccharides and include cellulose, colaminic acid (a polymer of *N*-acetylneuraminic acid), and polyuronides. The biochemical variety is well illustrated by pneumococci, each serotype of which produces a distinct polysaccharide.

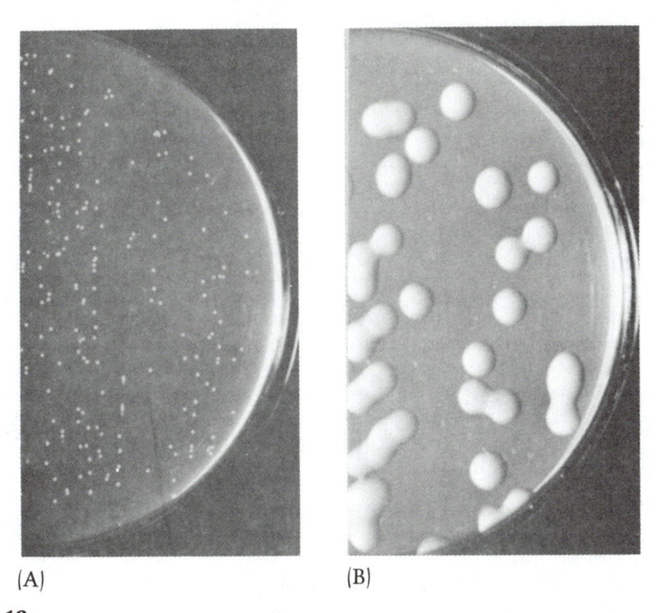

(A) (B)

Figure 19

Formation of capsular polysaccharide by bacteria. *Leuconostoc mesenteroides* streaked (A) on glucose medium and (B) on sucrose medium. The large size and mucoid appearance of the colonies on sucrose are due to massive synthesis and deposition of dextran around the cells. (From Stanier et al., 1986. Reprinted by permission of Prentice-Hall, Inc.)

ENERGY REQUIREMENTS FOR POLYMERIZATION

The energy requirements for the polymerization of the major classes of macromolecules for the reference *E. coli* cell (Chapter 1) are shown in Table 5. As discussed in Chapter 3, calculating the energy cost of synthesizing macromolecules has arbitrary aspects. To show how two sets of books can be kept, the values for nucleic acid and protein synthesis are shown both with and without the cost of activating their building blocks.

SUMMARY

1. Phospholipids are distributed asymmetrically among the leaflets of bacterial membrane bilayers. In Gram-negative bacteria, lipopolysaccharide is found in the outer leaflet of the outer membrane. Little is known about the mechanisms that ensure such asymmetry.
2. Many bacterial proteins are destined for insertion into the cell membrane, secretion into the medium, and, in Gram-negative bacteria, localization in the periplasm or the outer membrane. These translocations require passage from a hydrophilic to a hydrophobic environment. This passage is accomplished by a process that delays the

Table 5. Energy requirements for polymerization of the macromolecules in 1 g of cells[a]

Macromolecule	Amount of energy required (μmol ~P)[b]
From activated building blocks	
DNA: from dNTPs	136
RNA: from NTPs	236
Protein: from aminoacyl-tRNAs	11,808
Murein: in part from activated building blocks	138
Phospholipids: in part from activated building blocks	258
LPS	0
Polysaccharide (glycogen)	0
Total energy	12,576
From unactivated building blocks	
DNA: from dNMPs	336
RNA: from NMPs	1,516
Protein: from amino acids	21,970
Murein: in part from activated building blocks	138
Phospholipids: in part from activated building blocks	258
LPS	0
Polysaccharide (glycogen)	0
Total energy	24,218

[a]1 g of cells contains 961 mg macromolecules.
[b]These data are taken from summary tables in Chapters 2 and 3. They are used to calculate total energy and cost of polymerization of precursors (Table 2).

normal folding of a newly made peptide. These processes involve leader sequences and an export apparatus.

3. Most secreted proteins are synthesized with leader sequences that are cleaved upon maturation. The leader sequences help maintain the peptides in an unfolded conformation. In addition, the leader sequences interact directly with sites on the membrane, thereby allowing the translocation of the proteins.
4. It is not known with certainty whether the membrane is assembled at a few or many sites. The fluidity of the membrane makes it difficult to follow the fate of labeled precursors incorporated into membrane constituents.
5. The wall murein is assembled in a series of defined steps: synthesis of precursors, binding to the lipid carrier (undecaprenylphosphate), transfer across the membrane, and polymerization by addition to preexisting molecules. So-called penicillin-binding proteins located on the cell membrane are involved in these biosynthetic steps.
6. In *E. coli,* assembly of the wall murein takes place at many sites along the surface, apparently by the continuous elongation of preexisting murein chains. In some Gram-positive cocci, wall assembly takes place at the equator of the cells only.
7. Lipopolysaccharide, the characteristic component of the outer membranes found in Gram-negative bacteria, is made by separate pathways that independently synthesize the lipid A-core portion and the side chain polysaccharide. These constituents are then attached at the outer membrane.
8. Assembly of phospholipids and proteins into the outer membrane requires that the cell membrane be energized. Lipopolysaccharide is involved in the assembly of proteins into the outer membrane. It is likely that zones of adhesion between the cell membrane and the outer membrane are involved in the assembly process.
9. The wall lipoprotein, a characteristic protein of Gram-negative cells, is modified after synthesis by the addition of three fatty acids that help anchor it into the outer membrane. One third of the lipoprotein molecules are covalently bound to the underlying murein.
10. Flagella are assembled from the interior outward, component by component (basal body, hook, filament). The long filament is made by the extrusion of flagellin molecules through the core of the structure and their polymerization at the growing tip.
11. The assembly of pili differs from that of flagella because it does not take place by the extrusion of subunits through a core. These structures probably grow by addition at the basal end.
12. Bacterial capsules are synthesized either by the secretion of preformed polymers or by polymerization by extracellular enzymes.

STUDY QUESTIONS

1. Discuss possible mechanisms for the translocation of lipids across the cytoplasmic membrane and from the cytoplasmic membrane to the outer membrane.

2. Design a (preferably novel) experiment to determine whether a protein is inserted at one or many sites in the membrane.
3. What factors determine whether a protein will be secreted or retained in the interior of a bacterial cell?
4. Discuss the role and ultimate fate of the leader sequence of secreted proteins.
5. Discuss the main steps in murein biosynthesis.
6. In what ways does the assembly of the outer membrane of Gram-negative bacteria differ from that of the cytoplasmic membrane?
7. Contrast the mode of assembly of flagella and pili.
8. Discuss the features that characterize the synthesis of capsular polymers.

SUGGESTED READING

Topics in this chapter are covered in the following chapters from *Escherichia coli and Salmonella typhimurium: Cellular and Molecular Biology,* F. C. Neidhardt, J. L. Ingraham, K. B. Low, B. Magasanik, M. Schaechter and H. E. Umbarger (eds.). 1987. American Society for Microbiology, Washington, D.C.

Cronan, J. E., R. B. Gennis and S. R. Maloy. Cytoplasmic membrane. (Volume 1; Chapter 5)

Cronan, J. E. and C. O. Rock. Biosynthesis of membrane lipids. (Volume 1; Chapter 30)

Macnab, R. M. 1987. Flagella. (Volume 1; Chapter 7)

Park., J. T. Murein synthesis. (Volume 1; Chapter 42)

Rick, P. D. Lipopolysaccharide biosynthesis. (Volume 1; Chapter 41)

Other suggested readings include:

Koch, A. L. 1988. Biophysics of bacterial wall as a stress-bearing fabric. Microbiol. Rev. 52:337.

Randall, L. L., S. J. S. Hardy and J. R. Thom. 1987. Export of protein: a biochemical view. Annu. Rev. Microbiol. 41:507

Troy, F. A. H. 1979. The chemistry and biosynthesis of selected bacterial capsular polymers. Annu. Rev. Microbiol. 33:519.

5

Biosynthesis and Fueling

INTRODUCTION

The bacterial cell synthesizes the building blocks for construction of macromolecules from ingredients in its medium through a network of several hundred enzyme-catalyzed reactions. To emphasize their role in bacterial growth, we use a classification system that groups reaction pathways according to their metabolic products. BIOSYNTHETIC PATHWAYS lead from 12 metabolic intermediates called PRECURSOR METABOLITES (Figure 1) to building blocks. FUELING PATHWAYS lead from ingredients of the external medium to the metabolic needs—precursor metabolites, reduced pyridine nucleotides, energy, nitrogen, and sulfur—of the biosynthetic pathways (Figure 1 in Chapter 3).

The more conventional way to organize these many different reactions is to assign them to either CATABOLIC PATHWAYS or ANABOLIC PATHWAYS, depending on whether the pathway is primarily involved in degrading compounds into metabolically useful fragments (catabolism) or reforming these fragments into the building blocks of macromolecules (anabolism). In this nomenclature, reactions that fill both roles are called AMPHIBOLIC.

BIOSYNTHESIS

Elucidation of the pathways by which bacteria produce building blocks was very much dependent on the development of techniques for mutant isolation and analysis. This technology showed directly that particular enzymes were required to make a particular building block. By combining analysis of bacterial auxotrophic mutants with in vivo studies using isotope-labeled metabolites and in vitro studies on enzyme activities, researchers were able to uncover the several hundred biosynthetic reactions that produce amino acids, nucleotides, sugars, nucleotide–sugar derivatives, fatty acids, coenzymes, and enzyme prosthetic groups.

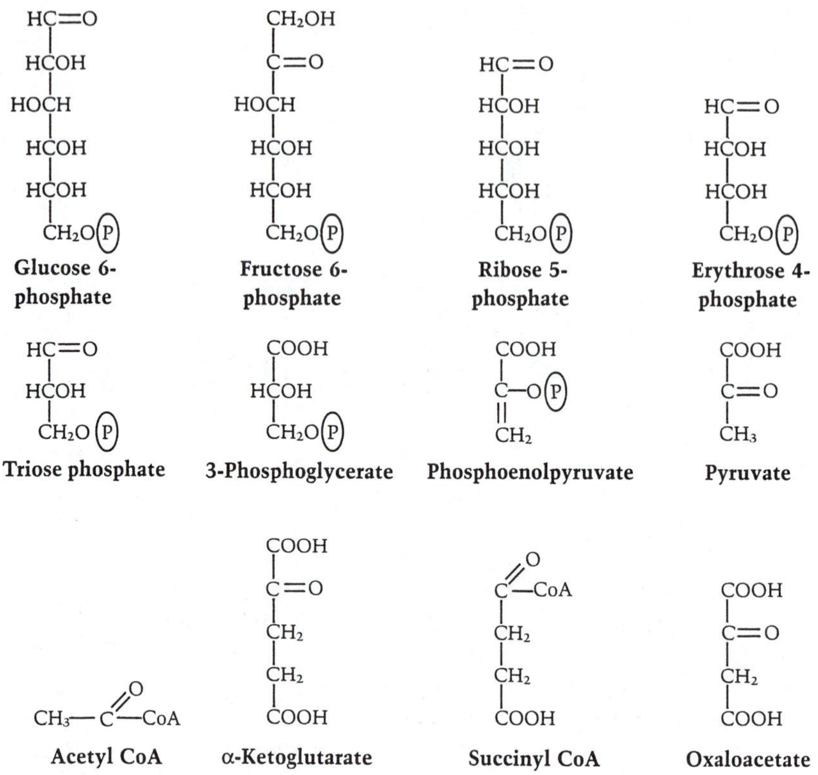

HC=O
HCOH
HOCH
HCOH
HCOH
CH2O(P)
Glucose 6-phosphate

CH2OH
C=O
HOCH
HCOH
HCOH
CH2O(P)
Fructose 6-phosphate

HC=O
HCOH
HCOH
HCOH
CH2O(P)
Ribose 5-phosphate

HC=O
HCOH
HCOH
CH2O(P)
Erythrose 4-phosphate

HC=O
HCOH
CH2O(P)
Triose phosphate

COOH
HCOH
CH2O(P)
3-Phosphoglycerate

COOH
C—O(P)
||
CH2
Phosphoenolpyruvate

COOH
C=O
CH3
Pyruvate

O
//
CH3—C—CoA
Acetyl CoA

COOH
C=O
CH2
CH2
COOH
α-Ketoglutarate

O
//
C—CoA
CH2
CH2
COOH
Succinyl CoA

COOH
C=O
CH2
COOH
Oxaloacetate

Figure 1

Structures of the 12 precursor metabolites.

This work, which reached its peak intensity in the 1950s and 1960s, was significant to all biology because, as it turned out, the biosynthetic pathways are essentially the same in all cells. To a first approximation, the same biochemical text serves the microbiologist, the botanist, and the zoologist. There are, of course, exceptions to the rule. A few building blocks are synthesized by different routes in different organisms. For example, the amino acid lysine is synthesized via one pathway (the DIAMINOPIMELATE PATHWAY) in bacteria, plants, and most algae, but it is synthesized by a completely different pathway (the α-AMINOADIPATE PATHWAY) in higher fungi and some algae. Other exceptions involve the biosynthesis of compounds unique to a given organism, and often these pathways are instructive indications of the special biochemical constraints imposed by a particular ecological niche. All of the 75–100 known building blocks, coenzymes, and prosthetic groups are synthesized from only 12 precursor metabolites (Figure 1) by reactions that employ energy, reducing power, and sources of nitrogen, sulfur, and single carbon units (Table 1). Regardless of which fueling reactions an organism employs, it synthesizes the same 12 precursor metabolites; they are the metabolic link between fueling and biosynthesis.

Table 1. Building blocks needed to produce 1 g of *E. coli* protoplasm

Building block	Amount present in *E. coli* B/r (μmol/g dried cells)	Cost of making 1 μmol of each of these building blocks (μmol/μmol)						
		Metabolites[a]	ATP	NADH	NADPH	1-C	NH_4^+	S
Protein amino acids								
Alanine	488	1 pyr	0	0	1	0	1	0
Arginine	281	1 αkg	7	−1	4	0	4	0
Asparagine	229	1 oaa	3	0	1	0	2	0
Aspartate	229	1 oaa	0	0	1	0	1	0
Cysteine	87	1 pga	4	−1	5	0	1	1
Glutamate	250	1 αkg	0	0	1	0	1	0
Glutamine	250	1 αkg	1	0	1	0	2	0
Glycine	582	1 pga	0	−1	1	−1	1	0
Histidine	90	1 penP	6	−3	1	1	3	0
Isoleucine	276	1 oaa, 1 pyr	2	0	5	0	1	0
Leucine	428	2 pyr, 1 acCoA	0	−1	2	0	1	0
Lysine	326	1 oaa, 1 pyr	2	0	4	0	2	0
Methionine	146	1 oaa	7	0	8	1	1	1
Phenylalanine	176	1 eryP, 2 pep	1	0	2	0	1	0
Proline	210	1 αkg	1	0	3	0	1	0
Serine	205	1 pga	0	−1	1	0	1	0
Threonine	241	1 oaa	2	0	3	0	1	0
Tryptophan	54	1 penP, 1 eryP, 1 pep	5	−2	3	0	2	0
Tyrosine	131	1 eryP, 2 pep	1	−1	2	0	1	0
Valine	402	2 pyr	0	0	2	0	1	0
RNA nucleotides								
ATP	165	1 penP, 1 pga	11	−3	1	1	5	0
GTP	203	1 penP, 1 pga	13	−3	0	1	5	0
CTP	126	1 penP, 1 oaa	9	0	1	0	3	0
UTP	136	1 penP, 1 oaa	7	0	1	0	2	0
DNA nucleotides								
dATP	24.7	1 penP, 1 pga	11	−3	2	1	5	0
dGTP	25.4	1 penP, 1 pga	13	−3	1	1	5	0
dCTP	25.4	1 penP, 1 oaa	9	0	2	0	3	0
dTTP	24.7	1 penP, 1 oaa	10.5	0	3	1	2	0
Lipid components								
Glycerol phosphate	129	1 triosP	0	0	1	0	0	0
Serine	129	1 pga	0	−1	1	0	1	0
$C_{16:0}$ fatty acid (43%)		8 acCoA	7	0	14	0	0	0
$C_{16:1}$ fatty acid (33%)		8 acCoA	7	0	13	0	0	0
$C_{18:1}$ fatty acid (24%)		9 acCoA	8	0	15	0	0	0
Average fatty acid	258	8.2 acCoA	7.2	0	14	0	0	0

[a]acCoA, Acetyl CoA; eryP, erythrose 4-phosphate; fruP, fructose 6-phosphate; gluP, glucose 6-phosphate; αkg, α-ketoglutarate; oaa, oxaloacetate; penP, ribose 5-phosphate; pep, phosphoenolpyruvate; pga, 3-phosphoglycerate; pyr, pyruvate; triosP, triose phosphate.

(*Table 1 continues*)

Table 1. (Continued)

Building block	Amount present in *E. coli* B/r (μmol/g dried cells)	Cost of making 1 μmol of each of these building blocks (μmol/μmol) Metabolites[a]	ATP	NADH	NADPH	1-C	NH_4^+	S
LPS components								
UDP-glucose	15.7	1 gluP	1	0	0	0	0	0
(CDP) ethanolamine	23.5	1 pga	3	−1	1	0	1	0
OH-myristic acid	23.5	7 acCoA	6	0	11	0	0	0
$C_{14:0}$ fatty acid	23.5	7 acCoA	6	0	12	0	0	0
(CMP) KDO	23.5	1 penP, 1 pep	2	0	0	0	0	0
(NDP) heptose	23.5	1.5 gluP	1	0	−4	0	0	0
(TDP) glucosamine	15.7	1 fruP	2	0	0	0	1	0
Peptidoglycan monomers								
UDP-*N*-acetylglucosamine	27.6	1 fruP, 1 acCoA	3	0	0	0	1	0
UDP-*N*-acetylmuramic acid	27.6	1 fruP, 1 pep, 1 acCoA	4	0	1	0	1	0
Alanine	55.2	1 pyr	0	0	1	0	1	0
Diaminopimelate	27.6	1 oaa, 1 pyr	2	0	3	0	2	0
Glutamate	27.6	1 αkg	0	0	1	0	1	0
Glycogen monomers								
Glucose	154	1 gluP	1	0	0	0	0	0
1-Carbon requirement								
Serine	48.5	1 pga	0	−1	1	0	0	0
Polyamines								
Ornithine equivalents	59.3	1 αkg	2	0	3	0	2	0

Other (small) molecules (less than 3% of cell dry weight)

Coenzymes: NAD, NADP, CoA, CoQ, bactoprenoid, tetrahydrofolate, cyanocobalamin, pyridoxal phosphate

Prosthetic groups: FMN, FAD, biotin, cytochromes, lipoic acid, thiamine pyrophosphate

Pool of unpolymerized monomers: average approximately 1% of amount in macromolecules

Biosynthetic pathways

Biosynthetic pathways differ markedly in complexity—some are linear, others branch or are interconnected. A simple pathway, consisting of three sequential enzymatic reactions, might be represented as

$$\underset{(PM)}{\text{Precursor metabolite}} \xrightarrow[(E_1)]{\text{Enzyme 1}} A \xrightarrow[(E_2)]{\text{Enzyme 2}} B \xrightarrow[(E_3)]{\text{Enzyme 3}} \underset{(BB)}{\text{Building block}}$$

where A and B are intermediate products in the pathway. Recognizing that these reactions are controlled as a physiological unit during growth, we might more simply represent the pathway as

$$PM \xrightarrow{3E} BB$$

Some pathways produce a building block that, in turn, is converted by a second pathway into another building block:

$$PM \xrightarrow{xE} BB_1 \xrightarrow{yE} BB_2$$

where x and y represent the number of enzymes in the two pathways. In some cases the pathway is branched:

$$PM \xrightarrow{xE} BB_1 \begin{cases} \xrightarrow{yE} BB_2 \\ \xrightarrow{zE} BB_3 \end{cases}$$

Most of the 12 precursor metabolites actually serve as the starting point for several pathways:

$$PM \begin{cases} \xrightarrow{xE} BB_1 \\ \xrightarrow{yE} BB_2 \\ \xrightarrow{zE} BB_3 \end{cases}$$

Finally, in many cases more than one precursor molecule is involved in the biosynthesis of a building block:

$$\left. \begin{matrix} PM_a \\ PM_b \end{matrix} \right\} \xrightarrow{xE} BB$$

Branching and interlocking of this sort are common among biosynthetic pathways. Building blocks that are produced from a common precursor are called a FAMILY. The ASPARTATE FAMILY, for example, consists of seven amino acids (asparagine, aspartate, diaminopimelate, isoleucine, lysine, methionine, and threonine) that are synthesized from the common precursor metabolite oxaloacetate (Figure 2).

The details of the various biosynthetic reactions are important for many purposes, including the biosynthetic solutions that have evolved for difficult problems in organic synthesis. A recent treatise (Neidhardt et al., 1987) on *E. coli* and *S. typhimurium* can be consulted for these

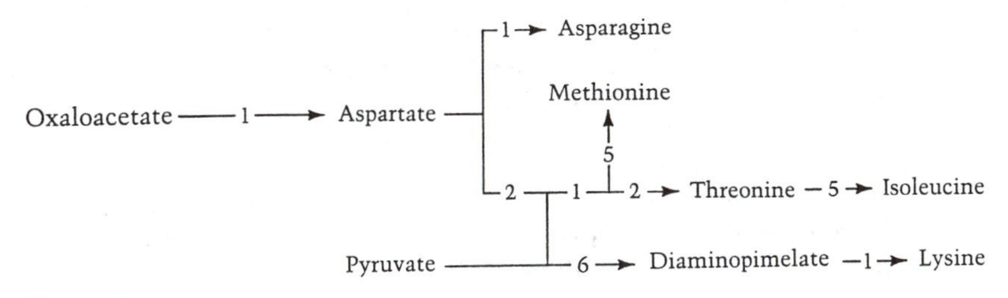

Figure 2

Pathways of biosynthesis of the aspartate family of amino acids. Numbers indicate how many enzymes catalyze various portions of the pathway.

details. In this chapter we shall concentrate on the role of biosynthetic reactions in cell growth and the metabolic resources required for these reactions. For instance, it is interesting to consider the resources needed to produce the building blocks to make 1 g of *E. coli.* By treating each pathway as a unit function, we can list its components (number of enzymes) and its metabolic costs (consumption of energy, reducing power, nitrogen, sulfur, and one-carbon units). The unit functions for most of the major biosynthetic pathways—those leading to building blocks for protein, RNA, DNA, phospholipids, lipopolysaccharides (LPS), peptidoglycan, and glycogen are shown in Figure 3.

Requirements for biosynthesis

We know enough about the biochemistry of *E. coli* and related bacteria to be able to draw a balance sheet of the cost of their principal constituents. The amount of each building block needed to make 1 g of bacterial cells is listed in Table 1. Also shown are the metabolic costs of making 1 μmole of each building block from its precursor metabolites. The cost of producing catalytically active small molecules—coenzymes, prosthetic groups, and related compounds—is not listed. Although their biosynthesis is vital to cell growth, they are synthesized in such small quantities that the metabolic costs incurred are insignificant. Nevertheless, inspection of their biosynthetic pathways is worthwhile, because one can gain some appreciation for the large number of enzymes necessary for this task and the consequent dedication of a significant part of the genome for encoding these enzymes (Neidhardt et al., 1987).

From the information in Table 1, one can sum up the total requirements of precursor metabolites, energy, reducing power, nitrogen, and sulfur necessary for the biosynthetic task of making 1 g of bacterial cells (Table 2). One can also calculate the costs in terms of energy and reducing power of synthesizing the cell's various components (Table 3).

Figure 3

Major bacterial biosynthetic pathways depicted as a unit function. A box at each pathway branch summarizes the number of enzymes, high-energy bonds (~P), reducing power ($NADH_2$ or $NADPH_2$), and other components that are used or produced. By convention, positive values indicate utilization; negative values, production. Costs are tallied as net change. For example, utilization of glutamine as an amino donor is charged as 1 ~P and 1 NH_4^+, the net cost of regenerating glutamine from the reaction product, glutamate. PEP, phosphoenolpyruvate. This figure and those tables derived from it are based on the presentations of metabolic economics by Fraenkel (1962), Umbarger (1977), and Stouthammer (1973).

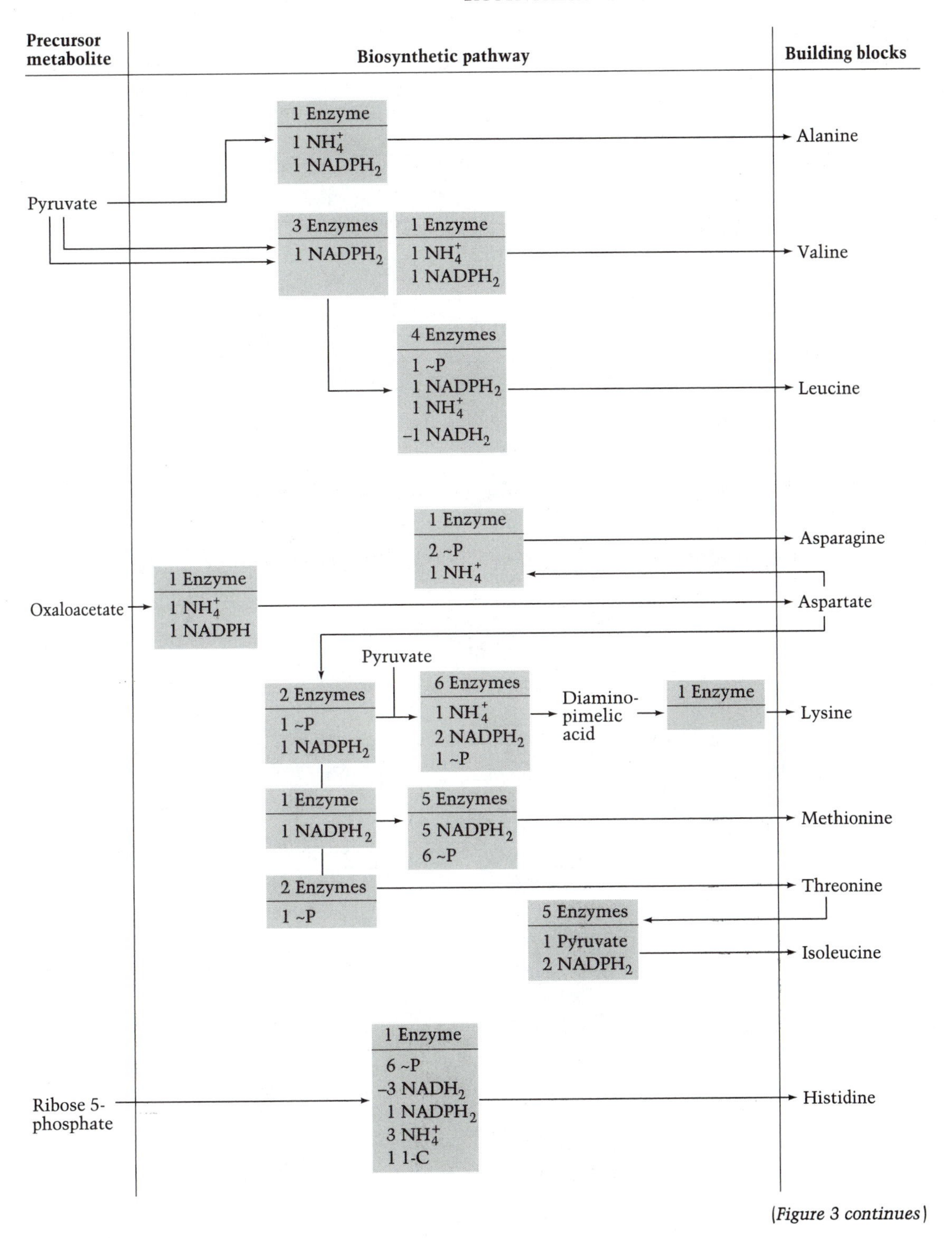

(Figure 3 continues)

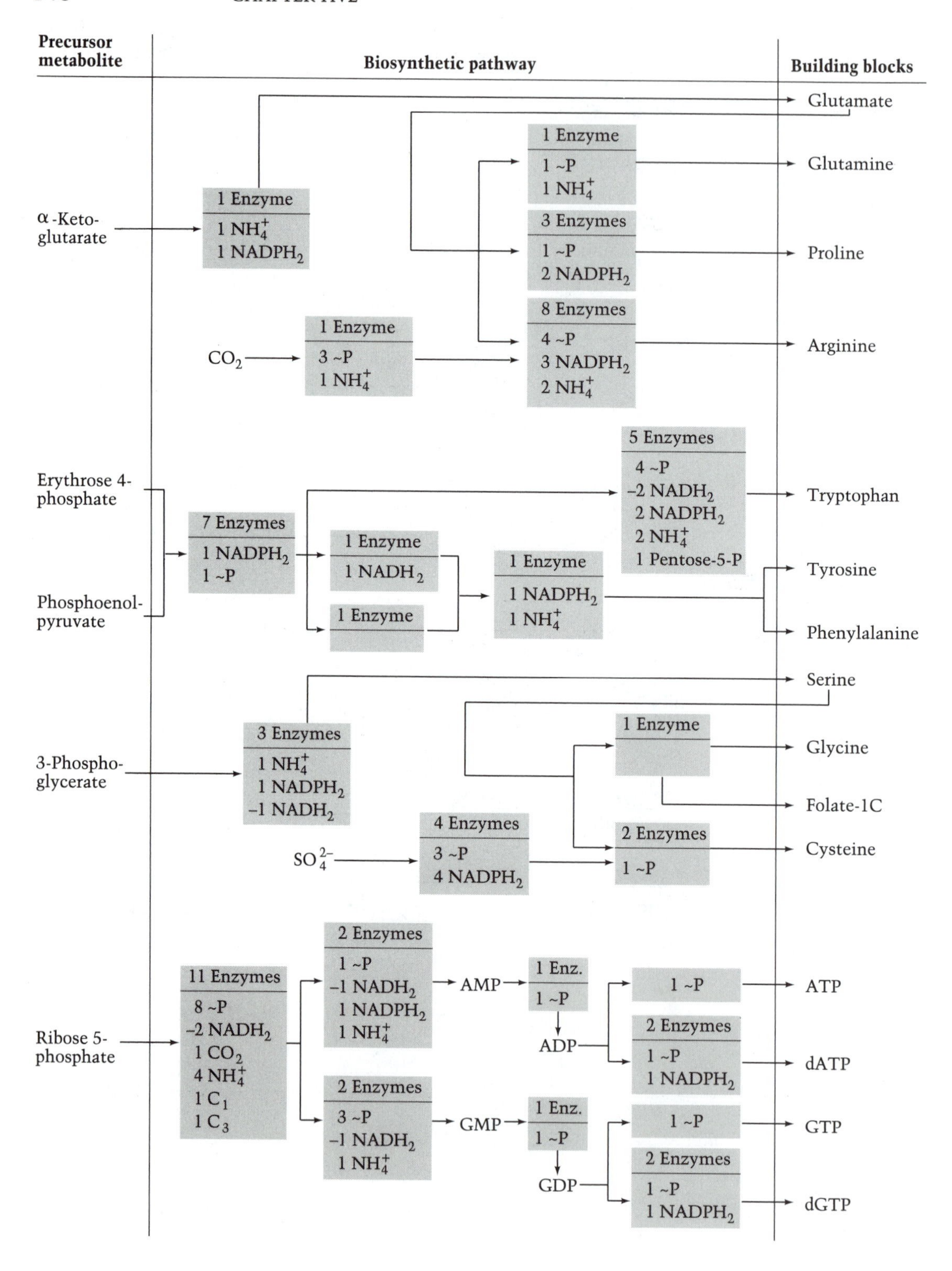

Precursor metabolite
Biosynthetic pathway
Building blocks
Glutamate
1 Enzyme
1 ~P
1 NH_4^+
Glutamine
α-Keto-glutarate
1 Enzyme
1 NH_4^+
1 $NADPH_2$
3 Enzymes
1 ~P
2 $NADPH_2$
Proline
8 Enzymes
4 ~P
3 $NADPH_2$
2 NH_4^+
Arginine
CO_2
1 Enzyme
3 ~P
1 NH_4^+
5 Enzymes
4 ~P
–2 $NADH_2$
2 $NADPH_2$
2 NH_4^+
1 Pentose-5-P
Tryptophan
Erythrose 4-phosphate
7 Enzymes
1 $NADPH_2$
1 ~P
1 Enzyme
1 $NADH_2$
1 Enzyme
1 $NADPH_2$
1 NH_4^+
Tyrosine
Phosphoenol-pyruvate
1 Enzyme
Phenylalanine
Serine
1 Enzyme
Glycine
3-Phospho-glycerate
3 Enzymes
1 NH_4^+
1 $NADPH_2$
–1 $NADH_2$
Folate-1C
4 Enzymes
3 ~P
4 $NADPH_2$
SO_4^{2-}
2 Enzymes
1 ~P
Cysteine
2 Enzymes
1 ~P
–1 $NADH_2$
1 $NADPH_2$
1 NH_4^+
AMP
1 Enz.
1 ~P
ADP
1 ~P
ATP
11 Enzymes
8 ~P
–2 $NADH_2$
1 CO_2
4 NH_4^+
1 C_1
1 C_3
Ribose 5-phosphate
2 Enzymes
1 ~P
1 $NADPH_2$
dATP
2 Enzymes
3 ~P
–1 $NADH_2$
1 NH_4^+
GMP
1 Enz.
1 ~P
GDP
1 ~P
GTP
2 Enzymes
1 ~P
1 $NADPH_2$
dGTP

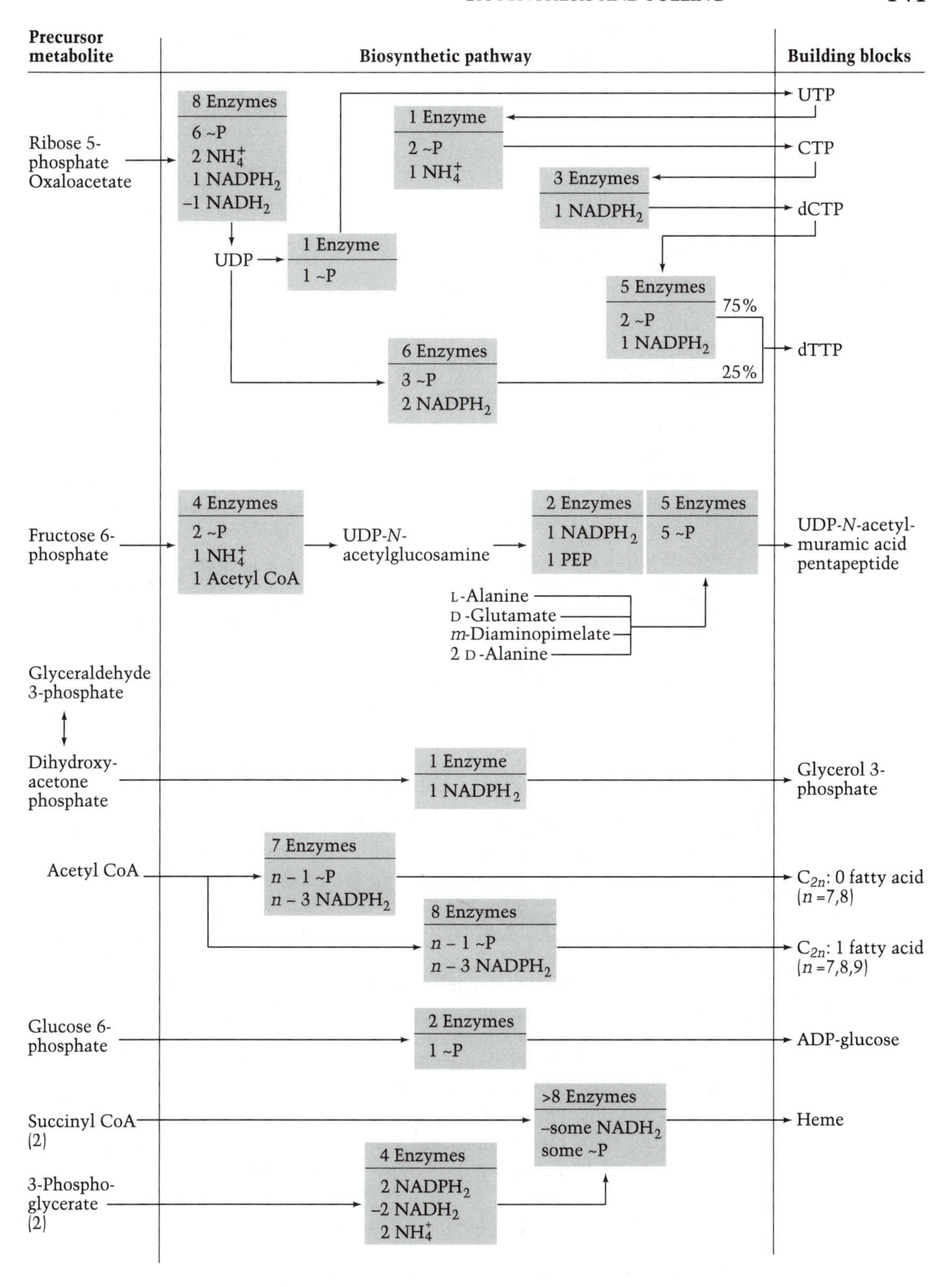

Precursor metabolite
Biosynthetic pathway
Building blocks
Ribose 5-phosphate
Oxaloacetate
8 Enzymes
6 ~P
2 NH_4^+
1 $NADPH_2$
–1 $NADH_2$
UDP
1 Enzyme
1 ~P
UTP
1 Enzyme
2 ~P
1 NH_4^+
CTP
3 Enzymes
1 $NADPH_2$
dCTP
5 Enzymes
2 ~P
1 $NADPH_2$
75%
6 Enzymes
3 ~P
2 $NADPH_2$
25%
dTTP
Fructose 6-phosphate
4 Enzymes
2 ~P
1 NH_4^+
1 Acetyl CoA
UDP-*N*-acetylglucosamine
2 Enzymes
1 $NADPH_2$
1 PEP
5 Enzymes
5 ~P
UDP-*N*-acetyl-muramic acid pentapeptide
L-Alanine
D-Glutamate
m-Diaminopimelate
2 D-Alanine
Glyceraldehyde 3-phosphate
Dihydroxy-acetone phosphate
1 Enzyme
1 $NADPH_2$
Glycerol 3-phosphate
Acetyl CoA
7 Enzymes
$n-1$ ~P
$n-3$ $NADPH_2$
C_{2n}: 0 fatty acid
(n=7,8)
8 Enzymes
$n-1$ ~P
$n-3$ $NADPH_2$
C_{2n}: 1 fatty acid
(n=7,8,9)
Glucose 6-phosphate
2 Enzymes
1 ~P
ADP-glucose
Succinyl CoA
(2)
>8 Enzymes
–some $NADH_2$
some ~P
Heme
3-Phospho-glycerate
(2)
4 Enzymes
2 $NADPH_2$
–2 $NADH_2$
2 NH_4^+

Table 2. Requirements for biosynthesis of building blocks

Precursor metabolite	Amount required[a] (μmol/g cells)	Precursor metabolite	Amount required[a] (μmol/g cells)
Glucose 6-phosphate	205	Succinyl CoA	—[b]
Fructose 6-phosphate	70.9	Oxaloacetate	1,786.7
Ribose 5-phosphate	897.7	~P	18,485
Erythrose 4-phosphate	361	$NADH_2$	(−3,547)
Triose phosphate	129	$NADPH_2$	18,225
3-Phosphoglycerate	1,496	1-C (accounted for as serine)	—[c]
Phosphoenolpyruvate	519.1	NH_4^+	10,180
Pyruvate	2,832.8	S	233
Acetyl CoA	3,747.8		
α-Ketoglutarate	1,078.9		

[a]Calculated from the information in Table 1 by multiplying the amount of each building block (first column of figures) by the molar requirement of precursor metabolites or other components to produce them.
[b]Succinyl CoA is used as a cofactor in the synthesis of building blocks, that is, succinate is later released from the biosynthetic intermediates. Therefore, there is a net expenditure of 1 ~P in the synthesis of lysine and methionine. Succinyl CoA is a precursor metabolite, however, in the synthesis of heme and hemelike compounds.
[c]The requirement for 1-C fragments in the synthesis of methionine, purines, and thymine has been arbitrarily accounted for by synthesizing the requisite number of serine molecules.

Assimilation of nitrogen and sulfur

None of the precursor metabolites contains nitrogen or sulfur. These elements enter the cell as constituents of a variety of organic or inorganic molecules or ions; but they always enter biosynthetic pathways in inorganic form—nitrogen as ammonium ion (NH_4^+) and sulfur as hydrogen sulfide (H_2S). Entry occurs in certain biosynthetic reactions (Figure 3).

Bacterial cells, like all cells, contain more nitrogen than sulfur or even phosphorus; nitrogen constitutes 14% of the dry weight of our reference bacterial cell. Like the nitrogenous compounds in bacterial cells, most utilizable organic sources of nitrogen contain this atom in the −3 oxidation state in amino groups, imino groups, or heterocyclic nitrogen atoms. These nitrogen sources are metabolized by the bacterial cells through pathways that yield NH_4^+ without changing the oxidation state of the nitrogen atom. As we note later, these pathways are usually under genetic control: when NH_4^+ is present in adequate concentration, the enzymes catalyzing these ammonia-yielding reactions are not synthesized.

Bacteria, as a group, also utilize inorganic sources of nitrogen: the most significant of these sources are NH_4^+, nitrate ion (NO_3^-), and dinitrogen gas (N_2). Probably all bacteria can utilize NH_4^+, but the ability to use NO_3^- and N_2 is more restricted. (Our reference cell can use NO_3^- only when it is growing anaerobically, and it cannot use N_2 under any conditions.) Those organisms that can assimilate NO_3^- reduce it to nitrite ion (NO_2^-) by ASSIMILATORY NITRATE REDUCTASE and then to NH_4^+ by ASSIMILATORY NITRITE REDUCTASE. In general, a bacterium that is able to utilize NO_3^- as a source of nitrogen for biosynthesis contains both assimilatory nitrate and assimilatory nitrite reductases—but not always.

Certain bacteria, including *E. coli*, can use NO_3^- for a totally different function. Nitrate can replace O_2 as the terminal electron acceptor in a respiratory process called ANAEROBIC RESPIRATION. The membrane-bound enzyme catalyzing this reaction is called DISSIMILATORY NITRATE REDUCTASE. It is synthesized only in the absence of oxygen and produces NO_2^- as a product. Under these conditions, *E. coli* also produces a nitrite reductase that reduces nitrite to NH_4^+ without generating ATP. Thus, under anaerobic conditions, *E. coli* can utilize NO_3^- as a source of nitrogen, but when growing aerobically, it requires a reduced nitrogen source.

In many natural environments, the amounts of usable nitrogen sources are insufficient to support optimal growth of organisms. Certain bacteria, however, have the capacity to reduce N_2 to NH_4^+, a process termed NITROGEN FIXATION. Because the earth's atmosphere contains approximately 80% N_2, these bacteria have an almost inexhaustible supply of nitrogen. (It would disappear with a half-life of about 20 million years were it not constantly replenished by denitrification; see the

Table 3. Costs of biosynthesis of cellular components from the precursor metabolites

Cellular component	Energy cost[a] (μmole ~P/g cells)	Reducing power cost (μmole NADPH/g cells)
Protein	7,287	11,523
RNA	6,540	427
DNA	1,090	200
Lipid	2,578	5,270
LPS	470	564
Murein	248	193
Glycogen	154	0
1-Carbon	0	48
Polyamines	118	0
Total	18,485	18,225

[a]Each nucleoside triphosphate is assumed to be made by consecutive reactions with ATP that consume 3 ~P per NTP produced. Formation of sugar–nucleotide derivatives are assumed to occur by direct reaction with the appropriate nucleoside triphosphate.

section in this chapter on anaerobic respiration.) Biological nitrogen fixation is a process that is unique to bacteria and that plays a key role in the global nitrogen balance. The process is catalyzed by a single enzyme complex, NITROGENASE, which is composed of two iron–sulfur proteins: AZOFERREDOXIN and MOLYBDOFERREDOXIN (which contains two molybdenum atoms in addition to the iron–sulfur centers).

Although the reduction of N_2 to NH_4^+ is exothermic, N_2 is an extremely unreactive molecule. Reduction to NH_4^+ requires considerable activation energy, which is supplied in the form of ATP. In vitro 6–15 moles of ATP are hydrolyzed for each mole of N_2 fixed; but in vivo the requirement is probably closer to 6. In any event, this is an extremely costly reaction in terms of expenditure of ATP. Nitrogenase is also remarkably sensitive to O_2. Thus, nitrogen-fixing bacteria have all evolved mechanisms to protect nitrogenase from O_2. Some, like *Clostridium pastorianum*, avoid the problem completely, because they are strict anaerobes that grow only in O_2-free environments. Others (e.g., *Rhodospirillum rubrum*) are facultative anaerobes that fix nitrogen only when the environment is anaerobic. Certain strict aerobes (e.g., *Azotobacter vinlandii*) maintain an anaerobic intracellular environment by respiring so actively that all available O_2 is utilized at the cell surface. Other aerobes have evolved specialized structures that maintain anaerobic interiors. In the case of some cyanobacteria, which are oxygenic phototrophs, nitrogen fixation occurs within certain thick-walled cells termed HETEROCYSTS. These cells exclude O_2, and, lacking the appropriate photosynthetic component (photosystem II), they do not produce it. The rhizobia (e.g., *Bradyrizobium japonicum*) can fix nitrogen by forming symbiotic relationships with plant roots. They invade the roots, thus causing the formation of nodules that become packed with rhizobia. In this environment, pO_2 is maintained at a low level compatible with nitrogenase activity by the action of an O_2-binding protein, LEGHEMOGLOBIN (Chapter 16).

Although there is considerable diversity in the way bacteria obtain ammonium ion, they incorporate it into organic constituents of the cell by only two pathways. Some bacteria possess one of the two pathways and others (e.g., *E. coli*) possess both. Ammonia can be assimilated (1) directly into glutamate via a reaction catalyzed by L-GLUTAMATE DEHYDROGENASE (GDH) or (2) indirectly by a cycle of reactions (GS-GOGAT) catalyzed by GLUTAMINE SYNTHETASE (GS) and GLUTAMATE SYNTHASE (sometimes termed glutamate-α-oxoglutarate aminotransferase, or GOGAT). The first pathway is an $NADPH_2$-specific reductive amination of the precursor metabolite α-ketoglutarate to produce glutamate (Figure 3). Ammonia becomes incorporated into the α-amino group of glutamate and, by subsequent reactions, it is transferred to the other nitrogenous constituents of the cell. In the second pathway—the GS-GOGAT system—the primary assimilation of NH_4^+ occurs during synthesis of glutamine from glutamate in the GS-catalyzed reaction. The GOGAT enzyme then catalyzes the transfer of the amide group of glutamine to the α-carbon of ketoglutarate, thereby generating two molecules of glutamate. The net reaction (α-ketoglutarate + $NH_4^+ \rightarrow$ glutamate) is the same in both cases.

The two primary assimilatory enzymes—GDH and GS—differ in two ways: (1) GS has a considerably higher affinity for NH_4^+ than does GDH and (2) the GS-catalyzed reaction utilizes ATP; the GDH-catalyzed reaction does not. Thus, at the expense of ATP, GS can assimilate ammonia when its concentration is low (1 m*M*). As might be expected, in organisms like *E. coli* that have both pathways, GS synthesis is repressed if ammonia is present in high concentrations. Indeed, GS synthesis is controlled by the same system—NITROGEN CONTROL—that regulates the ammonia-yielding fueling pathways mentioned earlier. The catalytic activity of GS is also subject to elaborate control, being inhibited in a cumulative way by each of the eight nitrogenous compounds derived from glutamine. Its activity is further modulated by POSTTRANSLATIONAL MODIFICATION a rather rare control mechanism in bacteria. A specific enzyme modifies GS by ADENYLYLATION (addition of an adenylate group to specific amino acid residues) and thus converts it to a less active form. The same enzyme deadenylylates GS, thereby reactivating it. The adenylyltransferase-deadenylylating enzyme fluctuates between its two activities in response to the ratio of the intracellular pools of glutamine and α-ketoglutarate (Chapter 11).

The major reaction by which sulfur is assimilated is catalyzed by *O*-ACETYLSERINE SULFOHYDROLASE. In this reaction, H_2S reacts with *O*-acetyl-L-serine to produce L-cysteine, acetate, and water; L-cysteine is the source, directly or indirectly, for the sulfur in most other sulfur-containing compounds in the cell. So, exogenous sources of sulfur must be converted to H_2S in order to be assimilated. Hydrogen sulfide is rare in aerobic environments because it is spontaneously oxidized by O_2, but it accumulates in anaerobic environments. So, not surprisingly, some strict anaerobic bacteria, including methanogens, use only H_2S as a sulfur source, but most bacteria, including all aerobes, can reduce oxidized sulfur compounds. The most common source of sulfur in laboratory media (but probably not in nature) is sulfate. Many bacteria, including *E. coli*, actively transport sulfate into the cell, where it is reduced. Reduction of sulfate requires that it first be activated by an energy-intensive series of enzyme-catalyzed reactions. Three high-energy phosphate bonds are utilized for each sulfate molecule reduced. Sulfate reacts with ATP, yielding adenosine-5′-phosphosulfate and pyrophosphate (P_i-P_i). A second molecule of ATP reacts with adenosine-5′-phosphosulfate, yielding ADP and adenosine-3′-phosphate-5′-phosphosulfate. In this form, sulfate is reduced to free sulfite, with thioredoxin serving as the electron donor. Sulfite is reduced to H_2S by SULFITE REDUCTASE, using $NADPH_2$ as the electron donor. Sulfate reduction is an expensive process, requiring the expenditure of three high-energy phosphate bonds and four molecules of $NADPH_2$ for each molecule of sulfate reduced. Four enzymes participate in the reduction.

Sulfate is probably not the major source of sulfur for microbial growth in nature. Certainly, this is the case in aerobic soil, where only a small percentage of the total sulfur occurs as sulfate and little, if any, occurs as elemental sulfur or H_2S. The bulk of sulfur is found as organic compounds (organic sulfates, amino acids, and other C-bonded sulfur compounds). Most organic sulfur compounds are metabolized to sulfate; sulfur-containing amino acids are probably assimilated directly. Although sulfate-yielding metabolism has not been studied in great detail, its regulation seems to differ from NH_4^+-yielding metabolism. The latter is under the repressive control of NH_4^+, i.e., synthesis of the enzymes of the pathway are decreased by NH_4^+, whereas enzymes catalyzing sulfate-yielding metabolism (e.g. sulfohydrolases) are induced by organic sulfates.

Assimilation of phosphorus

Assimilation of phosphorus differs from assimilation of nitrogen or sulfur in two important respects:

- Phosphorus is neither oxidized nor reduced in the assimilation process.
- Phosphorus, because it plays a central role in energy transduction, is assimilated in fueling pathways rather than in biosynthetic pathways.

Organic phosphates and phosphate ion both constitute sources of phosphorus for the bacterial cell. Phosphate ion enters *E. coli* by a specific as well as a nonspecific transport system. With a few exceptions, organic phosphate compounds do not permeate the cell membrane, rather they are hydrolyzed outside the cell or within the periplasm to phosphate ion. In *E. coli,* most hydrolysis of organic phosphates in the periplasm is catalyzed by a relatively nonspecific enzyme—ALKALINE PHOSPHATASE the synthesis of which is repressed when phosphate is present in the medium. Also present in the periplasm of *E. coli* is a 5′-nucleotidase that splits phosphate ion from nucleotides. Once inside the cell, phosphate can be incorporated into ATP by one of the several ATP-yielding reactions. The high-energy compound ATP is the major phosphate donor in the cell.

Nutritional diversity

In discussing biosynthesis we considered all the reactions that a bacterial cell needs in order to synthesize its building blocks from precursor metabolites. But not all these reactions are needed at all times. When building blocks are available in the environment, many bacteria can economize by using these preformed molecules rather than by synthesizing them (Chapter 11). Certain types of bacteria called AUXOTROPHS completely lack one or more biosynthetic pathways. The growth of these bacteria is dependent on the presence of relevant building blocks in the medium. The degree of auxotrophy or, conversely, the spectrum of biosynthetic competency is quite broad. It extends from PROTOTROPHS, like *E. coli,* which are complete biochemists able to synthesize all their building blocks and other small molecules from a single carbon substrate, to bacteria like the lactic acid bacterium *Leuconostoc mesenteroides*, which must be provided with all amino acids, several purines and pyrimidines, and 10 vitamins. Bacteria like *Shigella* spp. are intermediate in competency, being able to synthesize 19 of the 20 amino acids. Because all cells require the same building blocks, *there is a reciprocal relationship between biosynthetic capability and nutritional requirements.* As might be expected, bacteria that grow in environments that are relatively rich in organic material have tended to dispense with redundant biosynthetic pathways. Among such bacteria, we count those that are found in close association with the human body (e.g., *Staphylococcus aureus* and *Haemophilus influenzae*, each of which has a long list of growth requirements. In contrast, *E. coli* and other facultative body-dwellers require no preformed building blocks, probably because they must survive in the external environment as well as in association with humans. (Humans are more like auxotrophic bacteria than like prototrophic bacteria; human cells lack the ability to synthesize half of the 20 amino acids and all of the compounds called vitamins.)

Growth in rich media

The energy and reducing power required for biosynthesis imposes a heavy cost on the cell (Table 2) and, depending on the carbon/energy source available, may severely limit the growth rate. It is not surprising, therefore, to find that selective pressures significantly favor auxotrophy, that is, favor substitution of growth factor requirements for biosynthetic capability. These organisms can take almost full advantage of nutrients supplied to them (an auxotrophic mutant will outgrow its prototrophic parent in a rich medium). Even organisms that possess a full complement of biosynthetic pathways have evolved regulatory mechanisms that stop almost completely their endogenous biosynthetic pathways when building blocks are available in the medium. The growth rate of prototrophs usually increases when amino acids and other nutrients are added to a minimal medium. Our reference culture, *E. coli* B/r, doubles its growth rate when a large number of nutrients are added to a glucose minimal medium. The increase is even greater upon supplementation of a minimal medium containing a carbon source that is less readily utilized than glucose.

How great a savings can be made by the nutritional supplementation of a minimal medium? In principle, if every compound listed in the first column of Table 1 were supplied to the cells, all of the energy, reducing power, and precursor metabolites used in biosynthesis would be saved. In practice, the savings approach but do not reach this limit, because some building blocks (including nucleoside triphosphates and conjugated nucleotides) cannot permeate the cell membrane. But amino acids, nucleosides, sugars, aminosugars, fatty acids, and precursors of coenzymes and prosthetic groups do enter many bacterial cells, and the metabolic savings can be profound. In Table 4 the cost of producing building

Table 4. Costs of biosynthesis in a rich medium

Cellular component	Energy cost[a] (μmol ~P/g cells)	Reducing power cost (μmol NADPH/g cells)
Protein	0	0
RNA	1,890	0
DNA	300	0
Lipid	129	0
LPS	125	0
Murein	55	0
Glycogen	154	0
1-Carbon	0	0
Polyamines	118	0
Total	2,771	0

[a]The medium is assumed to contain the following utilizable nutrients (in addition to glucose and inorganic salts): 21 amino acids, ribonucleosides, deoxyribonucleosides, glycerol phosphate, fatty acids, ornithine, glucosamine, ethanolamine, heptose, and KDO. For each macromolecule, the number listed is the cost to produce the respective building blocks from substances in the medium.

blocks from compounds that can be supplied from the medium is calculated for each class of macromolecules. Comparing these results with those in Table 3 (the cost of synthesizing these cellular components from precursor metabolites), we see that 85% of the energy for biosynthesis (18,485 versus 2,771 μmol ~P/g cells) and all of the reducing power (18,225 μmol NADPH) can be spared.

This analysis neglects the savings due to nutritional supplementation that come from the reduced demand for precursor metabolites. In some cases this reduced demand will be highly significant, because, during growth on certain substrates, the formation of precursor metabolites is energetically costly. For example, during growth on acetate, formation of each mole of glucose 6-phosphate requires eight moles of high-energy phosphate bonds. Only one high-energy bond is required to make glucose 6-phosphate from glucose. Because this cost varies so strikingly with the nature of the substrate, we shall examine fueling reactions before completing our discussion of nutritional supplementation. Furthermore, some costs are associated with bringing nutrients into the cell, and before we can calculate the net impact of supplementation (see Table 4), these costs for cells growing on a single substrate must be compared with those for cells growing on a rich medium.

FUELING REACTIONS

The chemistry of biosynthesis has one marvelously simple aspect: virtually all biosynthetic pathways, polymerizations, and assembly processes are fundamentally the same in all bacteria. We have encountered some differences in biosyntheses, but these are trivial. Even the coordination of biosynthetic pathways is conceptually simple, because it involves adjustments in flow through essentially unidirectional pathways with few alternative routes to a given building block.

It is in fueling reactions that bacteria display their extraordinary metabolic diversity and versatility. Bacteria have evolved to thrive in almost all natural environments, regardless of the nature of available sources of carbon, energy, and reducing power, or pH, oxidation–reduction potential, and availability of electron acceptors. The collective metabolic capacities of bacteria allow them to metabolize virtually every organic compound on this planet (coal, diamonds, and many plastics constitute significant exceptions to this rule of metabolic infallibility). Regardless of the source of carbon and energy that they use or the environment in which they function, all kinds of fueling reactions must accomplish the same metabolic task—to produce precursor metabolites, reducing power, and high-energy phosphate in the precise proportions required for biosynthesis and polymerization. Because different substrates vary in their potential to yield these necessities for growth, the cell adjusts their proportions by modulating the relative flow through its various metabolic pathways.

Fueling pathways

All 12 precursor metabolites are synthesized by a series of reactions that are collectively called CENTRAL METABOLISM. In *E. coli* these reactions are divided into the EMBDEN-MEYERHOF-PARNAS (EMP) PATHWAY, which converts glucose 6-phosphate to pyruvate; the TRICARBOXYLIC ACID (TCA) CYCLE, which oxidizes acetyl CoA to CO_2, and the PENTOSE PHOSPHATE CYCLE, which oxidizes glucose 6-phosphate to CO_2. Central metabolism also includes linker reactions (such as pyruvate dehydrogenase and pyruvate formate-lyase, which produces acetyl CoA from pyruvate) that join these pathways, and reactions (such as those catalyzed by phosphoenolpyruvate synthase, which makes phosphoenolpyruvate from pyruvate) that enable a reversal of flow through these pathways. Finally, several reactions (such as the glyoxylate shunt) bypass certain portions of the pathways. Precursor metabolites are drawn off at various places in the fueling pathways. The EMP pathway produces six of them; the TCA cycle produces three more; the pentose phosphate cycle produces two; and the final one, acetyl CoA, is formed by a linker reaction. The interconnecting, reversing, and bypassing reactions are termed ANAPLEUROTIC REACTIONS; they replenish the pools of precursor metabolites drained by biosynthesis.

Other bacteria have additional or alternative central pathways, such as the ENTNER-DOUDOROFF PATHWAY, which replaces the EMP pathway in many pseudomonads. *Escherichia coli* uses enzymes of the Entner-Doudoroff pathway to metabolize gluconate but not glucose. The central fueling pathways of *E. coli* are outlined in Figure 4.

The central fueling pathways reduce great quantities of pyridine

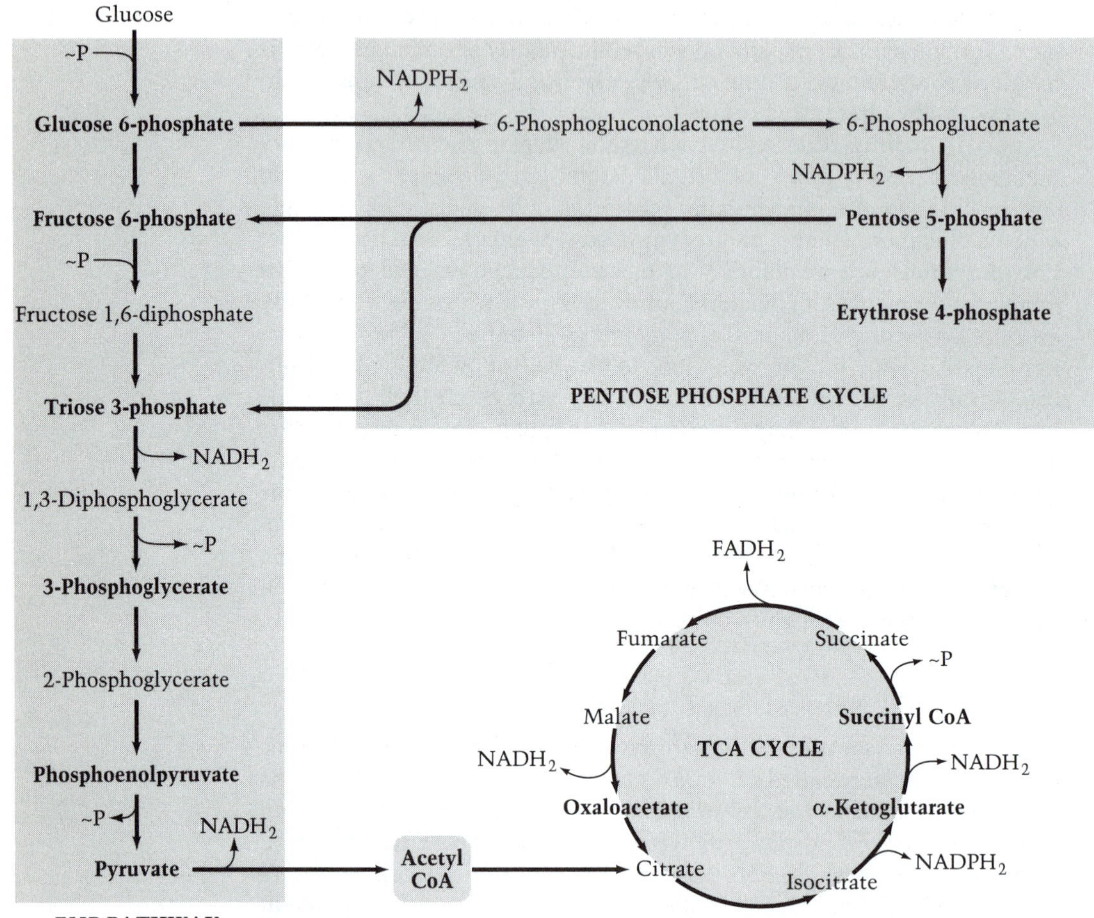

Figure 4

The central fueling pathways in *E. coli*. The three major pathways [Embden-Meyerhof-Parnas (EMP) pathway, tricarboxylic acid (TCA) cycle, and pentose phosphate cycle] are outlined to show their individual reactions, not to indicate how they operate during growth on any particular substrate. Anapleurotic reactions and peripheral pathways (including fermentative and respiratory pathways) are not shown. The reductive branch of the pentose phosphate pathway is summarized by the branched arrow connecting pentose 5-phosphate to the EMP pathway. The 12 precursor metabolites are in boldface.

nucleotides (NAD and NADP). Some reduced pyridine nucleotides (in the form of $NADPH_2$) are used for biosynthesis, but the rest (in the form of $NADH_2$) must be rapidly reoxidized or the central fueling pathways would come to a halt. Reoxidation of pyridine nucleotides occurs in two ways:

- In FERMENTATIVE PATHWAYS, reoxidation is accomplished directly by reducing organic metabolites.
- In RESPIRATORY PROCESSES, reoxidation is accomplished by electron

transport through a RESPIRATORY CHAIN to a terminal electron acceptor. This acceptor is usually O_2, but in some cases it can be oxidative ions, including nitrate and sulfate. In all respiratory processes, ATP is generated; so, for aerobes, reduced pyridine nucleotides are a source of metabolic energy.

These pathways are discussed below in the section dealing with the generation of ATP.

There is enormous flexibility in the way the central fueling pathways operate. They can function cyclically to degrade glucose completely to CO_2 and water, thereby producing large quantities of ATP and $NADH_2$. Under other conditions they can function unidirectionally to replace precursor metabolites drained off for biosynthesis. How the central fueling pathways operate depends on the nature of the substrates available for growth, which electron acceptors are present, and which preformed building blocks are supplied in the medium. For example, if no suitable electron acceptor is present, the reactions of the TCA pathway function solely to produce precursor metabolites. If oxygen or some other suitable electron acceptor is present, the TCA pathway oxidizes acetyl CoA to CO_2 and, coupled to the respiratory chain, generates most of the ATP for growth. Many adjustments throughout the central pathways are required to switch between these two modes of operation.

Almost any carbon source can be used by one bacterial species or another. Growth on substrates that are not intermediates in the central pathways is brought about by PERIPHERAL PATHWAYS that sometimes consist of many steps and that convert the particular substrate to an intermediate of a central pathway. Bacterial species are so diverse in their ability to utilize different substrates by peripheral pathways that these characteristic properties can be used for taxonomic purposes. This aspect of metabolic diversity is so vast that it is beyond the scope of this book. In the following sections we hope to emphasize the patterns, but not the details, of fueling diversity.

A useful approach to the complex world of the fueling reactions is to examine separately the need for their products. First we consider them solely as systems for generating precursor metabolites; then we examine the balance sheet to determine whether there are unmet needs for energy and reducing power.

Generation of precursor metabolites

Figures 5, 6, and 7 show the formation of each of the 12 precursor metabolites under three different conditions: aerobic growth on glucose, anaerobic growth (by fermentation) on glucose, and aerobic growth on malate. Features to note in these pathways include (1) the anapleurotic reaction needed in each situation; (2) the enzymes needed to reverse the flow of carbon in the pathways; and (3) the different yields of reducing power, energy, and CO_2 produced by growth on different substrates (Table 5).

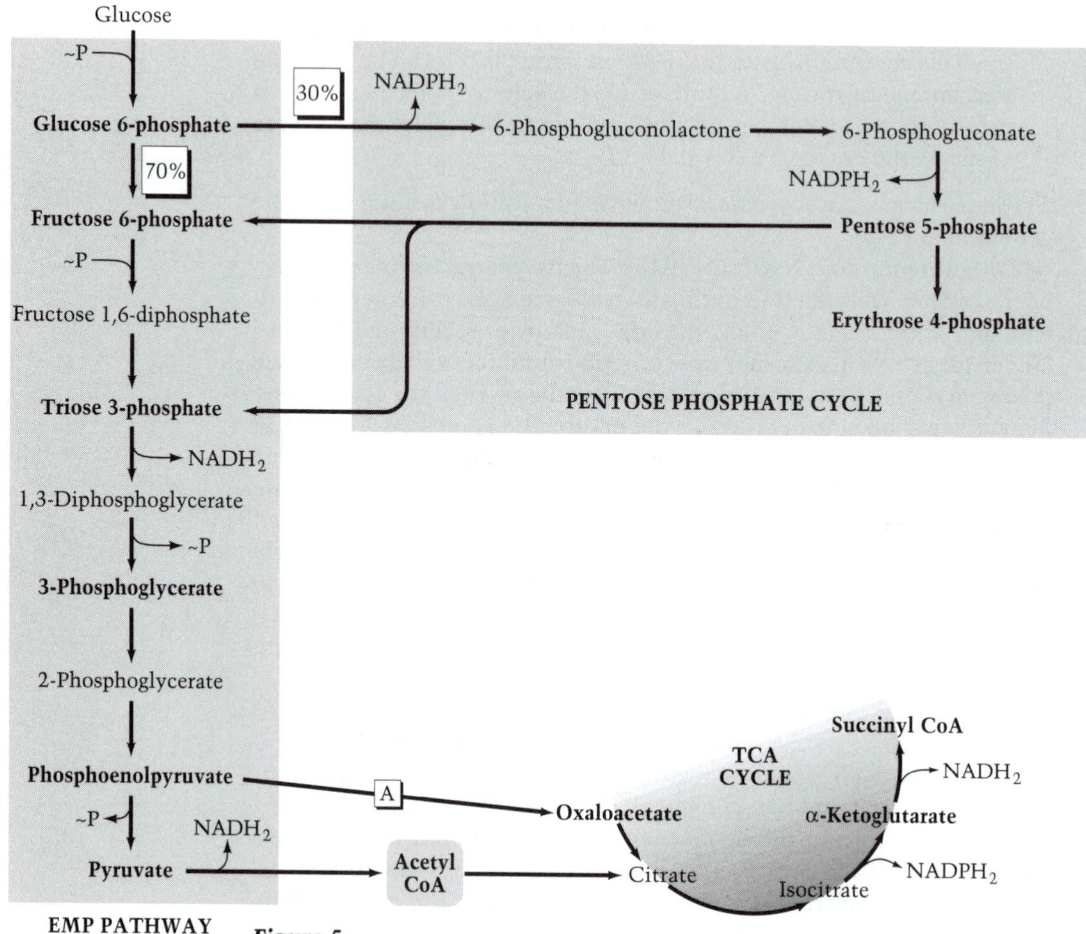

Figure 5

Formation of precursor metabolites during aerobic growth of *E. coli* on glucose. The major anapleurotic reaction, A (catalyzed by phosphoenolpyruvate carboxylase) forms oxaloacetate by carboxylation of phosphoenolpyruvate. Components of the TCA cycle function almost exclusively to provide three precursor metabolites, not as an energy-generating cycle. The 12 precursor metabolites are in boldface.

Knowing the costs of making each precursor metabolite and the amounts of them that are needed, one can calculate the total cost (yield) of fueling reactions for synthesizing 1 g of *E. coli* (Table 6). This information is notable in two respects:

- Quite different energy costs are associated with making precursor metabolites from the two substrates: from glucose, net energy is produced; from malate, net energy is utilized.
- In neither case is sufficient energy produced to meet the needs for synthesizing more cells from precursor metabolites.

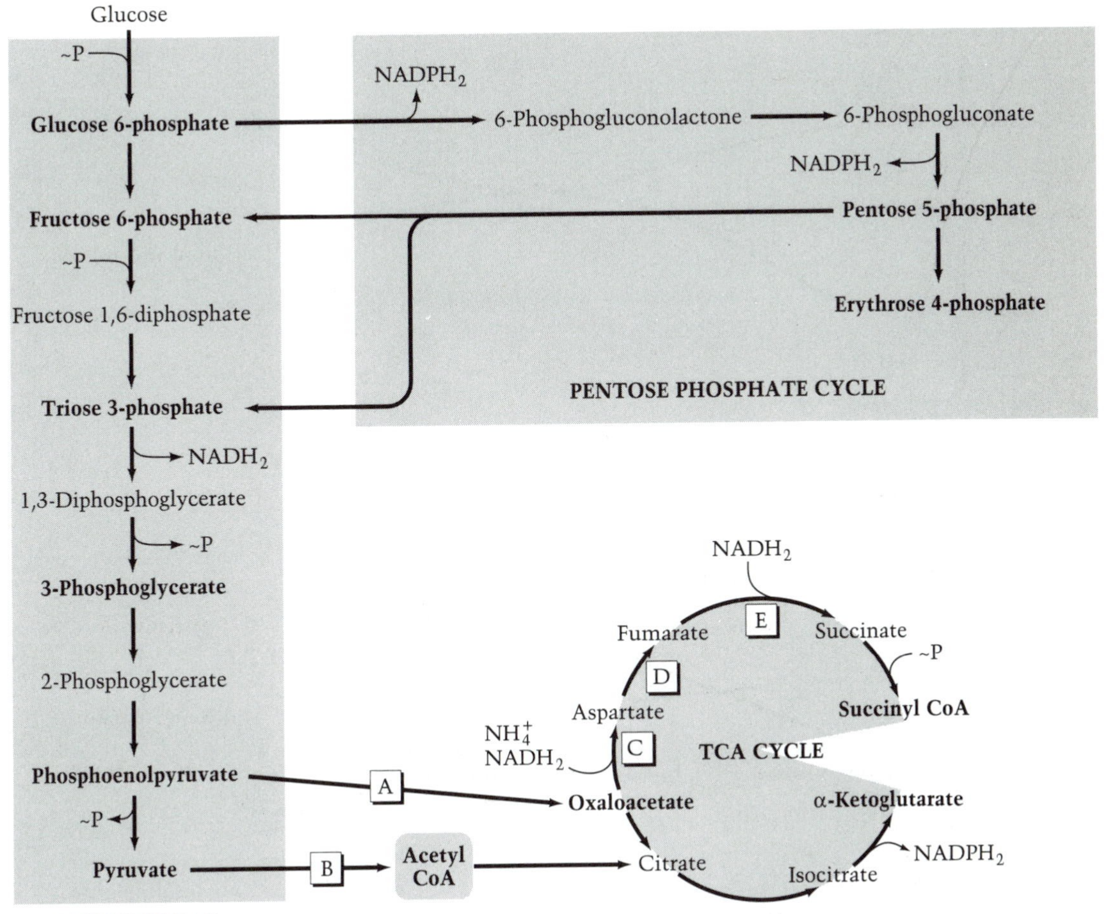

Figure 6

Formation of precursor metabolites during fermentative (anaerobic) growth of *E. coli* on glucose. The pentose cycle functions, although a lesser portion of glucose carbon flows through it during fermentation than during respiration. The major anapleurotic reactions are indicated by framed letters; the precursor metabolites are in boldface. As in aerobic growth (Figure 5), reaction A (catalyzed by the same enzyme) is the major replenishing reaction of the TCA cycle. Reaction B is catalyzed by pyruvate–formate lyase (rather than the aerobic pyruvate dehydrogenase); no $NADH_2$ is produced. Reaction C is a by-pass in which oxaloacetate undergoes reductive transamination at the expense of glutamate (formed from α-ketoglutarate by glutamate dehydrogenase). Aspartase, the enzyme catalyzing reaction D, produces fumarate by deaminating aspartate. Reaction E is catalyzed by an enzyme (fumarate reductase) different from the succinate dehydrogenase that functions in the opposite direction. The set of reactions that forms succinyl CoA constitute a reductive arm of the TCA cycle. The oxidative arm leads from oxaloacetate to α-ketoglutarate. Because there is no need anaerobically to oxidize this precursor metabolite, the enzyme α-ketoglutarate dehydrogenase is not formed under these conditions of growth.

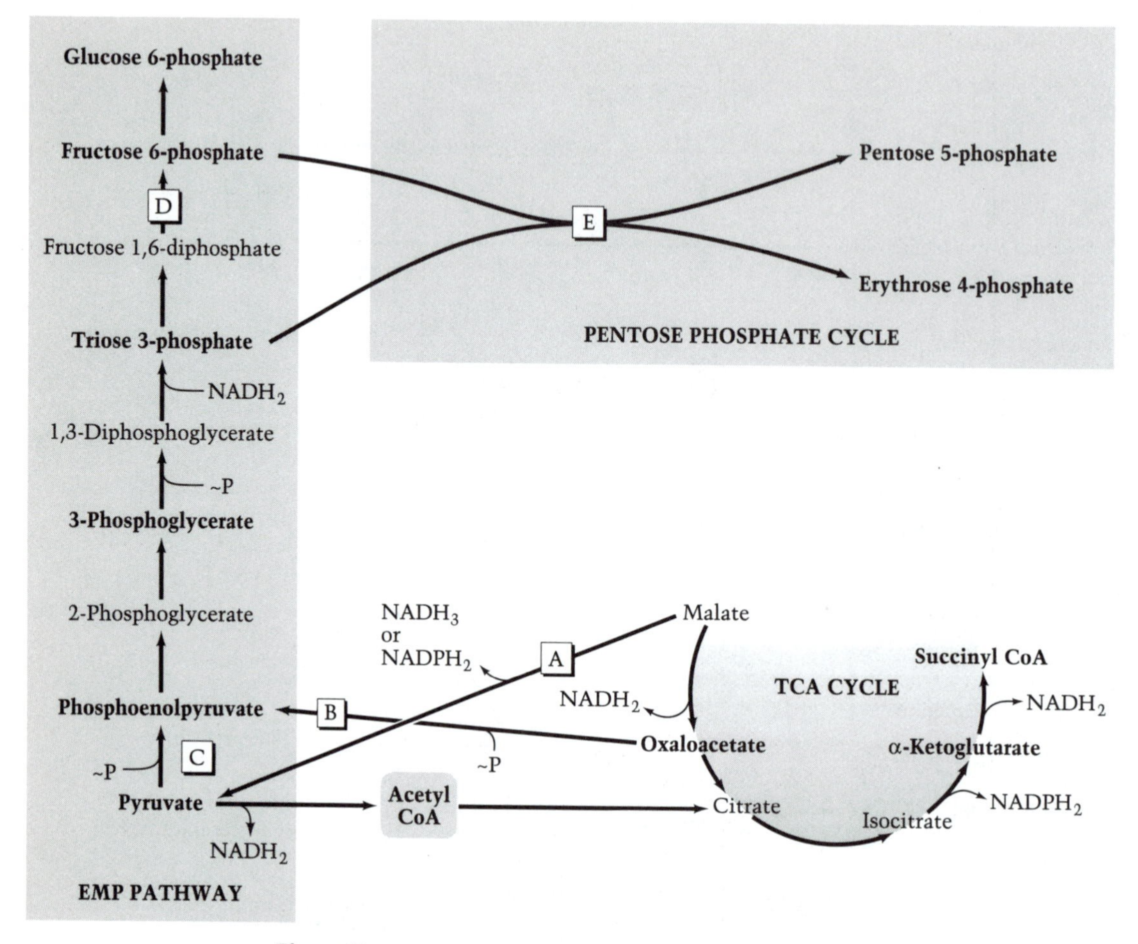

Figure 7

Formation of precursor metabolites during aerobic growth of *E. coli* on malate. Two routes lead from malate to the EMP pathway: reaction A (catalyzed by either of two malate enzymes) produces pyruvate and either $NADH_2$ or $NADPH_2$; reaction B (catalyzed by phosphoenolpyruvate carboxykinase) produces phosphoenolpyruvate after oxidation of malate to oxaloacetate and $NADH_2$. The pyruvate that is formed by reaction A can be converted to phosphoenolpyruvate by reaction C (catalyzed by phosphoenolpyruvate synthase), one of the two reactions needed to reverse the flow through the EMP pathway. The other is reaction D (catalyzed by fructose-1,6-bisphosphatase). Pentose and erythrose phosphates are formed by the anaerobic branch of the pentose phosphate cycle (catalyzed by reactions designated E).

That total is 42,703 μmol of ~P—24,218 for polymerization (Table 5 in Chapter 4) and 18,485 for biosynthesis (Table 3). The ways that this requirement is made up are discussed below.

Growth on a poor substrate like malate—one that places a heavy

Table 5. Amounts and costs of synthesizing sufficient precursor metabolites to make 1 g of *E. coli* from glucose or malate

Metabolite	Amount needed to produce *E. coli* B/r (μmol/g dried cells)	Cost of making 1 μmol of each of the metabolites (μmol/μmol)					Cost of making 1 μmol of each of the metabolites (μmol/μmol)				
		Glucose	~P	$NADH_2$	$NADPH_2$	CO_2	Malate	~P	$NADH_2$	$NADPH_2$	CO_2
Glucose 6-phosphate	205	1	1	0	0	0	2	4	0	0	−2
Fructose 6-phosphate	70.9	1	1	0	0	0	2	4	0	0	−2
Pentose 5-phosphate	897.7	1	1	0	−2	−1	1.7	3.3	0	0	−1.7
Erythrose 4-phosphate	361	1	1	0	−4	−2	1.3	2.6	0	0	−1.3
Triose phosphate	129	0.5[a]	1	0	0	0	1	2	0	0	−1
3-Phosphoglycerate	1,496	0.5	0	−1	0	0	1	1	−1	0	−1
Phosphoenolpyruvate	519.1	0.5	0	−1	0	0	1[e]	1	−1	0	−1
Pyruvate	2,832.8	0.5	−1	−1	0	0	1[f]	0	−1	0	−1
Acetyl CoA	3,747.8	0.5	−1	−2	0	−1	1	0	−2	0	−2
α-Ketoglutarate	1,078.9	1	−1	−3	−1	−1	2	0	−3	−1	−3
Succinyl CoA	—[b]	1[c]	−1	−4	−1	2	2	0	−4	−1	−4
Oxaloacetate	1,786.7	0.5[d]	0	−1	0	1	1	0	−1	0	0

[a]The values shown are those for the glycolytic pathway. By the pentose cycle, the cost is 0.6 glucose, 1.0 ~P, −0.6 CO_2, and −1.2 $NADPH_2$.
[b]Succinyl CoA is the precursor metabolite for tetrapyrroles (e.g., heme). It functions as a cofactor in the synthesis of major building blocks.
[c]The values shown are those for the TCA cycle functioning under aerobic conditions. Anaerobically, the cost is 0.5 glucose, 1.0 ~P, 1 $NADH_2$, and 1 CO_2.
[d]Formed by carboxylation of PEP.
[e]The PEP values assume formation from oxaloacetate.
[f]The pyruvate values assume formation by malic enzyme.

Table 6. Total cost of fueling reactions operating to produce precursor metabolites for 1 g of *E. coli* from glucose or malate

Substrate	Carbon source (μmol)	~P (μmol)	$NADH_2$ (μmol)	$NADPH_2$ (μmol)
Glucose	7,869	(−5,593)	(−16,965)	(−4,319)
Malate	15,109	7,152	(−16,965)	(−1,079)

energy burden on the cell—is usually slower than growth on a good substrate like glucose. This observation will be dealt with further in Chapters 8 and 12.

Generation of ATP

The cell needs energy for polymerization, biosynthesis, and sometimes formation of precursor metabolites. It needs additional energy for other purposes, such as bringing some substrates into the cell, keeping metabolites within the cell, maintaining a proper turgor pressure and internal pH, and motility. The magnitude of these additional needs can only be guessed. The total energy requirement is met by the fueling reactions through two general biochemical mechanisms—SUBSTRATE-LEVEL PHOSPHORYLATION and ELECTRON TRANSPORT.

In substrate-level phosphorylation, ATP is formed by a reaction between ADP and a phosphorylated intermediate of a fueling pathway. The central fueling pathways of *E. coli* contain three reactions (Figure 4) that yield substrate-level phosphorylations. They are those catalyzed by 3-phosphoglycerate kinase and pyruvate kinase in the EMP pathway and succinate thiokinase in the TCA cycle. Because these reactions participate in the pathway after glucose has been cleaved to produce two molecules of trioses, six molecules of ATP are produced from each glucose molecule by substrate phosphorylation. The 24 hydrogen atoms released during this process contribute to generating approximately 24 additional molecules of ATP by electron transport.

A theory to explain how electron transport could be coupled to the generation of ATP was proposed by Peter Mitchell in 1961. The succeeding three decades of research have established that his CHEMIOSMOTIC THEORY is almost certainly correct. Electron transport expels protons from the cell. Because the membrane itself is impermeable to H^+ and OH^- ions, the flow of protons out of the cell creates a difference in proton concentration (ΔpH) and electrical charge (membrane potential, $\Delta\psi$) across the cytoplasmic membrane of bacteria. The sum of these is a form of potential energy called PROTONMOTIVE FORCE, which can be used for a variety of purposes, including synthesizing ATP from ADP.

The electron transport chain is located in the cell membrane; the oxidation–reduction potentials of the compounds that constitute the chain are poised such that each member can be reduced by the reduced form of the preceding member. Thus, reducing power (as hydrogen atoms or electrons) can flow through the chain of carrier molecules to O_2 or some other terminal electron acceptor (NO_3^- or fumarate in the case of *E. coli*). The overall operation of the chain results in the transfer of hydrogen atoms (H) from an organic compound to the terminal electron acceptor. It is important to recognize that some members of the chain carry hydrogen atoms (a hydrogen atom is a proton + an electron) and others carry electrons. When reduction occurs by transfer of reducing power from a hydrogen-carrier to an electron-carrier, a proton (H^+) is released. Thus, the reduction of a hydrogen-carrier by an electron-carrier consumes a proton. If hydrogen-and electron-carriers alternate in the chain, the flow of reducing power through the chain can act as a PROTON PUMP. Protons are taken from the cytoplasm by reduction of a hydrogen-carrier on the inside of the membrane and released outside the cell by reduction of an electron-carrier on the outside of the membrane. One of the chains that functions in *E. coli* is shown in Figure 8. Pairs of hydrogen- and electron-carriers that transport protons out of the cell are called LOOPS. It is not clear, however, whether loops are the only way that protons are extruded. Some components might act directly as proton pumps that force protons out of the cell as reducing power flows through them.

Protonmotive force can be used to drive a variety of energy-linked processes, including (1) entrance of certain substrates into the cell against a concentration gradient; (2) maintaining the cell's turgor; (3) maintaining the cell's interior at a proper pH; (4) turning flagella; (5) driving a reverse flow of electrons through the respiratory chain to reduce NAD when the supply of $NADH_2$ is inadequate; and (6) generating ATP from ADP by a structure called F_1F_0 ATPASE. Some of these processes are shown schematically in Figure 9. The action of F_1F_0

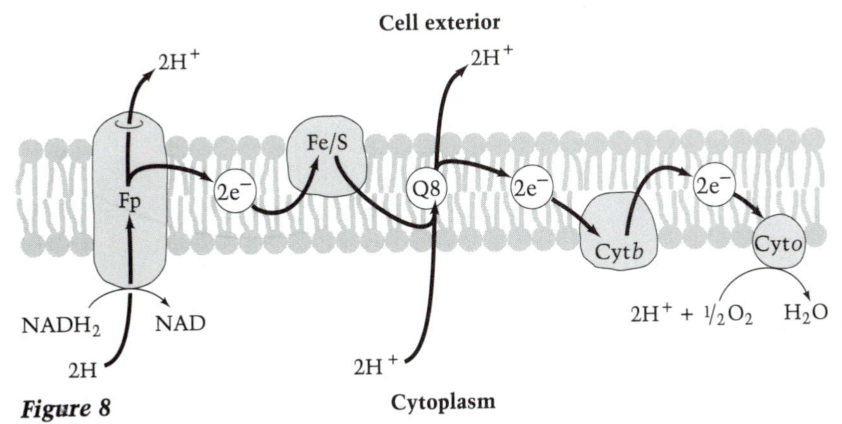

Figure 8

Functional organization of one of the electron transport chains in *E. coli*. This chain, which functions in environments with high pO_2, is composed of hydrogen carriers (flavoprotein, Fp; ubiquinone, Q8) and electron carriers (iron–sulfur protein, Fe/S; cytochrome *b*; cyt *b*). Cytochrome *o* (cyt *o*) transfers electrons to O_2, thereby reducing it. e^-, electron; H^+, proton.

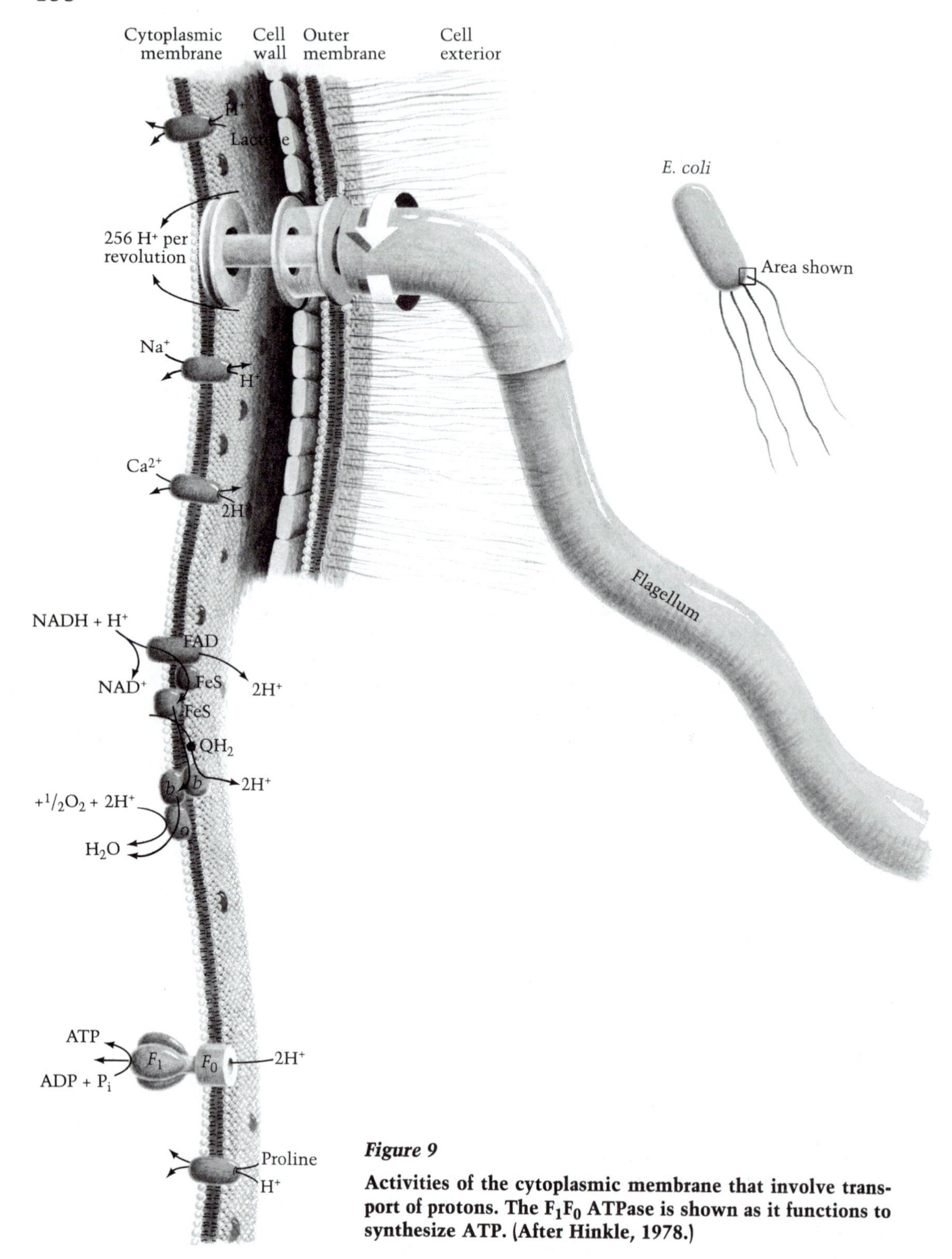

Figure 9

Activities of the cytoplasmic membrane that involve transport of protons. The F_1F_0 ATPase is shown as it functions to synthesize ATP. (After Hinkle, 1978.)

ATPase is reversible. It can use protonmotive force to synthesize ATP or ATP to increase protonmotive force. Both forms of energy (ATP and protonmotive force) are required for bacterial growth.

The F_1F_0 ATPase is a complex structure. It should really be considered to be a small cellular organelle. Electron micrographs of the cell membrane reveal small knobs that protrude into the cytoplasm. These are the components of the ATPase, termed F_1, and are easily removed from the membrane. In soluble form, F_1 retains ATPase activity (hydrolyzing ATP to ADP and P_i), but it cannot synthesize ATP. In situ F_1 is attached to the F_0 portion, which forms a proton-conducting channel through the membrane. Evidence indicates that one molecule of ATP is formed for each pair of protons that passes into the cell through the F_1F_0 ATPase.

How the F_1F_0 ATPase synthesizes ATP remains unresolved. Mitchell proposed that a phosphate ion binds to an active site in F_1 and that the two protons passing through the F_0 channel activate the phosphate by removing an oxygen atom from it, forming a molecule of water. The unattached phosphorus bond then binds directly to ADP, forming ATP. Others have proposed that the passage of protons changes the conformation of one of the proteins in the complex and this energy renders the synthesis of ATP thermodynamically feasible.

Generation of ATP during aerobic growth

In the pattern of aerobic respiration that we have discussed, the flow of reducing power from $NADH_2$ to O_2 expels four protons from the cell, sufficient to generate two molecules of ATP. The energy yield of respiration is usually described as the P:O ratio (the ratio of ATP formed per O atom consumed). In such a system the H^+:O ratio (protons expelled to O atoms consumed) is 4:1, and because the H^+:P ratio (protons reentering the cell to atoms of ATP synthesized) is about 2:1, the overall P:O ratio is about 2:1. This value of the P:O ratio is not universal among aerobic respiratory chains. In eucaryotic mitochondria the ratio is closer to 3:1, and in iron bacteria (see below) it is closer to 1:1.

Even in *E. coli* the P:O ratio varies with environmental conditions. As might be expected, oxidation of certain substrates yields less free energy, thus reducing the P:O ratio. Such is the case for L-α-glycerophosphate, D-lactate, and succinate. Hydrogen atoms from these substrates are donated directly to a quinone (Figure 8). This shunt results in the expulsion of only two protons and a P:O ratio of 1:1. The electron transport chain itself changes with conditions of growth. Anaerobically, menaquinone-8 is synthesized rather than ubiquinone-8. The level of aeration also affects the chain. The chain shown in Figure 8 is thought to function under conditions of high aeration. However, at low oxygen tension, cytochrome *d* rather than cytochrome *o* reduces O_2. Cytochrome *d* has a greater affinity for O_2, but the chain containing it probably has a P:O ratio of only 1:1.

So far we have talked about the way that organisms—called CHEMOORGANOTROPHS—derive energy from organic compounds. Another class of bacteria—termed CHEMOLITHOTROPHS—are able to oxidize inorganic compounds. They also generate ATP by electron transport. There are five classes of chemolithotrophs (Table 7): HYDROGEN BACTERIA, IRON BACTERIA, SULFUR BACTERIA, AMMONIA OXIDIZERS, and NITRITE OXIDIZERS. These bacteria generate ATP by transporting electrons from the reduced form of the inorganic compound to a terminal electron acceptor, O_2 (except for one sulfur bacterium that can use NO_3^-). In this respect ATP generation by chemolithotrophs and chemoorganotrophs (like *E. coli*) is similar, but there is an important distinction: the redox potential (E'_0) for the primary energy-yielding reaction of chemolithotrophs (with the exception of hydrogen bacteria) is lower than that for $NADH_2$ (footnote of Table 7). Thus, oxidation of these substrates cannot be coupled directly to the reduction of NAD. These bacteria use reverse electron transport (using protonmotive force to drive an electron transport chain in a thermodynamically unfavorable direction) to reduce NAD. Hydrogen bacteria, like chemoorganotrophs, couple substrate oxidation directly to the reduction of NAD. With the exception of the hydrogen bacteria, chemolithotrophs have a low P:O ratio.

Generation of ATP during anaerobic growth

The capacity to grow anaerobically is variable among bacteria. Some bacteria, such as most members of the genera *Pseudomonas, Streptomyces,* and *Bacillus,* and all members of the genus *Acinetobacter,* cannot grow in the absence of O_2. These bacteria are called STRICT AEROBES. ANAEROBES, such as members of the genus *Clostridium*, can grow only in the absence of oxygen; some of these are rapidly killed by oxygen. The

Table 7. Redox potentials of the primary energy-yielding reactions of chemolithotrophs

Class of chemolithotroph	Half-reaction of substrate (oxidation)[a]	E'_0 (volts)
Hydrogen bacteria	$H_2 \rightarrow 2H^+ + 2e^-$	−0.41
Sulfur bacteria	$H_2S \rightarrow S + 2H^+ + 2e^-$	−0.25
	$S + 3H_2O \rightarrow SO_3^{2-} + 6H^+ + 4e^-$	+0.005
	$SO_3^{2-} + H_2O \rightarrow SO_4^{2-} + 2H^+ + 2e^-$	−0.28
Iron bacteria	$Fe^{2+} \rightarrow Fe^{3+} + e^-$	+0.79
Ammonia oxidizers	$NH_4^+ + 2H_2O \rightarrow NO_2^- + 8H^+ + 6e^-$	+0.44
Nitrate oxidizers	$NO_2^- + H_2O \rightarrow NO_3^- + 2H^+ + 2e^-$	+0.35

[a]The oxidation of $NADH_2$ ($\frac{1}{2}$ NADH + $H^+ \rightarrow NAD^+ + 2e^- + 2H^+$) has an E'_0 of −0.32. The reduction of O_2 ($O_2 + 4H^+ + 4e^- \rightarrow 2H_2O$) has an E'_0 of +0.86.

FACULTATIVE ANAEROBES (of which *E. coli* is an example) can grow in the presence or absence of oxygen; generally, their metabolism differs depending on whether or not oxygen is present. But the lactic acid bacteria (including the genera *Streptococcus, Lactobacillus,* and *Leuconostoc*), are INDIFFERENT ANAEROBES. Regardless of the presence or absence of oxygen, they generate ATP by a fundamentally anaerobic process—fermentation.

The ability to utilize oxygen or to grow in its absence depends on the types of fueling pathways that an organism possesses, but the ability to withstand the toxic effects of oxygen depends on other factors:

- Does it have enzymes that are intrinsically sensitive to oxygen?
- Is it able to break down two highly toxic metabolic products of oxygen, hydrogen peroxide (H_2O_2) and superoxide ($O{\cdot}_2^-$)?

Most enzymes found in strict anaerobes are intrinsically sensitive to oxygen. Because even a brief exposure inactivates them, the study of enzymes from strict anaerobes like the METHANOGENS (methane-forming bacteria) must be carried out in chambers from which oxygen has been rigorously excluded. Curiously, even some aerobes produce certain oxygen-sensitive enzymes. For example, *Azotobacter* and cyanobacteria produce the highly oxygen-sensitive enzyme nitrogenase. As mentioned earlier, nitrogenase in *Azotobacter* spp. is protected by the organism's very high rate of respiration. This rapid utilization of oxygen creates an anaerobic pocket within the cell where nitrogenase resides. Cyanobacteria form specialized nitrogenase-containing cells, called HETEROCYSTS within which conditions are anaerobic.

Hydrogen peroxide or superoxide are inevitable by-products of certain oxygen-utilizing reactions, including those catalyzed by flavoproteins. Organisms that can tolerate oxygen have enzymes that destroy these potentially lethal agents. These protective enzymes include:

- CATALASE, which converts hydrogen peroxide to water and O_2 ($2H_2O_2 \longrightarrow 2H_2O + O_2$)
- SUPEROXIDE DISMUTASE, which converts superoxide to O_2 and hydrogen peroxide ($2O{\cdot}_2^- + 2H^+ \longrightarrow O_2 + H_2O_2$); the hydrogen peroxide is subsequently degraded by catalase

CHEMOTROPHS (organisms that use chemical energy for growth) generate ATP anaerobically by one or both of two modes of metabolism: ANAEROBIC RESPIRATION (in which the terminal electron acceptor of an electron transport chain is a compound other than O_2) and FERMENTATION (in which ATP is generated by substrate-level phosphorylation as a consequence of oxidizing one organic compound and reducing another).

Anaerobic respiration

A variety of compounds can serve as terminal electron acceptors in anaerobic respiratory chains. These include nitrate, sulfate, fumarate, trimethylamine-*N*-oxide, and selenate. Reduction of nitrate by some bacteria leads to a cascade of further anaerobic respirations, a process called DENITRIFICATION. The first product of this process is nitrite ion, which also serves as a terminal electron acceptor and becomes reduced to nitric oxide gas (NO). By a continuing process of accepting electrons, nitrous oxide gas (N_2O) and, finally, dinitrogen (N_2) are produced. Until humans began to make small amounts of N_2 as by-products of other chemical syntheses, this exclusively bacterial process accounted for all the N_2 in our atmosphere. In fact, the quantity of N_2 in our atmosphere remains almost constant, because the amount of nitrogen removed through fixation is roughly matched by the amount added through denitrification. The basic mechanism of ATP generation by anaerobic respiration, is the same as by aerobic respiration, but some of these alternative electron acceptors are less powerful oxidizing agents than O_2. In these cases, lesser amounts of ATP are generated per electron pair transported through the redox chain.

Fermentation

The other anaerobic mode of generating ATP is fermentation. The principles of fermentation are illustrated by considering HOMOLACTIC FERMENTATION (Figure 10), a relatively simple biochemical process carried out by certain lactic acid bacteria. By means of the EMP pathway, one molecule of glucose is cleaved and oxidized to yield two molecules of pyruvate. In this process, a net yield of two molecules of ATP is realized and two molecules of NAD are reduced. The two molecules of NAD are regenerated when pyruvate is reduced to lactate in a reaction catalyzed by lactate dehydrogenase. The substrates and products of the many bacterial fermentations are exceedingly diverse. Nonetheless, all fermentations share certain features of the homolactic fermentation.

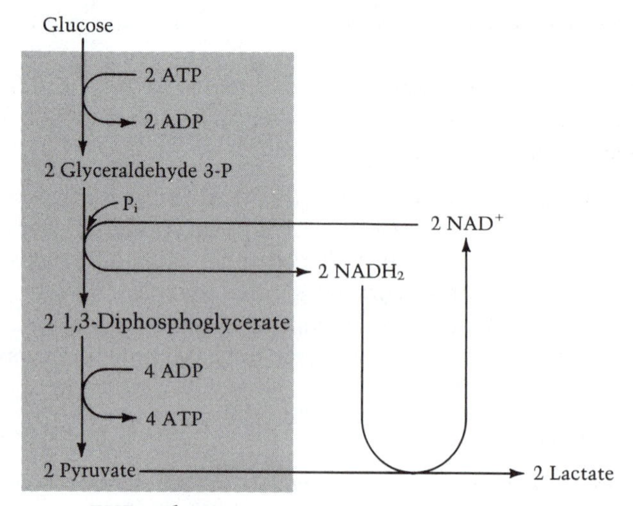

Figure 10

Homolactic fermentation.

- ATP is generated by substrate-level phosphorylation.
- Oxidative and reductive reactions occur, but a strict oxidation-reduction balance is maintained: the average oxidation state of the products and the substrate is the same (see Table 8). Some substrate carbon is assimilated into the cell, but this is a relatively small amount (see the last feature in this list), and the average oxidation state of carbon compounds in the cell is approximately that of the usual substrates of fermentation.
- For both oxidation and reduction to occur during fermentation, the substrates are usually at an intermediate state of oxidation. Hence, the substrates of fermentation are usually sugars.
- Because the yield of ATP from fermentation is relatively low (2 moles per mole of glucose in the homolactic fermentation versus about 28 in aerobic respiration of the same substrate), large quantities of substrate are used during fermentative growth; as a consequence, most carbon from the metabolized substrate can be recovered in the form of fermentative end products (Table 8).
- Most (but not all) pathways of bacterial fermentation involve pyruvate as an intermediate; the diversity of end products produced in various

Table 8. Parameters of the mixed acid fermentation of glucose by *E. coli*

Product	Moles formed/ 100 moles glucose fermented	Moles carbon	O/R value	O/R sum (O/R value × moles) +	−
Formate	2.4	2.4	+1	2.4	
Acetate	36.5	73.0	0		
Lactate	79.5	238.5	0	10.7	
Succinate	10.7	42.8	+1		
Ethanol	49.8	99.6	−2		99.6
2, 3-Butanediol	0.3	1.2	−3		0.9
CO_2	88.0	88.0	+2	166.0	
H_2	75.0	—	−1		75.0
Total		545.5		179.1	175.5

$$\text{Carbon recovery} = \frac{545.5}{100 \times 6} \times 100 = 91\%$$

O/R balance = 179.1/175.5 = 1.02[a]

[a]A number of methods can be used to calculate the oxidation–reduction balance of a fermentation. In the method used here, the O/R value of each product and of the substrate is calculated and multiplied by the number of moles of that product or substrate to give the O/R sum. When totaled algebraically, the total O/R of the products should equal that of the substrate used. In this fermentation, the substrate, glucose, has an O/R value of 0. Thus, the total of the positive O/R products approximately equals the negative O/R products. O/R value is calculated by assigning arbitrarily an O/R value of 0 to compounds with the empirical formulas of a carbohydrate $(CH_2O)_x$ and expressing each 2H in excess as a −1 value and each 2H shortage as +1. Thus, formate (CHOOH) falls two hydrogens short of being at the oxidation state of carbohydrate and is assigned an O/R value of +1. (Data from Wood, 1961.)

fermentations depends largely on the reactions by which pyruvate is metabolized and NAD is regenerated from $NADH_2$.

Some of the fermentative reactions by which pyruvate is metabolized and NAD is regenerated by various bacteria are shown in Figure 11. Not all of these reactions occur in the same bacterium. Some bacteria produce only a single fermentative end product. The HOMOFERMENTATIVE lactic bacteria, are one example; they produce only lactic acid. *Zymomonas* is another; it produces ethanol and CO_2, as fermentative products via the acetaldehyde pathway. Other bacteria produce many fermentative end products. For example, *Clostridium acetobutylicum* produces acetate, acetone, butyrate, butanol, CO_2, ethanol, and hydrogen.

Escherichia coli also carries out a relatively complex fermentation known as the MIXED ACID FERMENTATION because four of the seven end products are organic acids (Table 8). With the single exception of succi-

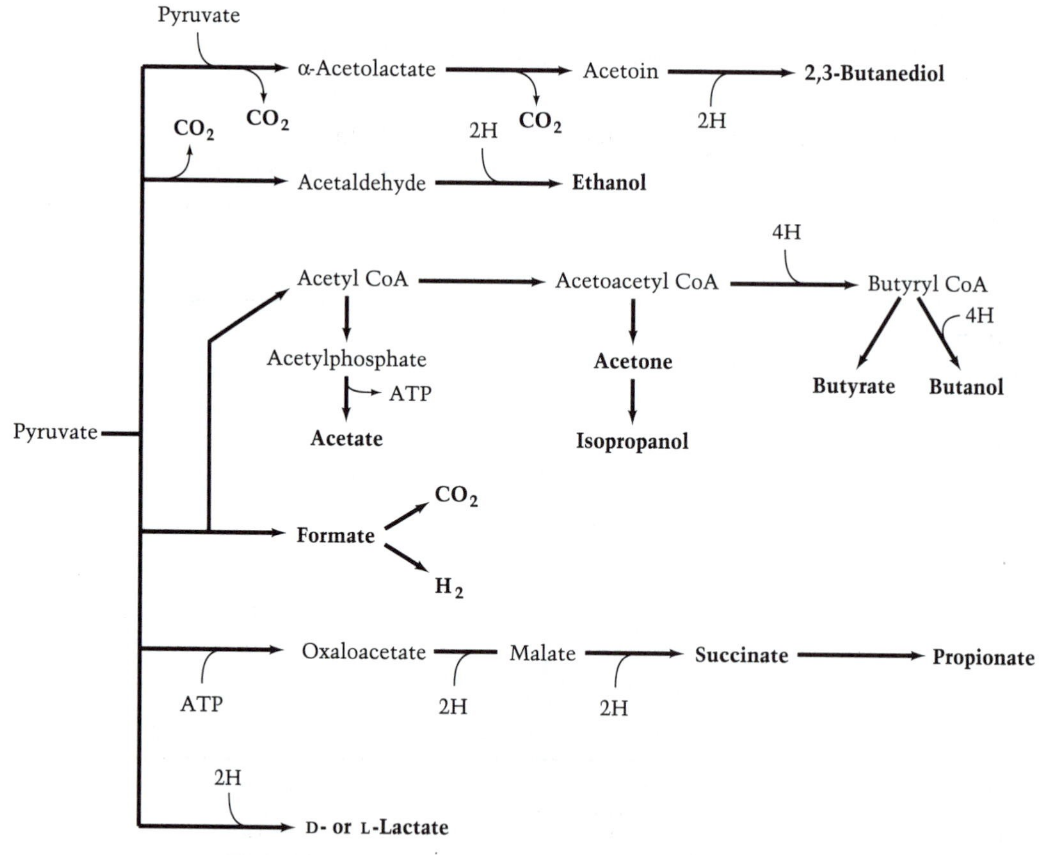

Figure 11

Fermentative end products that various bacteria form from pyruvate. End products are in boldface; reactions that oxidize $NADH_2$ are shown as H-utilizing reactions. (After Stanier et al., 1986.)

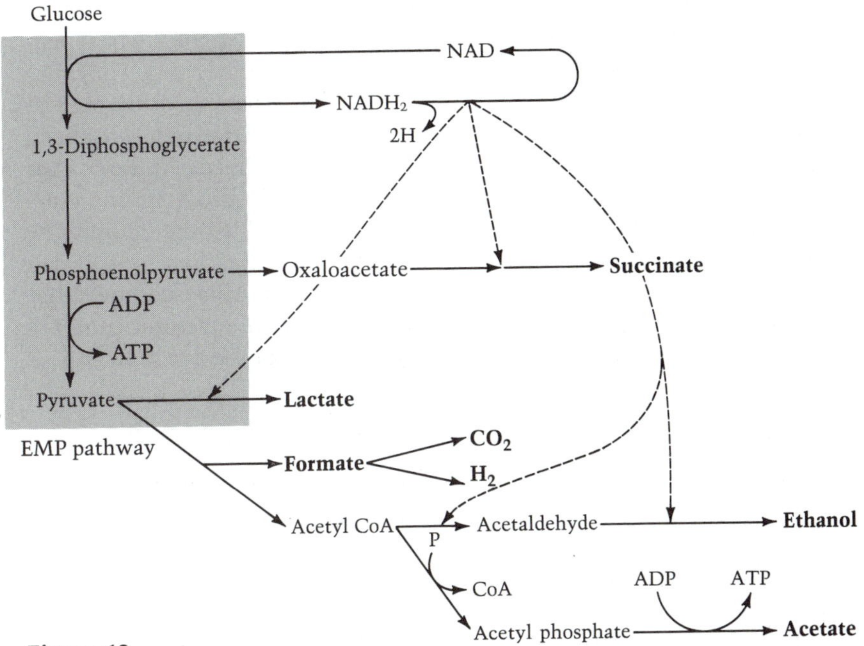

Figure 12

The mixed acid fermentation of *E. coli* and certain other enteric bacteria. The seven principal fermentation end products are in boldface. The dashed arrows indicate the four reactions that regenerate NAD from the $NADH_2$ formed in the EMP pathway. Certain enteric bacteria (including *Shigella, Yersinia*, and some species of *Salmonella*) mediate a mixed acid fermentation in which no gas is produced, because these bacteria lack the enzyme complex—formate hydrogenylase—that produces H_2 and CO_2 from formate.

nate, which is made from phosphoenolpyruvate, all of these end products are produced from pyruvate (Figure 12). The production of succinate involves a step that is not fermentative in the strict sense. Succinate is formed from fumarate, which serves as the terminal electron acceptor in an electron transport chain; so succinate is the product of anaerobic respiration.

There are nutritional advantages for an organism that carries out a complex fermentation like the mixed acid fermentation. Because they produce one product—ethanol—that is more reduced than carbohydrates and another—acetate—that is more oxidized, mixed acid fermenters are able to grow anaerobically on substrates like sorbitol (that are more reduced than sugars) and those like glucuronic acid (that are more oxidized) by adjusting the relative amounts of ethanol and acetate that are formed. In contrast, organisms that mediate a homolactic fermentation are limited to growth on sugars.

The prevalence of acids (succinic, lactic, formic, and acetic) among the products of the mixed acid fermentation accounts for the name of the fermentation and is the basis of a number of metabolic tests for certain enteric bacteria [*Aeromonas* (some species), *Escherichia*, *Photobacterium* (some species), *Proteus*, *Providencia*, *Salmonella*, *Shigella*, *Vibrio* (some species), and *Yersinia*] that mediate this type of fermentation. For example, McConkey's agar can be used to determine whether a strain is able to utilize a particular sugar for growth. McConkey's agar contains sufficient nutrients to support growth; it also contains a pH indicator (neutral red) and the sugar to be tested. When the strain to be tested is streaked on a petri plate containing McConkey's agar, it can grow whether or not it is able to ferment the sugar. Those colonies that do ferment the sugar become red, because cells in the anaerobic center of the colonies produce acids that change the color of the pH indicator.

Fueling reactions of phototrophs

The process by which light energy is utilized to drive fueling reactions is called PHOTOSYNTHESIS. Organisms that have this capacity—termed PHOTOTROPHS—include the higher plants, the algae, and certain bacteria. The type of photosynthesis carried out by higher plants, algae, and one group of bacteria (the cyanobacteria or blue-green bacteria) is termed OXYGENIC PHOTOSYNTHESIS because it is accompanied by the evolution of molecular oxygen. The other groups of bacterial phototrophs (the Rhodospirillaceae, or purple nonsulfur bacteria; the Chromatiaceae, or purple sulfur bacteria; and the Chlorobiaceae, or green bacteria) carry out ANOXYGENIC PHOTOSYNTHESIS; this process does not produce oxygen.

Generation of ATP / The major portion of the ATP in oxygenic photosynthesis and all the ATP generated in anoxygenic photosynthesis is synthesized by a process termed CYCLIC PHOTOPHOSPHORYLATION. In many respects, ATP generation by cyclic photophosphorylation and respiration are analogous. In both processes the flow of electrons and hydrogen atoms through an electron transport chain creates a protonmotive force that drives a membrane-bound ATPase to generate ATP. The processes differ with respect to the source of energy that causes the flow of electrons and protons through the chain. Respiration is driven by chemical energy; light energy drives cyclic photophosphorylation. The processes also differ with respect to the primary electron donors and acceptors of the chains. In respiration, as we have seen, there is an organic or inorganic substrate and the terminal electron acceptor is oxygen or some other oxidizing agent. In cyclic photophosphorylation, chlorophyll (in two different states) serves both as primary electron donor and as terminal electron acceptor. Light energy converts a molecule of chlorophyll from one of its two states (the ground state) to the other (the activated state). In its activated state chlorophyll is easily oxidized (it has a high negative E'_0) so it can readily donate electrons to the beginning of the chain; in its ground state it is easily reduced (i.e., it has a high positive E'_0, so it readily accepts electrons from the end of the chain).

The process of cyclic photophosphorylation is shown schematically in Figure 13. To accommodate the membrane-bound photosynthetic apparatus, photosynthesizing bacteria have an increased amount of cytoplasmic membrane. As a result, the membrane invaginates into the cytosol. The patterns thus created are characteristic of the various types of phototrophic bacteria. Some types produce rounded membrane vesicles; some produce regular layers of membrane that run parallel to the wall; others produce stacks of membranes that protrude into the cytosol at right angles to the wall; still others produce elaborate tubular structures that run through the cell (Stanier et al., 1986).

The photosynthetic apparatus can be divided into three functional components: the LIGHT-HARVESTING ANTENNA, the REACTION CENTER, and the ELECTRON TRANSPORT CHAIN. Light is absorbed by the light-harvesting antenna, which in the anoxygenic bacterial phototrophs is composed of carotenoids, bacteriochlorophyll, and proteins. In the oxygenic phototrophic bacteria, it is composed of carotenoids, chlorophyll a, and phycobiliproteins (proteins that are covalently bound to pigments, termed bilins, that are linear tetrapyrroles). The light-absorbing properties of these pigments determines the wavelength of light that supports phototrophic growth and therefore the environment in which various phototrophs are found. This action spectrum varies among individual phototrophs, but collectively they are able to use light energy extending from the near ultraviolet (~350 nm) to the infrared red range (~1,050 nm). For the remainder of our discussion of cyclic photophosphorylation, we shall focus on the particular reactions that occur in the most thoroughly studied nonoxygenic bacterial phototroph, *Rhodobacter sphaeroides.*

Light energy gathered by the light-harvesting antenna is transferred to the reaction center, where a molecule of bacteriochlorophyll initiates

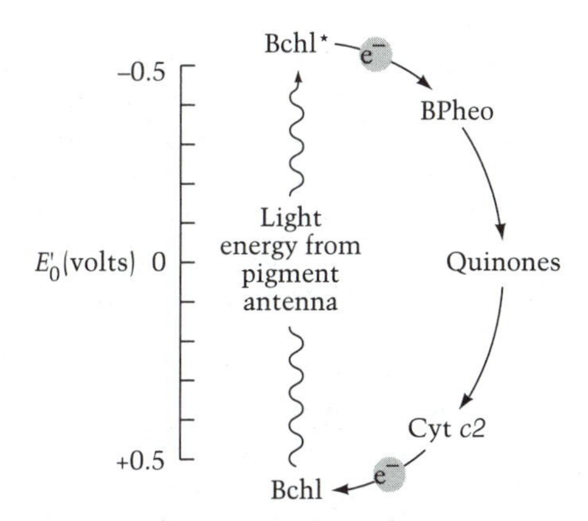

Figure 13

Cyclic photophosphorylation in bacteria. Bacteriochlorophyll (Bchl), at the expense of light energy from the pigment antenna, is raised to an activated state (Bchl*), which is readily oxidized by passing electrons through an electron transport chain composed in part of bacteriopheophytin (BPheo) and quinones to cytochrome *c*2, which reduces the oxidized bacteriochlorophyll.

flow through the electron transport chain. Only about 1 in 80 molecules of bacteriochlorophyll in the photosynthetic apparatus is located in a reaction center; the rest are in the pigment antenna. Each reaction center is composed of four molecules of bacteriochlorophyll, two molecules of bacteriopheophytin (a chlorophyll molecule that lacks Mg^{2+}), one atom of nonheme iron, and two molecules of quinone that form a complex with protein. The light-activated molecule of bacteriochlorophyll becomes oxidized as it donates an electron to a bacteropheophytin molecule. The electron is then passed to a ubiquinone·iron center from which it flows through the electron chain back to an oxidized molecule of bacteriochlorophyll. Thus, a molecule of activated chlorophyll is the primary electron donor for the electron transport chain, and an oxidized molecule of chlorophyll is the terminal electron acceptor. Like the electron flow of respiration, the flow of electrons through the photosynthesis chain generates a protonmotive force that is used to synthesize ATP by an F_1F_0 ATPase.

Generation of reducing power / By cyclic photophosphorylation, ATP is generated, but no reducing power is stored as $NADH_2$ or $NADPH_2$ – the form used in biosynthetic reactions. Members of the various groups of phototrophs reduce NADP in different ways, but in all cases two requirements must be met: (1) a source of reductant, either hydrogen atoms or electrons, and (2) a reducing agent that is sufficiently powerful (with an E'_0 more electronegative than −0.32; see Table 6) to reduce NADP.

As a source of reductant, the purple nonsulfur bacteria use organic compounds or H_2; the purple sulfur bacteria and the green bacteria use reduced sulfur compounds; the oxygenic cyanobacteria (like plants and algae) use water, producing O_2 as a product of oxidation. Purple bacteria obtain the NADP-reducing reductant much as chemolithotrophs do—by reverse electron transport—because no stable intermediate of their photosynthetic electron transport chain is a powerful enough reductant to reduce NADP. In contrast, the first stable intermediate in the chains of green bacteria and cyanobacteria can be used to reduce NADP. Such an electron flow is noncyclic—electrons flow from the oxidized molecule of chlorophyll to reduce NADP; the oxidized chlorophyll molecule is returned to its reduced state by accepting electrons from the source of reductant (Figure 14).

In the green bacteria, the source of reductant (a reduced sulfur compound) is a sufficiently powerful reducing agent to reduce oxidized chlorophyll directly. This reaction is not possible in the case of cyanobacteria, because water, which they use as a source of reductant, is a very weak reducing agent. Thus, the energy difference between oxidation of water and reduction of NADP is so great that it cannot be spanned by a single step of photoexcitation of chlorophyll. This activation is accomplished sequentially in two different reaction centers, called reaction center I (RCI) and reaction center II (RCII). Electrons from water are

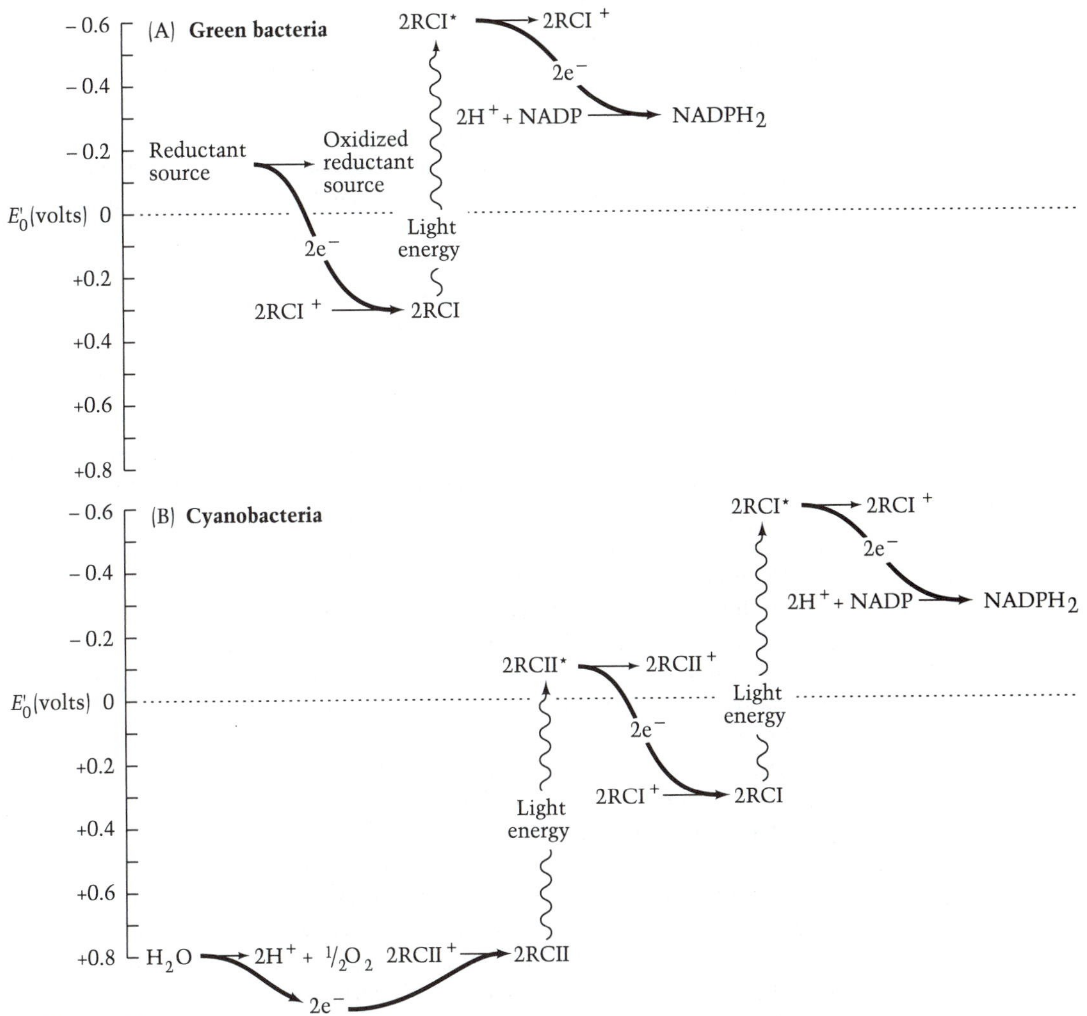

Figure 14

Comparison of the noncyclic flows of electrons that generate reducing power in green bacteria and cyanobacteria. (A) In green bacteria, electrons from the reductant source are passed to oxidized chlorophyll in reaction center I (RCI^+), thereby reducing it (RCI). Light energy from the pigment antenna converts the bacteriochlorophyll in reaction center I to an activated form (RCI^*) that is readily oxidized, donating its electron to an electron transport chain; the first stable intermediate of the electron transport chain is able to reduce NADP to $NADPH_2$. (B) In cyanobacteria, raising the electronegativity of reducing power of electrons in water sufficiently to reduce NADP requires two sequential photoexcitations of chlorophyll—first in reaction center II (RCII) and then in reaction center I (RCI). Thick arrows represent electron transport chains.

accepted by oxidized chlorophyll in RCII (E'_0 of about +0.86); excitation by light energy increases the reducing power of the chlorophyll to E'_0 of about −0.10. This value is insufficient to reduce NADP but adequate to generate protonmotive force by donating electrons to an electron transport chain for which oxidized chlorophyll in RCI is the terminal electron acceptor. Light excitation of RCI (the one used by cyanobacteria for cyclic photophosphorylation) raises it to a level of electronegativity that is capable of reducing NADP. Thus, electrons derived from the oxidation of water are made sufficiently electronegative to reduced NADP by two sequential photoexcitations—the first acting on RCII and the second on RCI. This process is called NONCYCLIC PHOTOPHOSPHORYLATION, because, in addition to reducing power, protonmotive force (and thereby ATP) is generated by this noncyclic flow of electrons.

Formation of precursor metabolites by phototrophs and chemolithotrophs / All phototrophs and chemolithotrophs can synthesize precursor metabolites from CO_2 (although under certain conditions, purple bacteria can also synthesize them from organic compounds).

The green bacteria assimilate (fix) CO_2 by employing reducing power (in the form of reduced ferredoxin) to reverse the normal direction of flow of carbon through the TCA cycle. This modified TCA cycle—called the REDUCTIVE TCA CYCLE—fixes CO_2 and forms six of the precursor metabolites; the other six are synthesized through connecting central fueling pathways.

Other phototrophs (including green plants and algae) and all chemolithotrophs fix CO_2 by the Calvin-Benson pathway. The single CO_2-fixing reaction of this cycle is catalyzed by RIBULOSEBISPHOSPHATE CARBOXYLASE, said to be the most abundant enzyme in nature. The product of this reaction is two molecules of 3-phosphoglycerate. The remainder of the cycle is composed of four enzyme-catalyzed reactions that regenerate a molecule of ribulosebisphosphate, the substrate of the CO_2-fixing reaction. The Calvin-Benson pathway does not produce ATP and reduced NADP, rather it utilizes them. All of these essential fueling products are generated in phototrophs by the processes described in the preceding two sections and in chemolithotrophs by aerobic respiration of the inorganic substrate.

SUMMARY

1. The building blocks for macromolecular polymerizations are made by biosynthetic pathways that form a network leading from 12 precursor metabolites to amino acids, nucleotides, sugars, sugar nucleotides, and other compounds that are needed. The pathways are similar in all organisms, but organisms differ in their complement of these pathways. All organisms require essentially the same set of building blocks; those that lack the pathways to produce certain building blocks must obtain them preformed from their environment.

2. In addition to precursor metabolites, biosynthetic pathways utilize large quantities of $NADPH_2$, ATP, and N, and lesser amounts of S. Nitrogen and sulfur are incorporated into metabolism in their reduced forms, NH_4^+ and H_2S. If these are not available in the medium, they must be obtained through pathways that degrade organic compounds or that reduce oxidized ions including NO_3^- and SO_4^{2-}. Some bacteria can utilize N_2 gas as a source of nitrogen for biosynthesis.
3. Fueling reactions of organotrophs consist of central and peripheral pathways that provide the precursor metabolites, $NADPH_2$, and energy needed for biosynthesis. The cell generates two forms of energy—~P and protonmotive force—that are interconvertible by the membrane-bound F_1F_0 ATPase. The central pathways (EMP, TCA, and pentose phosphate in *E. coli*) can function *cyclically* to degrade glucose and other substrates completely to CO_2 and water, thereby generating large quantities of energy and reducing power. They can also function unidirectionally to produce the 12 precursor metabolites. When intermediates of these pathways are drawn off to participate in biosynthesis, the continued functioning of the pathways is assured by anapleurotic reactions. Some anapleurotic reactions connect the central pathways, some permit the flow to be reversed, and some bypass portions of a pathway. As a result, growth is possible with any of several different compounds serving as the major substrate.
4. Fermentation and respiration pathways are coupled to the central pathways and regenerate NAD from the large quantities of $NADH_2$ produced in the central pathways. Respiration generates protonmotive force, which, by chemiosmosis, synthesizes ATP. Fermentative pathways regenerate NAD in reactions with organic compounds.
5. Peripheral pathways convert compounds that may be present in the environment to intermediates in the central pathways. Bacterial species differ markedly in the variety of compounds they can utilize as growth substrates.
6. Phototrophs generate protonmotive force and ATP by cyclic photophosphorylation. They generate reducing power through a noncyclic flow of electrons from their source of reductant (organic compounds, reduced sulfur compounds, or water, depending on the particular phototroph) to NADP. Most phototrophs, along with chemolithotrophs, synthesize precursor metabolites from CO_2 through the Calvin-Benson pathway. The green bacteria synthesize them through the reductive TCA pathway.

STUDY QUESTIONS

Unless otherwise specified, these problems refer to a culture of the reference strain of *E. coli* growing aerobically in glucose minimal medium at 37°C.

1. Considering only the carbon requirements for producing precursor metabolites, what fraction of metabolized glucose must be shunted through the pentose pathway?
2. Approximately 18,200 μmol of $NADPH_2$ are needed for biosynthesis of building blocks for 1 g of cells (Table 3). Only 4,319 μmol of $NADPH_2$ are produced incidentally in the formation of precursor metabolites (Table 6). How much additional glucose would have to be shunted through the pentose phosphate pathway to provide all the additional $NADPH_2$?
3. If the additional pentose molecules produced in generating $NADPH_2$ in Problem 2 were returned to the EMP pathway (in a reaction catalyzed by transketolase/transaldolase, three pentose 5-phosphate molecules form two fructose 6-phosphate molecules plus triose phosphate), what would be the new total glucose requirement to make the precursor metabolites?
4. Polymerization and biosynthesis require 24,200 and 18,500 μmol of ~P, respectively, to produce 1 g of cells (Tables 5 in Chapter 4 and Table 2 in this chapter). Only 5,600 μmol of ~P are produced incidentally in forming the precursor metabolites from glucose (Table 6). (a) If the ~P debit were made up exclusively by the combined operation of the EMP pathway and the TCA cycle, with all $NADH_2$ being used to generate ~P by oxidative phosphorylation, how much additional glucose would have to be consumed? (b) This strategy is not used by *E. coli*; instead, large quantities of glucose carbon are excreted as acetate, and much ATP seems to be generated through substrate level phosphorylation. Assume that this glucose is converted to acetate, generating 4 μmol of ~P per μmol of glucose, and that the $NADH_2$ produced is respired to generate additional ~P. How much glucose would be needed for the ~P debit?
5. The values in Table 4 ignore the effects of supplementation on the need for generating precursor metabolites. Calculate the *total* savings (in mol ~P per g of cells) by having all 20 amino acids provided ready-made. Assume that no amino acid biosynthetic reactions are needed (Table 1) and no precursor metabolites need be provided for these amino acids (Table 5).
6. Growth on malate (Figure 7; Tables 5 and 6) presents *E. coli* with a greater challenge for ~P production. Calculate how much malate would have to be consumed (beyond that needed for the precursor metabolites) to generate the ~P needed to produce 1 g of cells. Assume that malate is converted to pyruvate by malic enzyme and that the pyruvate is oxidized to acetyl CoA for complete oxidation in the TCA cycle. Ignore all costs for active transport.
7. Repeat the calculation of Problem 6, incorporating the additional assumption that entry of 1 μmol of malate costs 1 μmol of ~P, and the entry of 1 μmol NH_4^+ costs 0.5 μmol of ~P (see Problem 6).
8. Would you expect the addition of a full supplement of amino acids to be of different benefit to cells growing on malate than on glucose? Test your guess by calculating the benefit (in μmol ~P per g of cells).

SUGGESTED READING

Gottshalk, G. 1986. *Bacterial Metabolism*, 2nd Edition. Springer-Verlag, New York.
Harold, F. M. 1986. *The Vital Force: A Study of Bioenergetics*. W. H. Freeman and Co., New York.
Stanier, R. Y. , J. L. Ingraham, M. L. Wheelis and P. R. Painter. 1986. *The Microbial World*, 5th Edition. Prentice-Hall, Englewood Cliffs, New Jersey.
Neidhardt, F. C., J. L. Ingraham, K. B. Low, B. Magasanik, M. Schaechter and H. E. Umbarger (eds.). 1987. *Escherichia coli and Salmonella typhimurium: Cellular and Molecular Biology*. American Society for Microbiology. Washington, D.C.

6

Quest for Food

INTRODUCTION

In most natural environments, bacteria live in a state of chronic starvation. Finding food and concentrating it inside the cell is of enormous selective advantage in the procaryotic world. In this chapter we consider two related aspects of this metabolic challenge:

- how the bacterial cell transports substrates from its growth medium into its cytosol
- how some bacteria, by TACTIC RESPONSES are able to sense the locations of higher quality nutrients or more abundant supplies of nutrients and move toward them

The fraction of the bacterial genome dedicated to transport and tactic responses speaks eloquently to the metabolic importance of these functions. *Escherichia coli,* which is probably typical in these respects, has over 100 genetically distinct transport systems and over 60 genes that determine tactic responses.

TRANSPORT

The fact that bacteria have effective transport mechanisms can be inferred from the simple observation of how these organisms grow (Chapter 7). When inoculated into a favorable medium, bacteria grow at a constant rate until one nutrient in the medium is very nearly exhausted (i.e., until the concentration of the nutrient reaches the micromolar range); then growth stops abruptly. Even though most enzyme reactions slow at considerably higher concentrations of substrate, the metabolic reactions inside the bacterial cell proceed at full rate because transport mechanisms are able to maintain optimal intracellular concentrations of nutrients in spite of declining concentrations in the medium.

Transport systems are located in the cytoplasmic membrane, but to enter the cell, nutrients must also cross the other layers of the bacterial envelope—the murein layer and, in the case of Gram-negative bacteria, the outer membrane as well. With only a few possible exceptions, the murein layer is sievelike and offers no impediment to the passage of nutrients by diffusion. Nutrients can also diffuse across the outer membrane, but in a very different way—by passing through holes in the outer membrane that are formed by porins (Chapter 2). Thus, the concentration of an unbound nutrient in the periplasm must always be less than it is in the external medium; so the job of the transport systems located in the cytoplasmic membrane is greater than might be suggested by the concentration of nutrients in the growth medium.

Kinds of transport systems

Solutes enter a cell in several ways. A traditional and useful classification of these consists of simple diffusion, facilitated diffusion, active transport, and group translocation.

Simple diffusion / The membranes that surround bacterial cells are not barriers to all molecules. Some nutrients (for example, O_2, CO_2, NH_3, and water) simply diffuse through cytoplasmic membranes and the outer membrane of Gram-negative bacteria. In contrast, certain nonpolar compounds, including some highly toxic antibiotics can diffuse across cytoplasmic membranes but not across outer membranes, because the surface of the outer membrane is composed of a hydrophilic layer of carbohydrate, which is a barrier to hydrophobic solutes. The driving force of diffusion is the difference in concentration of the solute across the membrane (ΔC); the membrane's permeability to a solute is described by a PERMEABILITY CONSTANT (P) and is proportional to the total surface area (A) over which diffusion can occur. Thus flow occurs only down a concentration gradient, and its rate is proportional to ΔC, P, and A.

Only those few solutes that can penetrate a phospholipid membrane cross the cytoplasmic membrane by simple diffusion, but all essential nutrients (with a couple of possible exceptions) and waste products diffuse through the outer membrane. The diffusional properties of the outer membrane differ in two important respects from those of ordinary phospholipid membranes like the cytoplasmic membrane:

- The outer membrane (owing to the extended polysaccharide component of lipopolysaccharide) is impermeable to nonpolar solutes (a property that protects Gram-negative bacteria against a variety of antibiotics).
- The outer membrane is studded with porins (Chapter 2), through which all polar solvents enter or leave the cell.

The size and other properties of porins set the A and P terms of the diffusion equation, but no flow occurs unless a concentration gradient exists across the membrane. To establish a concentration gradient, Gram-negative bacteria maintain very low concentrations of nutrients within their periplasm, either by producing binding proteins that sequester certain nutrients and/or by actively pumping these and other nutrients across the cytoplasmic membrane into the cytoplasm. Some porins do not merely form holes in the outer membrane, rather they form channels that are specific for certain substrates (Chapter 2). In *E. coli,* for example, nucleosides and maltose pass the outer membrane through specific porins. The maltose porin is synthesized when maltose (or a maltodextran) is present in the growth medium. Maltose porin passes maltotriose 100 times faster than it passes the equal-sized trisaccharide, raffinose.

Facilitated diffusion / Facilitated diffusion is a process by which certain otherwise impermeable compounds can diffuse through a membrane. Stereospecific CARRIERS (transmembrane proteins that allow a specific compound to diffuse through the membrane) are said to participate in facilitated diffusion. A carrier that mediates facilitated diffusion (as the name implies) does not concentrate a substrate within the cytoplasm, it merely provides a mechanism by which a particular substrate to which the membrane is impermeable can diffuse down a concentration gradient into the cell. Think of the carrier as catalyzing a reaction for which substrate outside the cell is the reactant and substrate inside the cell is the product. Indeed, facilitated diffusion exhibits certain properties of enzyme-catalyzed reactions: it shows substrate specificity as well as saturation kinetics; and, increasing substrate specificity beyond a certain value does not increase the rate of the process, because all available proteins (enzymes or carriers) are in use.

COUNTERFLOW is a diagnostic property of facilitated diffusion systems in which a single carrier has two substrates. If the cell is preloaded with both substrates and then presented with an environment that contains only one of them, that substrate will enter the cell against a concentration gradient because both substrates compete for access to the carrier on the inner surface of the membrane while the single substrate has uncontested access to the carrier on the outer surface.

Although common among the transport systems of eucaryotes, facilitated diffusion is rarely encountered in procaryotes. Only glycerol is known to enter a number of species by facilitated diffusion; these species include *E. coli, S. typhimurium,* several species of *Pseudomonas, Klebsiella, Shigella, Bacillus,* and *Nocardia.* Indeed, in every bacterium in which it has been studied, glycerol was found to enter by facilitated diffusion. For this reason, the pattern of bacterial growth at the expense of this substrate is distinctive: rather than ceasing abruptly when glycerol disappears from the medium, growth slows down progressively as the concentration of glycerol in the medium declines.

Active transport / Active transport differs from facilitated diffusion in that a concentration gradient is established rather than exploited. Active transport resembles facilitated diffusion in that stereospecific, membrane-located carriers (sometimes termed PERMEASES) mediate the process, and saturation kinetics are seen. By active transport, the concentration of a substrate can be maintained at a level many fold higher than its concentration in the medium. For example, many amino acids and lactose are concentrated over 1,000 fold, galactose by a factor greater than 10^5, and K^+ by over 10^6. Obviously, the creation of a concentration gradient requires the expenditure of energy; the ways that metabolic energy is linked to active transport are discussed later in this chapter.

SECONDARY ACTIVE TRANSPORT

The processes of respiration and photosynthesis, as previously discussed, extrude protons from the cell, thus creating a pH gradient and a membrane potential that are collectively called PROTONMOTIVE FORCE (Chapter 5). Such proton extrusion can be considered to be a metabolism-linked active transport system, sometimes termed PRIMARY ACTIVE TRANSPORT. As was discussed earlier, hydrolysis of ATP by membrane-associated F_1F_0 ATPase also drives protons out of the cell; this, too, is primary active transport, and there are other forms as well. For example, certain amino acid decarboxylases act as primary active transporters; these membrane-associated enzymes couple the energy of decarboxylation to drive Na^+ ions out of the cell. The ion gradients created by primary active transport are a form of potential energy that can be used for a variety of metabolic purposes—including generating ATP (Chapter 5), causing flagella to turn (discussed later in this chapter), and moving molecules across the cytoplasmic membrane. Such movement of a molecule across the cytoplasmic membrane at the expense of a previously established ion gradient is called SECONDARY ACTIVE TRANSPORT.

There are three types of secondary active transport: symport, antiport, and uniport (Figure 1).

- SYMPORT is the transport of two substrates simultaneously in the same direction by a single carrier: if one of these substrates flows down its (previously created) concentration gradient, the other flows with it. If the driving force is an ion gradient (for example, a proton gradient; Figure 1A), either an oppositely charged ion or a neutral molecule can be brought into the cell by symport. In the former case, only the concentration gradient is diminished; in the latter, both the concentration gradient and the membrane potential are diminished. In this example, the product of primary active transport (a pH gradient) drives the symport system. Using other secondary active transport systems (see below), this primary gradient can create other ion gradients (for example, Na^+) to serve as driving forces for other symports.
- ANTIPORT is the simultaneous transport, mediated by a common carrier, of two materials in opposite directions (Figure 1B). Antiport processes are also gradient driven, and can be either electrogenic (involving a transfer of charge) or neutral.
- ACTIVE UNIPORT (Figure 1C) is a flow of ions driven directly by an ion gradient. (Facilitated diffusion can be considered as passive uniport of a neutral compound.)

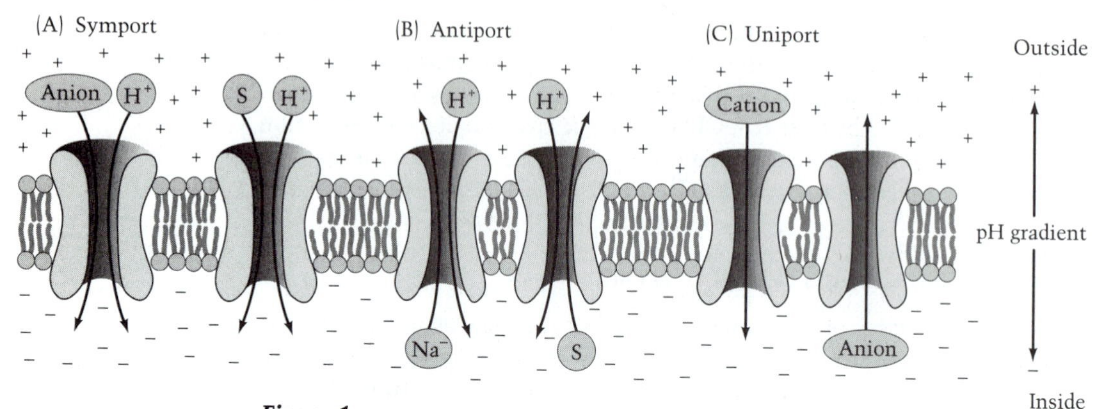

Figure 1

Secondary active transport systems. (A) Symport reactions. The pH gradient created by primary active transport drives (on the left) an electroneutral symport of an anion with a proton, and (on the right) an electrogenic symport of an uncharged solute (S) together with a proton. (B) Antiport reactions. The pH gradient drives (on the left) the electroneutral exchange of a cation for a proton and (on the right) the electrogenic exchange of an uncharged solute (S) for a proton. (C) Uniport reactions. The pH gradient drives a cation into the cell (left) or an anion out (right). (After Rosen and Kashet, 1978.)

The most thoroughly studied example of secondary active transport and possibly the most thoroughly studied bacterial transport system is galactoside permease, the symport system that brings lactose and certain other β-galactosides into *E. coli.* This system, encoded by the *lacY* gene, consists of a single 46,500-Da transmembrane protein that can bind a proton and a molecule of a β-galactoside; protonmotive force, either its ΔpH or its Δψ component, can drive the two molecules into the cell. The elegant simplicity of galactoside permease can be illustrated by a revealing experiment. When purified galactoside permease is added to a suspension of LIPOSOMES (spherical membrane structures that form spontaneously when phospholipid is added to water), it is incorporated into the artificial membrane. Then, when a pH gradient is created by adding acid to the suspension, β-galactosides are actively transported into these membrane vesicles.

Some secondary active transport mechanisms are not driven directly by protonmotive force but indirectly by another ion gradient formed at the expense of protonmotive force. The sugar melibiose and the amino acid glutamate enter *E. coli* by such a system. They flow into the cell by symport with sodium ions; the Na^+ gradient that drives this flow is formed by a H^+–Na^+ antiport system.

Shock-sensitive transport systems

If Gram-negative bacteria are suspended in a buffered 20% sucrose solution containing ethylenediaminetetraacetic acid (EDTA), centrifuged, and rapidly resuspended in 0.5 m*M* $MgCl_2$ at 0°C, the proteins normally located in the periplasm leak into the medium. This sudden change in osmotic strength of the external environment—or OSMOTIC SHOCK, as it is called—apparently damages the outer membrane, which is also the outer barrier of the periplasm. A number of periplasmic enzymes are released along with a group of proteins, termed BINDING PROTEINS. These proteins bind tightly (with binding constants in the range of $10^{-7}M$) and specifically to certain nutrients. The first binding protein, one that interacts specifically with SO_4^{2-}, was discovered by A. Pardee in 1966. Since that time, a number of others that bind to other ions, amino acids, sugars, and various other compounds have been identified. Although binding proteins are clearly not carriers, because they are free in the periplasm rather than being inserted into the cytoplasmic membrane, they do play a vital role in certain transport systems. (Some binding proteins also play a role in chemotaxis, as we note later in this chapter.) We have already discussed the indirect role they play in the diffusion of certain solutes through the outer membrane by maintaining low concentrations of the free forms of these solutes in the periplasm. They also play a direct role in certain active transport systems located in the cytoplasmic membrane. These systems are said to be SHOCK-SENSITIVE, because osmotically shocked cells lack or have greatly diminished activities of the binding proteins. Transport systems that remain in osmotically shocked cells—for example, secondary active transport systems—are called SHOCK-INSENSITIVE.

The complexity of shock-sensitive transport systems stands in stark contrast to the relative simplicity of the secondary active transport systems. The apparently typical shock-sensitive systems that bring histidine (Figure 2) and maltose into *S. typhimurium* and *E. coli*, respectively, are made up of three distinct proteins in addition to the periplasmic binding protein. Two of these proteins span the membrane, presumably forming a channel through which histidine passes. The third, HisP, which is attached to the membrane on its inner surface, is thought to be an energy transducer. Not surprisingly, the way that

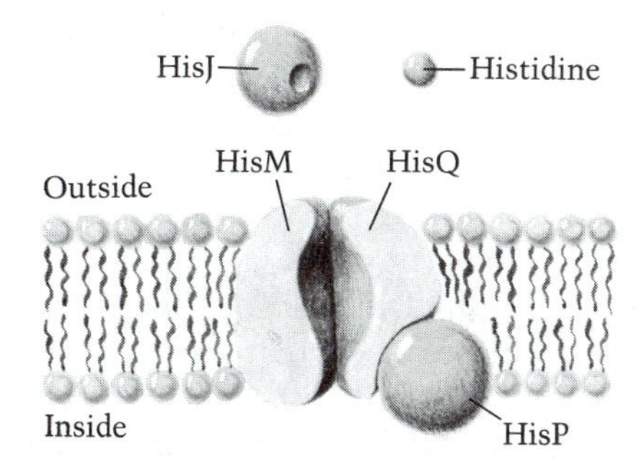

Figure 2

The shock-sensitive histidine transport system of *Salmonella typhimurium*. The participating proteins and their membrane locations are shown. (After Hengge and Boos, 1983.)

shock-sensitive systems function is not completely clear, but a plausible explanation for their action is the following:

- When the substrate, histidine, enters the periplasm, it binds tightly to the binding protein, HisJ.
- HisJ passes the substrate to sites on the transmembrane proteins, HisM and HisQ.
- HisP hydrolyzes a high-energy phosphate bond, the energy of which changes the conformation of HisM and HisQ such that the bound molecule of histidine is released on the membrane's inner surface.

The energy source has been thought to be ATP, but it may be some other compound with a high-energy phosphate bond, possibly acetylphosphate.

In addition to histidine and maltose, branched-chain amino acids, oligopeptides, ribose, β-methylgalactoside, and phosphate enter *E. coli* through shock-sensitive transport systems.

Group translocation

GROUP TRANSLOCATIONS are mechanisms that chemically alter substrates to impermeable derivatives as they cross the cytoplasmic membrane. These mechanisms are not active transport, because a concentration gradient is not established. But the same metabolic function is accomplished: the concentration of the substrate derivative inside the cell is greater than the concentration of the free substrate outside the cell. Entry of a substrate by group translocation rather than by active transport usually constitutes a net savings of metabolic energy. Although the transport-related reaction requires the expenditure of a high-energy phosphate bond, the derivative is one that would have to be formed in any case, because it is identical to the product of the first intracellular metabolic reaction. Derivatization has the same metabolic costs regardless of whether it occurs during or after the entry of the compound into the cell, and the normal costs of active transport are saved by group translocation. In view of the energy savings associated with group translocation, it is not surprising that such mechanisms are most frequently encountered in strict and facultative anaerobes.

The best established of the group translocations is the PHOSPHOTRANSFERASE SYSTEM (PTS) by which certain carbohydrates are brought into some bacteria (Figure 3). The system is somewhat complex. At least three proteins (EI, HPr, and EII) are needed to transport and phosphorylate any of the so-called PTS carbohydrates, and a fourth protein (enzyme III, EIII) is additionally required for a subset of these carbohydrates. These proteins form a chain of carriers that transfer a high-energy phosphate group from phosphoenolpyruvate to the incoming carbohydrate. The same two proteins—enzyme I, (EI) and a small, heat-stable protein (HPr)—are the first carriers in all PTS chains. Substrate specificity of any particular chain resides in the membrane-bound enzymes II, each of which can recognize a series of structurally related carbohydrates;

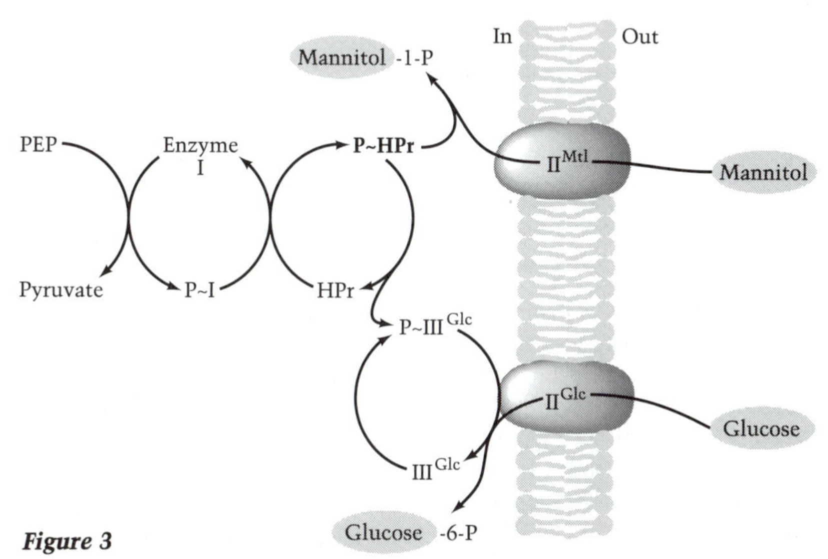

Figure 3

Schematic representation of the phosphotransferase system (PTS). Enzymes I and HPr participate in all PTS chains. Of the many different enzymes II found in *E. coli*, only two are shown. II^{Mtl} is specific for mannitol, and II^{Glc} is specific for glucose. III^{Glc} participates in the glucose chain and certain others, including the sucrose chain. P~1, P~HPr and P~III^{Glc} are phosphorylated forms of enzyme I, HPr, and III^{Glc}, respectively. PEP, phosphoenolpyruvate. (After Postma and Lengeler, 1985.)

although a bacterium produces only one species of enzyme I and of HPr, it produces many different enzymes II. Enzymes III are obligate penultimate members of certain PTS chains. They are hydrophobic carriers associated with the membrane and are less specific than enzymes II. *Escherichia coli* is known to produce at least four enzymes III. A minority of PTS chains in *E. coli* contain an enzyme III, but they are the rule in PTS chains in Gram-positive bacteria.

As mentioned earlier, PTS transport is energy conserving; so, because fermentations yield fewer ATPs than respirations, one might expect to find PTSs predominantly associated with fermentative organisms. As a generalization, this expectation is true. The strict aerobes *Azotobacter, Micrococcus, Mycobacterium,* and *Nocardia* do not possess PTSs, although *Arthrobacter pyridinolis* does. The facultative anaerobes known to possess PTSs include *Escherichia, Salmonella, Staphylococcus, Photobacterium,* and *Vibrio*; the strict anaerobes include *Clostridium* and *Fusobacterium.*

In addition to PTSs, other less well established examples of group translocation have been proposed. A coenzyme A transferase system mediated by acyl:CoA-synthase is presumed to play the same role in the uptake of fatty acids as PTSs do in the uptake of certain carbohydrates. Similarly phosphoribosyltransferases, which catalyze the class of reactions

$$\text{Purine or pyrimidine base} + \text{PRPP} \xrightarrow{\text{phosphoribosyltransferase}} \text{Nucleoside monophosphate} + P_i\text{-}P_i$$

are thought to constitute a group translocation mechanism for the uptake of adenine, guanine, hypoxanthine, xanthine, and uracil.

Uptake of specific substrates

Of the various transport mechanisms, almost all (including facilitated diffusion, shock-sensitive active transport, and secondary active transport) participate in the uptake of one or another sugar or sugar alcohol by *E. coli* (Figure 4). Interestingly, the uptake of a particular sugar does not follow a pattern: for example lactose is taken into *E. coli* by proton symport but into *Staphylococcus aureus* by a PTS.

Amino acids are transported by shock-sensitive or secondary active transport systems. In *E. coli*, 14 different transport systems are dedicated to bringing amino acids into the cell (Table 1). From an examination of the amino acid transport systems of bacteria, several patterns emerge. Some of the systems transport a group of amino acids with similar struc-

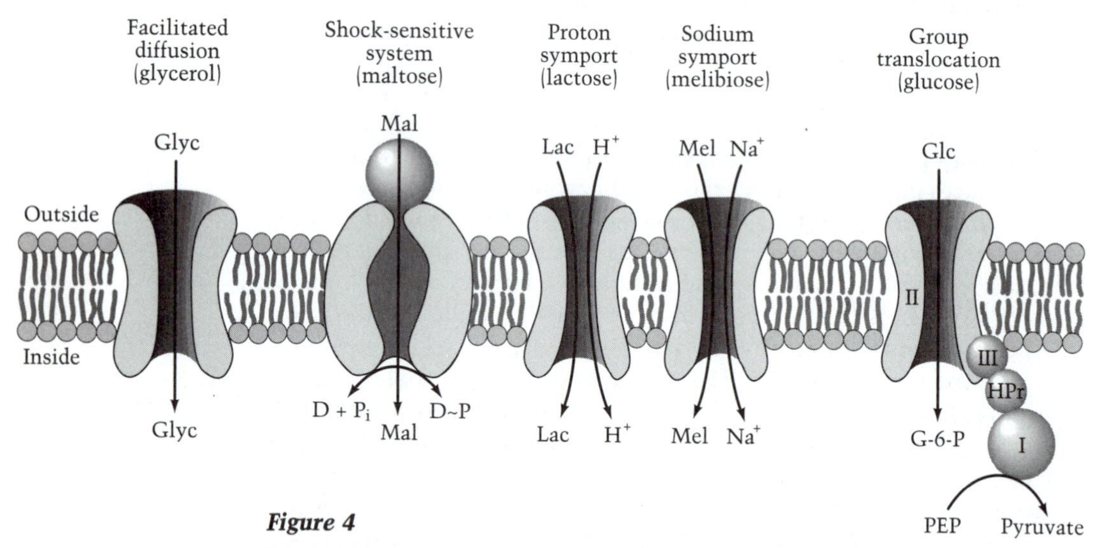

Figure 4

Representative carrier-mediated mechanisms for the uptake of sugars and sugar alcohols by *E. coli*. Glycerol (Glyc) enters by facilitated diffusion. Maltose (Mal) enters by a shock-sensitive system dependent on a binding protein (sphere), two transmembrane proteins (ovals), and an energy-transducing protein (curved, double-headed arrow) that hydrolyzes a high-energy phosphate donor (D~P). Lactose (Lac) enters by proton symport and melibiose (Mel) by a Na^+ symport of secondary active transport. Glucose (Glc) enters by a PTS through the mediation of enzyme I (I), HPr, enzyme II (II), and enzyme III (III); the intracellular product of the process is glucose 6-phosphate (G-6-P). (After Dills et al., 1980.)

Table 1. Transport systems for amino acids in *Escherichia coli*

1. Glycine-alanine	8. Cystine
2. Threonine-serine	9. Asparagine
3. Leucine-isoleucine-valine	10. Glutamine
4. Phenylalanine-tyrosine-tryptophan	11. Aspartate
5. Methionine	12. Glutamate
6. Proline	13. Histidine
7. Lysine-ornithine-arginine	14. Cysteine (probably)

tures. Often these have subsystems specific for only one of the amino acids. This apparent redundancy serves a purpose. One of the systems has high affinity (affinity constant in the submicromolar range) and low flow; the other has low affinity (affinity constant 10 times or more greater) and high flow. Each has obvious advantages in particular environments. In addition to amino acid transport systems, *E. coli* has a variety of transport systems for cofactors, various ions, and metabolic intermediates.

In the preceding discussion, we have emphasized the role of transport systems in bringing substrates into the cell. They play other roles as well. One of them is to keep metabolites in the cell. Endogenously synthesized metabolites, such as amino acids for which transport mechanisms exist to capture exogenous supplies, constantly leak out and are pumped back. Other transport systems have the function of pumping certain materials out of the cell. The transposon Tn*10* (see Chapter 9), which confers resistance to the antibiotic tetracycline, does so by encoding a transport system that pumps the antibiotic out of the cell.

MOTILITY AND TAXES

In spite of their small size and lack of a nervous system, some bacteria exhibit what appears to be purposeful behavior. They swim toward more favorable locations, including those with a better supply of nutrients and away from unfavorable locations, for example, those that contain toxic materials. These responses are called TAXES (singular: TAXIS). Not all species of bacteria are capable of locomotion, but of those that are, probably all are capable of taxis. As best we know, the sole purpose of bacterial motility is taxis—not just random movement.

A variety of different mechanisms of motility are found among the bacteria. Many species, representing Gram-positive and Gram-negative eubacteria as well as archaebacteria, swim by means of flagella that occur in different numbers and patterns on their surfaces. The spirochetes propel themselves with an AXIAL FILAMENT, a fin like structure that runs the length of the cell and is powered by flagella lying inside the outer membrane. Certain Gram-negative species, including representatives of the cyanobacteria, *Cytophaga*-like bacteria, and the myxobacteria move only when they are in contact with a solid surface; then they move by a mysterious process called GLIDING MOTILITY.

The types of taxes exhibited by bacteria are quite varied. Moving toward certain chemicals or away from others is called CHEMOTAXIS. If the attracting chemical is O_2, the process is called AEROTAXIS,. Moving toward light—even discriminating light of different wavelengths as some phototrophs do—is called PHOTOTAXIS. Some bacteria exhibit MAGNETOTAXIS: they move along magnetic lines of force, the presumed way that these microaerophilic bacteria find their way into the semianaerobic silts at the bottom of bodies of water (Chapter 17). This localization is possible because representatives of this kind of bacterium in the northern hemisphere swim toward the north magnetic pole and those in the southern hemisphere swim toward the south magnetic pole; they swim along lines of force that enter the Earth. Bacteria can also respond tactically to the nonnutrient environment. *Escherichia coli,* for example, swims away from alkaline and acidic environments toward more neutral ones; it seeks warmer regions within the temperature range of 20° to 37°C; and it avoids intense blue light.

Escherichia coli and *S. typhimurium* have been intensively studied with respect to their motility and chemotaxis. These studies have provided a rich body of information that explains, at least in broad outline, how these organisms swim and how they direct it to their nutritional advantage.

We begin by considering the various classes of taxis-deficient mutations, because they identify the functional modules that make up the process. The largest class (almost 60 genes are involved) affects the structure of the flagellum. A small second class affects the flagellar motor; strains with this class of mutation make a normal flagellum but are unable to turn it. The third class affects the chemotactic response to stimuli; strains with these mutations can swim, but they might swim in any direction—not necessarily in a favorable one.

Mechanics of flagella-mediated motility

The bacterial flagellum (described in detail in Chapter 2) is functionally analogous to a propeller attached to a motor. Its helical filament is the propeller; its basal body is the motor with associated bearings. (The hook probably functions as a universal joint.) The motor, which is embedded in the cytoplasmic membrane, is turned by a flow of protons driven by protonmotive force, as has been shown very directly: the flagella attached to empty cell envelopes will turn at full rate when the buffer outside the cell is adjusted to about four pH units less than that inside. Approximately 1,000 protons flow through the cell envelope during each revolution of the motor. This amount represents a minuscule expenditure of energy—less than 1% of the cell's energy budget is spent for motility. Although a flow of protons drives the flagellar motors of enteric bacteria and most other bacteria, other ions are sometimes used.

The flagella of alkalophilic bacteria are turned by a flow of sodium ions.

Enteric bacteria, being peritrichous (Figure 5), bear a number of flagella spread somewhat randomly over their surface, an arrangement that seems inconsistent with directional movement of the cell. One might imagine that such an arrangement would push the cell evenly from all directions, and that no net progress would be made in any direction. But this is not what happens. When the flagella turn in the counterclockwise direction (as viewed by looking outward from the cell) all the individual flagella coalesce into a bundle (Figure 6A) that functions like a composite propeller. It drives the cell evenly through the medium at a relatively rapid rate: flagellated bacteria swim at the rate of 10–20 μm/sec. Put in relative terms, this rate is equivalent to about 10 body lengths a second, a speed that would correspond to about 40 mph for a human being. This condition of smooth swimming, which specialists in the field of bacterial chemotaxis call a RUN, does not continue indefinitely. After a brief period, the extent of which is determined by whether or not the cell is being tactically attracted, the direction of flagellar rotation reverses.

As soon as the flagella begin to turn in the clockwise direction, the flagellar bundle flies apart and the cell TUMBLES without making net progress in any direction. In some still mysterious way, the turning of all flagella on a cell is coordinated: when one switches its direction of turning, they all switch. The period of tumbling is always quite brief—less than a second—but it changes the direction in which the cell swims; the next run might proceed in any direction. For taxis to occur, the tumble is as essential as the run.

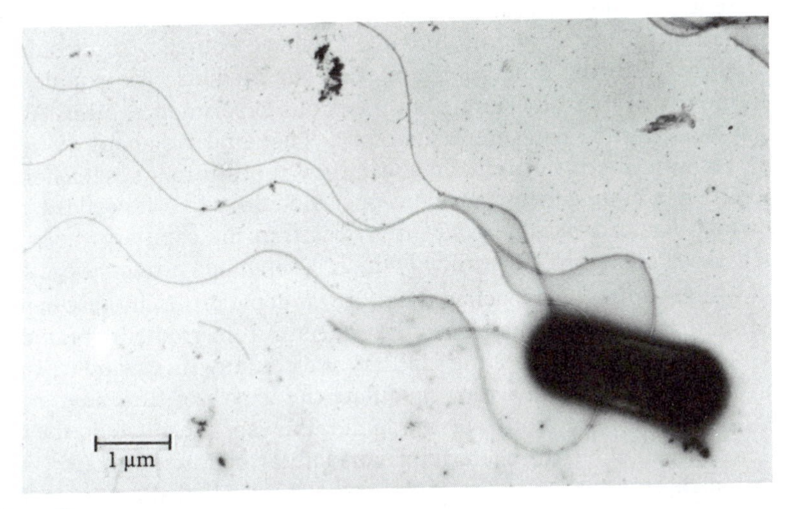

Figure 5

Electron micrograph of *Salmonella typhimurium* showing the distribution of flagella on its surface. (Courtesy of R. M. Macnab.)

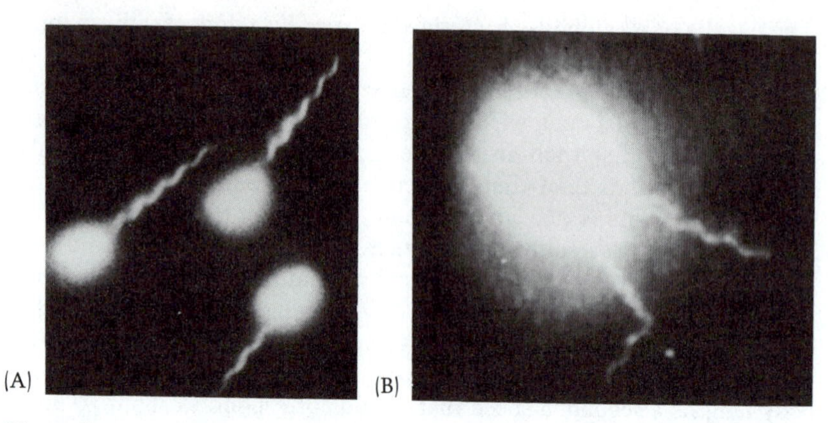

Figure 6

Darkfield micrographs of *Salmonella typhimurium*. (A) When all the flagella turn in the counterclockwise direction, the individual flagella coalesce into a bundle that moves the cell evenly through the medium. (B) When the flagella turn in the clockwise direction, the bundle flies apart and the cell tumbles aimlessly. (From Macnab and Ornston, 1977; Khan et al., 1978.)

Chemotaxis: A result of a biased random walk

When, with the aid of an ordinary bright-field microscope, one observes a bacterial cell swimming through a homogeneous medium (when not undergoing chemotaxis), individual flagella cannot be resolved; but the cell's movement reveals how they must be turning. Each cell runs for about a second, then it tumbles for about a tenth of a second, then it begins a new run in a randomly set direction. Each cell follows an apparently aimless zigzag path, so there is not net migration of the population in any direction. But if a localized source of nutrients is introduced into the otherwise homogeneous medium, the population will move toward it. By simple but ingenious experiments, Julius Adler made quantitative measurements of this phenomenon about 25 years ago. He introduced nutrient-filled capillary tubes into a suspension of bacteria in a homogeneous nutrient-poor medium and followed the rate at which bacteria became concentrated within the capillary tubes. By such experiments, he determined which compounds attracted bacteria and which compounds repelled them. It was apparent from these experiments that bacteria could sense the concentration gradient created as the nutrient diffused out of the capillary and could swim toward it. They do this in a curious way: they modulate the length of time that a run continues. When swimming up the gradient toward the nutrient, the run is lengthened; when the bacterium swims in the opposite direction, it is shortened. This swimming pattern—sometimes called a BIASED RANDOM WALK—is not the quickest way to get to the source of nutrient, but it inevitably gets you there. Daniel E. Koshland, a specialist in this field of research, described the process as "taking giant steps in the right direction and small ones in the opposite direction."

Nature and function of sensors and signal transducers

How does a bacterial cell sense whether the concentration gradient is increasing or decreasing so that it can adjust the length of its runs appropriately? There are two possibilities:

- It could sense at any one moment the difference in concentration between its leading and trailing pole.
- It could sense whether the concentration of nutrient in its environments increases or decreases with time of swimming.

The first possibility can almost be excluded on theoretical grounds: the bacterial cell is so short and the detectable gradient is so shallow (a change of only one part in 10^4 over a cell length) that the concentration difference across the cell would amount to a change of only a few molecules in a reasonable sample volume. The validity of the second possibility was established by observing how bacteria in a uniform, thoroughly mixed nutrient environment respond when the nutrient concentration is abruptly changed by dilution or addition of more nutrient. In other words, the experiments asked, How do cells respond if nutrient concentration is changed with time, not distance? The results of such experiments were quite clear. When concentration was abruptly increased, the cells swam smoothly (in various directions) for several seconds. When concentration was abruptly decreased, the cells tumbled (Figure 7). Thus, the cell must sense whether it is moving in a favorable or unfavorable direction by making repeated measurements of nutrient concentration, constantly comparing the most recent measurement with the previous one. The cell measures, remembers, measures again, and compares.

The cell responds to a *change* in nutrient concentration, not to its absolute value, so it must constantly *adapt* to higher concentrations in order to be able to detect still higher ones. Indeed, ADAPTATION is a fundamental aspect of sensory perception; our own visual sense depends on similar principles of detecting differences in light intensity at various absolute levels of intensity.

Our understanding of how bacteria sense nutrient concentration, remember what it used to be, and convert this information to a signal that changes or maintains the direction of flagellar rotation is progressing very rapidly as a result of research being done in a small number of highly productive laboratories. Already the broad outlines are available. This dramatic progress is eloquent testimony to the power of mutant analysis. By observing which functions are lost in particular mutants, the researchers can deduce the number of participating proteins and their behavioral role.

A set of transmembrane proteins called METHYL-ACCEPTING CHEMOTAXIS PROTEINS (MCPs) detects and measures concentrations of chemotactic compounds in the medium. As their name implies, these proteins become methylated in response to changes in concentration of attractants or repellants in their environment, and, during the period of adaptation, they return gradually to their original state of methylation. A

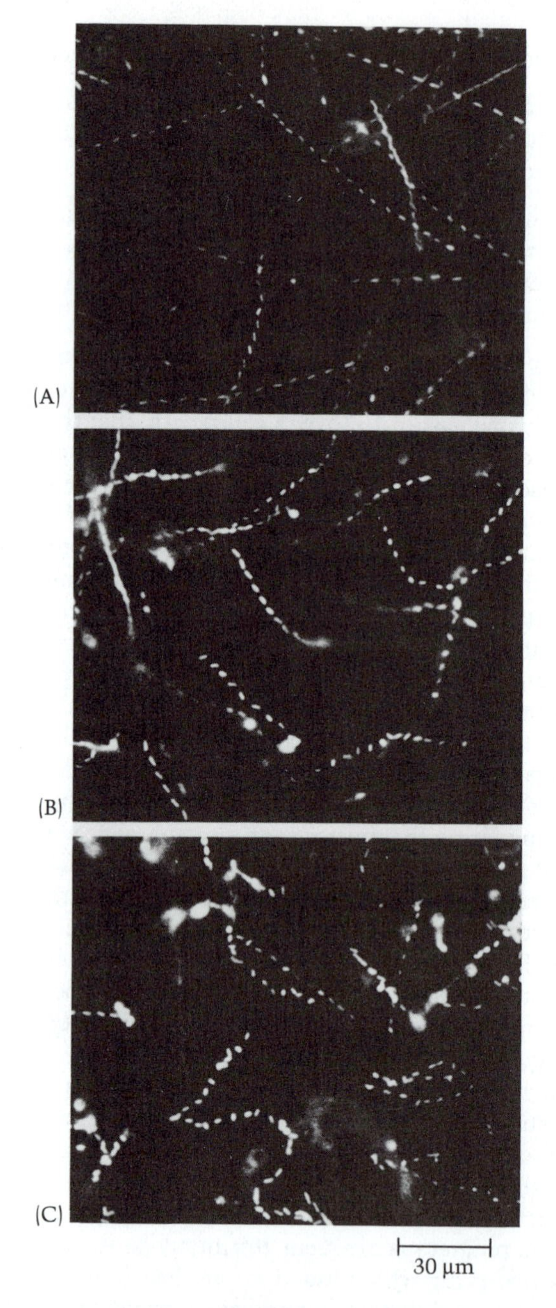

Figure 7

Motility tracts of *Salmonella typhimurium* taken in the time interval of 2 to 7 sec after subjection of the culture to a sudden (200 msec) change in attractant (serine) concentration. (A) Concentration increased from 0.0 to 0.76 m*M*. Smooth linear tracts. (B) No change in concentration. Some changes in direction. Bright spots indicate tumbling or nonmotile bacteria. (C) Concentration decreased from 1 to 0.24 m*M*. Frequent tumbles and erratic changes in direction. (Photographs were taken in darkfield with a stroboscopic lamp operating at five pulses per second.) (From Tsang et al., 1973.)

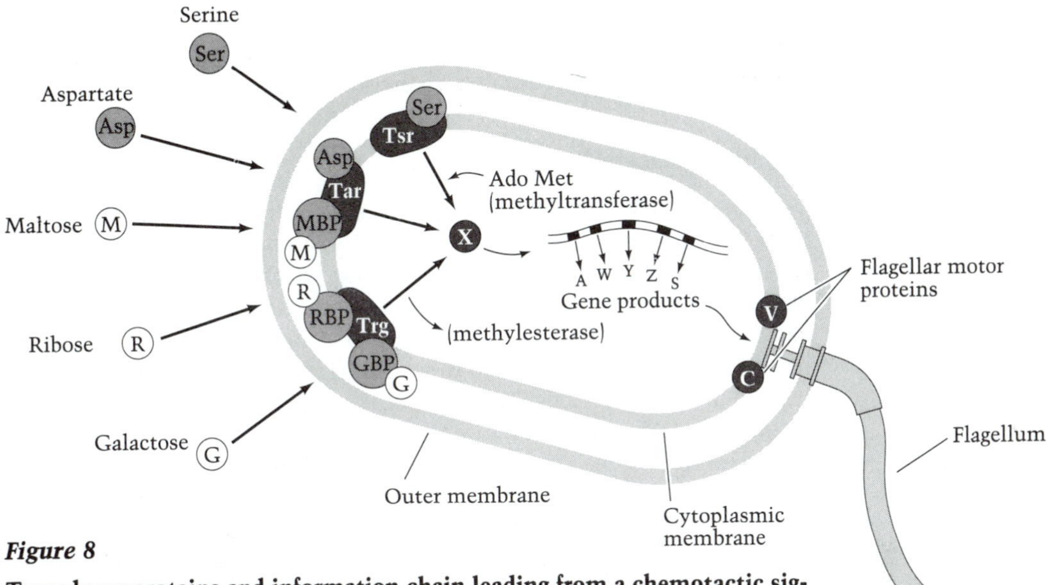

Figure 8

Transducer proteins and information chain leading from a chemotactic signal to control of the flagellar motor. (After Taylor, 1983.)

high level of methylation signals prolonged runs; a low level increases the frequency of tumbles.

From our discussion so far, it is apparent that MCPs must lie within an information pathway leading from environmental change to control of rotation of the flagellar motor (Figure 8). Change in the concentration of attractants in the medium determines the state of methylation of the MCPs; the state of their methylation determines how long the motor continues to turn in a counterclockwise direction. There are three major MCPs—Tsr, Tar, and Trg—that respond to different attractants. A particular chemoattractant, either in a free state or bound to a periplasmic binding protein, attaches to specific sites on the periplasm-exposed surface of its particular MCP. For example, free aspartate attaches to Tar; maltose also attaches to Tar, but only when complexed to maltose-binding protein. Attachment of a free or complexed attractant most probably induces a conformational change in an MCP, exposing sites on its cytoplasmic surface that can now become methylated by a chemotactic-specific methylase. If the concentration of attractant does not increase, a chemotactic-specific methylesterase gradually removes these methyl groups. Thus, the presence of a chemoattractant in the medium is converted into an intracellular signal registered as a methylated MCP. This signal is passed through about five intracellular proteins to a switching mechanism on the flagellar motor, causing it to prolong or diminish the time of counterclockwise rotation. How the signal is passed through the intracellular information chain and how it directs the rotation of the flagellar motor is not yet clear, but the proteins in the chain have structural similarities to proteins in two-component regulatory systems (see Chapter 13), a finding suggesting that these proteins pass information down the chain, possibly by one protein phosphorylating the next.

The MCPs are not the only kind of proteins that detect chemotactic signals. Enzymes III of the PTS, by some still-mysterious mechanism, detect the presence of certain carbohydrates.

Other means of locomotion

One may ask how bacteria with flagella arranged in some way other than peritrichous make their random walk to a source of nutrients. Doubtless there are many variations on the enteric bacterial theme, but obviously formation and dissolution of a flagellar bundle cannot occur in bacteria with only a single flagellum. *Rhodobacter sphaeroides,* which has a single flagellum inserted medially (Figure 9), swims by rotating it in a clockwise direction. This organism does not change the direction of flagellar rotation, rather it reorients its swimming pathway by ceasing to rotate its flagellum. When it does, the flagellum coils tightly back on itself, and the cell is reoriented by the forces of Brownian motion. When the flagellum begins to turn again in the clockwise direction, it reforms a helical shape and the bacterium swims away on a new course. Although the details differ, the pattern of chemotactic swimming by *R. sphaeroides* is like that of the enteric bacteria—swimming for a few seconds in a straight line and then reorienting to swim in a new direction.

Rhodobacter sphaeroides illustrates a second principle of flagellar motion: torsional stress contributes to determining the helical waveform of the filament. The effect is particularly extreme in the case of *R. sphaeroides.* When torsional stress is applied, the filament assumes the helical shape necessary for motility; when stress is removed, the filament assumes the shape of a tight coil. Probably the amplitude, wavelength, and handedness of the filaments of enteric bacteria change when their direction of rotation changes.

Pseudomonas putida has a tuft of flagella (usually 5 to 7, with a range of 1 to 12). Like the rotation pattern seen in *E. coli,* counterclockwise rotation of the flagella causes them to form a bundle that propels the cell smoothly. When flagellar motion reverses, the bundle flies apart; the cells do not tumble, but they do reorient. Thus, like the chemotactic response of *E. coli,* chemotaxis of *P. putida* depends on modulating the periods of counterclockwise and clockwise rotation of flagella. Occasionally, when flagellar rotation switches from counterclockwise to clockwise, the synchronization is so precise that the flagellar bundle remains intact for a brief period and the cell swims backward. This response is the normal one of flagellar reversal of some polar monotrichous bacteria.

Although the helical waves that progress down the length of spirochetes suggest a unique type of bacterial motility, it is, in fact, merely a form of flagellar motility. Rather than extending into the medium, the flagella of spirochetes are contained within the periplasm, where they aggregate into a composite structure—the axial filament (Figure 10). As these flagella—(sometimes called ENDOFLAGELLA or PERIPLASMIC FLAG-

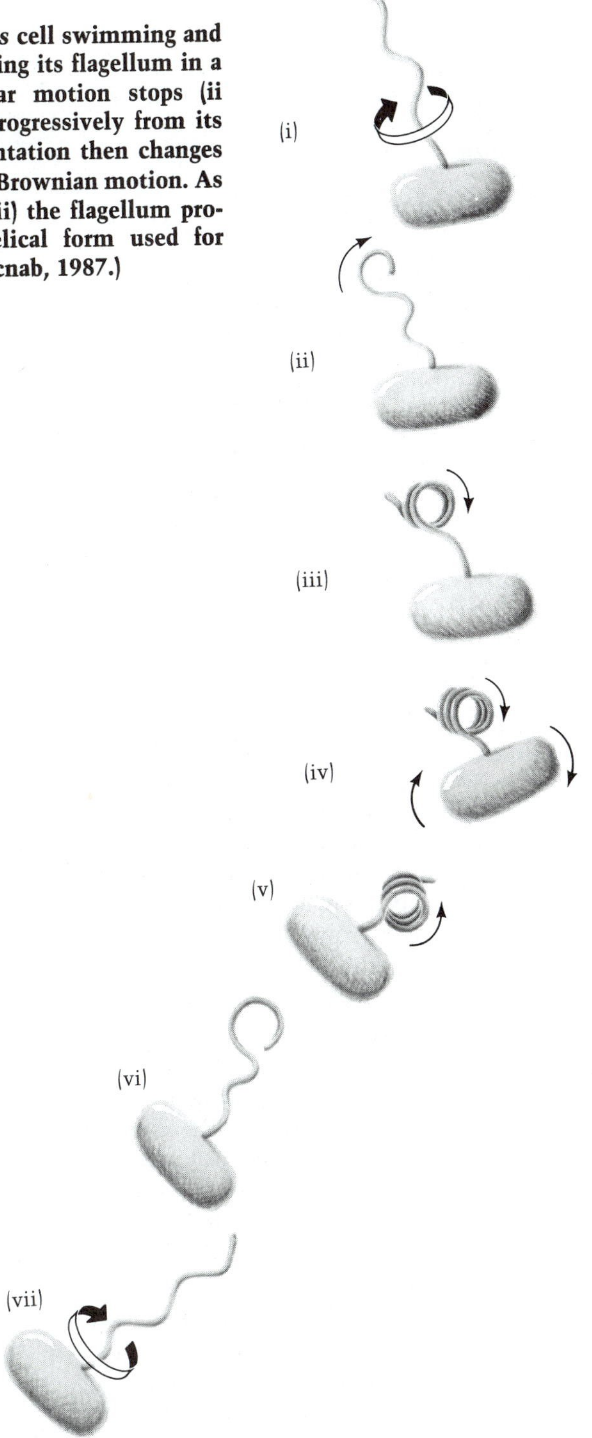

Figure 9

Sketch of a *Rhodobacter sphaeroides* cell swimming and reorienting. A cell swims (i) by turning its flagellum in a clockwise direction. When flagellar motion stops (ii through iv), the flagellum relaxes progressively from its distal end into a tight coil. Its orientation then changes randomly (iv) under the influence of Brownian motion. As rotation recommences (v through vii) the flagellum progressively assumes the normal helical form used for propulsion. (After Armitage and Macnab, 1987.)

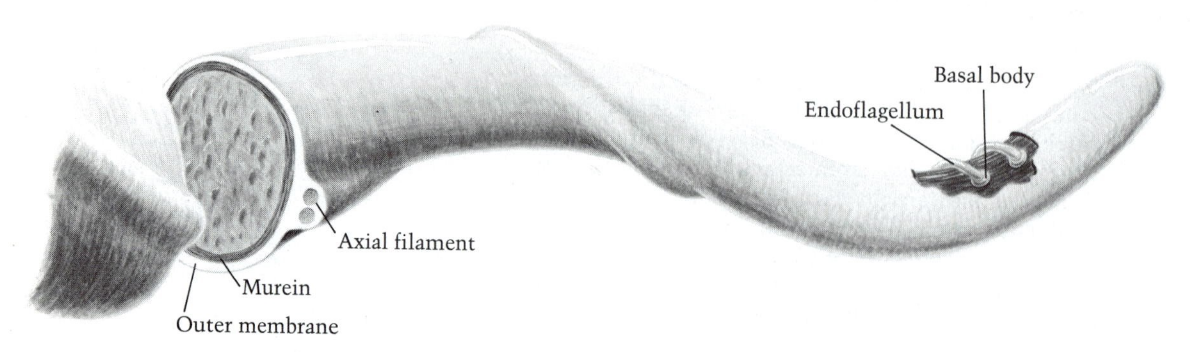

Figure 10

Sketch of a spirochete showing the arrangement of the axial filament within the periplasm. (After Holt, 1978.)

ELLA—turn, they produce the helical waves that move the cell. Spirochetal motility seems to be particularly suited to viscous media. Whereas ordinary flagellar motility slows as viscosity increases, and stops completely at about 60 centipoise (cP), spirochetal motility becomes more rapid up to values of 500 cP.

Gliding motility is yet another major means of bacterial locomotion. The mechanism for this phenomenon remains unexplained, but it almost certainly is not dependent on flagella because no such structures have been revealed by careful searches employing electron microscopy. However, gliding motility does resemble flagella-based motility in being driven by protonmotive force. Possibly this form of bacterial movement has more than one physical basis, because the patterns of gliding motility differ among the various bacterial groups that employ it. Some strains of *Myxococcus* move no faster than 1 to 2 μm/min (about the rate of continental drift!), whereas the speed of some strains of the cyanobacterium—*Oscillatoria,* can exceed 600 μm/min.

Gliders also differ with respect to their content of SULFONOLIPIDS (*N*-fatty-acyl-2-amino-3-hydroxyisoheptadecane-1-sulfonic acids), which are present in the outer membrane of some of them. These unusual lipids are completely missing from or rare in myxococci, but they constitute as much as 20% of the total membrane lipids of some cytophagas. Without doubt, the motility of some cytophagas depends on the presence of sulfonolipids in their outer membrane, for the following reasons. Certain mutants of *Cytophaga johnsonae* that are deficient in motility and synthesis of sulfonolipids regain both abilities when they are supplied with specific biosynthetic intermediates. It also seems significant that when *C. johnsonae* is in contact with a solid surface (the state in which gliding motility is possible), it produces sulfonolipids that differ from the ones it produces when suspended in liquid. The former contain more highly polar fatty acids than do the latter.

Knowing that sulfonolipids are essential for at least some forms of gliding motility does not tell us much about the mechanism of the process, but some hints come from the pattern of movement of small particles that become attached to the surface of gliding bacteria. Attached particles move to and fro on the surface of cytophagas at about the same rate (ca. 120 mm/min) as the cell glides across a solid surface (Figure 11). Anoxia stops the movement of both the particles and the cell. These observations have lead to the model that sites on the cell to which particles attach move within the fluid outer membrane along tracts fixed to the rigid murein framework. The cell glides when one or more sites moving in the same direction attaches to the substratum.

Chemotaxis of gliding bacteria

Although the mechanism remains a matter of rank speculation, the pattern and ecological significance of gliding taxes is richly developed. One of the more interesting examples is the diurnal tactic responses of *Beggiatoa* in mixed microbial mats with cyanobacteria. *Beggiatoa* are sulfur-oxidizing bacteria that prosper at interfaces between aerobic and anaerobic environments, where two of its nutrients—O_2 and H_2S—are both available. (H_2S is spontaneously oxidized by O_2.) *Beggiatoa* becomes concentrated in this zone and, indeed, creates it as a consequence of its phobic tactic responses to O_2 and H_2S (Figure 12). (In this narrow zone, the concentrations of both nutrients are decreased by the metabolic activity of the *Beggiatoa.* Increased concentrations of O_2 above the zone and increased concentration of H_2S below it trap *Beggiatoa* there.) However, these two taxes, by themselves, would lead to disastrous consequences to *Beggiatoa* living in microbial mats that contain cyanobacteria. These O_2-producing (and gliding phototactic) photo-

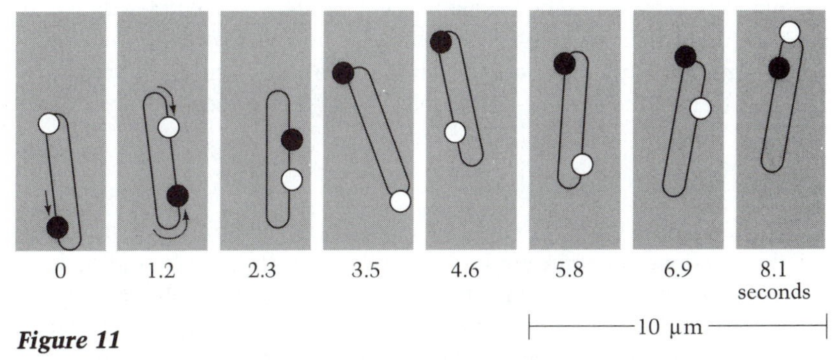

Figure 11

Movement of two polystyrene latex spheres, 0.26-μm-diameter. The gray box is a spatial reference point; the numbers are seconds after observations were begun. The black sphere was moving downward at time 0, while the cell was moving in the opposite direction. The sphere then looped around the lower pole and moved up the right side. At times (for example, between 3.5 and 4.6 sec), movement of one of the spheres would stop, or (for example, between 1.2 and 2.3 sec) the spheres would move in opposite directions. (After Lapidus and Berg, 1982.)

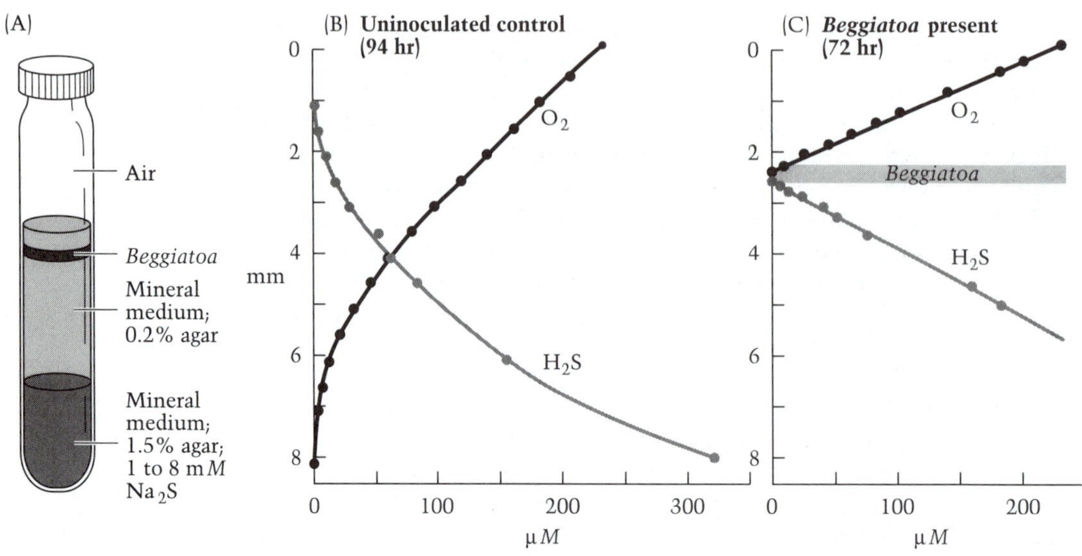

Figure 12

Tactic responses of *Beggiatoa*. (A) Sulfide gradient culture. (Courtesy of D. C. Nelson.) (B, C) Oxygen and sulfide concentrations as a function of depth in artificial gradients prepared in tubes as shown in A. (B) Uninoculated control. (C) In the presence of *Beggiatoa*, which concentrates itself in the shaded area. (After Nelson et al. 1986.)

trophs arrange themselves at the surface of microbial mats, where they have access to light. At night when the cyanobacteria are not producing O_2, the *Beggiatoa* migrate to the surface of the mat, following the rising aerobic–anaerobic interface. If *Beggiatoa* responded phobically only to O_2 and H_2S, they would be trapped away from H_2S at the mat surface in the morning when oxygenic photosynthesis begins. However, *Beggiatoa* are also phobically tactic to light, so they are able to migrate back into the mat to find the aerobic–anaerobic, O_2–H_2S interface.

SUMMARY

1. All substances that enter or leave the bacterial cell must pass through the envelope. The murein layer, being an open molecular mesh, does not constitute a significant barrier to the passage of solutes, but the basic structures of the membrane layers do. Both the cytoplasmic membrane and the outer membrane (in Gram-negative bacteria) contain proteins that facilitate the passage of solutes, because only a few nutrients—including water, O_2, and NH_4^+—are able to pass directly through the lipid portion of these structures. Highly nonpolar solutes can pass through the lipid portion of the cytoplasmic membrane, but the lipopolysaccharide content of the outer membrane makes it relatively impermeable to this class of compounds.
2. Small-molecular-weight solutes diffuse through holes in the outer

membrane formed by trimeric proteins called porins. Some of these are specific for certain sugars, nucleosides, or ions.

3. Solutes pass through specific proteins in the cytoplasmic membrane by facilitated diffusion, active transport, or group translocation. Facilitated diffusion, rarely found in bacteria, but commonly encountered in eucaryotic microorganisms, is driven by a concentration gradient of the solute and is mediated by transmembrane protein carriers that catalyze the passage of specific solutes across the membrane.
4. Active transport mechanisms maintain high intracellular concentrations of certain solutes, allowing the bacterial cell to grow rapidly even in media containing low concentrations of essential nutrients. Active transport is mediated by relatively simple secondary active transport mechanisms (symport, antiport, or active uniport) or by more complex multiprotein systems, called shock-sensitive transport because they require the participation of periplasmic binding proteins that are lost following cold osmotic shock.
5. Group translocation accomplishes the purposes of active transport by derivatizing nutrients to impermeable forms as they enter the cell. Phosphotransferase systems (PTSs) are multicomponent systems that phosphorylate certain carbohydrates as they enter the cell. The PTSs are energy conserving, because they usually render unnecessary the energy-utilizing first step of metabolism that would otherwise occur inside the cell.
6. Motile bacteria are capable of a number of selectively advantageous taxes—chemotaxis, phototaxis, magnetotaxis, aerotaxis. They also can move toward regions of more favorable temperature and pH. They move by one of two quite different mechanisms—flagella-mediated motility or gliding motility. There are many variations on the theme of flagellar motility, including how the flagella are arranged and whether they form a composite propulsive structure or function independently. In all cases the flagella turn, thereby moving the cell. In one group of Gram-negative bacteria—the spirochetes—flagella are contained within the periplasm, forming a helical ridge on the cell surface. When these endoflagella turn, propulsive waves pass down the ridge. The mechanism(s) of gliding motility which occurs only when the cell is in contact with a surface is not well understood. In some cases, gliding is dependent on the presence of certain unusual sulfonolipids in the cell's outer membrane. Gliding might depend on the movement of attachment sites on the outer membrane.
7. Flagella-mediated chemotaxis, the best studied of the tactic responses, proceeds as a random walk. Bacteria swim smoothly for a few seconds, then stop, reorient, and set out in a new direction. Smooth swimming is extended when the cell proceeds in a favorable direction and is shortened in an unfavorable direction. Enteric bacteria swim smoothly when their several flagella turn counterclockwise; they reorient (tumble) when flagella turn in the opposite

direction. The flagellum of *Rhodobacter sphaeroides* turns only clockwise; this cell reorients when the flagellum stops turning. In the case of enteric bacteria, chemotaxis has been shown to be directed by temporal comparisons of concentrations of chemoattractants. Transmembrane proteins, called MCPs, detect these concentrations by becoming methylated when attractants bind to them. Adaptation, a property common to all sensory responses, occurs as the MCP is demethylated.

STUDY QUESTIONS

1. Compare the relative concentrations in the medium, the periplasm, and the cytosol of a nutrient entering a bacterial cell by secondary active transport. Why should such a relationship exist? What would these relative concentrations be if the nutrient were entering the cell by a PTS?
2. Compare the growth curves of a bacterium in a glucose minimal and a glycerol medium. Explain the physiological basis for the differences.
3. What are the physiological advantages of group translocation over active transport?
4. What physiological purposes does active transport serve that facilitated diffusion could not?
5. Why is adaptation an essential feature of sensory perception?
6. Why has the biased random walk rather than setting the best course and proceeding on it, evolved as the fundamental pattern of bacterial chemotaxis?

SUGGESTED READING

Burchard, R. P. 1981. Gliding motility of prokaryotes: Ultrastructure, physiology, and genetics. Annu. Rev. Microbiol. 33:497.

Koshland, D. E., Jr. 1981. Biochemistry of sensing and adaptation in a simple bacterial system. Annu. Rev. Biochem. 50:765.

Postma, P. W. and J. W. Lengeler. 1985. Phosphoenolpyruvate: Carbohydrate phosphotransferase system of bacteria. Microbiol. Rev. 49:232.

7

Growth of Cells and Populations

INTRODUCTION

Bacteria, with some notable exceptions, do not have obligatory life cycles. As long as the environment is propitious, most bacteria, just like many other unicellular life forms, reproduce continually as vegetative cells. Bacteria undergo developmental changes such as spore formation only when the environment changes and not, as in higher plants or animals, as the result of an inexorable program of biological differentiation. Thus, when bacteria are not actually growing, they are poised to do so.

The hallmark of bacterial growth is the speed with which it can take place. The entire process of synthesizing a cell of *E. coli* from glucose and minerals, as occurs during growth in minimal medium, takes only about 40 minutes at 37°C and can be accomplished in half that time in a rich medium. Although some bacteria grow quite slowly, requiring hours for a cell to synthesize a new one, rapid growth is characteristic of many species, particularly those that have to compete in the environment for scarce or intermittently available nutrients.

UNRESTRICTED GROWTH

In a nutritionally and physically appropriate medium, a bacterial culture will grow at a characteristic rate. We say that growth is UNRESTRICTED as long as the concentration of nutrients does not become limiting and that of toxic compounds does not reach an effective level. Growth becomes restricted by changes in the environment imposed either by the growth of the bacteria themselves or by external alterations in chemical or physical conditions. Among the more common perturbations are exhaustion of nutrients, lessened availability of oxygen, and accumulation of acids and other products of metabolism.

During unrestricted growth, cells in culture increase at a rate proportional to the number of cells present at a particular time: the more cells there are, the more cells are made. Thus, unrestricted growth mimics an autocatalytic first-order chemical reaction, a response that can be anticipated from the fact that bacteria are self-duplicating units or factories to make more bacteria. In mathematical terms,

$$dN/dt = kN \tag{1}$$

in which N is the concentration of cells (number of cells per unit volume), t the time, and k a constant of proportionality called the SPECIFIC GROWTH RATE CONSTANT. Equation (1) states that the change in the number of cells over time depends on how many cells are present and on an intrinsic property of the organisms—how fast they can grow and multiply in that medium. The dimensions of the specific growth rate k are reciprocal time, usually expressed as reciprocal hours, or hr^{-1}. Integration of Equation (1) between the limits of 0 and t and N_1 and N_2 gives

$$\ln(N_2/N_1) = k(t_2 - t_1) \tag{2}$$

and, converting to logarithms to the base 10,

$$\log_{10} N_2 - \log_{10} N_1 = k(t_2 - t_1)/2.303 \tag{3}$$

Equations (2) and (3) describe a straight line ($y = ax + b$). Cultures in a state of growth that obey these equations are said to be GROWING EXPONENTIALLY.

In practice, if one determines the number of bacterial cells present at various times and plots the log of that number, one gets a straight line. From the slope of the line, we can calculate the specific growth rate k of the culture, which is the most encompassing parameter of how fast a particular bacterium grows in a particular environment. For example, if a culture contains 10^3 cells at t_1 and 10^8 cells 6 hours later, the specific growth rate k is $(8 - 3)\ 2.303/6$, or $1.92\ hr^{-1}$.

The rate of growth of a bacterial culture is often described by the time required for the number of cells to increase by a factor of 2, or the DOUBLING TIME or GENERATION TIME, g. The relationship between g and k can be established by returning to Equation (2), since when $N_2 = 2N_1$, $t_2 - t_1$ becomes equal to g. Substituting values for t and N into Equation (2), we obtain

$$g = \ln 2/k = 0.693/k \tag{4}$$

In the example used above, $g = 0.693/1.92 = 0.36$ hour; in other words, the culture doubles every 21.7 minutes. For maximal growth rates of other bacteria, see Table 1.

Growth can also be considered in terms of cell multiplication (remember the ditty "when cells divide, they multiply"?). At each division, one bacterium becomes two, these become four, and the series 2, 4, 8, 16, . . . ,2^n describes the increase in cell numbers. Thus

$$N_2/N_1 = 2^n \tag{5}$$

or

$$\ln N_2/N_1 = n \ln 2 \qquad (6)$$

where n is the number of generations required to produce $N_2 - N_1$ cells. From Equation (2)

$$n \ln 2 = k(t_2 - t_1) \qquad (7)$$

and

$$k = n \ln 2/(t_2 - t_1) \qquad (8)$$

which, in words, says that the specific growth rate is a function of the number of times the average cells will divide in a given period of time.

Balanced growth

When unrestricted growth has occurred for some time, all cell constituents begin to increase by the same proportion over the same interval. In other words, the time it takes to double the number of cells will be the same as the time it takes for these cells to double their content of DNA, ribosomes, individual enzyme molecules, and so on. This special condition has been called BALANCED GROWTH and, although it can be approximated for a considerable time in the laboratory, it does not usually persist for long in a natural environment. Under natural conditions, growth is usually unbalanced for most bacteria.

Working with cultures in balanced growth has three practical advantages. In the first place, samples taken over different periods of time vary only by the extent to which the culture has grown between these times, but are otherwise identical. Thus, as long as the culture is in balanced growth, it does not matter when it is sampled (as long as the time of

Table 1. Maximum recorded growth rates for certain bacteria in complex media

Organism	Temperature[a] (°C)	Doubling time (hours)
Vibrio natriegens	37	0.16
Bacillus stearothermophilus	60	0.14
Escherichia coli	40	0.38
Bacillus subtilis	40	0.43
Pseudomonas putida	30	0.75[b]
Vibrio marinus	15	1.35
Rhodobacter sphaeroides	30	2.2
Mycobacterium tuberculosis	37	≈6
Nitrobacter agilis	27	≈20[b]

Source: From Stanier et al. (1986).
[a]Cultures were grown at or near their respective optimal temperature.
[b]Growth in synthetic medium.

sampling is recorded and related to some growth measurement). In the second place, the RELATIVE RATE OF SYNTHESIS of any and all cellular components of the culture becomes known just by measuring the growth rate. In other words, if a culture doubles every 30 minutes and is in balanced growth, its DNA, ribosomes, and so on double every 30 minutes. Note that to convert this relative value to an absolute one (e.g., femtograms/bacterium/minute), a single measurement of cellular content of a given constituent suffices (a "single point on the curve"). As an example, if you determined once that the DNA content of an *E. coli* cell is 26 femtograms in a balanced growth culture that is doubling every 40 minutes (or if you looked up this value in Table 1), you would know that the average rate of synthesis of DNA is 0.65 femtograms/bacterium/minute. The third advantage of working with cultures in balanced growth is that this is the most reproducible physiological state of a bacterial culture. Thus, experiments done with cultures in balanced growth in different laboratories can usually be compared directly.

Balanced growth also requires that the mean cell size remain constant, a condition that might at first glance appear paradoxical because, as they grow, individual cells increase in size and eventually divide. Indeed, balanced growth refers to the *average behavior of cells in a population,* not to that of individual cells. Bacteria do not usually grow and divide synchronously, thus the population represents the average behavior of all individual cells.

Measurement of growth

In the laboratory, the definition of bacterial growth is operational and depends on the property that is measured. Thus, growth may be estimated as the increase in cell mass, the number of bacteria, or any cellular constituent. Also related to growth are the rates of disappearance of nutrients and accumulation of metabolites. As long as a culture is in balanced growth, all these measurements are equivalent and any of them can be used to estimate the rate of growth. Given a choice, convenience wins out, and the measurements most often used are those that can be carried out rapidly and cheaply. The most popular one is the measurement of the turbidity of a culture in liquid medium, which can be determined nearly instantaneously using a spectrophotometer. There are many instances, however, where such a measurement is not practical, for example, if the medium itself is cloudy.

Often one wishes to determine the number of living cells in the population rather than the total bacterial biomass. Counting living cells is based on the ability of a single viable bacterium to produce a colony on solid media. To carry out a VIABLE COUNT, one dilutes the sample, spreads a measured aliquot of the dilution on an agar plate, incubates the petri dish, and counts the number of individual colonies that develop. The technique has potential errors that can be difficult to avoid. For example, if the bacteria are clumped together, the number of colonies reflects the number of clumps and not individual cells. Also, the true time of sampling may be difficult to determine because bacteria may continue to divide during the dilution process.

In addition to these operational types of errors, there is an actual SAMPLING ERROR; that is, the number of bacteria in parallel samples differs by chance because the organisms are randomly distributed in the suspension. The sampling error is relatively small when the mean number of bacteria per sample is large. However, the countable number of colonies per petri dish is usually in the low hundreds, beyond which colonies run together and are hard to count. In this range, the sampling error is appreciable. An estimate of the size of the error can be obtained from the fact that the standard deviation is approximately the square root of the mean number of colonies per petri dish. Thus, if individual agar plates were spread with identical samples of the same suspension and the mean number of colonies resulting is 100 per plate, the standard deviation would be ±10, or 10.0% of the mean. If the mean number of colonies per plate were 625, the standard deviation (±25) would be 4.0% of the mean. It may help to remember that, when dealing with a randomly distributed population, about two-thirds of the samples would give counts within plus or minus one standard deviation. In skilled hands, the errors in making dilutions and plating can be smaller than the sampling error.

Most growing bacterial cultures contain some dead cells. With laboratory strains of *E. coli* growing under standard conditions, the proportion of dead cells is usually small and the viable count is a direct index of growth. A TOTAL COUNT, which includes dead as well as live bacteria, can be carried out in one of two ways. The number of cells in a sample can be determined by counting them under the microscope in a counting chamber. This is a tedious procedure that can be circumvented with an electronic particle counter. Both methods are used for counting other particles, for example, white blood cells or animal cells in culture.

A COUNTING CHAMBER is a glass slide with a central depression of known depth (for bacteria, usually 0.02 mm), the bottom of which is ruled into squares of known area. The depression is filled with a bacterial suspension, covered with a rigid cover slip, and allowed to stand until the cells have settled to the bottom. The number of cells per unit area (here proportional to volume) is counted under a microscope, and the total number of cells in the suspension is calculated. An analogous device used for counting blood and other animal cells is called a HEMOCYTOMETER.

ELECTRONIC COUNTING is based on the principle that the electrical conductivity of a bacterial cell is less than that of a saline solution. An electronic particle counter consists of two chambers connected through a small hole (for counting bacteria the hole is usually 30 μm in diameter). Each chamber is provided with an electrode, so that the electrical conductivity across the hole can be monitored. Then bacteria are added to one of the chambers and pumped to the other. Each time a bacterium passes through the hole, the conductivity of the circuit decreases momentarily, causing a pulse of voltage; the number of pulses can be tallied electronically. Thus, by knowing the volume of the suspension pumped through the hole, one has a rapid measurement of the concentration of bacteria in the suspension. The size of the voltage pulse is proportional to the size of the bacterium traversing the hole (actually, to the volume of solvent displaced). With proper electronic circuitry, a measure of the size distribution as well as of the number of bacteria can be obtained. The device counts hundreds of thousands of bacteria per sample rapidly and accurately, but it is expensive and often available only in specialized laboratories.

The most direct way of measuring the DRY WEIGHT of cells of a culture is to harvest them by filtration or centrifugation, wash them with distilled water to remove adhering components of the medium, dry them at 105°C, and weigh them. Such measurements are time consuming and require the sacrifice of considerable culture volume. Dry weight determinations are not suitable for routine monitoring of a growing culture.

The dry weight of cells of a culture can be estimated TURBIDIMETRICALLY, a technique based on the light-scattering properties of bacterial cells. This technique is so commonly used that it deserves some discussion. If a parallel beam of light is passed through a suspension of bacteria, it is diminished in intensity by the extent to which it is absorbed by the bacteria and scattered by them. Most bacteria are virtually colorless, thus they absorb a negligible amount of light. Therefore, changes in light transmission through a bacterial suspension are largely due to scattering.

The amount of light scattered is proportional to the ratio of particle size to wavelength of incident light. Therefore, the shorter the wavelength, the more sensitive the measurement. The ability to make sensitive measurements is an issue in physiological experiments, which often are carried out with relatively dilute cultures. Two sorts of measurements can be made. We can measure the scattered or the unscattered light. If the scattered light is measured, the photodetection device is located at an angle other than 180° (usually 90°) to the incident light. The device used for such measurement is called a NEPHELOMETER; it is highly sensitive but specialized and not often available. More commonly, unscattered light is measured using a more readily available COLORIMETER or SPECTROPHOTOMETER (whose photodetection device is located 180° to the incident beam).

A colorimeter or a spectrophotometer is an instrument designed to measure ABSORBANCE (A), that is, the logarithm of the ratio of the incident (I_0) to the transmitted (I) light through nonparticulate solutions. Absorbance is related to the concentration of solute by the Lambert-Beer law:

$$\log(I_0/I) = \epsilon l c = A \tag{9}$$

where ϵ is the extinction coefficient, l the length of the light path in the liquid, and c the concentration of the solute. Thus, when one knows the extinction coefficient and the light path and measures A in the colorimeter, one can determine the concentration (c) of the solute. However, bacteria are in suspension, not in solution, and a few other considerations apply. The parameter A—or the OPTICAL DENSITY, as it is usually called—is here largely a measure of light scattering. Therefore, the value of A depends on the precise geometry of the light path and will vary with the design or alignment of the instrument. A particular instrument can be calibrated for the bacterium being studied by determining the dry weight of a suspension and measuring its optical density. The result of such a calibration is shown in Figure 1. The optical density is proportional to the dry weight of the culture only at low densities. Deviations at high densities are a consequence of multiple scattering of a single ray. Once a calibration curve has been made, a rapid and reliable determination of dry weight of the culture can be made simply by measuring the optical density and reading the dry weight from the curve. If a culture is too dense, it must be accurately diluted before measuring its optical density.

GROWTH OF CULTURES

If an ordinary bacterium, say, a staphylococcus, a pseudomonad, or a coliform, is inoculated into a suitable medium in the evening, the culture will show considerable growth by the next morning. If the culture medium is nutrient broth, the culture aerated, and the temperature of incubation 37°C, the density of bacteria will become about 1 mg/mL bacterial biomass. By this time, the culture will have stopped growing and is said to be in the STATIONARY PHASE OF GROWTH. Had the growth of the

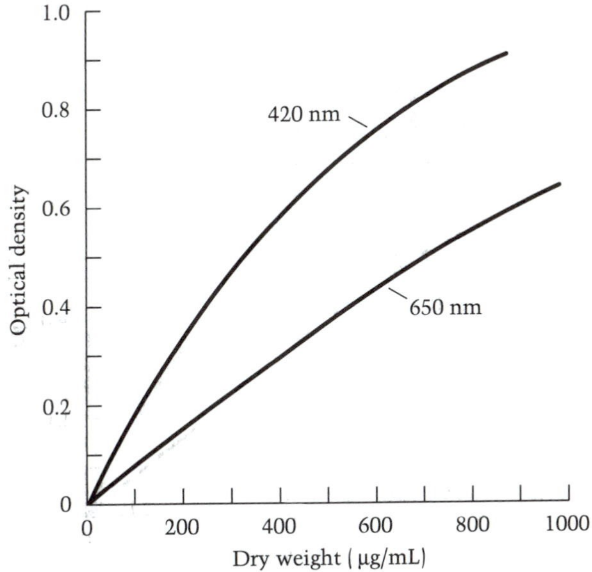

Figure 1

The relationship between the optical density and the biomass (dry weight) of a culture. Note that the relationship is linear at low bacterial densities. Also note that measurements are more sensitive with light of shorter wavelength.

culture been measured, the plot shown in Figure 2 would have been obtained. What made the bacteria in the culture stop growing? The reason that cultures cannot go on growing forever is that, as they grow, cells use nutrients from the medium and secrete waste products into it. Eventually, some nutrient becomes exhausted or some waste product reaches toxic concentrations, so continued balanced growth is no longer possible. When the culture depletes a single limiting nutrient from a defined medium, the transition from the exponential to the stationary phase may be surprisingly abrupt (Figure 2). However, if a culture is growing in a complex medium, several nutrients may become limiting in sequence or the medium may gradually become acidic. In this case, the transition to the nongrowing condition is more gradual.

During the transition from the exponential to the stationary phase, growth becomes UNBALANCED because the syntheses of some cellular components slow down or stop before others. During the stationary phase, cells do not have a constant composition and therefore differ chemically with the time of sampling. For example, *E. coli* cells in the stationary phase continue actively synthesizing DNA after their protein synthesis has decelerated. Thus, their DNA:protein ratio will increase with time. Cells in the stationary phase are smaller than those in the exponential phase, because cell division continues after the synthesis of most macromolecules has slowed down (see Chapter 14).

If a culture in exponential growth is diluted into fresh medium of the same composition, it will continue to grow exponentially. On the other hand, if a culture in the stationary phase is transferred to fresh medium, it will not start growing right away. Growth is delayed for a period of time that depends on the organisms and the particular medium. This period is called the LAG PHASE. An extreme example is seen with spore-forming bacteria such as *Bacillus subtilis* (see Chapter 16). Such organisms sporulate during the stationary phase, and their growth in fresh medium can only resume after the lengthy process of spores germina-

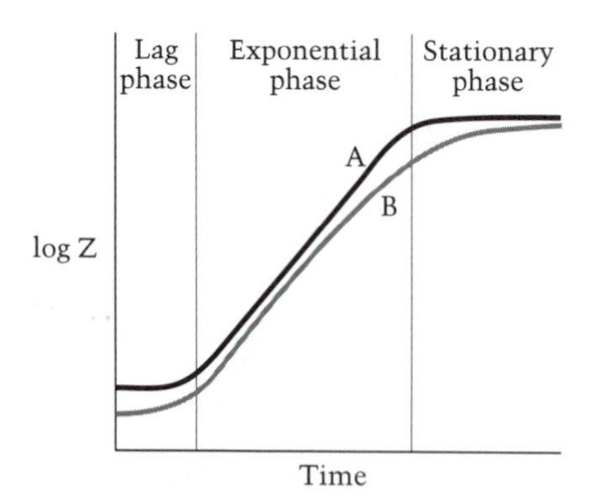

Figure 2

Growth and growth phases of a culture of *Salmonella typhimurium*: lag, exponential, stationary. Curve A: Culture grown in a glucose-minimal medium. Curve B: Culture grown in nutrient broth. *Z* is any readily measurable property of the culture, such as its turbidity.

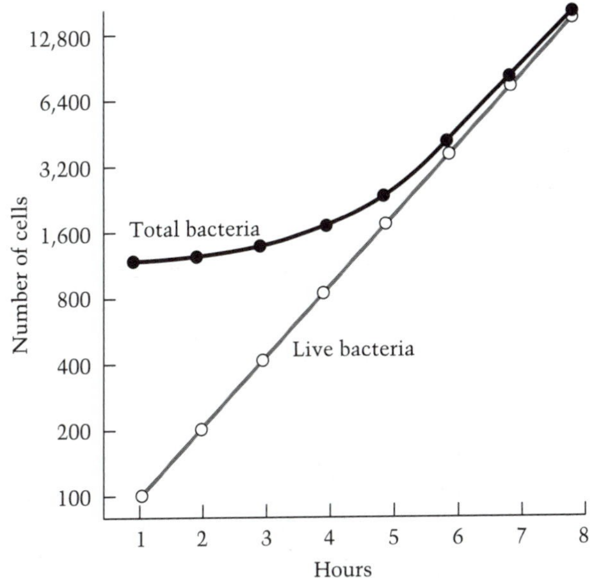

Figure 3

The measurement of live and total cells. If a population consisted initially of 100 living cells and 900 dead cells and growth is measured by estimating the total number of cells, an apparent lag will be seen. If living cells only are measured, there is no lag. This may happen when a 24-hour culture of *E. coli* has been transferred to fresh medium. (After Mandelstam and McQuillen, 1973.)

tion. The rearrangements that take place in *E. coli* during the stationary phase are less obvious, but the need for physiological readjustment is manifested in this species also (Chapter 13).

It follows that the length of the lag phase may be extremely variable. For example, the longer a culture remains in the stationary phase, the more profound the rearrangements necessary to begin growing again, and the longer the lag phase. In some instances, the lag phase is more apparent than real. Frequently, bacteria in the stationary phase die at rates characteristic of the species and the medium. Note that if the new culture is inoculated with a mixture of living and dead cells, a misleading result is obtained from turbidimetric measurements of cell mass because dead cells also contribute to turbidity (Figure 3). In this case, a viable count would reveal that the lag in growth of the living cells is very short.

GROWTH YIELD

Assume that the culture depicted in Figure 2 stopped growing because a needed component of the medium was depleted. If that component had been present in half the concentration, the final yield of bacterial biomass would have been half. This linear relationship holds for nearly all nutrients over wide ranges of concentration. Thus, a medium that contains all nutrients in excess except for one supports a final crop of bacteria that is directly proportional to the concentration of the limiting nutrient. This relationship can be stated as

$$B_f - B_0 = Y(c_0 - c) \qquad (10)$$

where B_f is the final concentration of bacteria, B_0 the initial concentration of bacteria, c_0 the concentration of the limiting nutrient initially present in the medium, c the concentration of the limiting nutrient when growth stops, and Y the YIELD COEFFICIENT. The parameter Y is unitless, being the ratio of the dry weight of cells produced per unit weight of the limiting nutrient. For aerobes, the value of Y is usually about 0.5 for the carbon source—say, glucose; but Y may be 100 times greater for a required amino acid or vitamin.

The linear relationship between the concentration of a limiting nutrient and the growth yield can be used to determine the concentration of a particular nutrient. For example, an organism that requires methionine for growth can be used to assay the concentration of methionine in a sample. A medium that supplies all nutrients except methionine is prepared. The yield coefficient for methionine is determined by adding it in various known amounts to the medium. Knowing the yield coefficient and the amount of growth with a certain amount of the sample, we can calculate the amount of methionine present in the sample. This procedure is known as the BIOASSAY and was used extensively in the 1940s and 1950s to determine concentrations of amino acids, vitamins, and other nutrients. Bioassays were used to estimate the potency of vitamin preparations and the nutritional value of foodstuffs. Today, because better analytical methods are available, this technique is no longer extensively used, but it remains a valuable research tool in the study of biosynthetic intermediates.

The yield coefficient has been used to make intelligent guesses about the energy metabolism of bacteria. Consider this syllogism: (a) The ATP requirement for polymerization of macromolecules is remarkably constant among bacteria (Chapter 3). (b) The amount of macromolecular material is proportional to the growth yield. (c) Therefore, the growth yield is an index of ATP generation. Because ATP generation varies among different fermentations, growth yield has been used to deduce the probable pathway of fermentation used by a bacterial culture. The results are highly variable, however, and this approach is now mainly of historical interest.

In practice, measurements of Y_{ATP} have been restricted to cultures that generate ATP by fermentation, because only for these are the conditions for which the ATP yield per mole of substrate known with precision. Much less is known about aerobic growth, although an interesting relationship was described by Andersen and von Meyenburg (1980). They found that the SPECIFIC OXYGEN CONSUMPTION—Q_{O_2} (millimoles oxygen consumed per hour per milligram dry weight)—in cultures of *E. coli* does not vary significantly with the growth rate (Figure 4). The values for

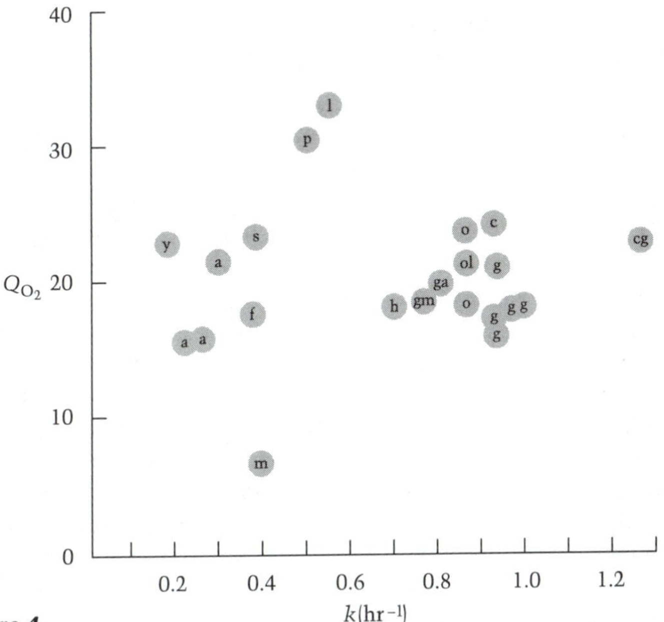

Figure 4

The effect of the growth rate (k) on the specific oxygen consumption (Q_{O2}) of batch cultures of *E. coli*. Cultures were grown in a mineral salts medium with various carbon sources. Letters indicate the carbon sources used: a, acetate; c, casein hydrolysate; f, fumarate; g, glucose; h, galactose; l, lactate; m, mannose; o, glycerol; p, pyruvate; s, succinate; y, glycolate. The units of Q_{O2} are mmol/hr/mg dry weight of cells. (After Andersen and von Meyenburg, 1980.)

growth with mannose as a carbon source are low and those for pyruvate and lactate are high. But these are exceptions (for which plausible rationalizations were presented) that do not affect the following conclusion: If we assume that ATP generated per mole of oxygen consumed does not vary with the growth rate (and there is no experimental or theoretical basis for questioning it), then *the total energy available for doubling the cell mass decreases in proportion to the growth rate.*

How can we interpret the finding that the specific oxygen consumption does not vary with the growth rate? If the energy required for polymerization is constant (Chapter 5), then cells growing at low rates (which make smaller amounts of macromolecules per unit time) have energy available for other purposes. The most obvious demand for this extra energy is to synthesize the monomers required for polymerization. To grow fast, bacteria are provided with preformed building blocks; thus, they are spared the need for biosynthesis of such monomers (Chapter 11). At low rates, the cells must make their own building blocks, a demand requiring extra energy. Under the conditions of the Andersen-von Meyenburg experiments, the growth rate appears to be set by the energy requirement for the synthesis of monomers.

EFFECT OF CONCENTRATION OF NUTRIENTS ON THE GROWTH RATE

An examination of Figure 2 shows that the culture grew at a constant rate over the greater part of the growth period in spite of the fact that substrate was being increasingly consumed. The growth rate remained constant and decreased to zero rather abruptly when substrate was depleted. Therefore, growth rate is a function of substrate concentration only at very low substrate concentrations. This relationship is not easy to measure: If the concentration of substrate is low enough to affect the growth rate, it will be rapidly utilized even by moderately sparse bacterial cultures. Monod (1942) determined this relationship using techniques that allow growth measurements at low bacterial densities. He used a nephelometer (see earlier) to measure the turbidity in a sensitive manner. His results have been generally confirmed more recently using electronic counting and more sensitive turbidimetry. From his measurements, Monod proposed that the relationship between growth rate and substrate concentration followed first-order kinetics, where some component of the system becomes saturated (limiting) at low substrate concentrations. It has the form

$$k = k_{max}c \mathbin{/} (K_s + c) \qquad (11)$$

where k is the specific growth rate at a concentration c of limiting nutrient; K_s is a constant that is numerically equal to the substrate concentration at which $k = k_{max}/2$. If growth is measured during unrestricted growth, the estimate of growth rate obtained gives the value of k_{max}.

The relationship found by Monod between growth rate and substrate concentration has the same form as that between the velocity of an enzyme-catalyzed reaction and the concentration of the substrate for that reaction, that is, the Michaelis-Menten equation:

$$v = v_{max}S \mathbin{/} (K_m + S) \qquad (12)$$

where v is the velocity of the reaction, S the substrate concentration, and K_m the Michaelis-Menten constant.

The full applicability of Equation (11) has been questioned by some investigators. For example, as shown in Figure 5, Shehata and Marr found deviations from the prediction of the equation at high substrate concentrations. This result suggests that the limiting nutrient—in this case, glucose—may not be metabolized by the same pathway at all rates of growth. Thus, cultures growing at different rates even in the same medium may differ in basic aspects of their physiology. This point is discussed in detail in Chapter 15.

CONTINUOUS CULTURE

In the laboratory, a culture can be maintained in balanced growth by diluting it at intervals with fresh medium. If this is carried out fre-

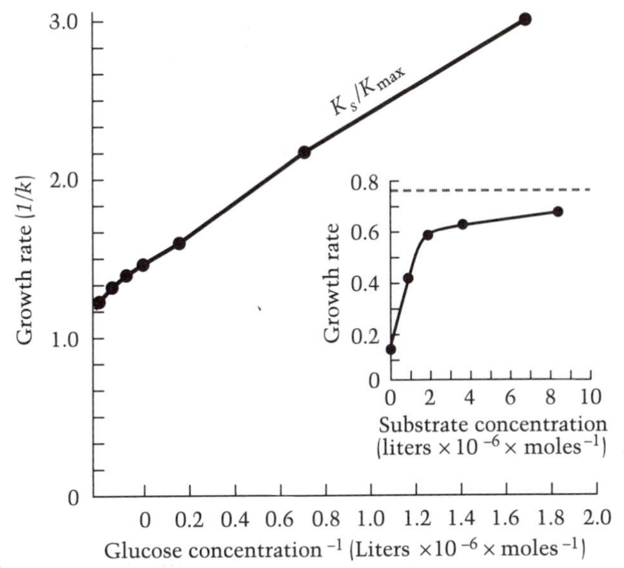

Figure 5

Effect of limiting concentrations of glucose on the growth of *E. coli* in a chemostat. A plot of the inverse of the growth rate (1/*k*) against the inverse of the glucose concentration yields a straight line whose slope is K_s/K_{max} (see Equation 11). The inset shows the direct plot of growth rate against the substrate concentration. (After Shehata and Marr, 1970.)

quently (e.g., once per doubling time), then the culture will fluctuate in density over a factor of 2 and will remain in the exponential phase of growth (Figure 6). As long as the operator is willing to carry out such dilutions, the cells will grow in an unrestricted manner. The tedium involved in making periodic dilutions over a long period of time can be avoided by automation. Thus, a culture vessel can be outfitted with a device to monitor the density of the bacterial culture, for example, by constantly measuring its optical density. If the optical density exceeds the setting, the device automatically allows the flow of fresh medium from a reservoir. If the vessel is equipped with an overflow syphon, the volume of the culture will remain constant, as will the bacterial density. Such an apparatus is called the TURBIDOSTAT.

What if one wishes to grow the same bacteria at *different* growth rates? One can still use a turbidostat, as long as nutrients are added at a concentration below that which supports maximal growth (Figure 5, inset). However, this relationship limits the usefulness of the apparatus, because to grow the cell at submaximal rates, the bacterial density will be low. For many experiments, the mass of bacteria will be too low for certain analysis. To grow bacteria at submaximal growth rates *and* at high concentrations, a different apparatus is used—the CHEMOSTAT. Both turbidostats and chemostats permit CONTINUOUS CULTURE of bacteria ("bugs on tap") and are used for a variety of genetic and physiological experiments. They also have applications in industrial fermentations.

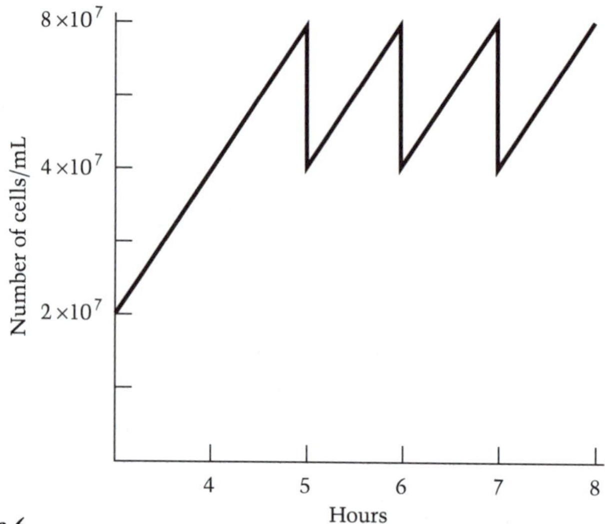

Figure 6

The effect of periodic dilution on a culture in the exponential phase of growth. A culture growing with a doubling time of 1 hour reached a titer of 5×10^7 cell/mL after 5 hours. Thereafter, it was diluted twofold with fresh prewarmed medium once after every doubling. The resulting saw-tooth pattern shows that exponential growth can be maintained for a long time by this simple expedient.

A sketch of a chemostat is shown in Figure 7. In principle, it is a remarkably simple apparatus, consisting of a culture vessel that can be kept at a desired temperature and, for the growth of aerobes, that can be aerated. Fresh medium is added at a rate set by a system, such as a metering pump, that regulates the flow rate. The volume is kept constant by removing medium through the overflow device at the same rate as fresh medium is added. Vigorous mixing is employed to ensure rapid equilibration of the fresh medium with the contents of the culture vessel. Details in the design of chemostats vary considerably with regard to volume of the culture chamber, the manner in which unwanted organisms are kept out, and the device that regulates the addition of fresh medium. In reality, commercially available chemostats can be quite complex. The important properties of a chemostat are (1) the rate of addition of fresh medium (per volume), which determines the growth rate in the culture vessel; (2) the density of bacteria in the culture vessel which is constant and is determined by the concentration of a limiting nutrient. For a chemostat to function properly, the bacterial density should not exceed one that allows balanced growth in batch cultures. This condition is achieved by making an essential nutrient limiting. In practice, this condition is met by using a minimal medium and limiting the concentration of glucose, ammonia, phosphate, or a needed amino acid.

The principal difference between a chemostat and a turbidostat is

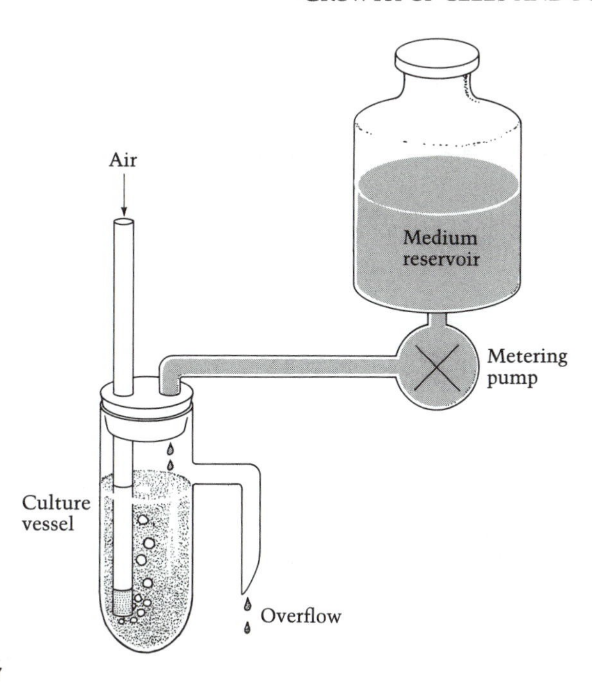

Figure 7

A chemostat. Fresh medium from a reservoir is fed at a constant rate to the culture vessel through a metering pump. A constant volume of culture is maintained by means of an overflow, which removes culture fluid at the same rate as fresh medium is added. Not shown is a device for rapidly mixing the culture vessel contents. The size of the culture vessel may vary from a few milliliters to thousands of gallons.

that in a turbidostat, the growth rate is determined *internally*, by properties of the bacteria themselves. At any given temperature, the growth rate can only be changed by using different media. In the chemostat, a single medium can be used and the growth rate is adjusted *externally* by the rate of flow.

By studying Figure 8, one can see how the chemostat works. The chemostat operates in a region of the growth curve where (like that shown in the inset of Figure 5) the growth rate (k) varies with the concentration of limiting substrate (c). The reason is that only a small amount of nutrient is added from the reservoir per unit time. Because this small amount is rapidly utilized, the effective concentration of the limiting nutrient in the culture vessel is very small. How small depends on K_S (Equation 11), the affinity of the bacteria for the limiting nutrient. In the course of running a chemostat for long periods of time, mutants with greater affinity will be selected (see Figure 9). As expected, the chemostat has proved useful in studies of bacterial mutagenesis and evolution.

The chemostat is a self-correcting system. We can illustrate that fact by the following scenario. Let us assume that it is operating at a steady state. When we increase the rate of addition of fresh medium, the rate of

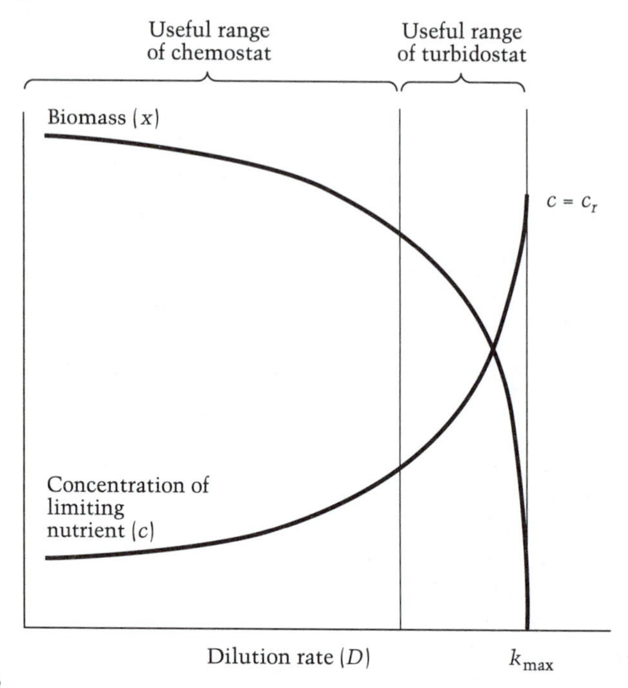

Figure 8

The relationship between biomass (x), the concentration of limiting nutrient (c), and the dilution rate (D) in continuous culture. The useful ranges of the turbidostat and the chemostat are indicated.

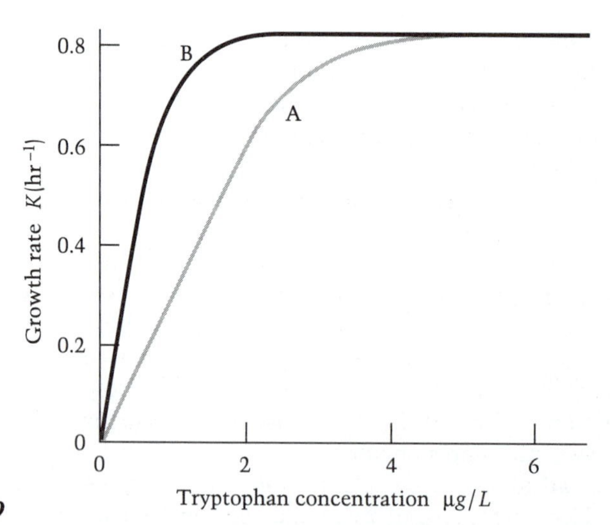

Figure 9

The relationship between growth rate and concentration of a limiting substrate (tryptophan) for the original tryptophan-requiring mutant of *E. coli* (curve A) and a strain selected for more efficient uptake by growth in a chemostat (curve B). (Modified from Novick and Szilard, 1950.)

loss of cells increases transiently, exceeding their rate of formation in the vessel. As a consequence, the density of cells in the vessel decreases. The less dense culture will utilize the limiting nutrient at a slower rate, thus allowing the concentration of nutrient in the vessel to increase. Consequently, the growth rate increases until it is sufficient to match the rate of loss of cells through the overflow. When the rate of addition of fresh medium decreases, the opposite series of events ensues. Thus, growth rate of the culture adjusts to match the rate of loss from the vessel and the culture density remains constant.

Consider now the mathematical basis for the functioning of a chemostat. Let V be the volume of the culture in the vessel and f the flow rate, measured in units of culture volume per hour; then the MEAN RESIDENT TIME (MRT), or the average time that a cell spends in the culture vessel, is

$$\mathrm{MRT} = V/f \qquad (13)$$

The reciprocal of MRT is the DILUTION RATE (D). If a chemostat is operating at a steady state, the number (N) or mass (x) of cells removed through the overflow is matched exactly by the number or mass of cells produced in the culture, that is

rate of production of cells through growth =
rate of loss of cells through overflow

The rate of production of cells through growth of a culture can be described as

$$dx/dt = kx - xf/V \qquad (14)$$

and

$$dx/dt = kx - Dx \qquad (15)$$

In the chemostat the number of bacteria does not change, so dx/dt is zero, and

$$kx = Dx \qquad (16)$$

or

$$k = D \qquad (17)$$

Equation (17) is a restatement of the fact that the growth rate (k) of a culture in a chemostat is determined by the flow rate and the volume of the culture vessel; it is exactly equal to the dilution rate. But how a chemostat works requires an understanding of the fact that at low nutrient concentration the growth rate is related to the nutrient concentration. As shown in Equation (11), $k = k_{\mathrm{max}}c/(K_s + c)$. Solving for c yields

$$c = K_sD/(k_{\mathrm{max}} - D) \qquad (18)$$

thus giving us a fundamental relationship between substrate concentration (c) in the growth vessel and the dilution rate (D).

Let us now consider the relationship between cell mass and nutrient utilization. In a steady state the substrate concentration in the culture vessel remains constant; therefore,

substrate added from the reservoir =
substrate used for growth + substrate lost through overflow

or

$$Dc_r = dc/dt + Dc \tag{19}$$

where c_r is the nutrient concentration in the reservoir and dc/dt the rate of utilization of the nutrient. Substituting $(dx/dt)(dc/dx)$ for dc/dt—recall that $dx/dt = kx$ (Equation 1) and $dc/dx = 1/Y$ (from Equation 10)—we have

$$Dc_r = kx/Y + Dc \tag{20}$$

Because in the steady state $D = k$,

$$x = Y(c_r - c) \tag{21}$$

which is the second fundamental equation of the chemostat. From Equations (18) and (21), the relationships shown in Figure 8 can be plotted. Several important conclusions can be drawn from a consideration of this figure. First, over a wide range of dilution rates (D), the density of the culture (x) remains nearly constant, and this is the range where a chemostat is most conveniently operated. At high dilution rates, where x changes rapidly with D, the chemostat becomes unstable because small inadvertent changes in D can cause wash out. Obviously, no chemostat can be run when D exceeds k_{max}, the unrestricted growth rate in the medium used. If a culture is to be grown continually at its maximal rate, a turbidostat must be used.

Is there a lower limit to the dilution rate at which a chemostat can be operated? The answer depends on the nature of the limiting nutrient. If the limiting nutrient is the source of energy, growth ceases at low dilutions and the cell are washed out. On the other hand, if the limiting nutrient is an amino acid or some other specific precursor in macromolecular synthesis, chemostats can be operated at dilution rates leading to a mean residence time of several days or perhaps weeks. The reason for this disparity is discussed below. Note one technical detail: at very low dilution rates, nutrient from the reservoir dribbles into the culture vessel as a small volume per minute. Unless the capacity of the culture vessel is very large, continuous addition becomes a problem. Continuous addition from the reservoir is a necessary condition for a device to fit the definition of a chemostat. If the addition is intermittent, the cells will not be growing continuously but will alternate between feeding and starvation. The consequence may be subtle, but physiologically important. Thus, there can be practical limits as to how slowly one can run a chemostat *sensu stricto*.

MAINTENANCE ENERGY

Why does a chemostat not operate at very low dilution rates when the limiting nutrient is the source of energy for growth? The reason is that, under any growth condition, a certain amount of energy is used for essential processes other than those leading to increase in mass. Some of these processes take precedence, and growth can take place only if their demand is met first. This energy demand for essential (non–growth-related) processes is called MAINTENANCE ENERGY.

What are the processes that require maintenance energy? We count among them the maintenance of a potential across the cytoplasmic membrane and the transport of certain solutes; the constant hydrolysis and resynthesis of certain macromolecules—termed TURNOVER; and cell motility. In addition, for reasons that are not clear, bacteria synthesize certain polysaccharides at a nearly constant absolute rate (femtograms/bacterium/minute). When the carbon source becomes limiting, the proportion of energy diverted to this synthesis becomes so great that cells cease growing.

Maintenance energy can be described as follows:

$$dx/dt = Y \cdot dc/dt - ax \quad (22)$$

where a is the SPECIFIC MAINTENANCE RATE. Typical values for a of *E. coli* growing at 37°C are 0.02–0.03 hr^{-1}. This amount of energy is a small proportion of the available energy for cultures growing at rates greater than 1.0 hr^{-1}, but it becomes a major proportion in slow-growing cultures. In practice, another parameter—MAINTENANCE COEFFICIENT—is often used. It is related to the specific maintenance rate a by the equation

$$m = a/Y_G \quad (23)$$

where m is the maintenance coefficient and Y_G the yield coefficient for substrates used for growth. For further details, consult Pirt (1965).

BALANCED GROWTH IN PRACTICE

When we go from theory to practice, terms related to balanced growth, such as *unchanging environment* and *steady state,* must be used as approximations; in practice, they really mean "presumed unchanging environment" or "approximate steady state." Thus, the first of these terms can be defined only empirically—by finding out to what density a culture can grow before the medium is no longer "essentially" the same. This point is a subtle one. Some cell properties change early in the growth of a culture, long before the increase in mass slows detectably. This phenomenon was very striking in a study of lysogenization of *Salmonella typhimurium:* in broth cultures at densities lower than 10^8 cells/mL, a constant fraction of cells was lysogenized under standardized conditions. However, when the density exceeds 10^8 cells/mL, the lysogenization frequency dropped sharply, despite the fact that the rate of mass increase remained unchanged for at least one more doubling time.

As a consequence of often imperceptible changes, it is unsafe to assume that all properties of the cells in culture remain invariant unless the culture is at a sufficiently low density that growth is a long way from slowing down. Reference is often made in the literature to cultures harvested in "early" or "late" log phase, of even in "mid-log"; such arbitrary statements are disturbing because they give the impression that no great care was taken to establish balanced growth conditions. There is no assurance that the growth rate was not changing at the time of sampling. Unless growth is monitored throughout a physiological experiment, the results may not be reproducible, whether in your own or somebody else's laboratory. In fact, a physiological experiment done with a poorly characterized culture *is all but useless.*

A good rule is not to start experiments that relate to growth physiology unless both optical density and cell number have been shown to increase exponentially by a factor of 10 or more. If a large inoculum was used, this condition may require several dilutions of the culture. There are two reasons for advocating such stringent criteria. First, many readings are necessary to make sure that the definitive growth rate has been reached. Second, the final adjustment of parameters such as cell dimensions and the number of ribosomes requires several generations (Chapter 15). As it turns out, *E. coli,* related enterics, and pseudomonads can commonly be safely used at densities up to 10^8 cells per milliliter.

GROWTH-RELATED BIOCHEMICAL AND RADIOCHEMICAL MEASUREMENTS

How can we study the rate of synthesis of individual chemical constituents of a bacterial culture? To use purely chemical means is sometimes difficult because growing bacterial cultures are relatively dilute and large samples are required. For this reason, radiochemical assays are often used. They not only are much more sensitive and require smaller samples, but they also give unique information about "new" versus "old" molecules. The main issues to consider are the choice of label, its time of addition, and the time of sampling.

The choice of label

A radioactive precursor may be a GENERAL LABEL and be used for the biosynthesis of many different kinds of molecules. Examples are [^{14}C]glucose or [^{32}P]phosphate, each of which is incorporated into a variety of large or small molecules. Which product of biosynthesis will become labeled depends on the bacterium and the composition of the medium. When [^{14}C]glucose is added to heterotrophic bacteria growing in a minimal medium with glucose as the only carbon source, all carbon-containing compounds will be radioactively labeled. As discussed in Chapters 11 and 12, the endogenous synthesis of many precursors is often suppressed if such compounds are present in the medium in utilizable form. If such precursors are added in a non-radioactive form, their products of metabolism will not become labeled by the added [^{14}C]glucose. For example, if methionine is present in excess in the medium, cellular methionine will not, in general, acquire radioactivity from [^{14}C]glucose.

Other compounds can be used as SPECIFIC LABELS: amino acids, purines, or pyrimidines, and certain sugars added to the medium can be used for more specific biosyntheses than glucose. Thus, [^{3}H]thymidine labels DNA; [^{35}S]methionine, proteins; [^{14}C]diaminopimelic acid, murein; or [^{14}C]heptose, the lipopolysaccharide of Gram-negative bacteria. Other compounds are not so selective and serve as labels for more than one product. Thus, labeled adenine ends up in both RNA and DNA. Most useful specific labels, such as the ones mentioned, are compounds that are end products of simple metabolic pathways and are not used for multiple pathways.

In some cases, the rate of uptake from the medium is too slow to permit efficient labeling of cellular constituents. In such cases, it is convenient to use auxotrophic double mutants that are deficient in the biosynthesis of a given precursor and that also have efficient transport mechanisms. An example is the use of mutants requiring the murein component, diaminopimelic acid, which wild-type *E. coli* takes up poorly. Such mutants readily incorporate exogenous diaminopimelic acid, and their murein can be labeled accordingly. It must be remembered that in order to grow such mutants, diaminopimelic acid must be added to the culture. The addition of the compound in the medium dilutes the isotopic species and lowers the specific radioactivity. If the resulting specific radioactivity is too low to be effective, the culture can be washed free of diaminopimelic acid just before adding the labeled compound. Should this procedure not be convenient, the level of unlabeled diaminopimelic acid must be decreased to the minimum needed to ensure balanced growth.

These considerations also apply to heavy isotopes. The purpose of using precursors containing the heavy atoms ^{2}H, ^{13}C, or ^{15}N is to physically separate "new" from "old" molecules. As in the classical Meselson-Stahl experiment, "heavy" molecules can be physically separated from "light" by density gradient centrifugation. Heavy isotope-containing precursors are often used in conjunction with radioactive precursors.

Experiments concerning the synthesis of macromolecules are often carried out by combining antibiotics with radioactive labeling. A number of antibiotics are known to inhibit the synthesis of one class of macromolecules specifically without initially altering that of others (Table 2). For example, rifampin is used in studies of the different classes of RNA (see below).

Steady-state labeling

One cannot just add a labeled compound to a culture and expect a useful result. Some thought must go into realizing what happens to the radioactive compound. Thus, metabolic end products are not the only substances that become labeled by the addition of a radioactive precursor. Invariably, the added radioactive precursor must enter a pool of metabolites that themselves become labeled (Figure 10). Thus, thymidine must first become thymidine triphosphate and proline, prolyl-tRNA,

Table 2. Some commonly used antibiotics

Antibiotic	Macromolecular synthesis inhibited (step involved)	Mode of action
Penicillins, cephalosporins	Murein (elongation)	Covalently binds murein biosynthetic enzymes (penicillin-binding proteins)
Cerulenin	Fatty acids (elongation)	Inhibits condensing enzymes
Chloramphenicol	Protein (translation of ribosome)	Inhibits peptidyltransferase
Fusidic acid	Protein (ribosome movement)	Inhibits elongation factor G
Rifampin	RNA (initiation of synthesis)	Binds RNA polymerase
Streptolydigin	RNA (elongation)	Binds RNA polymerase
Nalidixic acid	DNA (supercoiling)	Binds DNA gyrase
Mitomycin C	DNA (elongation)	Cross-links DNA (induces breaks)

both of which will also become radioactive. For many macromolecular biosynthetic pathways, the metabolic pool (low-molecular-weight precursors) in bacteria is relatively small. Therefore, if a labeled precursor is added in excess and growth proceeds over a long period of time, the bulk of the radioactivity will be in the macromolecular product and only a small proportion in the metabolic pool.

With *E. coli,* it is commonly found that the content of an amino acid pool is equivalent to the consumption of that amino acid during 10–20 seconds of protein synthesis at 37°C. One minute after the addition of [^{35}S]methionine, the pool has been renewed 5–10 times and its specific activity will be very similar to that of the medium. After 10 or more minutes, the protein synthesized during the labeling period will have a methionine specific activity that is almost equal to that of the medium. For even longer periods, the pool effect becomes negligible, and the radioactivity in the acid-soluble material will be a good measure of the amount of protein synthesized after the addition of [^{35}S]methionine.

Radioactivity in the metabolic pool can be removed by suitable techniques. For example, treatment with 5% trichloroacetic acid precipitates proteins and nucleic acids but leaves amino acids and nucleoside phosphates in soluble form. Recovering the precipitate by filtration allows one to measure the radioactivity in the macromolecular fraction. Conversely, the radioactivity in the acid-soluble fraction allows one to estimate the size of the metabolic pool.

If a radiolabeled precursor has been present in the medium for several previous generations, the amount incorporated into cellular material

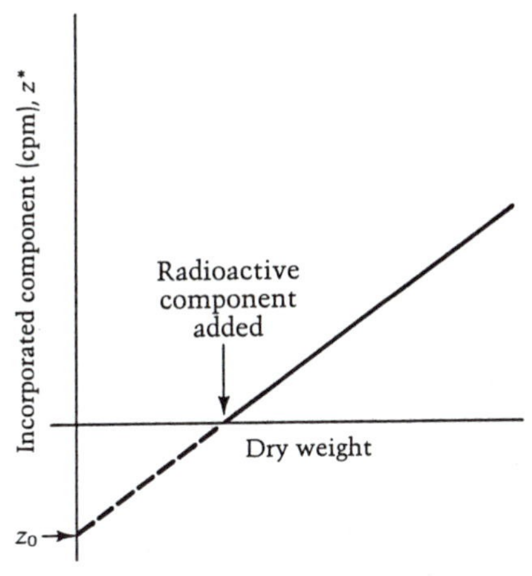

Figure 10

Differential plot of incorporation of a radioactive compound. By plotting counts per minute (cpm) incorporated into biomass (z^*) versus the mass of the culture (dry weight), we obtain a straight line. Extrapolation to 0 mass gives an estimate of z_0, the amount of cell component (measured as radioactivity) present when the radioactive component was added to the medium.

can be used to determine the growth rate by plotting the logarithm of the radioactivity in the cells versus time. This procedure is no different from other methods used to calculate the growth rate. In many experimental situations, the radioactive label is added at the beginning of the experiment. Now the incorporated radioactivity is an index of the amount of the labeled constituent synthesized after addition of the radiolabel, not of the total amount of that material. Under these conditions,

$$dz/dt = k(z^* + z_0) \tag{24}$$

where z is the total amount of a cellular component, z^* the label incorporated into that component, and z_0 the amount of the component when label was added. Hence, a plot of log z^* versus time yields not a straight line but a curve of decreasing slope (approaching the growth curve asymptotically). The value of z_0 can be estimated from a differential plot in which z^* is plotted against the mass of the culture (Figure 10). Extrapolation of this curve to 0 mass gives z_0 as the negative intercept. Plotting log $(z^* + z_0)$ versus time yields a straight line with slope equal to $k/2.303$. Using chemical analysis, the amount of component z can be related to counts of radioactivity (e.g., cpm per micromole).

Pulse labeling

When growth becomes unbalanced, the rates of labeling of individual compounds change with time (Figure 11) and do not give direct information about rates of biosynthesis. At any given time, the synthetic rate can be estimated from tangents to the curve, but this is not a very

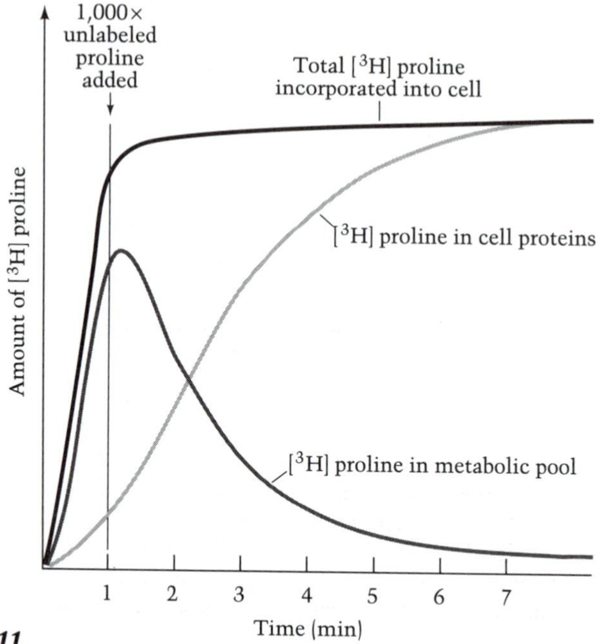

Figure 11

The incorporation of [^{3}H]proline during a pulse-chase experiment. The labeled precursor was added to the culture at 0 time. After 1 minute, a 1,000-fold excess of unlabeled proline was added, effectively decreasing the specific activity in the medium and "chasing" the label compound.

accurate measurement. The instantaneous rate of biosynthesis can be determined by PULSE LABELING, which consists in the addition of radioactive precursor for a short period of time. For example, [^{3}H]-thymidine can be added for a 60-second interval to samples of the culture and the amount of radioactivity in the DNA determined. The result can be expressed as label incorporated into DNA per minute. If the number of cells or the amount of cell mass are determined at the same time, the result can be expressed as *radioactivity incorporated into DNA/cell/minute,* which is a representation of the rate of DNA synthesis per cell.

Pulse labeling gives such direct information regarding rates of biosynthesis only if two conditions are met: (1) The metabolic pool is very small. If the pool constitutes an appreciable proportion of the label incorporated, its contribution must be determined and a correction must be made. (2) The final product is metabolically stable. In bacteria, most macromolecules are stable and, under growth conditions, are not degraded to any great extent. The most notable exception are molecules of messenger RNA, which are broken down as the same rate as they are made (see below for a discussion of the kinetics of labeling of mRNA).

A variation on the theme is the so-called PULSE-CHASE TECHNIQUE. Assume that you wish to measure the kinetics of flow of a precursor through a metabolic pool. If a pulse of short duration (say, 1 minute) is followed by the addition of a large amount of the same compound in the nonradioactive form, the radioactivity with be "chased" into the end products of biosynthesis. As shown in Figure 11, radioactivity in the pool material will first rise, then decrease, as the labeled substrate exits the pool to become the final product. Note that curve C need not be measured directly but can be obtained by subtraction of B from A.

When carrying out a kinetic experiment with a labeled precursor, it becomes important to correct for unequal recovery among samples. Imagine, for example, that you wish to determine the rate of synthesis of a protein by measuring the rate of incorporation of [^{3}H]methionine. However, the recovery of that protein may vary from sample to sample, in which case the amount of tritium present would give an imprecise estimate. A simple way out of this problem is to use two labels. For example, if a culture is labeled for long periods with [^{3}H]methionine and pulse-labeled with [^{35}S]methionine, the rate of incorporation is given by the ^{35}S:^{3}H ratio, regardless of the accuracy in recovery of the protein in each sample.

Labeling kinetics of messenger RNA

Can labeling techniques be used to determine the amount and rate of synthesis of unstable products? So far we have considered the labeling kinetics of stable macromolecules such as DNA or proteins. How do the synthesis and degradation of unstable products affect the kinetics of incorporation of a radioactive precursor? We will use as an example labeling with radioactive guanine of mRNA, the most abundant of the unstable macromolecules of the cell. As shown in Figure 12, if guanine is incorporated into both stable (rRNA, tRNA) and unstable products (mRNA), its rate of incorporation (a) will depend mainly on the rate of synthesis (b) of the stable RNAs. The reason is that the difference between the rates of synthesis (c) and degradation (d) of mRNA will be equal to the growth rate. Over a short period of time—say, under 1 minute—the difference between these rates will be small (the amount of growth occurring will be small) and the amount of degradation will nearly cancel out that of synthesis. Notice that this may introduce a hidden complication: If the synthesis of stable RNA were inhibited, the incorporation of precursors would greatly decrease, thereby giving the erroneous impression that *all* synthesis of RNA had stopped.

On further examination, it becomes apparent that the direct proportionality between guanine uptake and the synthesis of stable RNAs holds only if labeling is carried out over a long period. Initially, guanine will be incorporated into both components at their characteristic rates of synthesis (b and c). Only after sufficient time has passed for the label in mRNA to achieve a steady state does the contribution of label to the pool (P) from mRNA degradation equal the rate of its withdrawal for its synthesis. Therefore, it is possible to determine the rate of mRNA synthesis after pulse labeling with a suitable precursor, such as tritiated guanine. Such measurements depend on a number of parameters, especially the

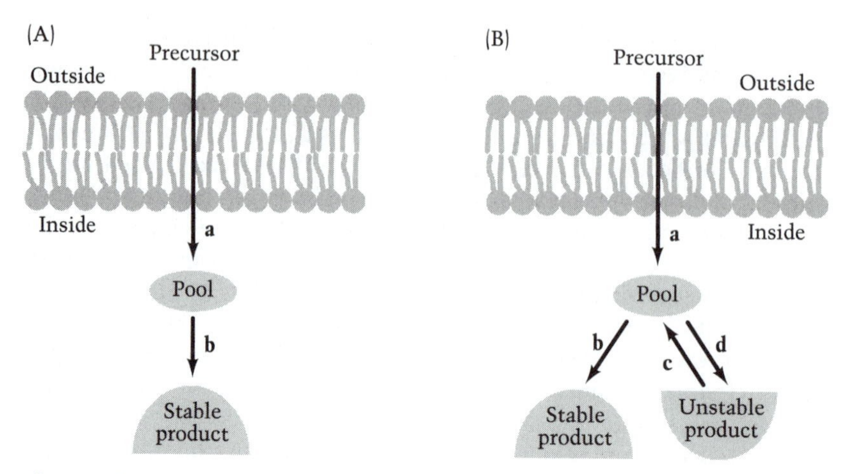

Figure 12

The incorporation of labeled precursors into macromolecules. (A) Incorporation into a stable product (e.g., proline into proteins). The rate of uptake into the cell (a) will be equal to that of synthesis (b). (B) Incorporation into both a stable product and an unstable product (e.g., uracil into RNA). The rate of uptake into the cell (a) will be equal to that of synthesis (b) of the stable species (e.g., rRNA) as long as the unstable species (e.g., mRNA) is made at the same rate (c) as it is degraded (d).

kinetics of labeling of pool material—in this case, the guanine nucleotide GTP. Notice that such measurements can only give average values, because they do not take into account the fact that individual mRNA molecules may vary in their rates of synthesis and degradation.

It should be apparent that direct labeling techniques are not an easy way to estimate the rate of synthesis and degradation of mRNA, or the proportion of total RNA that is mRNA. Fortunately, a so-called RUN-OUT EXPERIMENT allows a simpler estimation of these values. A culture of growing bacteria is labeled with a radioactive RNA precursor. At the same time, an antibiotic that inhibits the *initiation* of new molecules of RNA, but not their on-going synthesis, is added. The drug most usually employed is RIFAMPIN. The resulting "up and down" curve (Figure 13) is full of information: the ascending part of the curve reflects the rate of synthesis of all RNA species. The time required for the peak value to be reached is approximately the average time required to finish ongoing synthesis of RNA molecules initiated before the experiment. This value can be used to calculate the average length of RNA molecules. The ratio between the peak value reached and the eventual plateau represents the fraction of total RNA to stable RNA. The difference between the two values is the proportion of unstable RNA. Although run-out experiments do not give absolute values, they are useful for comparing different cultures. For example, Pato and Von Meyenburg (1970), using this approach, concluded that the chain growth rate of RNA was nearly the

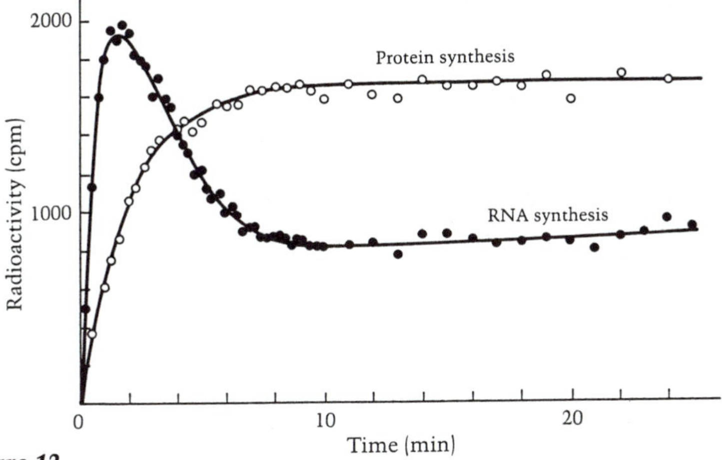

Figure 13

Residual RNA and protein synthesis in the presence of rifampin and nalidixic acid. [³H]uridine and [¹⁴C]proline were added at the same time as the drugs. (After Pato et al., 1973.)

same for cultures of *E. coli* growing at different rates (see Chapter 15 for details).

SUMMARY

1. When growing in a medium in which all essential nutrients are present in nonlimiting concentrations—a condition termed unrestricted growth—the rate of increase of bacterial biomass mimics an autocatalytic, first-order chemical reaction: the rate of increase of mass is proportional to the amount of mass present at any particular time.
2. Under conditions of unrestricted growth, the rate of increase in cell number can be predicted from the equation

   $$\log_{10} N_2 - \log_{10} N_1 = k(t_2 - t_1)/2.303$$

   where N is the number of cells, t is the time, and k the specific growth rate constant.
3. The value of k or the related parameter g (doubling time) is sufficient to describe completely the growth rate of a bacterial culture. These parameters are related by the equation $k = \ln 2/g$.
4. After a period of unrestricted growth, balanced growth ensues, that is, all components of the biomass increase at the same rate. As a consequence, any measurement of biomass can be used to compute the growth rate.
5. Upon exhaustion of an essential nutrient or the accumulation of toxic products of metabolism, growth of a bacterial culture ceases. The culture is now said to leave the exponential phase of growth and

enter the stationary phase. On transfer of such a culture to a fresh medium, growth is delayed for a period—termed the lag phase of growth.

6. The final crop of cells that a medium can support is linearly related to the concentration of the growth-limiting nutrient. The constant of proportionality is called the yield coefficient, Y.
7. Bacterial cultures can be grown continuously in a turbidostat or a chemostat by feeding medium to a culture and withdrawing a corresponding volume of culture at the same constant rate. If the rate of dilution of the culture by this process does not exceed the unrestricted growth rate of the bacterium, the culture density remains constant.
8. A certain portion of the energy derived from the utilization of nutrients—termed the energy of maintenance—does not contribute directly to growth but is used for other processes needed for survival. As a consequence, continuous cultures of bacteria are unable to growth when the limiting nutrient is added too slowly.
9. Growth is balanced as long as the bacterial density does not exceed about 10^8 cells/ml. Constant growth rate is the most useful criterion of balanced growth, and its determination is the sine qua non for reproducible experiments in bacterial growth physiology.
10. Labeling of macromolecules with radioactive precursors is a useful and often simple way of determining relative amounts and rates of synthesis of cellular constituents.

STUDY QUESTIONS

1. The following data were obtained using a culture of *Salmonella typhimurium*. A sample (100 mL) of a growing culture was collected, resuspended in distilled water, centrifuged again, dried at 105°C, and weighed. This sample weighed 103.5 mg. A second sample was taken at the same time. The A_{420nm} of this culture and dilutions of it were measured with the following results. Note that the numbers in the dilution columns indicate milliliters of culture plus milliliters of uninoculated medium.

Dilution	A_{420nm}	Dilution	A_{420nm}
10 + 0	1.017	5 +5	0.670
9 + 1	0.972	4 + 6	0.568
8 + 2	0.919	3 + 7	0.445
7 + 3	0.844	2 + 8	0.300
6 + 4	0.763	1 + 9	0.147

By taking a number of samples of the same culture of *S. typhimurium* growing in a minimal medium with glucose as the sole source of carbon and energy, the investigator obtained the following data. At the time of inoculation (0 time), the glucose concentration was 0.2% (w/v). Assume that the culture stopped growing because glucose was exhausted in the medium.

Time of sampling (min)	A_{420nm}	Time of sampling (min)	A_{420nm}
0	0.091	140	0.577
20	0.092	160	0.732
40	0.111	180	0.902
60	0.152	200	1.001
80	0.208	220	1.080
100	0.303	240	1.080
120	0.430		

a. Plot dry weight and log of dry weight in the form of a growth curve on millimeter paper.
b. Plot the data on semilog paper.
c. Calculate k.
d. Calculate g.
e. Calculate Y.
f. Calculate Y_m (grams cell dry weight per mole substrate utilized).
g. Assuming the value of $Y_{ATP} = 10.5$, calculate the number of moles of ATP produced per mole of glucose metabolized.
h. To obtain a stationary culture containing 200 μg cells/mL, what concentration of glucose should the medium contain?
i. If the average dry weight of a *S. typhimurium* cell in the exponential phase is 1×10^{-12} g, how many cells per milliliter are in the culture at 240 minutes?

2. A glucose-limited chemostat is operated at a dilution rate of 0.4 hr^{-1}. If K_s for glucose is 1×10^{-7}, the glucose yield constant (Y) is 0.48; and if the same medium supports growth in batch (regular) cultures with $k = 0.9$ hr^{-1}, what concentration of glucose should be added to the reservoir to maintain a steady-state cell density in the growth vessel of 600 μg/mL? (b) After resetting the flow rate of the chemostat, you find that the steady-state concentration of glucose is 2×10^{-7} *M*. At what dilution is the chemostat now operating?

3. Assume that a glucose-limited chemostat is operating at a steady state. (a) List three changes that would increase the steady-state cell density in the growth vessel. (b) What changes could be made to increase the growth rate (k)?

4. A technician has isolated a mutant of *E. coli* that has lost a certain function. When mixed with its parent and grown at low dilution in a glucose-limited chemostat, the mutant persists and the parent disappears. How could this phenomenon be explained?

SUGGESTED READINGS

Gerhardt, P., et al. 1981. *Manual of Methods for General Bacteriology*. American Society for Microbiology, Washington, D.C.

Kubitschek, H. E. 1970. *Introduction to Research with Continuous Cultures*. Prentice-Hall, Englewood Cliffs, New Jersey.

Maaløe, O. and N. O. Kjeldgaard. 1966. *Control of Macromolecular Synthesis*. W. A. Benjamin, New York.

Monod, J. 1942. *Recerches sur la croissance des cultures bacteriennes*. Hermann et Cie, Paris.

Monod, J. 1949. The growth of bacterial cultures. Annu. Rev. Microbiol. 3:371.

Novick, A. 1955. Growth of bacteria. Annu. Rev. Microbiol. 9:97.

8

The Effects of Temperature, Pressure, and pH

INTRODUCTION

Not being aggregated into tissues the way plant and animal cells are, bacteria must deal with their environment individually and on intimate terms. Most animal cells are afforded the environmental luxury of being bathed in an unchanging isotonic nutrient solution maintained at an optimal temperature and pH. Plant cells are subject to greater environmental stress, but their multicellular organization also provides them with some protection from their environment. In contrast, environmental stresses have immediate impact on a bacterial cell, so we ought not to be surprised to learn that bacteria have mechanisms to help them cope with a changing and sometimes hostile environment. Possibly because of the constant selective pressures of environmental challenge, bacteria as a group have evolved remarkable tolerance to extreme environments. Some species of bacteria can grow under dramatic circumstances: above the boiling point or below the freezing point of water; in saturated brine solutions; at pH values as low as 1.0 and as high as 11; and at the extreme hydrostatic pressures of the deepest oceans. Some bacteria are able to grow only in these extreme environments. One mechanism by which bacteria deal with a hostile environment was discussed in Chapter 6: by a tactic response, they leave that environment and seek out a better one. The range of environments that allow bacterial growth and how bacteria deal with environmental changes are discussed in this chapter.

EFFECT OF TEMPERATURE ON BACTERIAL GROWTH

Although bacteria grown in the laboratory are usually incubated at a precisely maintained temperature (our reference bacterial cell described in Chapters 1–5 was grown at 37°C), most species have the capacity to grow over a range of temperature—usually one spanning about 40 Celsius degrees. To inquire how temperature affects the rate of bacterial growth within this range, we might start by recalling how it affects the rate of chemical reactions. The Swedish chemist Arrhenius discovered in the last century how temperature affects the rate of a chemical reaction and found it to be described by the equation

$$\nu = Se^{-\Delta E^{*}/RT} \quad (1)$$

in which ν is the velocity of the reaction, S is a constant, ΔE^{*} is the activation energy of the reaction, R is the gas constant, and T is the temperature in K. In its logarithmic form, the equation becomes

$$\ln \nu = (-\Delta E^{*}/R)(1/T) + S \quad (2)$$

Thus, the logarithm of the velocity of a chemical reaction is a linear function of the reciprocal of absolute temperature; the line has a negative slope $(-\Delta E^{*}/R)$, from which the value of the activation energy can be calculated. If this type of plot (frequently termed an ARRHENIUS PLOT) is made for bacterial growth rate (k) rather than for chemical reaction rate, a somewhat different response is seen (Figure 1). In a midrange of temperature, normal chemical kinetics seem to apply: a straight line, like one describing an isolated chemical reaction, is obtained. The slope of this line (times the gas constant, R) is called the TEMPERATURE CHARACTERISTIC—a name designed to make clear that this parameter, unlike activation energy, has no precise physical referent. This linear region of the plot extends over the NORMAL RANGE of temperature. Above and below the normal range, growth rate is less than the value predicted by extrap-

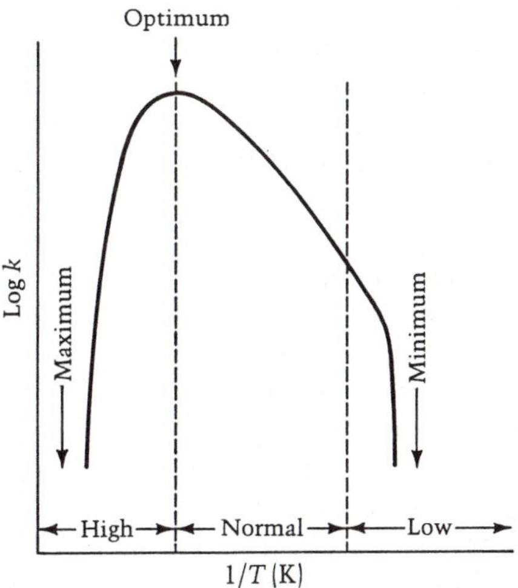

Figure 1

General form of an Arrhenius plot of bacterial growth. The cardinal temperatures (maximum, optimum, and minimum for growth) and the growth temperature ranges (high, normal, and low) are shown.

olation of the Arrhenius relationship. In the LOW RANGE, the slope of the plot increases, becoming vertical at the MINIMUM TEMPERATURE FOR GROWTH. At high temperature, the plot becomes vertical at the MAXIMUM TEMPERATURE FOR GROWTH. The temperature at which growth rate is maximal is called the OPTIMUM TEMPERATURE FOR GROWTH. Collectively, these three temperatures (minimum, maximum, optimum) are called the CARDINAL TEMPERATURES; they are useful in describing the reponse of a species to growth temperature.

Although the absolute values of the defining parameters (temperature characteristics and cardinal temperatures) vary widely among species, the general form of the Arrhenius plot of growth rates is typical for all bacteria studied. Certain information contained in the generalized Arrhenius plot of bacterial growth rate is broadly applicable to all bacteria and, therefore, worthy of emphasis.

- Bacteria do not simply grow incrementally more slowly at high and low temperatures; there are precise temperature limits above and below which growth of any particular strain cannot occur.
- Growth rate decreases rapidly at temperatures above the optimum, so the optimum and maximum temperatures are not far apart.

An Arrhenius plot of the strain of *E. coli* analyzed in Chapter 1 is shown in Figure 2. Its maximum temperature of growth is approximately 48°C; its optimal temperature is 39°C; its minimal temperature is 8°C (not shown in plot); and its normal temperature range extends from about 21° to 37°C. The richness of the medium affects growth rate at all temperatures, but it has no effect on the temperature characteristic. Neither does it have any effect on the optimum temperature of growth. The effect of the medium on the maximum and minimum temperatures of growth varies with strain and species. Enriching the growth medium with amino acids, purines, pyrimidines, and vitamins does not change the minimum temperature of growth of *E. coli,* but enrichment has a marked effect on the maximum temperature of growth, increasing the maximum from about 43° to almost 48°C. The effect of enrichment on the maximum and minimum temperatures of growth is quite variable among different species. Commonly, the maximum temperature of growth is increased, but sometimes—for example with some species of *Bacillus*—the minimum temperature of growth is decreased. In those cases that have been studied, extension of the temperature range of bacteria is nutrient specific. In other words, high or low temperatures appear to block specific biosynthetic pathways, thereby rendering a bacterium dependent on the availability of exogenous building blocks in order to grow at the extremes of its temperature range. The effect of exogenous methionine on the maximum temperature for growth of K-12 strains of *E. coli* illustrates this phenomenon. In the absence of methionine, growth stops at 45°C; and in the range of 40° to 45°C, growth rate is limited by the availability of the amino acid. At these temperatures, the activity of the first enzyme (homoserine succinyltransferase) in the methionine biosynthetic pathway is reversibly inhibited.

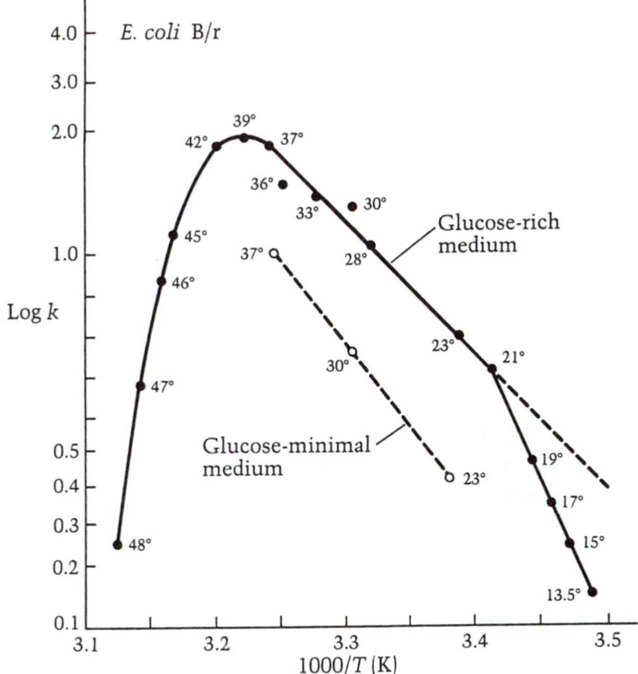

Figure 2

Arrhenius plot of growth rate of *E. coli* B/r. Individual data points are marked with corresponding degrees Celsius. *E. coli* B/r was grown in a rich glucose medium and a glucose-minimal medium. The units of the growth constant k are reciprocal hours (hr^{-1}). (After Herenden et al., 1979.)

Ratkowski et al. (1982) made the interesting observation that the square root of growth rate is a linear function of growth temperature over the normal and low temperature range (Figure 3). Although this relationship has no theoretical basis and fails by extrapolation to predict the minimum temperature of growth, it should be useful in predicting intermediate growth rates from limited experimental data. Failure to predict the minimum temperature of growth shows that the square root relationship with growth rate does not hold in the extreme low range.

The values of cardinal growth temperatures vary widely among bacteria. Some bacteria can grow at temperatures slightly above the boiling point of pure water and others can grow at temperatures as low as −10°C. Indeed, bacterial growth seems to be limited only by the availability of liquid water, the upper limits being set by the boiling point and the lower being set by the bacterium's ability to withstand the osmotic pressure required to maintain water as a liquid below the freezing point of pure water. Although there is no direct relationship between the maximum and minimum temperatures of growth, most bacteria grow over a range of approximately 40 Celsius degrees (Figure 4). If the maximum and minimum temperatures for growth of various bacteria are plotted against each other, they fall on a straight line, the intercept of

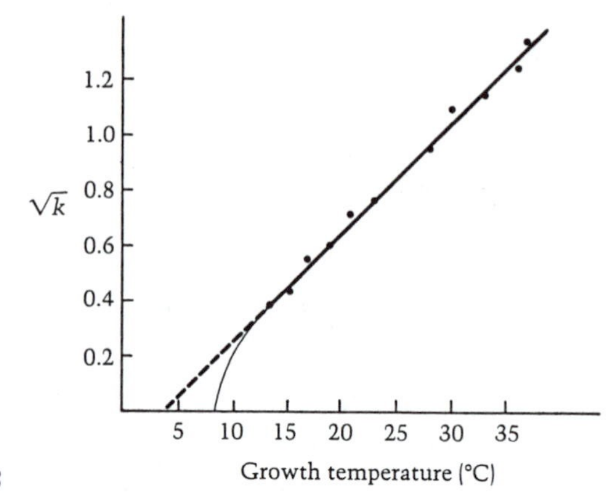

Figure 3

Plot of the data from Figure 2 according to the method of Ratkowski et al. (1982). The square root of the growth rate *k* is plotted against growth temperature. The data points closely fit a straight line, but extrapolation of the line to zero (at 3.5°C) does not accurately predict the actual minimum temperature of growth (8°).

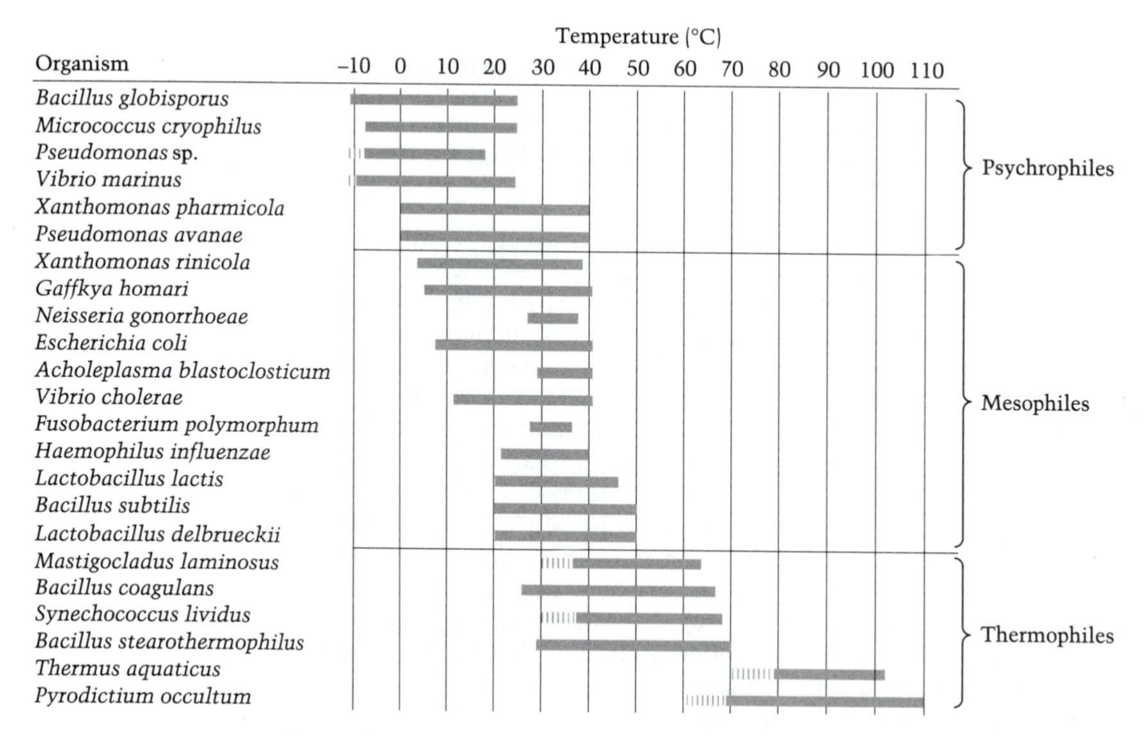

Figure 4

Temperature ranges for growth of certain bacteria. Bars ending in dashed sections indicate that the minimum temperature of growth has not been determined.

which on the maximum temperature axis is approximately 40°C. Some points lie below the line, a result indicating that certain bacteria grow only over a more limited range of growth temperatures, but very few lie significantly above the line, this result indicating that the ability to grow over a range significantly greater than approximately 40 Celsius degrees is rare. The reasons for this natural restriction of the growth-temperature range is not obvious, because the maximum and minimum temperatures permitting growth of a particular bacterium appear to be determined by different metabolic factors. Studies on TEMPERATURE-SENSITIVE (*ts*) MUTANTS that have more restricted growth-temperature ranges than their parents have, show that the maximum and minimum temperatures of growth are set by different genetic determinants. Those *ts* mutants—termed HEAT-SENSITIVE (*hs*)—that have a decreased maximum temperature of growth do not have an altered minimum temperature of growth. COLD-SENSITIVE (*cs*) MUTANTS in which the minimum temperature of growth is increased do not have altered maximum temperatures of growth.

On the basis of their temperature range of growth, bacteria are frequently divided into three broad classes: those that grow at 50°C or above, THERMOPHILES; those that grow best at approximately 37°C, MESOPHILES; and those that can grow at 5°C or below, PSYCHROPHILES or PSYCHROTROPHS. Thermophiles are defined by their maximum temperature of growth, mesophiles by their optimum, and psychrophiles by their minimum, these definitions reflecting the interest of bacterial physiologists rather than any fundamental principle.

To the bacterial physiologist, the observed effects of temperature on bacterial growth pose two important questions:

- Does growth temperature affect the physiological state and the chemical composition of bacteria?
- What biochemical factors set the temperature limits of growth of bacteria?

Effect of temperature on cell composition and physiology

That growth temperature can profoundly affect the physiological state of bacterial cells can be illustrated by a series of simple experiments. If the incubation temperature of an exponentially growing culture of *E. coli* is suddenly shifted within the normal (Arrhenius) temperature range, exponential growth continues at a rate characteristic of the new temperature. But if temperature shifts are made between the normal range and either the high or the low range, growth proceeds at transitional rates before exponential growth at a rate characteristic of the new temperature begins. Transient growth periods are most marked in shifts to or from the low range. Shifting a culture of *E. coli* from 37° to 12°C causes it to stop growing for approximately 4 hours before exponential growth begins again. The reverse shift in temperature causes the culture to grow at a markedly reduced rate for approximately two doublings before full rate is attained. These experiments suggest that within the normal temperature range, cellular reactions remain coordinated simply by changes of enzyme activity, but full growth rate in the high or low range requires that composition of the cell be altered. Detailed analysis of the effect of growth temperature on the cellular content of 133 different proteins (constituting 70% of the cell's protein mass; Herenden et al., 1979) lends support to this hypothesis. The concentrations of very few proteins change appreciably if growth temperature is varied within the normal range. But growth in the high or low range is accompanied by large changes (25-fold or more) in the steady-state level of a number of proteins. These changes in level are brought about very rapidly; within a minute after a shift in temperature from, say, 30° to 42°C, rapid changes occur in the synthesis of many of the 1,000 or so readily observable cellular proteins. There is a gradient of response extending from some proteins that are hyperinduced transiently 100-fold to some that for a time cease to be made. Proteins that are induced transiently by an increase in temperature are repressed by a decrease in temperature and vice versa.

The response of *E. coli* to a shift in temperature above the normal Arrhenius range is analogous in many ways to the HEAT-SHOCK RESPONSE that exists throughout the living world. Cells of all animals, plants, and microbes studied so far display a transient synthesis at high rates of approximately two dozen proteins, when exposed to temperatures above the normal temperature for growth. In *E. coli,* 24 proteins are induced by a shift-up in temperature; of these, 20 are under the control of a single gene, *htpR,* which encodes a protein called a sigma factor (Chapter 13), which confers specificity on RNA polymerase. An RNA polymerase that contains the *htpR* product transcribes only heat-induced genes. Mutant strains that are altered in *htpR* and, therefore, are unable to mount a heat-shock response die after a short period of growth at the elevated temperature. Apparently, a higher concentration of one or more of these proteins is needed for the cells to be able to grow at high temperature. The heat-shock proteins seem to be involved in a bewildering array of cellular processes, and therefore it is likely that there is a multitude of requirements for them (Chapter 13).

The fatty acid composition of the phospholipids in bacterial membranes also varies with growth temperature, apparently in order to preserve a more or less constant degree of membrane fluidity. In most bacteria—indeed, in most poikilothermic organisms—the proportion of unsaturated fatty acids (those with double bonds) in phospholipids increases as the temperature decreases (Figure 5). This change maintains fluidity despite decreasing temperature because the melting point of lipids decreases as their proportion of unsaturated fatty acids increases. The differential melting characteristics are due to the fact that saturated fatty acids line up in quasi-crystalline arrays, whereas unsaturated fatty acids have "kinks" that prevent close packing. Other bacteria, notably the aerobic sporeformers, maintain optimal fluidity of their membranes by modulating the proportion of branched chain fatty acids and the molecular weight of the fatty acids.

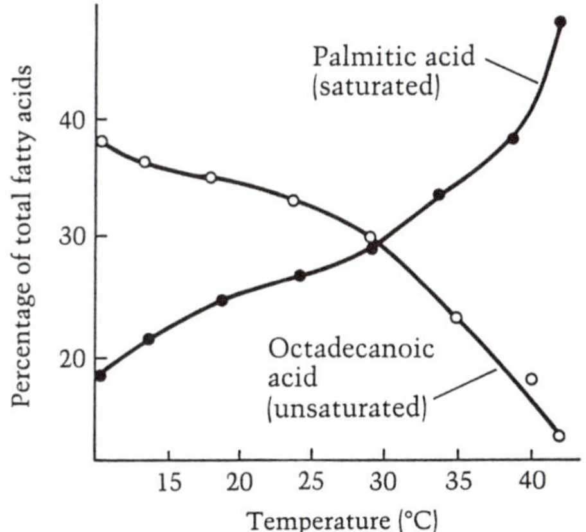

Figure 5

Effect of growth temperature on the fatty acid composition of *E. coli* ML30. Cells were harvested during exponential growth in a glucose-minimal medium. Results are presented as the percentage of the total saturated fatty acid fraction (by weight) that is the saturated fatty acid palmitic acid and the unsaturated fatty acid octadecanoic acid. (After Marr and Ingraham, 1962.)

Factors that determine the upper temperature limits of growth

A number of studies have established the fact that the maximum temperature for growth of most bacteria is set by the stability of proteins. The primary structure of proteins dictates their stability, their susceptibility to turnover, and their functional activity, be it catalytic or structural. How much of the size and complexity of a protein is required for heat stability, as contrasted with functional activity, remains unknown, but certainly heat stability is a particularly sensitive property of a protein. It seems to be more readily changed by mutations affecting primary structure than is the functional activity itself. Several lines of experimental evidence support this contention.

The most revealing data are those of Langridge (1968) who examined 54 randomly altered mutant forms of the enzyme β-galactosidase from *E. coli.* This set was generated by isolating a number of nonallelic amber mutations in the gene encoding β-galactosidase and then introducing a suppressor mutation that introduced a new amino acid (serine) at the site of the amber codon. Thus, he was able to obtain a series of mutationally altered enzymes without selecting for loss of catalytic activity. He found that over 70% of these mutant proteins showed distinct loss of heat stability (Table 1), whereas only one showed detectable loss of catalytic function.

These results suggest that most naturally occurring mutations are selected against because they decrease heat stability rather than destroy function. It follows that a microorganism evolving in the absence of the challenge of elevated temperature would contain very few thermostable proteins. A variety of observations supports this contention. Most mesophiles contain very few thermostable enzymes, and many marine psy-

Table 1. Frequency of half-life classes at 57°C of 54 serine-substituted β-galactosidases

Half-life classes[a] (min at 57°C)	Frequency
10	16
10–20	5
20–30	4
30–40	3
40–50	4
50–60	1
60–70	1
70–80	1
80–90	1
90–100	3
100–110	15

Source: From Langridge (1968).
[a]Wild-type enzyme has a half-life of 104 minutes.

chrophiles, which are protected from the challenge of even moderate temperatures, fail to grow above 20°C. Geneticists have come to accept, as a matter of course, that missense mutations have a high probability of decreasing the heat stability of the product protein.

Loss of function at low temperature

The chemical basis for loss of function at high temperature is self-evident: those chemical bonds that maintain the proper secondary and tertiary structure of proteins become weakened at elevated temperatures, changes resulting in denaturation and loss of function. Loss of function at low temperature is more difficult to explain, because most chemical bonds are strengthened as temperature decreases; but hydrophobic bonds weaken at low temperature because of physical changes in the structure of the solvating water.

The selective pressure for growth at low temperature is quite different from that for growth at high temperature. We have discussed evidence indicating that exposure to the challenge of high temperature is essential if a microorganism is to retain its ability to grow at high temperature, but such does not seem to be the case at low temperature. Organisms, like enteric bacteria that grow almost exclusively at 37°C, still maintain the ability to grow at temperatures as low as 8°C.

The study of cold-sensitive mutants has provided examples of loss of function at low temperature. The typical growth response of a cold-sensitive mutant of *E. coli*, as affected by temperature, is shown in Figure 6. At 37°C the mutant grows almost as fast as its parent, but as temperature is lowered, the growth rate of the mutant decreases more rapidly than that of the parent, and it stops growing completely at approximately 20°C, rather than at 8°C.

Although cold sensitivity is as frequent a consequence of mutation as heat sensitivity, mutations conferring cold sensitivity are restricted to a smaller number of genes. Cold-sensitive mutants provide a means of analyzing the biochemical basis for the minimum temperature of growth. Because a single genetic change increases the minimum temperature of growth, the gene's altered protein product is the biochemical determinant of the minimum temperature of growth of the mutant strain. Among different species of bacteria, different independent functions cease at the minimal temperature of growth. Thus, there is not a single cause determining the minimum temperature for growth. Although a study of cold-sensitive mutants does not reveal the actual lesions that cause wild-type microorganisms to stop growing at low temperature, it gives information about the types of lesions that can prevent growth.

One type of lesion alters the sensitivity of allosteric proteins to their small molecule effectors (Chapter 11) and hence distorts their regulation. Cold-sensitive histidine mutants are an example. In the absence of histidine, these mutants have a minimum temperature of growth approximately 12 Celsius degrees higher than that of their parent; but in the presence of histidine, parent and mutant grow at identical rates at all temperatures. The functional block at low temperature lies in the first reaction of the histidine pathway—that catalyzed by phosphoribosyl-ATP pyrophosphorylase. This enzyme is sensitive to feedback inhibition

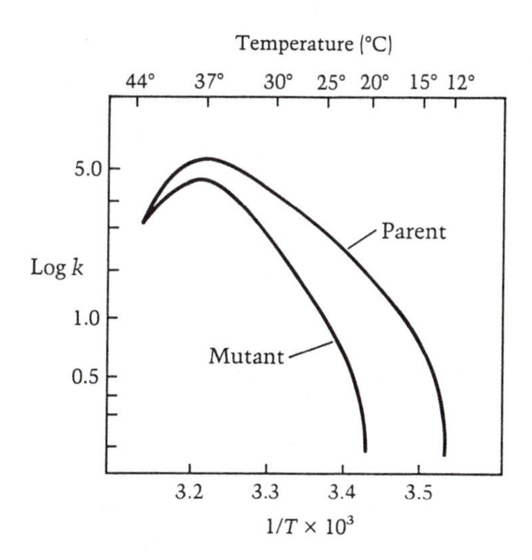

Figure 6

Arrhenius plot of the specific growth rate (k) of an *E. coli* cold-sensitive mutant. Both the mutant and its parent were grown in a minimal medium. The ordinate is the log of the specific growth rate and the lower abscissa is the reciprocal of the absolute temperature times 1,000.

by the end product of the pathway—histidine. The mutant enzyme is almost 1,000-fold more sensitive to feedback inhibition than the wild-type enzyme. Moreover, both mutant and wild-type enzyme are approximately 10 times more sensitive to feedback inhibition at 20°C than at 37°C. Thus, the increased sensitivity of the mutant enzyme to feedback inhibition, plus the increased sensitivity imposed by low temperature, creates conditions whereby the intracellular concentration of histidine sufficient to prevent its own further biosynthesis is insufficient to allow biosynthesis of proteins. As a consequence, the mutant cannot grow at 20°C in the absence of exogenous histidine.

If the biochemical explanation of cold sensitivity of histidine biosynthesis has general significance, the degree of inhibition of many other allosteric proteins might be expected to change with temperature. Considerable evidence indicates that this is the case, but the direction of the change is not predictable. Some allosteric proteins become more sensitive to inhibition as temperature is decreased; others become less sensitive. Such effects most probably reflect fundamental properties of allosteric proteins, even in organisms that do not undergo large temperature shifts. As an example, we can consider the effect of temperature on the inhibition of fructose-1,6-bisphosphatase from rat liver by adenosine monophosphate (AMP). This enzyme is dramatically more sensitive to AMP at low temperature. In the absence of AMP, the effect of temperature follows normal chemical kinetics: the Arrhenius plot is completely linear over the temperature range of 46° to 2°C; but in the presence of low levels of AMP the reaction is selectively inhibited at low temperature (Figure 7). Adenosine monophosphate at 10^{-4} M has no inhibitory effect on the enzyme at 46°C, but at 20°C this concentration inhibits enzyme activity by more than 90%. So even allosteric enzymes from homothermic organisms are subject to temperature-sensitive inhibition by effectors. It is not surprising that mutations that alter the sensitivity of allosteric proteins to their effectors have a high probability of being expressed as a cold-sensitive phenotype.

Another major class of cold-sensitive mutants contains mutations affecting an assembly process. Some mutants are unable to synthesize ribosomes at low temperature because they are unable to assemble the ribosomes properly. The mutations involved lie in structural genes encoding ribosomal proteins.

Cold-sensitive mutants of both the regulatory and assembly type probably share the same biochemical basis: a change in primary structure plus the weakening of hydrophobic bonds causes the protein to have a slightly altered conformation at low temperature. If activity of that protein is sensitive to conformational change, it might lose function at low temperature. The sensitivity of allosteric proteins to their effectors is known to be affected by conformation, and one would expect that the assembly of ribosomal proteins into ribosomes would be dependent on a precise conformational state.

Other aspects of protein activity are also affected. When shifted to a temperature slightly below the minimum for growth, polysomes do not

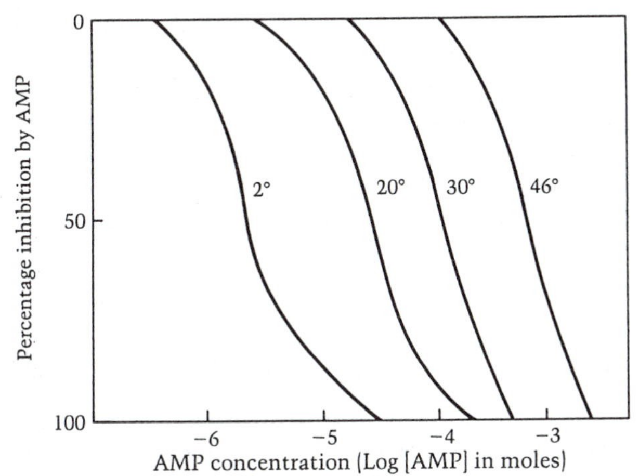

Figure 7

Inhibition of fructose 1,6-bisphosphatase by various concentrations of AMP at various temperatures (°C). (After Takita and Pogell, 1965.)

form in *E. coli,* probably as a consequence of a conformational change in a ribosomal protein or an initiation factor. As has been suggested, this defect might set the minimum temperature for growth of the organism. But other bacteria are able to assemble polysomes at their minimum temperature for growth. It seems that one can generalize about the sorts of protein changes that set the minimum temperature of growth of bacteria but not about the particular protein that is affected.

Lipids have also been implicated as the cause of the cessation of growth below a minimum temperature. The scenario runs as follows. We know that the fraction of unsaturated fatty acids in membranes increases with decreasing growth temperature; possibly a point is reached beyond which higher levels of unsaturated fatty acids cannot occur. That theoretical point would be the minimum temperature of growth. Although logically attractive, the fatty acid scenario is refuted by experiments: by physiological or genetic manipulation, cultures can be caused to grow near their minimum temperature of growth with a fatty acid composition typical of the culture growing at its optimum temperature for growth. We can only conclude that (at least in those cases that have been studied) temperature modulation of fatty acid composition is a physiological nicety that must (because of its widespread occurrence) confer subtle selective advantages; it is not essential for growth at a low temperature.

LETHAL EFFECTS OF TEMPERATURE

Bacteria may be killed by exposure to high temperature, by freezing, and by sudden chilling. Although some bacteria die when held at temperatures above the freezing point and below the minimum for growth, there is no evidence that such killing is caused specifically by low temperature. Most probably, it is the usual decline in viability that occurs in the absence of growth.

Lethal effects of high temperature have been studied extensively, because sterilization by heat is simple and effective. In spite of the enormous variations among bacteria with respect to intrinsic susceptibility to heat, the importance of the physiological state of the organism, and the protective effects of components of the medium, certain generalizations can be made:

- The rate of killing of bacteria is directly related to their moisture content.
- The rate of killing of bacteria, in general, follows exponential kinetics.

Although deviations (sometimes major ones) occur, a plot of the logarithm of the number of surviving cells in a population exposed to a lethal high temperature is usually a linear function of the time of exposure. Thus, sterility is an absolute state, but the design of lethal heat treatments to attain it are based on probability calculations, even when the initial population and the slope of the killing curve are known precisely. The common indexes of susceptibility to killing by heat of a particular suspension of bacteria are:

- THERMAL DEATH POINT (TDP): the lowest temperature that results in the organism's being unable to reproduce after 10 minutes or some other fixed exposure time
- THERMAL DEATH TIME (TDT or F value): the time of exposure to a particular temperature necessary to sterilize a definite concentration of organisms in a particular medium
- D VALUE: the time of exposure necessary to reduce the number of survivors by a factor of 10

Often the logarithm of both D values and F values are linear functions of temperature over relatively narrow ranges. The slope of the curve relating the logarithms of F values to temperature, that is, the number of degrees required for the line to traverse one log cycle, is termed the Z VALUE.

Susceptibility of bacteria to freeze-killing, like their susceptibility to heat-killing, depends on their physiological state. In addition, the rate at which the culture is chilled markedly affects the number of bacteria that survive the treatment. The degree of killing by freezing can be separated into two components: the killing that occurs at the time of freezing (IMMEDIATE EFFECT) and the slow loss of viability that occurs in the frozen state (STORAGE EFFECT). The immediate effect is to kill a certain portion of the population. Thus, the logarithm of the number of survivors of a particular suspension of bacteria is usually a linear function of the number of times it has been frozen. The rate of loss of viability of a culture held in the frozen state is usually related to the temperature at which it is held.

A number of compounds are remarkably protective when added to the medium in which a bacterial suspension is frozen. They include glycerol, dimethylsulfoxide, dimethylacetamine, dimethylformamine, *N*-methylpyrrolidinone, acetamide, and polyvinylpyrrolidone. Other compounds that confer somewhat less protection include milk proteins, meat extract, sucrose, glucose, and lactose.

Bacteria that make ice

It is ice formation, not low temperature itself that kills microorganisms. With this in mind, it might be surprising to learn that some bacteria have the ability to stimulate ice formation. Ice always melts at 0°C, but it does not necessarily form at that temperature because ice crystals need a template on which to grow. Preformed ice acts as a template, as do lattices that form spontaneously at low temperature in liquid water. But such lattices are so small that their equilibrium temperature is approximately −40°C: above that temperature the spontaneous ice lattice melts in pure water. Many heterogeneous substances and mechanical treatments like scratching a solid surface also can serve as templates, causing a nucleation event for the formation of ice. But most heterogeneous materials that cause ice nucleation are not effective above −5°C. Certain bacteria, however—called ICE-NUCLEATING BACTERIA—are an exception. They cause nucleation at this temperature and slightly above by producing a specific protein (INA) located in their outer membrane. Ice-nucleating bacteria, which belong to the species *Pseudomonas syringae, P. fluorescens, P. viridiflava, Erwinia herbicola,* and *Xanthomonas campestris,* are all epiphytes (grow on the surface of plants). One may wonder how these bacteria benefit from producing INA. There are two possible answers:

- They cause dew to form on the plant surface, thereby providing the bacterium with a source of the water.
- By causing water to freeze just below 0°C, plant cells are killed, thereby providing the surviving bacteria with a source of nutrients.

The protein INA has captured the attention of biotechnologists because it promises so many applications:

- Mutant strains of ice-nucleating bacteria that cannot form INA might be spread on frost-sensitive crops, thereby displacing the normal epiphytic flora and preventing frost damage on nights when the temperature dips only a few degrees below 0°C.
- Seeding clouds with INA-producing bacteria could cause rain.
- Seeding snow-making machines would allow artificial snow to be made at higher temperatures.
- INA could be used to make smooth ice cream without stirring.
- INA could serve as an amplifier to detect certain organisms. In theory, one molecule of INA is detectable by its ability to cause a large quantity of water to freeze.

Cold shock

Many bacteria exhibit a curious susceptibility to death when a culture is rapidly chilled. This phenomenon, known as COLD SHOCK, can, if unappreciated, cause huge errors in certain procedures, such as the making of dilutions prior to determining the number of viable cells in a growing culture. For example, the rapid chilling of cultures of *E. coli* from 37° to 5°C can kill over 90% of the population. Rapidly growing cultures are most susceptible to cold shock. Slow chilling of a culture protects it completely, as does the presence of glycerol. In general, Gram-positive bacteria are more susceptible to cold shock than are Gram-negative bacteria, for unknown reasons.

EFFECT OF OSMOTIC PRESSURE ON GROWTH

Any difference in solute concentration across the cytoplasmic membrane establishes an OSMOTIC PRESSURE, because water molecules tend to equalize their concentration across the membrane, through which they pass freely. Depending on the sense of the solute gradient, the cytoplasmic contents will tend to shrink (PLASMOLYZE) or expand. Either condition is potentially disastrous to animal cells, but, within reasonable limits, only sudden changes in the osmotic environment are a threat to plant or bacterial cells, because they have a greater capacity to adjust their internal osmolarity. Bacterial cells maintain an internal osmolarity in excess of that of their environment: they maintain a somewhat constant turgor pressure because the cell membrane is contained within the rigid cell wall. This turgor pressure is thought to accomplish two cellular purposes:

- It obviates the necessity for precise control of the elaborate mechanisms that would be needed to match internal osmolarity exactly and rapidly to the external one.
- It provides the driving force for growth-associated expansion of the cell wall.

All bacteria examined, with the exception of mycoplasmas (see below) and halobacteria (archaebacteria with somewhat flexible cell walls that live in very high saline environments), maintain a turgor pressure. In some species, turgor pressure is quite high. Possibly because they have a thicker and, therefore, stronger murein layer, Gram-positive bacteria maintain a higher turgor pressure [5 to 22 atmospheres (atm)] than that of Gram-negative bacteria (0.8 to 5 atm). The fact that bacteria can maintain a constant turgor pressure in spite of a changing osmotic environment poses two physiological questions:

- How does the bacterial cell adjust its internal osmolarity?
- What effect does a changing internal osmotic environment have on cellular functions?

Adjustment of the cell's internal osmolarity

Bacteria respond to an external hyperosmotic stress by increasing the internal concentration of only a few solutes, sometimes called COMPATIBLE SOLUTES. These include K^+, certain amino acids (glutamate, glutamine, proline, γ-aminobutyrate, alanine, glycinebetaine, and other fully *N*-methylated amino acid derivatives), and sugars (sucrose, trehalose, and glucosylglycerol).

Of these, K^+ plays the quantitatively major role (Figure 8). As external osmolarity is increased, its concentration within the cell (along with the concentration of a monovalent counterion) rises sufficiently to maintain turgor, up to external values of osmolarity as high as 1.2 os*M*. When the osmolarity of the medium is increased, K^+ is pumped into the cell; when it falls, K^+ leaves the cell. A simple observation shows that the signal to pump K^+ into the cell or allow it to leave is turgor pressure, not external osmolarity: high concentrations in the medium of compounds like glycerol and ethanol, which pass freely through membrane, do not cause K^+ to accumulate inside the cell.

Two amino acids, glutamate and glutamine, have a special role in regulating osmotic pressure. The intracellular concentration of glutamate contributes in a major way to maintaining the cell's turgor, particularly in Gram-negative bacteria. The intracellular concentration of glutamate in *E. coli* is directly proportional to the osmolarity of the medium, rising to values at high osmolarity that make up over 90% of the cell's pool of free amino acids. In contrast to intracellular K^+, gluta-

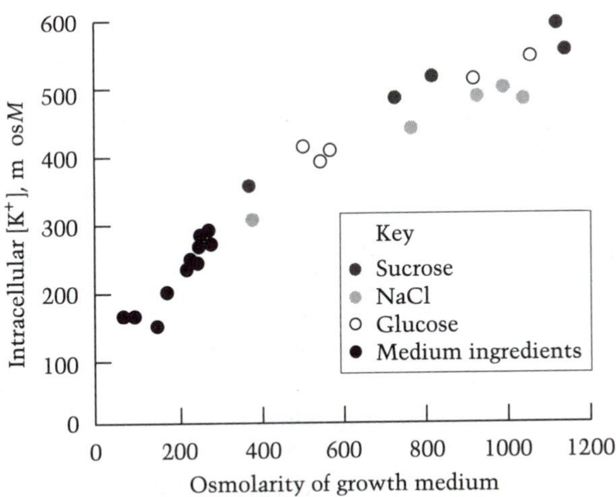

Figure 8

Effect of the osmolarity of the growth medium (milliosmolar) on the intracellular concentration of K^+ (milliosmoles per liter of cell water). Osmolarity of the medium was set by the elements indicated in the key. (After Epstein and Schultz, 1965.)

mate pools are adjusted by synthesis, not transport, as can be demonstrated by the fact that normal adjustments of the glutamate pool occur with cells growing in a glutamate-free medium. In general, the glutamate pools of Gram-positive bacteria are less responsive to osmotic stress; they respond more slowly and to a lesser extent. In extreme cases represented by species of *Planococcus* and *Staphylococcus*, glutamate pools do not respond at all to osmotic stress; in these organisms glutamine and/or alanine play the equivalent role.

Effect of high osmolarity on cellular functions

The system of increasing internal osmolarity to maintain a constant turgor pressure has obvious limits. Sooner or later, high internal osmolarity will inhibit some essential enzyme activity and growth will stop. For example, enteric bacteria grow extremely slowly in media that contain more than about 0.65 *M* NaCl. But not all solutes are equally harmful. Certain compounds, called OSMOPROTECTANTS, are able to ameliorate the deleterious effects of high osmolarity. Some osmoprotectants, notably proline and glycinebetaine, are remarkably effective. For example, *S. typhimurium* grows very slowly in a minimal medium containing 0.8 *M* NaCl, but if 1 m*M* proline or the choline derivative glycinebetaine is added, growth rate is increased two- to fourfold. How does the cell acquire osmoprotectants? *Salmonella typhimurium* can synthesize proline, but the elevated internal pools of proline that protect it from high osmotic environments are achieved exclusively by transporting it into the cell. So, for this organism and most other Gram-negative species as well, proline protection depends on its being available in the medium. In contrast, when exposed to high osmotic environments, Gram-positive bacteria can protect themselves by synthesizing more proline when they are exposed to high osmotic environments. In those bacteria that have been studied, glycinebetaine is accumulated by transport.

Although there are some differences among bacteria with respect to the effectiveness of osmoprotectants, these compounds are active in most species. They even protect plants from osmotic damage. In spite of their dramatic effects, the mode of action of osmoprotectants is far from clear. They might merely be exceptionally benign compatible solutes, or they might play an actively protective role, interacting with proteins and thereby protecting the proteins from osmotic destruction.

The capacity of choline to protect enteric bacteria from high osmolarity illustrates the biological importance and selective value of osmoprotectants. Choline itself is not an osmoprotectant, but it can be converted to the active compound—glycinebetaine—through a two-step oxidative pathway in which glycinebetaine aldehyde is the intermediate. Synthesis of the enzymes that catalyze this pathway as well as the transport system that brings its substrate—choline—into the cell is stimulated in a high osmotic environment. Glycinebetaine is not further metabolized. Almost certainly, the only cellular function of the choline-to-glycinebetaine pathway is to make this osmoprotectant; it is the only route of its biosynthesis by *E. coli. Salmonella typhimurium* lacks this pathway, so choline affords no osmoprotection to this species.

An exception: The mycoplasmas

The mycoplasmas, being wall-less, constitute a major exception to the generalization that bacteria maintain a nearly constant turgor pressure because of the strong wall. Mycoplasmas seem to maintain their cytoplasm in a state nearly isosmotic with their external environment by actively pumping sodium out of the cell with a Na^+-translocating ATPase. If mycoplasmas are deprived of an energy source (and, therefore, are unable to pump out Na^+), they will swell and lyse when suspended in an isosmotic solution of NaCl.

EFFECT OF HYDROSTATIC PRESSURE ON GROWTH

Bacteria are able to withstand and flourish at the highest hydrostatic pressure on the planet—those found in the deepest parts of the ocean. Because water flows readily into bacteria, high hydrostatic pressure cannot crush them; but it can inhibit certain chemical reactions by preventing the formation of the activated state, the obligatory intermediate of any chemical reaction. If the specific molecular volume of the activated state exceeds that of the reactants, high hydrostatic pressure will slow or even stop the reaction. In contrast, if the specific molecular volume of the activated state is less than that of the reactants, the reverse result obtains: the reaction is speeded up. Without knowing the effect of high hydrostatic pressure on the rate of a reaction, the relative molecular volumes of the activated state and reactants cannot be predicted—in the case of enzyme-catalyzed reactions, the direction of the volume change can be different for the same reaction catalyzed by different enzymes. This fact is illustrated by comparisons of the enzyme fructose bisphosphatase from rainbow trout and abyssal rat-tail fish. The trout, which lives in shallow water, has an enzyme that is inhibited by high hydrostatic pressure; in contrast, the enzyme from the rat-tail fish is stimulated. Thus, natural selection is able to affect the molecular volume of the activated state of an enzyme-catalyzed reaction, so one might anticipate that various bacteria would respond differently to hydrostatic pressure, depending on the environment in which they evolved. Such is indeed the case. In the deep ocean, one class of bacteria—called BAROPHILES—can grow better at elevated hydrostatic pressure than at atmospheric pressure. There are even bacteria—called OBLIGATE BAROPHILES—that grow only under hydrostatic pressures that exceed 1 atm.

Ordinary bacteria that probably have not been challenged by high hydrostatic pressure during their evolution are, nevertheless, remarkably tolerant to such pressure. *Escherichia coli* grows well at hydrostatic pressures as high as 300 atm. In stark contrast, most yeasts cannot grow at pressures that exceed 8 atm—a fact that makes bottled champagne possible.

EFFECT OF pH ON GROWTH

What strategy have bacteria devised for coping with different pHs? Most species can grow over a wide pH range, although many enzymes upon which microbial growth depends function only within a narrower range of pH. These simple facts suggest that bacteria deal with wide-ranging changes in the pH of their growth medium by maintaining their internal pH near a fixed optimal value. A specific example illustrates this point. *Escherichia coli* grows well near neutrality, at pH values from 6.0 to 8.0; it can grow, albeit more slowly, at pH values of one pH unit or so above or below this range. Thus, with respect to its response to external pH, *E. coli* is called a NEUTROPHILE. But regardless of the value of external pH, within the cell pH is maintained at a value quite close to 7.6–a remarkable homeostasis. This strategy of dealing with changes in external pH is widespread, possibly universal, among bacteria, although the actual values of internal pH vary considerably. The ACIDOPHILE *Thiobacillus ferrooxidans,* which grows best at pH 2.0, maintains its internal pH close to 6.5 over its pH growth range, which extends from 2 to 8. At the other extreme, the ALKALOPHILE *Bacillus alkalophilus,* which grows best at pH 10.5, maintains an internal pH of 9.

Thus, when a bacterium grows at pH values lower than its internal value, its cytoplasm is more alkaline than its environment, but when it grows at higher values, its cytoplasm is more acidic. At first glance, this fact seems to contradict what we learned about chemiosmosis in Chapter 5. How can the chemiosmotic cyclic flow of protons (out of the cell during electron transport and back in through the membrane-bound ATPase to generate ATP) occur in the range of pH over which the cytoplasm is more acidic than the environment? The answer to this perplexing question lies in the fact that protonmotive force is made up of two interconvertible components: ΔpH and $\Delta\psi$ (membrane potential), either of which has the capacity to drive protons through the membrane-bound ATPase to generate ATP.

We also have to address the more mechanistic question of how ΔpH is exchanged for $\Delta\psi$ when an organism grows in media with a pH value higher than the internal value. This question is not completely answered, even for the well-studied *E. coli,* but certain requirements are obvious. The cell must bring protons into the cell and simultaneously create a membrane potential. Most probably, this task is accomplished by either a proton/K^+ or a proton/Na^+ antiport. There is some experimental support for and against either possibility, but the weight of evidence suggests that the proton/K^+ antiport system plays the major role.

Growth at high pH introduces a further physiological complication. If K^+ ions must be exchanged for protons, how is it possible to maintain an appropriate turgor pressure, which we believe to be largely dependent on the intracellular concentration of K^+? We do not know. The physiology of growth at high pH values is obviously complex. It involves an interplay of respiratory proton pumping, cation–proton exchange, and transport of K^+ into the cell that are not yet fully understood.

SUMMARY

1. An Arrhenius plot of the growth rate of any bacterial culture has a characteristic shape that defines three temperature ranges—high, normal, and low. At both temperature extremes, the plot becomes vertical at the maximum and minimum temperatures for growth. Temperature shifts within the normal temperature range require minimal adjustments of cellular composition before growth can proceed, but major changes occur when bacteria are shifted to or from temperatures outside the normal range.
2. Most bacteria are able to grow over a temperature range of about 40 Celsius degrees. Some can grow slightly above the boiling point of water, and some can grow at temperatures as low as −10°C.
3. Thermal instability of proteins sets the maximum temperature of growth of most strains of bacteria. Those factors setting the minimum temperature of growth derive from the weakening of hydrophobic bonds that occurs at low temperature.
4. Various species of bacteria maintain a characteristic constant turgor pressure over relatively wide changes in the strength of their osmotic environment. Turgor pressure is maintained by adjusting the intracellular concentration of certain "compatible solutes", principally K^+ and, to a lesser extent, glutamate, glutamine, and alanine.
5. Some compounds—notably, proline and glycinebetaine—are called osmoprotectants because, when they are present in the growth medium, many bacteria are able to tolerate higher osmotic environments.
6. Most bacteria can withstand extremely high hydrostatic pressures. Some, called barophiles, grow better at high hydrostatic pressures; others, called obligate barophiles, can grow only if hydrostatic pressure exceeds 1 atm.
7. Bacteria maintain a nearly constant internal pH that can be quite different from the pH of their medium. Depending on the pH of their medium, the cytoplasm of a bacterium might be either more alkaline or more acidic than its medium. The relative contribution of ΔpH and $\Delta\psi$ to protonmotive force shifts as external pH changes.

STUDY QUESTIONS

1. Present a plausible explanation for the observation that different strains of bacteria are able to grow over the entire range of temperatures at which water is liquid but few if any individual strains are able to grow over a range greater than 40 Celsius degrees.
2. What changes in the composition of the cytoplasm of a growing bacterium would occur if it were shifted to a medium with a higher osmolarity and a higher pH?
3. How might the kinetic properties of enzymes from a barophile and an obligate barophile differ?

SUGGESTED READING

Booth, I. R. 1985. Regulation of cytoplasmic pH in bacteria. Microbiol. Rev. 49:359.

Csonka, L. N. 1989. The physiological and genetic responses of bacteria to osmotic stress. Microbiol. Rev. 53:121.

Harold, F. M. 1986. *The Vital Force: A Study of Bioenergetics.* W. H. Freeman, New York.

Ingraham, J. L. 1987. Effect of temperature, pH, water activity and pressure on growth. Chapter 97 in Volume 2 of Neidhardt, F. C., J. L. Ingraham, K. B. Low, B. Magasanik, M. Schaechter, and H. E. Umbarger (Eds.). 1987. *Escherichia coli and Salmonella typhimurium: Cellular and Molecular Biology.* American Society for Microbiology, Washington, D.C.

Jannasch, H. W. and C. D. Taylor. 1984. Deep-sea microbiology. Annu. Rev. Microbiol. 38: 487.

9

Genetic Adaptation: The Genome and Its Plasticity

INTRODUCTION

In Chapter 8 we noted that bacteria as a group have successfully colonized every conceivable niche on this planet, including some that from the human perspective seem inimical to life. This ubiquity of bacteria is the result of both genetic and physiological adaptation. PHYSIOLOGICAL ADAPTATION, which is the change in gene expression in the cells of a population in response to some change in the environment, is the subject of Chapters 11–13. GENETIC ADAPTATION, which is the overgrowth of a mutant cell to become the predominant type in the population, is the subject of this and the following chapter.

The bacterial genome, which encodes all the structures and functions discussed in the previous eight chapters, can undergo rapid changes. In fact, changes can be so rapid that on occasion microbiologists are challenged to decide whether an observed change in some property of a bacterial culture is the result of physiological or genetic processes.

Genetic adaptation occurring within a given population on a short time scale can be called MICROEVOLUTION—a subject of interest for many microbial geneticists. As more is learned about the structure, physiology, and activity of procaryotic cells, curiosity is growing about the MACROEVOLUTION of these cells. That is, microbiologists are seeking answers to a number of fundamental questions about the past and future of microbes. For example, how did bacteria originate? They are clearly adept at adjusting to a variety of environments over which they can exert only minimal control. How did they get that way? Is their development completed, or are they still undergoing evolutionary change as the environment of this planet gradually changes, through natural processes and the influence of human activity?

In this chapter we summarize some of the main features of the bacterial genome, its organization and evolution, and the means by which variation in existing characteristics can occur and by which entirely new properties can be acquired. The chief generators of diversity in the bacterial genome are mutations. We shall see how mutations produce changes in genes, creating mutant alleles, how they can add DNA to the bacterial genome, and how they can rearrange its order. Finally, we shall consider the major facilitator of genetic plasticity, transposable genetic elements.

THE BACTERIAL GENOME

General features

The sole known function of the bacterial genome is to serve as a replicating repository of the information encoded in the sequence of bases in its DNA. In this respect the bacterial genome is like that of all other cells. In many other respects, however, it differs markedly. We have already noted some differences in molecular architecture (Chapters 2 and 3). Other unique aspects of the bacterial genome include

- the absence of a diploid complement of genes
- the use of nearly the entire genome for coding or regulation, without redundant stretches of DNA
- the absence from the chromosome of histones, the basic proteins of eucaryotic chromosomes
- the tendency for bacterial genes encoding related functions to be linked in multicistronic transcriptional units termed OPERONS
- the colinearity of bacterial genes with their protein products (i.e., the absence of introns in bacterial genes)

Other noteworthy, but not necessarily unique features of the procaryotic genome are

- the existence of accessory genetic units—plasmids—in addition to the single chromosome
- the occurrence with about equal frequency of coding sequences on each of the two strands of the chromosome

Organisms other than bacteria have haploid genotypes, but only for part of their life cycle (the germ cells of higher plants and animals, for example) or, as in the case of yeast, as one of alternative states of their genome. Haploidy, however, is a hallmark of the procaryotic life form, and a few observations about its definition and significance are in order.

How can we call bacteria haploid when—as we have seen repeatedly (Chapters 1, 2, and 3) and shall have occasion to note later (Chapters 14 and 15)—during rapid growth a bacterial cell may have more than one nucleoid and multiple chromosomal equivalents of DNA? The answer is that ploidy is a functional concept that refers to the inheritance of genetic material. A DIPLOID CELL is one that has two complete copies of its genome (i.e., two of each of its chromosomes) obtained at some time in past history from the zygotic fusion of two parental haploid cells. Upon division, a diploid cell passes to each daughter cell a copy of each pair of homologous chromosomes. As a consequence, if a diploid cell is HETEROZYGOUS for a particular gene, having one allele of this gene on one chromosome of a pair and a different allele on the other, *each of its offspring cells will be similarly heterozygous.* A haploid bacterial cell may contain multiple nucleoids during rapid growth, but simply because cell division has not kept up with nucleoid formation—the rapidly growing bacterial cell is COENOCYTIC (multiple nucleoids cohabiting the same cell). This cell has identical alleles on each chromosome. And, should a mutation occur in a gene on one of the chromosomes (and not on the other), at cell division one of the offspring will carry the mutant allele and one the wild-type allele. In a haploid organism, no matter how many physical copies of the genome exist, descendant cells will eventually be homozygous. Consequently, RECESSIVE MUTATIONS are always phenotypically expressed in bacteria, at least within a few divisions of the mutational event. In contrast, in a diploid cell, recessive alleles are (by definition) never expressed until a mating of some sort results in a homozygous state, with the recessive mutant allele on both chromosomes.

Haploidy has at least two consequences for the plasticity of the bacterial genome. First, any mutational change has immediate consequences, for better or worse, and therefore is immediately subjected to the pressure of natural selection. Second, haploidy helps bacteria use genetic switches in the promoters of genes as a way to control their function. As we see later in this chapter, species of *Salmonella* employ a device of this sort—called PHASE VARIATION—to alternate their surface antigenic structure by switching genes on and off by a quasi-mutational process. Control of gene expression in this way is known to occur in some diploid eucaryotic cells (in the mammalian immune system) but is necessarily a more complex process because of the problem of the unswitched allele on the other chromosome.

Patterns of gene arrangement

Is the current structure of the bacterial genome a historical accident or are there forces selecting a particular order and arrangement of the genes on the chromosome? A close look at a well-studied genome—that of *E. coli*—reveals that there are some patterns in the arrangement of its genes. First, as already noted, genes of closely related function (i.e., those encoding enzymes in the same metabolic pathway) are in many instances contiguous on the chromosome and are cotranscribed. Of 900 mapped genes shown on the seventh edition of the *E. coli* genetic map (Bachmann, 1987), 260 can reasonably be inferred to be organized into 75 operons. Clustering of genes of related function is found in other bacteria as well.

A second pattern, discerned quite recently, is the tendency for genes (operons) with high frequencies of transcription to be oriented with respect to the origin of replication in a manner that minimizes collisions between the transcribing RNA polymerase and the replication apparatus (Figure 1). Recall that one strand of DNA runs 3′ to 5′ in the clockwise direction around the chromosome (let us call it the clockwise strand) and the other runs 3′ to 5′ counterclockwise (the counterclockwise strand). When the clockwise strand is the sense strand for a given gene, that gene is transcribed clockwise; genes for which the counterclockwise strand is the sense strand are transcribed counterclockwise. Recall also that the two replication forks initiate DNA synthesis at the origin (near minute 83.5) and proceed in opposite directions around the circular chromosome and meet at the terminus (near minute 31). Highly active genes are found almost exclusively on the clockwise strand if they are located on the half of the chromosome extending clockwise from the origin to the terminus, and on the counterclockwise strand if they map to the segment extending counterclockwise from the origin to the terminus. Through this arrangement, RNA polymerase and the replication apparatus move in the same direction through these busy genes.

A third pattern, less pronounced than the one just considered, is a distinct tendency for genes of related function and of presumed related

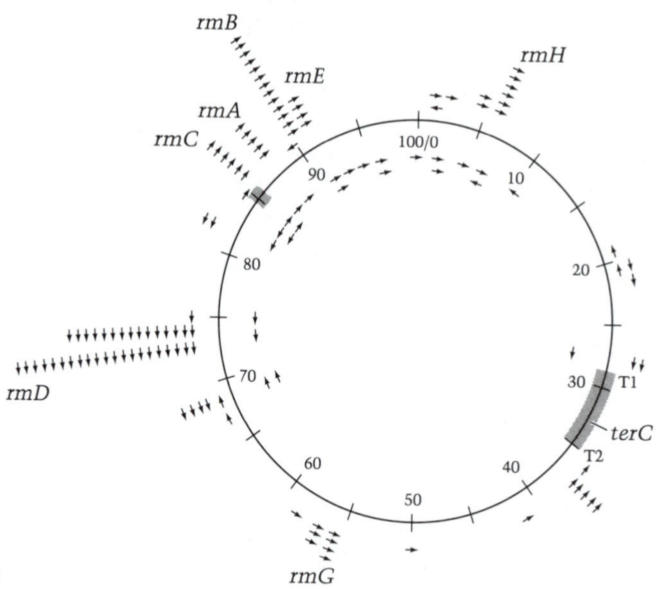

Figure 1

Arrangement of genes on the two strands of DNA of the *E. coli* chromosome. The figure shows the direction of transcription (and hence the sense strand) for only highly expressed genes. Each arrow outside the circular map refers to a gene encoding a part of the protein-synthesizing machinery; the seven rRNA (*rrn*) operons are labeled. Each arrow inside the map refers to a gene necessary for DNA and/or RNA synthesis. Multiple arrows at one location indicate the presence of multiple genes closely spaced on the chromosome. The origin of replication (*oriC*) is located at 83.5 minutes; the terminus (*terC*) is the region from 28 through 34 minutes. (Adapted from Brewer, 1988.)

evolutionary origin (i.e., genes exhibiting HOMOLOGY) to be clustered in four groups separated by 90° (25 map units) on the circular chromosome. This observation has led to the hypothesis that the chromosome of *E. coli* might have evolved by two successive duplications of a smaller chromosome (Riley and Anilionis, 1978).

Finally, in a rapidly growing cell, genes located near the point of origin of replication are present in greater numbers than those near the terminus, because cells with replicating chromosomes contain more copies of that region of the chromosome. Thus, the location of a gene relative to the origin of replication can affect its level of expression in *E. coli* by a factor as large as 2. Other regulatory factors (Chapters 12 and 13) exert much larger effects, but for genes that might be operating near the maximum possible frequency for bacterial promoters, such as ribosomal genes, a two-fold factor might be important. Indeed, such very active genes (genes for ribosomal RNA and ribosomal proteins and for heat-shock proteins) tend to be located on the *oriC*-proximal portion of the chromosome.

Although gene location is highly conserved, it is difficult to rationalize this observation further. The linkage maps of *E. coli* strains K-12 (Figure 2), B, and C, *Salmonella typhimurium, S. abony, Shigella dysenteriae, Citrobacter freundii, Klebsiella pneumoniae,* and *Enterobacter aerogenes* are quite similar. The detailed maps of *E. coli* strain K-12 and *S. typhimurium* are virtually identical except for a region of approximately 10% of the chromosome in which the gene order in *S. typhimurium* is the inverse of that in *E. coli.* The similarity in gene order between these organisms is remarkable in view of the fact that they have diverged to the point where recombination between them is infrequent. Gene order on the chromosomes of more distantly related bacteria varies widely (Figure 2); the orders for *Bacillus subtilis, Pseudomonas aeruginosa, Acinetobacter calcoaceticus, Streptomyces coelicolor,* and *Rhizobium japonicum* differ greatly from that of enteric bacteria and from one another.

The arrangement of the genes that encode the enzymes of tryptophan biosynthesis illustrates the principle of conservation and variation of gene arrangement (Figure 3). These genes are arranged very similarly in all members of the enteric group that have been examined, but this arrangement has not been found in any other bacterium, nor do other species seem to have gene orders that are similar to one another. Still, certain patterns of clustering can be discerned (e.g., the conserved order E-C-B-A) even in the least related organisms. It would be interesting to know whether this pattern represents CONVERGENT EVOLUTION (independent formation of the same, functionally optimal structure) or DIVERGENT EVOLUTION (retention of an ancestral structure, but with alternative solutions to a common function).

On the basis of a comparison of the nucleotide sequence of ribosomal RNAs, several proposals regarding the pathway of bacterial macroevolution have been offered in recent years. One such proposal (Woese, 1987) is presented in Figure 4; an alternative proposal has been presented by Lake (1988).

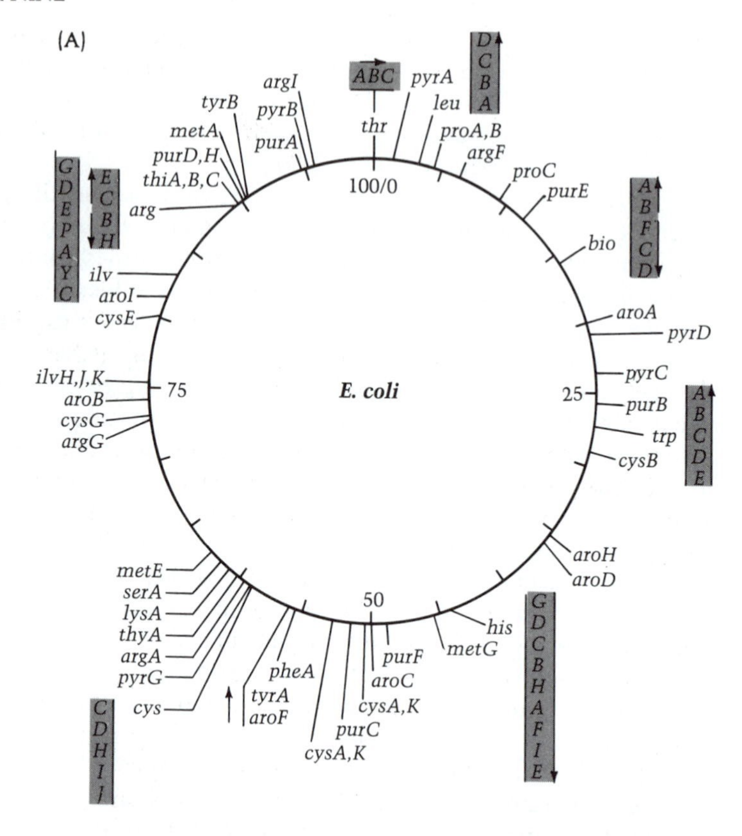

Figure 2

Maps of selected genes on the chromosomes of *Escherichia coli*, *Bacillus subtilis*, and *Pseudomonas aeruginosa*. Genes encoding enzymes in biosynthetic pathways are shown. Gene designations of biosynthetic pathways are *arg*, arginine; *aro*, aromatic amino acids; *bio*, biotin; *cys*, cysteine; *gly*, glycine; *gua*, guanine; *his*, histidine; *ilv*, isoleucine-valine; *leu*, leucine; *lys*, lysine; *met*, methionine; *phe*, phenylalanine; *pro*, proline; *pur*, purine; *pyr*, pyrimidine; *ser*, serine; *thi*, thiamine; *thr*, threonine; *thy*, thymine; *trp*, tryptophan; *tyr*, tyrosine. (A) The map of *E. coli* is divided into 100 units, which correspond to the number of minutes required for a chromosome from an Hfr cell to enter an F$^-$ cell at 37°C. (The map of *Salmonella typhimurium* is similarly divided and the gene arrangement is almost identical.) Genes arranged in operons are shown joined by a horizontal or vertical bar. The arrows indicate direction of transcription. Ninety of a total of 900 mapped genes are shown. (B) The map of *B. subtilis* is divided into 360°; 55 of 360 mapped genes are shown. (After Henner and Hoch, 1980.) (C) The *P. aeruginosa* chromosome is divided into minutes on the basis of the time taken for the chromosome to enter a recipient cell by FP2-mediated conjugation. FP2 is inserted in donor strains at 0 minutes. Late markers have been located by a conjugation system mediated by plasmid R68.45. Although the chromosome has been demonstrated genetically to be circular, the total length of the chromosome in minutes is not yet established. The location of 41 of 95 mapped genes is shown. (After Royle et al., 1981.)

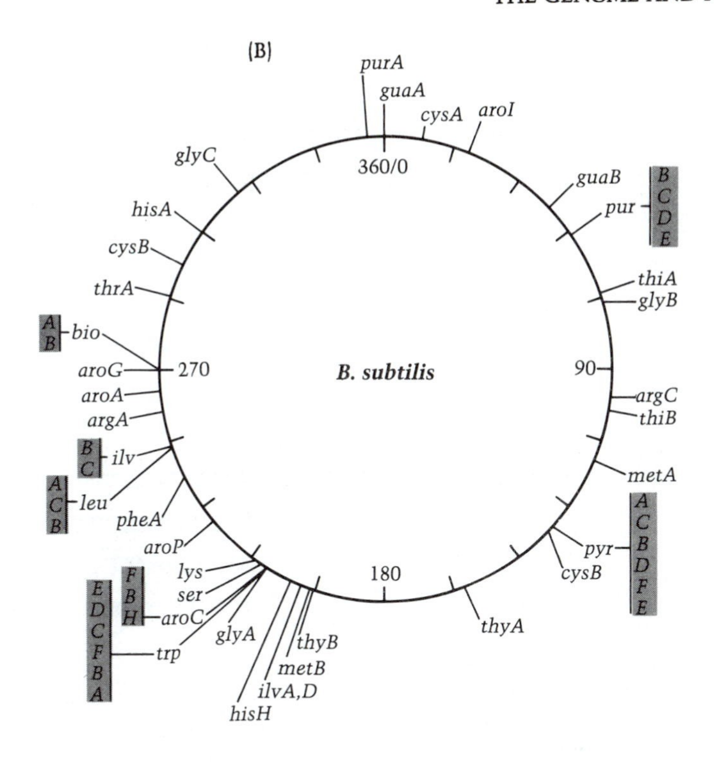
(B)
purA
guaA
cysA
aroI
360/0
glyC
guaB
pur
B
C
D
E
hisA
cysB
thrA
thiA
glyB
A
B
bio
aroG
270
B. subtilis
90
aroA
argC
argA
thiB
B
C
ilv
metA
A
C
B
leu
pheA
A
C
B
D
F
E
pyr
aroP
cysB
lys
F
B
H
ser
180
aroC
E
D
C
F
B
A
trp
glyA
thyB
thyA
metB
ilvA,D
hisH

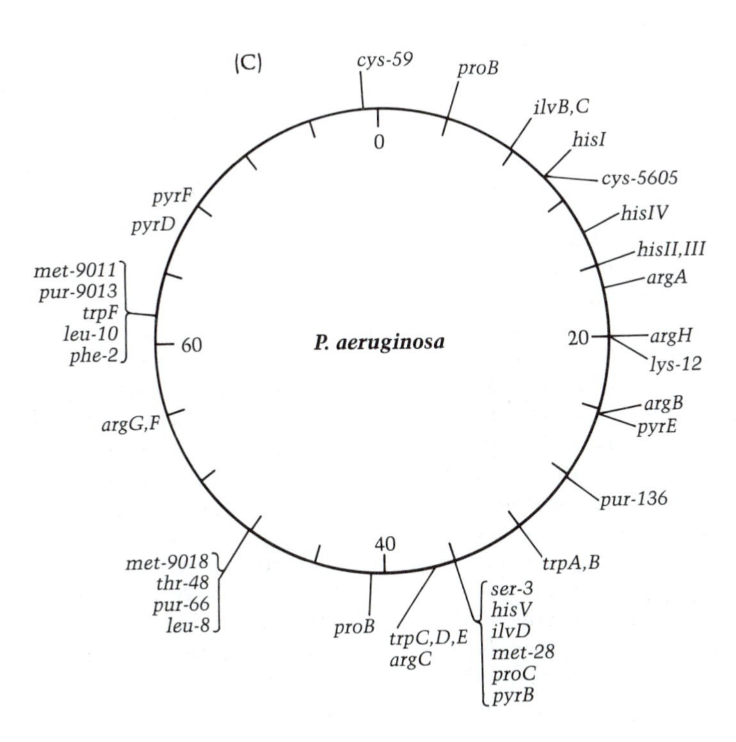
(C)
cys-59
proB
ilvB,C
hisI
0
cys-5605
pyrF
pyrD
hisIV
hisII,III
argA
met-9011
pur-9013
trpF
leu-10
phe-2
60
P. aeruginosa
20
argH
lys-12
argB
pyrE
argG,F
pur-136
40
trpA,B
met-9018
thr-48
pur-66
leu-8
proB
trpC,D,E
argC
ser-3
hisV
ilvD
met-28
proC
pyrB

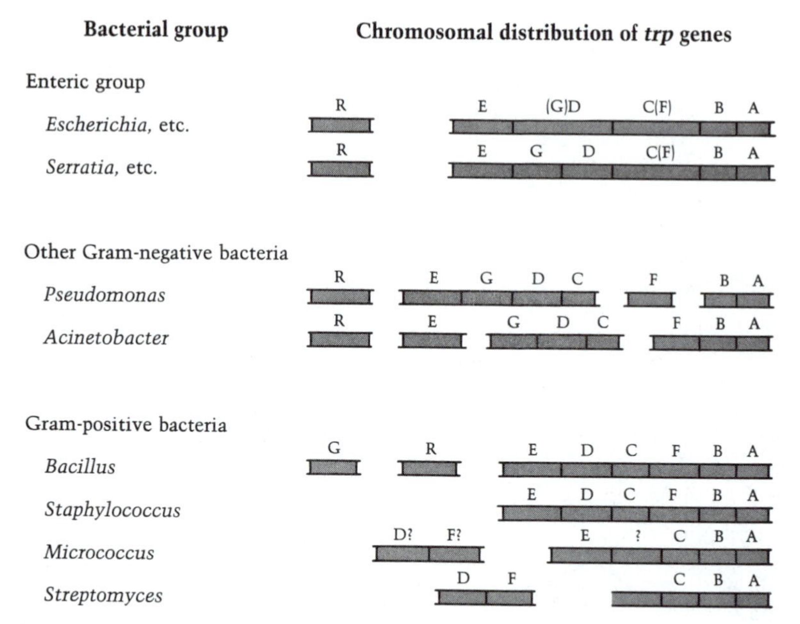

Figure 3

Arrangement of the genes (*trp*) encoding tryptophan biosynthesis in various bacterial species. Continuous double lines indicate contiguous regions of DNA. Capital letters represent the genes (*trpA* through *trpG*) encoding enzymes that catalyze the various biosynthetic steps and the gene *trpR* that regulates their expression. Gene termini are indicated by vertical lines. Letters in parentheses represent other functions also catalyzed by the protein product of the gene designated by the letter without parentheses. (After Crawford, 1975.)

Figure 4

Phylogenetic trees determined from rRNA sequence comparisons. (A) Universal phylogenetic tree. A matrix of evolutionary distances was calculated from an alignment of representative 16S rRNA sequences from each of the three kingdoms shown. This matrix was used to construct a distance tree based on those nucleotide base positions represented in all sequences in the alignment in homologous secondary structural elements. Line lengths on the tree are proportional to calculated distances. The branching order within each kingdom is correct to a first approximation only. (B) Eubacterial phylogenetic tree based on 16S rRNA sequence comparisons. An alignment was constructed from one representative sequence from each of the eubacterial phyla together with an archaebacterial consensus sequence. Using only those positions represented in all sequences in the alignment, the investigators generated an evolutionary distance matrix from which a distance tree was constructed. Branch lengths on the tree are proportional to calculated distances. For those phyla in which additional 16S rRNA sequences are available, the known sequence variation of the group has been calculated and is indicated by the shaded wedges. (From Woese, 1987.)

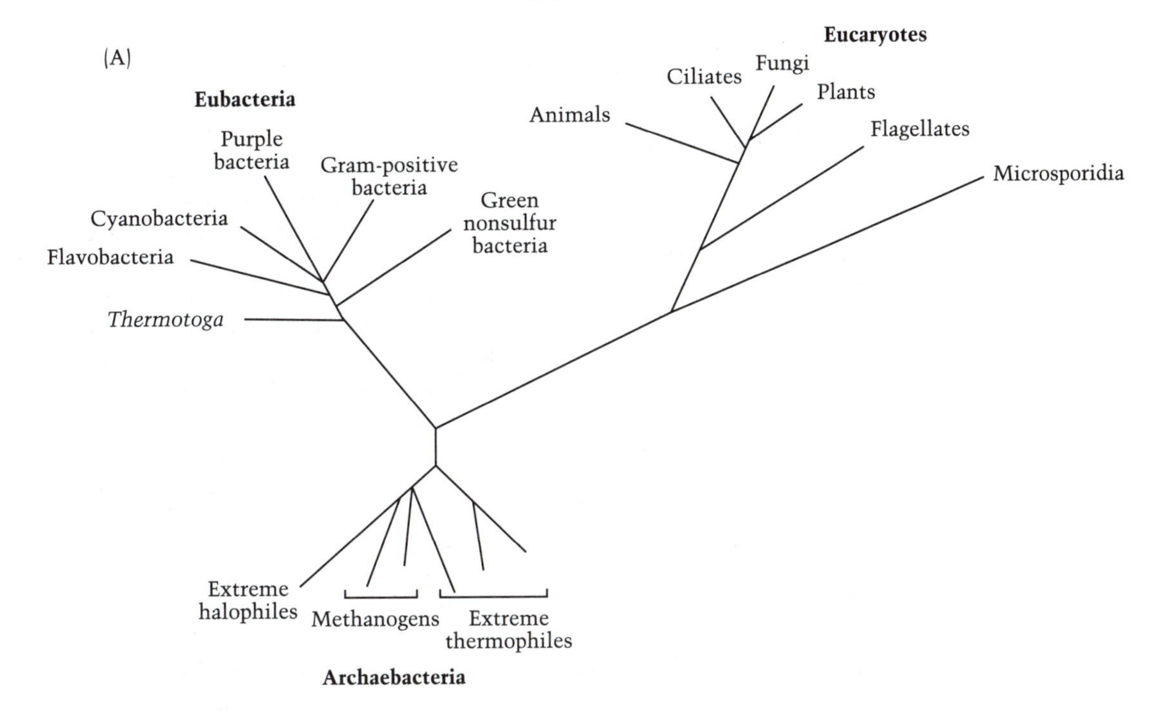
(A)
Eubacteria
Purple bacteria
Gram-positive bacteria
Green nonsulfur bacteria
Cyanobacteria
Flavobacteria
Thermotoga
Eucaryotes
Ciliates
Fungi
Plants
Animals
Flagellates
Microsporidia
Extreme halophiles
Methanogens
Extreme thermophiles
Archaebacteria

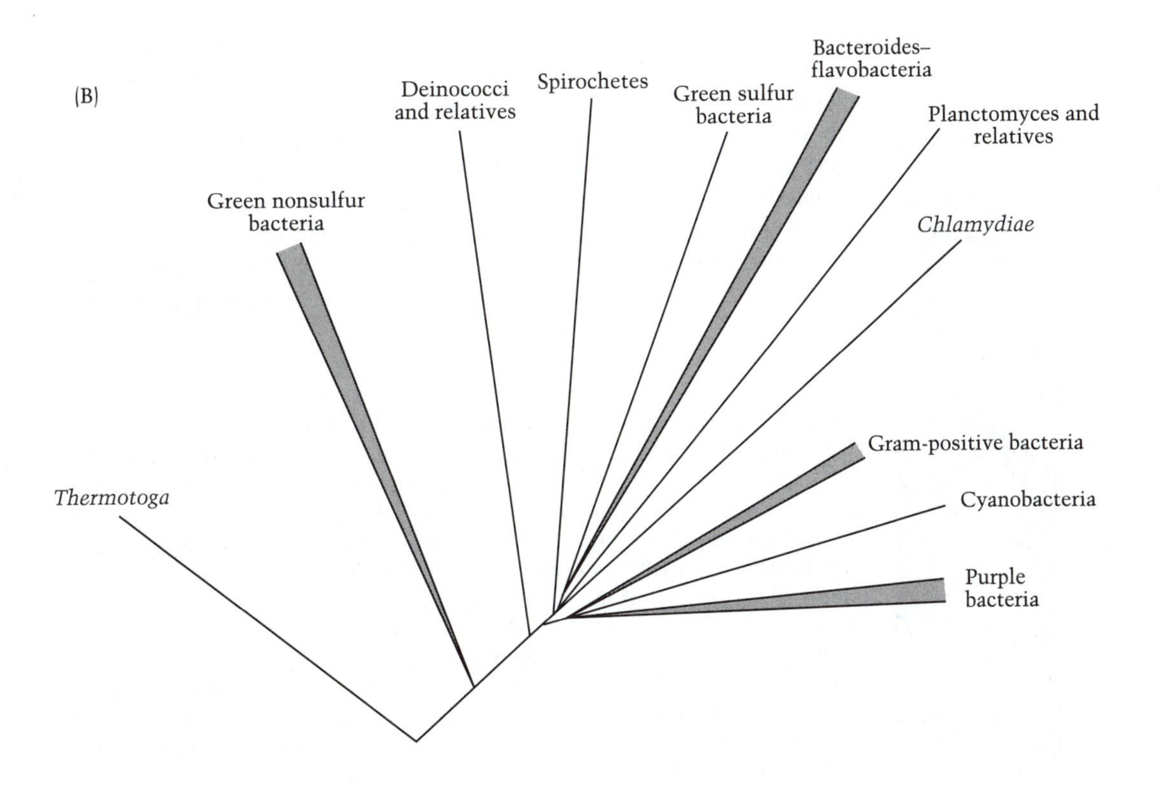
(B)
Deinococci and relatives
Spirochetes
Green sulfur bacteria
Bacteroides–flavobacteria
Planctomyces and relatives
Green nonsulfur bacteria
Chlamydiae
Gram-positive bacteria
Thermotoga
Cyanobacteria
Purple bacteria

PLASMIDS AS ACCESSORY GENETIC ELEMENTS

Definition and detection

Although it is correct to say that the genome of procaryotes is contained on a single chromosome, in fact, most bacteria—Gram-positive and Gram-negative alike—carry a surprising number of different accessory genetic elements in addition to their single chromosome. PLASMIDS are autonomously replicating, extrachromosomal elements composed of circular, double-stranded DNA. They vary in size from 1 to 200 kbp of DNA (bacterial chromosomes are typically 1,000–5,000 kbp). Most plasmids have little or no nucleotide sequence homology with the cell chromosome and can, in this sense, be regarded as foreign to the cell. Moreover, plasmids rarely encode functions that are critical for cell growth or survival in the laboratory.

A single organism may harbor several distinct plasmids, each with its own replication system. The presence of plasmids is easily revealed by subjecting DNA from cell extracts to electrophoresis through an agarose gel, which acts as a molecular sieve. Because DNA molecules move through such a gel as a function of their size and shape, small circular plasmid DNAs migrate much more rapidly than does the chromosomal DNA. It is common to find as many as half a dozen plasmids in strains freshly isolated from natural environments (Figure 5). After prolonged cultivation in the laboratory, strains tend to lose some or all of their plasmids, a fact that provides a clue to important aspects of plasmid biology. The presence of plasmids is, if anything, a burden under laboratory conditions, which generally permit rapid growth; cells that have by chance inherited less than a full complement of plasmids have a selective advantage and outgrow their fellows. In nature, however, the opposite is true: cells with plasmids seem to be favored over those without. Presumably the advantages conferred by plasmids operate when factors other than the simple availability of food determine growth and

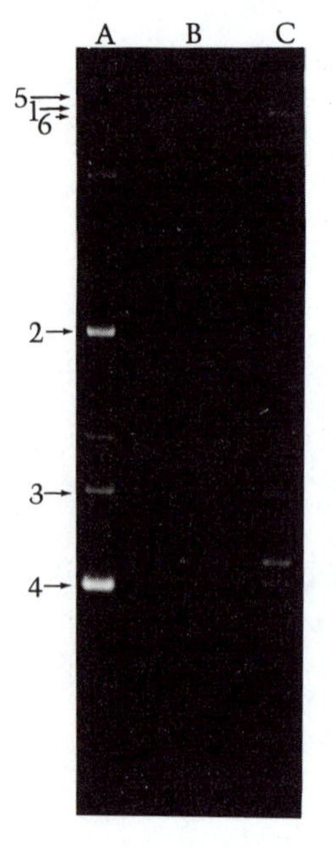

Figure 5

Agarose gel electrophoresis of plasmid DNA from various bacteria. Plasmid DNA from cultures of three organisms was extracted and partially purified by density gradient centrifugation and precipitation to separate it from chromosomal DNA. The plasmid DNA from each organism was then analyzed by agarose gel electrophoresis. The DNA bands are made visible as fluorescing complexes with ethidium bromide. Smaller plasmids migrate faster (toward the bottom of the gel). Lane A: *Enterococcus faecalis*, strain 39-5; lane B: *E. faecalis*, strain DS16; lane C: *Escherichia coli*, strain VA517. The several DNA bands visible in lane A are produced by six plasmids, two of which produce different bands (as a result of nicking of the normally supercoiled circular DNA). The two bands in lane B are two different plasmids. The multiple (>8) bands in lane C are DNA of multiple small plasmids carried by this strain. (Courtesy of D. B. Clewell.)

survival. In the following sections, we shall encounter many examples of situations in which a particular feature of the environment favors the plasmid-carrying cell.

Conjugative plasmids and DNA mobilization

All plasmids contain the requisite sequences and products to govern their own replication. Some plasmids have the additional ability to direct their own transfer from cell to cell. These CONJUGATIVE PLASMIDS carry a set of genes—called *tra* genes—that encode products promoting intimate cell contact and transfer of the plasmid across a conjugation bridge from the donor cell to the recipient. Conjugative plasmids also may facilitate the transfer of nonconjugative plasmids that happen to cohabit the same donor cell; or, as in the case of the F plasmid (Chapter 10), they may mobilize the cell's chromosome for transfer to a recipient cell.

Mobilization usually involves INTEGRATION (covalent insertion) of the conjugative plasmid into the DNA molecule being mobilized. In some cases mobilization may not depend on integration; possibly transient single-strand connections between plasmid and mobilized DNA are responsible. Also, integration need not always result in transfer of the hybrid molecule to a recipient cell. Some plasmids have a high frequency of integration into the chromosome, and, once integrated, can be maintained indefinitely in this fashion during subsequent cell divisions. Plasmids that can either replicate autonomously or be maintained as part of the chromosome are known as EPISOMES.

In *E. coli*, cells carrying integrated F plasmids are called Hfr cells, for **H**igh **f**requency **r**ecombination.

Variety and classification

Plasmids come in various sizes. They differ not only in the number and nature of the genes that they carry but also in the mode of control of their replication. Small plasmids may be no more than a few million daltons (coding for only a few proteins), large ones a hundred times this size. (In all cases, however, they are small with respect to the bacterial chromosome.) In general, small plasmids exist in multiple copies per cell, whereas large ones are often present as a single copy.

The vast variety of plasmids that occur in bacteria are classified on the basis of the organism in which they were discovered and the INCOMPATIBILITY GROUP into which they fall, that is, on the basis of their ability to coexist in the same cell. The phenomenon of incompatibility cannot yet be fully explained in mechanistic terms, but it can be easily described: certain closely related plasmids cannot stably coexist in the same cell. If two incompatible plasmids are introduced into a cell, only one will be passed to progeny cells. A single bacterial cell can harbor a large number of different plasmids if they all fall into *different* incompatibility groups. Some plasmids occur in multiple copies in a cell (the number can easily reach many dozen for some small plasmids), but the copies are identical. If a change that affects its compatibility occurs in one of the copies, then either it or its unchanged partners will not be inherited. Incompatibility has been chosen as the primary criterion of relatedness, and therefore plasmids that are quite different in phenotype are grouped together if they are incompatible. Table 1 lists a number of bacterial plasmids and certain of their properties, including functions they encode.

The plasmids of one incompatibility group, IncP in the enteric bacteria grouping or IncP1 in the pseudomonad group, deserve an additional comment. Whereas most plasmids can exist in only a limited number of closely related bacteria (NARROW HOST RANGE PLASMIDS), IncP plasmids can exist in almost any Gram-negative species (BROAD HOST RANGE PLASMIDS). They also are capable of self-transfer among members of this broad group and even have the capacity to mobilize chromosomal genes of a number of bacterial species at low frequency (10^{-8} per cell). Isolated variants of certain IncP plasmids have a greatly increased ability to mobilize chromosomal genes.

Are broad host range plasmids a potent force in bacterial evolution? Surely it is improper to believe that the interspecies and intergenus transfer of genes by IncP-like plasmids is creating one large melting pot of Gram-negative genomes. In nature, such transfer is presumably rare. But it is clear that the potential for cross-species and even cross-genus and cross-kingdom gene transfer does exist. This possibility must be a serious consideration in evaluating genetic engineering ventures, because a set of specially engineered genes placed in one bacterial species could end up being transferred to others.

Functions encoded by plasmids

All plasmids must have whatever genes are necessary for self-replication, including an origin of replication and, usually, a protein that specifically regulates replication. Conjugative plasmids must carry additional genes for mediating transfer into a recipient cell. Many plasmids carry considerable additional genetic information, much of which may be important to the bacterial cell in natural environments. The variety of cellular properties now known to be plasmid-encoded is very large, and the list continues to grow as more is learned about the molecular genetics of species beyond the well-studied enteric group. These properties include the production of toxins, pili, and other adhesins; resistance to many antimicrobial agents; production of BACTERIOCINS (toxic proteins that kill some other bacteria); production of siderophores for scavenging Fe^{3+}; and many functions of unknown biochemical nature that are required to colonize various niches within mammalian hosts with or without causing overt disease.

In short, many of these plasmid-encoded features could be called SOCIAL FUNCTIONS, i.e., properties that contribute not to the process of growth per se but to the ability of the cell that harbors the plasmid to

Table 1. Properties of six bacterial plasmids[a]

Incompatibility group	Host range	Plasmid	Original host	Phenotype[b]
IncFI	*Escherichia coli* *Proteus morganii* *Proteus vulgaris* *Salmonella typhimurium* *Rhizobium lupini*	F	*E. coli*	Tra^+ Dps (Fl, F2, F17, MS2, μ2) Phi (T3, T7, Q11)
		ColV-K94	*E. coli*	Tra^+ Dps (F1, F2, μ2, MS2, M13)
	Enterobacter cloacae *Proteus mirabilis*	R453	*Proteus morganii*	Tra^+ Dps (F1, F2, MS2) Phi (λ) Ap, Cm, Sm, Sp, Su, Tc
IncFII		R100	*Shigella flexneri*	Tra^+ Fi^+ (F) Dps (F1, F2, MS2, M13) Cm, Fa, Sm, Sp, Su, Tc, Hg
IncA-c	*Providencia*	R480	*Providencia*	Tra^+ Fi^- (F) Dps (PRDI, PRRI)
IncP	Gram-negative bacteria	RP1	*Pseudomonas aeruginosa*	Ap, Km, Tc

[a]Selected from over 400 entries in a table compiled by A. E. Jacobs, J. A. Shapiro, L. Yamamoto, P. I. Smith, S. N. Cohen, and D. Berg describing plasmids studied in *Escherichia coli* and other enteric bacteria, p. 607, in Bukhari, et.al., 1977.
[b]Tra, Ability to promote self-transfer; Dps, plasmid encodes sensitivity of host to phages listed in parentheses; Phi, plasmid encodes resistance of host to phages listed in parentheses; Fi^+, fertility inhibition of F, produces product that inhibits expression of *tra* genes on F. Antibiotic resistance: Ap, ampicillin, Cm, chloramphenicol; Fa, fusidic acid; Km, kanamycin; Sm, streptomycin; Sp, spectinomycin; Su, sulfanylamide; Tc, tetracycline; Hg, mercury resistance.

adapt to its environment. Recall that *occupancy* and *resistance to damaging agents* were two of the environmental demands on bacteria that we identified at the start of Chapter 1. A third demand mentioned was coping with alternating feast and famine, that is, adjusting to the intermittent nature of nutrient availability in natural environments. The pseudomonads are particularly adept at utilizing any one of an enormous number of organic molecules (some as exotic as camphor) for carbon and energy. It turns out that many of the substrate-degrading capabilities of pseudomonads are the result of enzymes encoded by plasmids carried by these cells in their natural environment (soil).

Plasmids that include genes conferring resistance to antibiotics are of great significance in human and veterinary medicine. These plasmids are called R-FACTORS (R = resistance), and their resistance genes encode enzymes that inactivate or degrade specific classes of antibiotics, or change the permeability of the cell to them. Many R-factors encode resistance to multiple antibiotics, making effective chemotherapy of certain diseases an increasingly difficult problem. This problem is compounded by the broad host range of many R-factors. Both pathogenic and nonpathogenic bacteria can therefore serve as reservoirs of R-factors. The heavy use of antibiotics in animal husbandry to increase the rate of growth of beef cattle and swine and to control diseases in chicken factories has raised the fear that these animals are serving as special environments conducive to the evolution and amplification of large populations of R-factors that could spread to human or animal pathogens.

MUTATIONS

It is time now to see how the bacterial genome mutates and evolves under potent influences from the environment. A MUTATION is a change in the base sequence of DNA. Frequently mutations alter the cell's PHENOTYPE (observable characteristics). Mutations can therefore be classified either by the nature of the change in DNA or the change in phenotype. Table 2 lists classes of bacterial mutants on the basis of the phenotypic change they produce.

Table 2. Phenotypes of bacterial mutants

Class	Mutational defect
Auxotrophic	Defect in a biosynthetic pathway—growth of the mutant is dependent on the presence in the medium of the end product of the pathway (amino acid, purine, pyrimidine, vitamin, etc.)
Carbon source	Defect in a metabolic pathway of a compound that serves as a carbon or energy source—the mutant cannot grow if carbon sources prior to the block are the only ones present
Nitrogen source	Defect in a catabolic pathway that yields ammonia—the mutant cannot grow if nitrogen sources prior to the block are the only ones present
Cryptic mutants	Mutants that have lost a particular function but retain the intracellular enzymatic activities to catalyze the reactions of the function; usually such mutants have lost a permease
Conditionally expressed	
Temperature-sensitive	Gene product cannot function or be synthesized correctly (temperature-sensitive synthesis) at the restrictive temperature; functions or is synthesized at near-normal levels at permissive temperature
Heat-sensitive	High temperature is restrictive, low temperature is permissive (usually 42° and 30°C, respectively, in the case of *E. coli*)
Cold-sensitive	Low temperature (20°C) is restrictive, high (37°C) is permissive
Osmotic remedial	Permissive and restrictive conditions are set by the osmotic strength of the medium
Streptomycin remedial[a]	Permissive medium contains streptomycin, restrictive does not

[a]Other aminoglycosides inlcuding kanamycin and neomycin also suppress certain missense mutations.

Table 3. Macrolesions of DNA

Type	Molecular change[a]
Deletion	abcdefghi ⟶ abc-ghi[b]
Duplication	abcdefghi ⟶ abcdef-defghi
Inversions	abcdefghi ⟶ abc-fed-ghi
Translocations (insertions)	abcdefghi ⟶ uvw-def-xyz

[a]Letters are meant to indicate genes on double-stranded DNA.
[b]-indicates improper junction.

In molecular terms, mutations can be conveniently divided into two broad classes: MICROLESIONS, in which a single base pair has been altered, and MACROLESIONS, in which more extensive changes have occurred. Macrolesions include deletions, duplications, inversions, translocations, frameshifts, and insertions (Table 3). Microlesions include transitions, transversions, and frameshifts.

Deletions

DELETION MUTATIONS (losses of segments of DNA) occur at a significant rate during growth and account for approximately 12% of spontaneously occurring mutations. It can be assumed, therefore, that this process plays a significant role in creating genetic diversity in bacterial populations in nature.

The obvious consequence of a deletion mutation is complete loss of a portion of DNA. Because this DNA cannot be regained by subsequent mutation, deletions are characterized as mutations that do not revert spontaneously or by mutagenic treatment. But lack of a detectable reversion rate is not limited to deletion mutations.

A less obvious consequence of a deletion is formation of an IMPROPER JUNCTION (Table 3), that is, the fusion of two previously separate regions of DNA. Sometimes such fusions join part of one gene to the regulatory region of another gene. Such fusions would alter the regulation of the gene in question. This phenomenon can be put to practical use in the laboratory. By fusing the regulatory region of a gene to the *lacZ* gene of *E. coli,* one can study its mechanism of regulation by assaying β-galactosidase activity by a sensitive and convenient procedure. In this case the *lacZ* gene is said to be a REPORTER GENE.

Duplications

DUPLICATIONS (the formation of additional copies of chromosomal segments) occur with remarkably high frequency. Values in the range of 10^{-4} per cell doubling have been obtained for individual *E. coli* genes—

that is, a given gene is duplicated in one cell in every 10,000. Thus, a significant fraction of the cells in a culture can be expected to carry duplications somewhere on the chromosome. Once a duplication is formed, a large region of homology is created, a condition that can lead to further amplification of these genes (Figure 6). On the other hand, duplications are highly unstable, being lost by homologous recombination (Chapter 10) between duplicate segments (Figure 6).

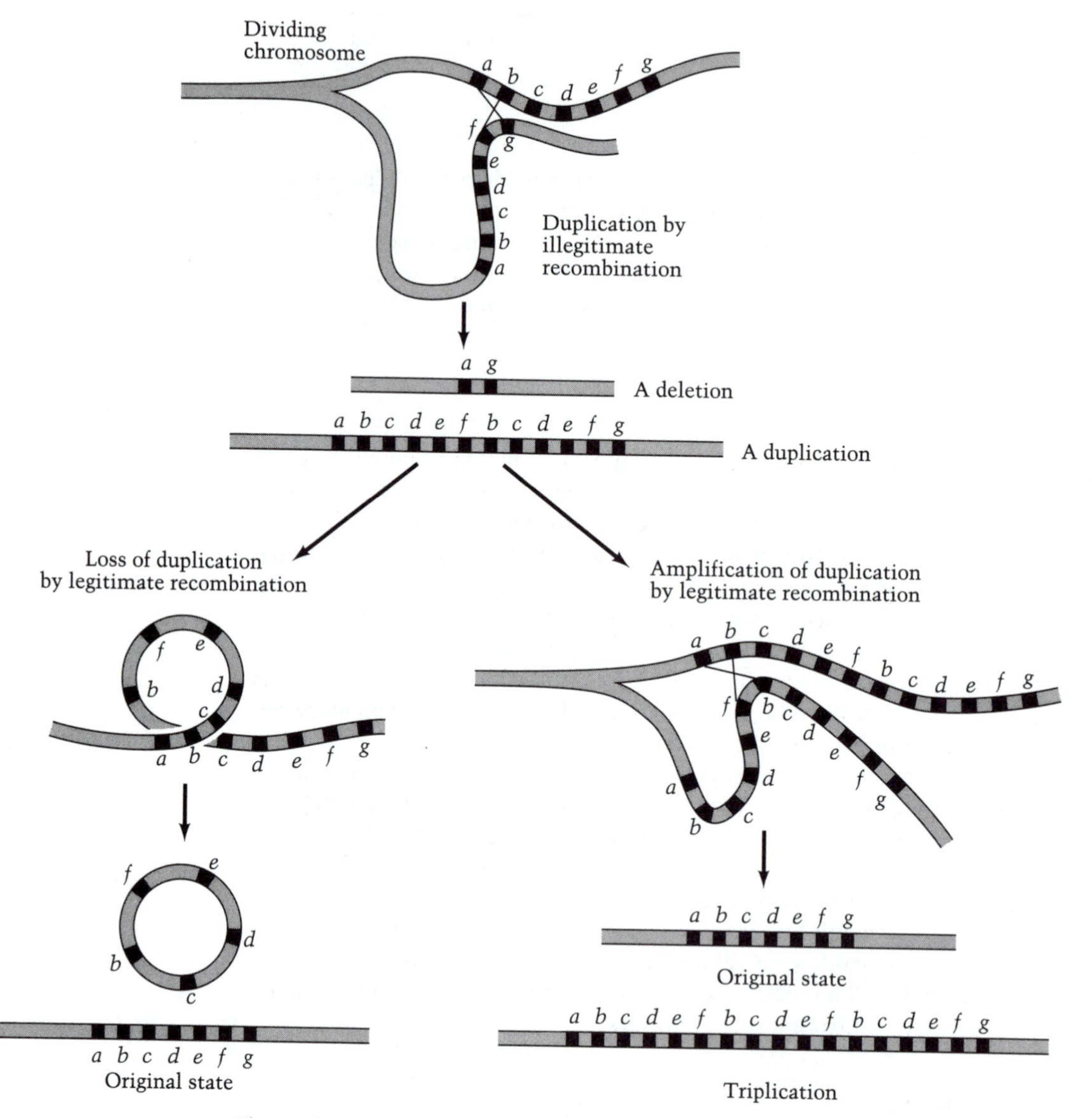

Figure 6

Probable mechanisms for the generation, loss, and amplification of duplication mutations. (After Anderson and Roth, 1977.)

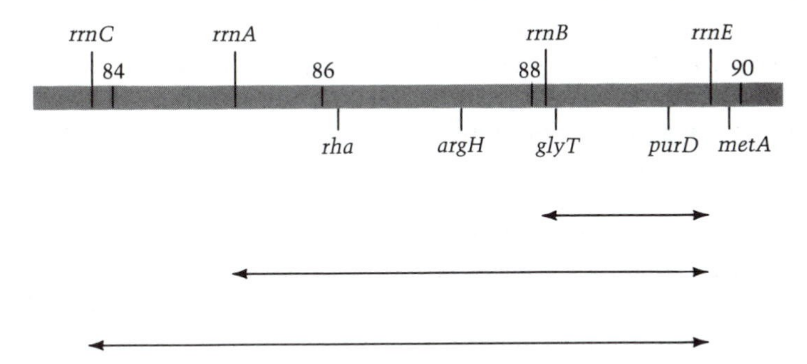

Figure 7

Extent of duplications in the region of the chromosome of *E. coli* that contains genes encoding ribosomal RNA. A set of 25 duplications of *glyT* were selected and their extent was determined by genetic and physical techniques. The majority were shown to terminate within operons encoding ribosomal RNA (*rrnA*, *rrnB*, *rrnC*, and *rrnE*), evidence that these duplications were formed by homologous recombination (Chapter 10) between these redundant genes. The bar represents a region of the chromosome of *E. coli*: numbers along the bar refer to chromosome map units (minutes). In addition to those encoding ribosomal RNA, certain other genes (including *glyT*) are shown. Double-headed arrows indicate the extent of the duplicated regions. (After Hill et al., 1977.)

The mechanism by which duplications are formed is considerably less obvious than the ones by which they are amplified or lost. In some cases it may involve nonhomologous recombination between sister chromosomes. However, in strains that carry a mutation in *recA* (a gene required for homologous recombination) and are therefore unable to undergo homologous recombination, the frequency of formation of duplications is decreased approximately 1,000-fold. Thus, most duplications must form between preexisting homologous regions of the chromosome. This amplification certainly occurred in the genes that encode ribosomal RNA. Four of the seven sets of these genes are clustered in the region of the *E. coli* chromosome between 83 and 90 units (Figure 7). In media that support growth at a high rate (i.e., growth conditions that call for large numbers of ribosomes), cells accumulate duplications that extend between homologous regions of these gene clusters. These duplications are lost when such cultures are transferred to poorer media in which optimal growth does not require large numbers of ribosomes (Chapter 14).

If duplications are so readily lost, how can strains with duplications be isolated? One answer is found in the GENE DOSAGE EFFECT, by which duplication can lead to the synthesis of greater amounts of the products of the duplicated genes. In fact, duplication-carrying clones can be fairly readily isolated by selection in the laboratory. One simply arranges conditions that render the amount of a specific gene product growth-rate limiting. Clones of *E. coli* with duplication of the *lac* operon and others with duplication of the gene for glycyl-tRNA synthetase have been isolated in this way.

There are a number of interesting physiological, genetic, and evolutionary consequences of duplications. Duplications are the only means by which the amount of the cell's resident DNA can be increased other than by addition of foreign DNA. Such mutations are very likely, therefore, to play important roles in evolution. Indeed, as we have noted, the arrangement of genes on bacterial chromosomes provides some evidence for duplication in the relatively recent evolution of bacterial chromosomes.

The duplications that reveal the most about evolutionary processes are those that occur when selection pressures require that a new gene be created with a function similar to that of an existing gene. For example, the tRNA for glycine reads the codons GGA and GGG; by mutation the $tRNA_{Gly}$-encoding gene can be altered to read the arginine codons, AGA and AGG. Such a strain would grow extremely poorly, because its capacity to read the glycine codons would be inadequate. Instead, selection for ability to read both sets of codons yields strains that carry duplicate $tRNA_{Gly}$ genes; one copy makes a tRNA that recognizes GGA and GGG, the other makes one that recognizes AGA and AGG (Hill et al., 1969). Given the high frequencies of duplications in bacteria, such double events can occur with detectable frequency.

Duplications do not cause any loss of function if they extend beyond a gene or an operon, so there is no upper limit to their length, but they do create an improper junction that can result in gene fusion. The existence of a duplication-generated improper junction creates the unusual situation in which a large duplication can be transferred from one cell to another by physically transferring only the small piece of DNA that includes the improper junction (Figure 8). This apparent anomaly occurs as a consequence of crossovers between the legs of a replicating chromosome at regions on either side of the improper junction.

Inversions

INVERSIONS (reversal of gene order) presumably occur spontaneously in bacteria. We make this statement on the basis of the fact that the only significant difference in the organization of the chromosomes of *E. coli* and *Salmonella typhimurium* is the reversal of the gene sequence in one region of the chromosome. Systematic studies on spontaneous formation of inversions are still few in number, but interesting information is being learned about the rules governing inversion in *E. coli*. For example, inversions are more readily isolated if the inverted segment includes the terminus (*ter*) region of the chromosome. Also, certain genetic elements (see the section in this chapter on insertion sequences and transposons) can generate inversions as well as deletions during their translocation to different regions of the chromosome.

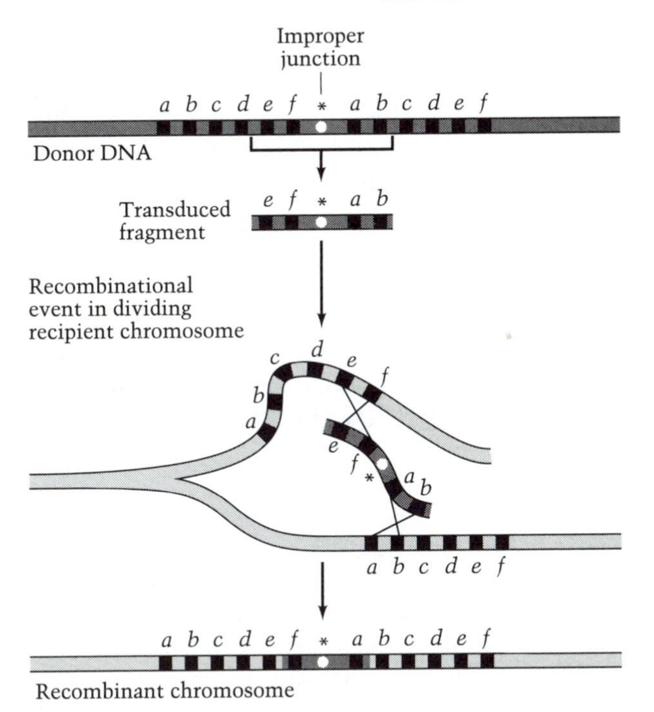

Figure 8

Mechanism of transduction of duplications larger than the phage genome. The asterisk indicates the improper junction; see text for details. The fragment of DNA is introduced into the recipient cell by a pseudovirus in the process known as transduction (Chapter 16).

Translocations (Insertions)

TRANSLOCATION, which is the movement of a fragment of DNA from one region of the chromosome to another, occurs rarely in bacteria. But genetic elements called insertion sequences and transposons have the special property of replicative translocation, whereby a copy appears at a new location on the chromosome and the original copy remains intact. This event results in an INSERTION MUTATION, which usually inactivates the gene within which the element has been introduced. The properties of these mutagenic elements are discussed in a later section.

Microlesions

Changes in a single base pair in which one purine is substituted for another purine, or one pyrimidine for another pyrimidine, are called TRANSITION MUTATIONS; substitutions of a purine for a pyrimidine and vice versa are termed TRANSVERSION MUTATIONS (Figure 9). Additions or losses of one or two base pairs within coding regions produce FRAMESHIFT MUTA-

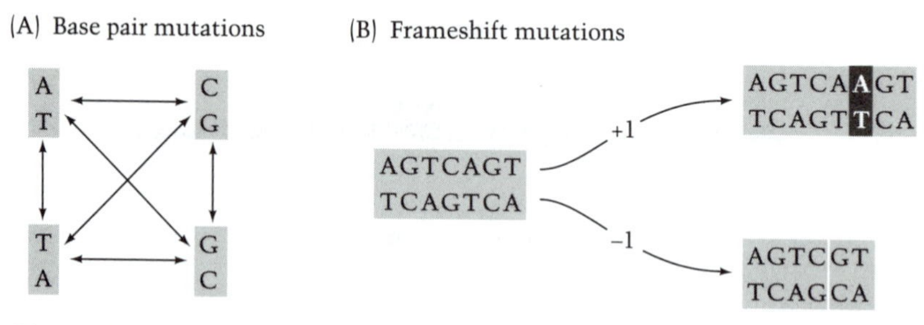

Figure 9

Classes of microlesion mutations. (A) Base pair mutations. Horizontal and vertical arrows indicate transversions; diagonal arrows indicate transitions. (B) Frameshift mutations illustrated by possible changes in a length of double-stranded DNA.

TIONS because, not being divisible by three, they shift the reading frame of the ribosome as it traverses the mRNA made from the mutated gene. (Recall that bases in mRNA are read in groups of three.)

There are several possible consequences of a base-pair substitution, depending on the specific circumstances. There is no effect whatsoever if the gene product is a protein and the substitution produces an alternate codon for the amino acid encoded by the wild-type allele. For example, as stated earlier, the AGA and AGG codons both encode arginine (Figure 10 in Chapter 3). A transition mutation resulting in a change in the purine in the third position would have no effect on the protein product. Base substitutions that cause a different amino acid to be inserted into the protein product—mutations termed MISSENSE MUTATIONS—might or might not have a significant effect on the activity of the protein, depending on the extent to which the affected amino acid changes the properties of the protein. Most missense mutations do not detectably affect the catalytic activity of an encoded enzyme; temperature lability of the protein is more frequently affected, as is susceptibility to proteolytic attack.

The base-substitution mutations that generate a nonsense codon—therefore termed NONSENSE MUTATIONS—almost always lead to production of a nonfunctional gene product, because truncated proteins are usually nonfunctional. Several tRNA species can be mutationally altered so that they read, albeit poorly, a nonsense codon. This situation results in the cell being able, at low frequency, to produce a functional protein with an amino acid inserted at the position of the nonsense codon. The change in the tRNA gene is called a SUPPRESSOR MUTATION, because it suppresses the effect of the primary nonsense mutation. Misreading of the nonsense codon where it appears as an appropriate termination signal occurs in such a cell, but at a frequency so low that it is not fatal.

Frameshift mutations change the register and hence the coding properties of the portion of the gene that is transcriptionally distal to the site of the mutation. This miscoded region is usually so different from the normal sequence that frameshift mutations almost always render the gene product completely inactive. Nonsense codons are frequent among the new codons generated by this mutational shift in reading frame. Thus, truncated proteins are a common product of frameshift mutations.

MUTAGENESIS

Mutations occur spontaneously because the process of replicating the chromosome is not completely free of errors. In spite of proofreading during replication (Chapter 3), errors do occur, albeit at very low frequency during normal growth in a benign environment. The rate of formation of spontaneous mutations, termed the MUTATION RATE, is calculated as the number formed per cell doubling, according to the formula

$$a = \frac{m}{\text{cell generations}} = \frac{m \ln 2}{n - n_0}$$

where a is the mutation rate and m is the number of mutations formed as the number of cells increases from n_0 to n.

The spontaneous mutation rate can be increased by certain mutant alleles, termed MUTATOR GENES. Although the molecular mode of action of most mutator genes remains unexplained, one of them encodes a defect in DNA polymerase, and it is assumed that other mutator genes also affect some aspect of the replication process, making it more susceptible to error.

Environmental conditions, including the presence of low levels of certain agents that induce mutations (MUTAGENS), can be evaluated by their effect on mutation rate, but mutation rate alone is not a good index of efficacy of most mutagenic treatments used in the laboratory. Many mutagens do not depend on chromosomal replication to be fully active. The frequency of mutant cells in the surviving population is a better index of the effect of a mutagenic treatment.

In the natural environment, mutagenesis is brought about by

- replication errors
- chemicals
- irradiation
- insertion of transposons or similar DNA sequences

Mutagenic chemicals (Figure 10) include BASE ANALOGUES, which are structural analogues of the four DNA bases that become incorporated into DNA and cause mispairing when the DNA containing them is replicated; DNA-REACTING MUTAGENS, which directly alter DNA bases, thereby changing their complementarity properties; and INTERCALATING AGENTS, which are flat molecules that can intercalate between base pairs in the central stack of the DNA double helix, thereby distorting the structure and causing subsequent replication errors.

Various types of radiation are mutagenic. X rays cause chromosomal breaks; attempts by bacteria to repair such breaks often lead to various

Class	Examples	Structure	Mode of action
Base analogue	2-Aminopurine → 2,6-Diaminopurine	2-aminopurine ring (N, NH, NH_2, N, N)	Can pair with thymine or cytosine
	5-Bromouracil 5-Chlorouracil		Can pair with adenine or guanine
Chemical modifier of DNA	Nitrous acid		Deaminates DNA bases
	Hydroxylamine		Hydroxylates 6 amino groups of cytosine
Alkylating agent	Nitrosoguanidine → Ethyl methane sulfonate Methyl methane sulfonate Diethyl sulfonate	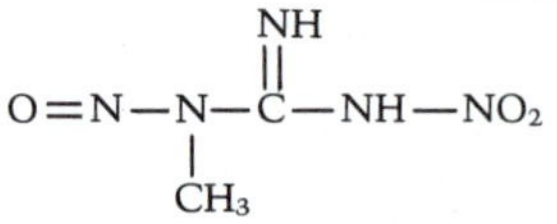 $O{=}N{-}N(CH_3){-}C({=}NH){-}NH{-}NO_2$	Alkylate purines
Intercalating agent	Acridine orange ICR 191 → Proflavine	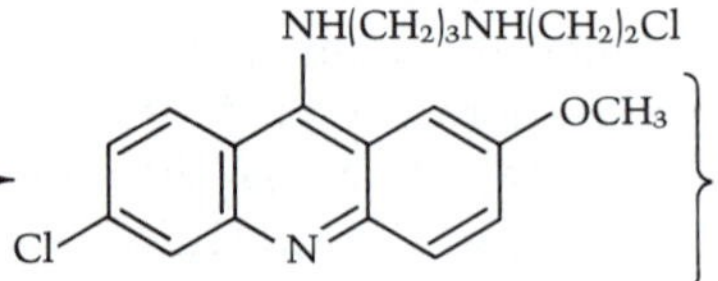 Acridine ring with $NH(CH_2)_3NH(CH_2)_2Cl$, OCH_3, Cl, N	Cause frameshift mutations

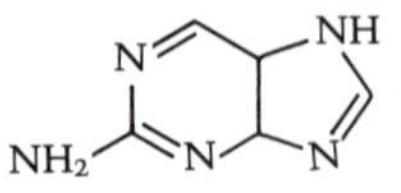

Figure 10

Classes and examples of mutagenic chemicals. Nitrosoguanidine is the trivial name of *N*-methyl-*N*′-nitro-*N*-nitrosoguanidine.

macrolesions. Ultraviolet (UV) light, an environmentally prevalent agent, catalyzes the joining of adjacent pyrimidines, principally thymine, at positions 4 and 5. The cell's repair system (SOS system, Chapter 13) is then activated to excise these PYRIMIDINE DIMERS. The process of repair of these single-strand gaps, which are first extended by nucleases, is less accurate than is normal chromosomal replication. This ERROR-PRONE REPAIR causes mostly transversions but also brings about some frameshifts and transitions.

Finally, a very special class of mutator elements, the transposable genetic elements, deserve extensive consideration because of the enormous contribution they make to the plasticity of the bacterial genome.

TRANSPOSABLE GENETIC ELEMENTS

Within the bacterial chromosome and the accessory genetic material (the plasmids) reside elements that have the capacity to TRANSLOCATE in a special way to new locations in the genome. These elements are collectively known as TRANSPOSABLE ELEMENTS; they include four general types—insertion sequences (IS), transposons (Tn), invertible elements, and the prophage (Chapter 10) of a particular virus called Mu. Note that transposition occurs within a single cell; it is unrelated to conjugative transfer of DNA between cells.

All transposable elements consist of segments of DNA that translocate by duplication: one copy of the element appears at a new site in the genome and the other copy remains at the original site. Insertion at the new site involves a staggered cut in the target DNA; after ligation of the transposed element, the gaps due to the staggered cuts are filled by synthesis directed by the opposite strand, a process resulting in short segments of directly repeated nucleotides bordering the transposed element. These DIRECT REPEATS of the target DNA, by the nature of their origin, differ from site to site for any given transposable element. The insertion in a new site is often mutagenic, because the element usually interrupts the continuity of a gene and destroys its function. Most transposable elements encode strong translational stop signals near their ends. Therefore, the mutations they cause are highly POLAR; that is, if they are inserted in a gene within an operon, the expression of all genes of that operon distal to the one inactivated by insertion of the element is greatly reduced.

Insertion sequences and transposons are distinguished by size and by the amount of information they encode. INSERTION SEQUENCES are small transposable elements (800–2,000 bp) and encode only the ability to transpose. TRANSPOSONS are larger transposable elements (more than 2,000 bp long) and encode a variety of enzymes in addition to those that catalyze transposition (Table 4). The structures of all known transposons have certain common features (Figure 11). At their ends are short IS-like regions, almost exactly duplicated at the ends of any particular transposon. These REPEATS, as they are called, lie either in the same orientation with respect to each other (DIRECT REPEATS), or, as is more commonly the case, in the opposite orientation (INVERTED REPEATS). Note that these repeated segments are in addition to the direct repeats in the target DNA. A variety of enzyme activities are encoded in the central region (the CORE) between the repeated ends. Many of these have the capacity to inactivate a specific antibiotic, by phosphorylation, acetylation, or adenylylation. Other transposons encode membrane proteins that cata-

Table 4. Properties of seven transposable elements

Designation	Size (base pairs)	Direct repeat at target (base pairs)	Core encodes	Terminal repeats (base pairs)
IS*1*	768	9	—	30, inverted
IS2	1,327	5	—	32, inverted
IS3	1,400	3–4	—	32, inverted
Tn*1*	5,000	5	Ampicillin resistance	38, inverted
Tn5	5,700	9	Kanamycin resistance	1,400, inverted
Tn9	2,600	9	Chloramphenicol resistance	768 (IS*1*), direct
Tn*10*	9,300	9	Tetracycline resistance	1,400 (IS*10*), inverted

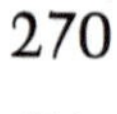

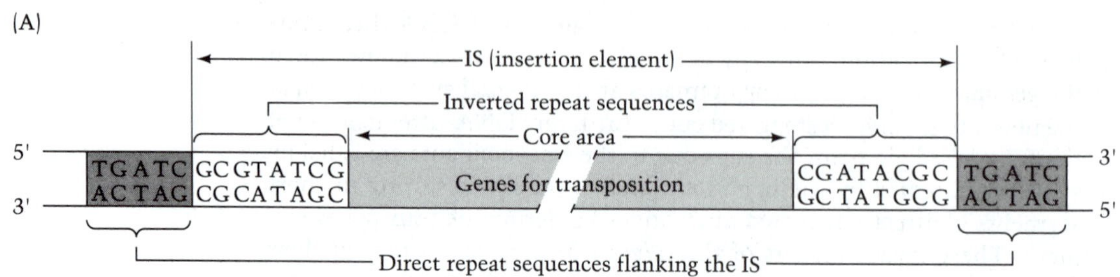

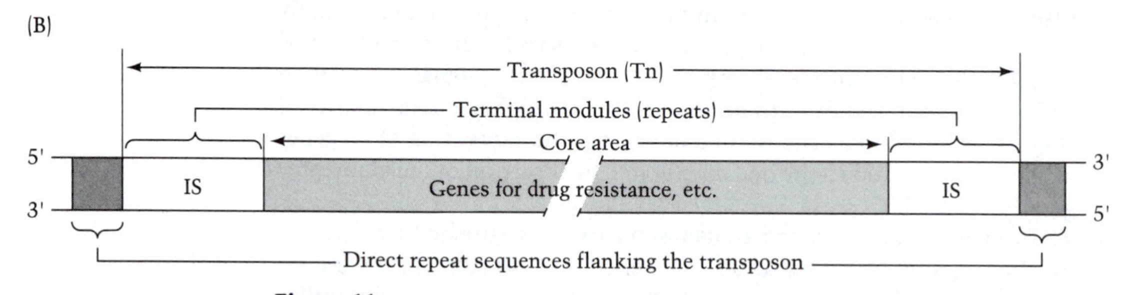

Figure 11

Generalized structure of transposable elements. (A) An insertion sequence (IS) element. (B) A transposon (Tn). Features of specific elements are given in Table 4.

lyze the active efflux of antibiotics from the cell or encode enzymes that catalyze the catabolism of a substrate. The antibiotic-resistance genes carried by R-factors are often found within transposons. Thus, transposons probably play a major role in the distribution of antibiotic resistance in natural bacterial populations.

The process of transposition of insertion sequences and transposons occurs unimpeded in mutant cells that cannot manage "normal" (*recA*-mediated) homologous recombination (Chapter 10). It is clear that these transposable elements carry sufficient information to encode the proteins necessary and sufficient for their own transposition. There is specificity with respect to the sites into which transpositions occur. In certain cases (e.g., Tn*5* and Tn*10*), the apparent specificity is low, because insertions occur at so many sites that it appears they can take place in most, if not all, genes. But more detailed analysis has shown that insertion occurs with greatest frequency at certain sites within a gene. The insertion of other transposons is highly specific; Tn*7* inserts with detectable frequency only at a single site on the chromosome of *E. coli* and at a limited number of sites on the chromosomes of other bacteria. Besides causing mutations, the introduction of transposons into the bacterial genome creates regions of homology that facilitate additional genetic rearrangements.

How do transposable genetic elements enter cells? The most common vehicles are viruses (phages) and conjugative plasmids. Once in, the transposable element has a good chance of translocating to the cell's chromosome. Transposons Tn*5* and Tn*10*, and presumably other transposons, inhibit their own translocation through the mediation of a soluble protein product that they encode. As a consequence, translocation frequency is highest when a transposon is freshly introduced into a cell. Later, buildup of the controlling protein reduces the frequency of subsequent transposition.

The discovery of transposons explained the ubiquity and mobility of genes encoding resistance to antibiotics. The antibiotic resistance genes carried by R-factors that are embedded within transposons can be amplified by tandem duplications on the plasmid and can hop to other plasmids or to the chromosome in the same cell. The rapid evolutionary development of multiple drug resistance plasmids since 1960 is a predictable result of natural selection resulting from the properties of transposons and plasmids and the intensive use of antibiotics in human and veterinary medicine.

The discovery of transposable elements also paved the way to solution of a very long-standing problem in bacterial genetics—PHASE VARIATION. For decades it had been known that populations of some bacterial species have the ability to switch from one phenotype to another, from making flagella of one antigenic type to making flagella of another type. For the bacteria, the switching made sense, for it occurred under circumstances that enabled them to evade the immunological response of the human or animal host during infection. What was disturbing to microbi-

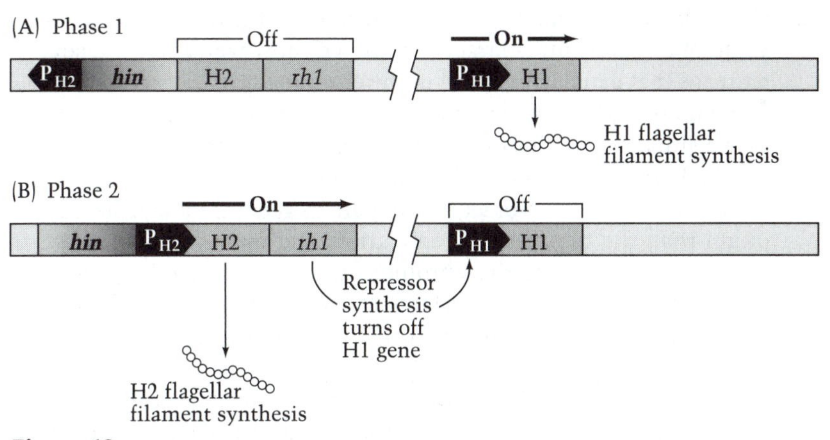

Figure 12

Mechanism of flagellar antigenic phase variation in *Salmonella typhimurium*. The two genes encoding flagellins, H1 and H2, are in separate locations on the chromosome. The promoter, P_{H2}, for the H2 gene, is within an invertible element (heavy arrow), which is bounded by inverted repeat sequences and contains the gene, *hin*, encoding the DNA invertase, Hin. The H2 operon also contains a gene, *rh1*, encoding a repressor protein that acts on the promoter, P_{H1}, of the flagellin H1 gene. (A) In phase 1, the only active flagellin gene is H1, because P_{H2} is oriented away from the H2 gene. (B) In phase 2, P_{H2} is oriented toward H2, leading to the synthesis of flagellin H2 as well as repressor Rh1, which ensures that the H1 gene is turned off.

ologists was that the population switch occurred at a speed that seemed too high to be caused by selection operating on random mutational variation in the population. Studies on the molecular mechanisms of transposition, and of related site-specific recombinational processes (Chapter 10), eventually contributed to the discovery of transposon-like INVERTIBLE ELEMENTS. These segments of DNA code for the proteins necessary to carry out transposition, but in this case the proteins include a DNA INVERTASE, an enzyme capable of bringing about the inversion of the element without changing its location on the chromosome. An invertible element that produces a DNA invertase called Hin, controls two flagellar protein genes, one directly and one indirectly; consequently, only one is ON and the other OFF at a given time (Figure 12). Because the inversion is brought about as a specific, enzyme-catalyzed reaction, the high frequency of phase variation (as high as once per 1,000 cells per generation) is readily understood as a very special form of mutational event.

SUMMARY

1. In addition to the physiological adaptability of individual bacterial cells, populations of bacteria change genetically in response to environmental pressures. Genetic adaptation occurs in several ways, but all involve natural selection operating on populations of bacteria that have acquired mutations.

2. The rationale for the detailed arrangement of genes in the chromosome of a bacterium is largely unknown; however, there are several patterns that can be discerned in the chromosome of enteric bacteria. Coregulated genes of a pathway are more often than not clustered into a single transcriptional unit, an operon. Related genes are commonly found either 90° or 180° apart on the circular chromosome. Genes with high rates of expression are in general located nearer the origin than the terminus of replication, and most are transcribed in the same direction as the chromosome is replicated. The chromosomes of relatively closely related species and genera display a conserved gene order. More distantly related species display little conservation.

3. A broad outline of phylogenetic relatedness of bacteria can be offered from a study of the detailed molecular structure of conserved genes, such as those for ribosomal RNA.

4. Most bacteria carry accessory genetic elements in the form of small, dispensable DNA molecules called plasmids. Plasmids commonly carry genes that encode functions contributing to the ability of the bacterial cell to cope with its natural environment. They are typically responsible for properties related to colonization of surfaces and to defense against harmful agents, but they may also encode auxiliary enzymes of fueling reactions. Besides vertical transmission through partition to daughter cells, some plasmids are capable of horizontal, conjugal transfer between cells.

5. Mutations occur frequently enough in the large natural populations of bacterial species to provide allelic heterogeneity in virtually each gene. This condition permits continued evolution of protein products to optimize their properties and the nature of their regulation. Duplications can provide "experimental" copies of genes, thereby enabling the accumulation of microlesions in one of the copies while retaining another copy for its unmodified and perhaps essential function. This process can account for the formation of entirely new genes with protein products having properties and functions related to, but distinctly different from, those of the original genes.
6. Mutations occur spontaneously as uncorrected errors in replication and as inversions and deletions brought about by intrachromosomal recombinational events. Mutation frequency is increased by a large number of environmental agents, both physical and chemical, that either damage components of DNA or interfere with accurate replication.
7. Mutations are brought about also by naturally occurring genetic elements with the property of replicative transposition, that is, the ability to move from one location on a chromosome to another location on the same or a different DNA molecule. The simplest of these transposable elements are insertion sequences, which encode functions related only to their own transposition. More complex elements, the transposons, encode other functions, typically antibiotic resistance. Insertion of a transposable element within the coding segment of a gene inactivates that gene. Transposons provide a rapid means for the evolution of plasmids with novel arrangements and combinations of genes and are therefore potent forces in the microevolution of bacteria.
8. Other changes in DNA are brought about by segments of DNA called invertible elements. These transposon-like segments of DNA produce DNA-invertase which brings about the inversion of the segment without changing its location on the chromosome. One of these elements is responsible for the control of flagellar genes in *S. typhimurium*, bringing about phase variation by determining which of two alternative genes will be active.

STUDY QUESTIONS

1. What general features of the bacterial genome can be related to the ability of these organisms to grow rapidly?
2. What findings indicate that the arrangement of genes on the bacterial chromosome is not random?
3. Define each of the following names of genetic elements, making clear the distinctions among them: phage, prophage, plasmid, and episome.
4. On the basis of your understanding of plasmids as accessory genetic elements in bacteria, predict which of the functions listed below are likely to be encoded on plasmids and which on the chromosome.

Explain the basis of each of your predictions.

- synthesis of valine
- degradation of superoxide anion
- synthesis of an adhesin
- production of a protein toxic to human cells
- synthesis of an enzyme of the TCA cycle
- synthesis of $tRNA_{fMet}$

5. Show how the various mutational changes might be involved in the evolution of a new, complex function in bacteria: acquisition of the ability to synthesize, through a six-enzyme pathway, a new amino acid structurally related to threonine.
6. What difficulties do you perceive in the task of defining "the normal rate of mutation" in bacteria.
7. Define each of the following terms, making clear the distinctions among them: transposon, insertion sequence, transposable element, R-factor, and invertible element.
8. Evaluate the concept that mutational changes are random events, not directed to occur in particular genes by environmental factors.
9. Explain to your brother/sister/significant other/friend why an individual bacterium with two copies of its single chromosome is defined as haploid.

SUGGESTED READING

Topics in this chapter are covered in the following chapters from *Escherichia coli and Salmonella typhimurium: Cellular and Molecular Biology,* F. C. Neidhardt, J. L. Ingraham, K. B. Low, B. Magasanik, M. Schaechter and H. E. Umbarger (eds.). 1987. American Society for Microbiology, Washington, D.C.

Bachmann, B. J. Linkage map of *E. coli* K-12, Edition 7. (Volume 2; Chapter 53)
Craig, N. L. and N. Kleckner. Transposition and site-specific recombination. (Volume 2; Chapter 62)
Eisenstadt, E. Analysis of mutagenesis. (Volume 2; Chapter 59)
Riley, M. and S. Krawiec. Genome organization. (Volume 2; Chapter 56)
Sanderson, K. E. and J. A. Hurley. Linkage map of *Salmonella typhimurium.* (Volume 2; Chapter 54)

Other suggested readings include:

Bacterial Evolution

Lake, J. A. 1988. Origin of the eukaryotic nucleus determined by rate-invariant analysis of rRNA sequences. Nature 331:184.
Woese, C. R. 1987. Bacterial evolution. Microbiol. Rev. 51:221.

Insertion Sequences and Transposons

Berg, D. E. and M. M. Howe (eds.). 1989. *Mobile DNA.* American Society for Microbiology, Washington, D.C.

10

Genetic Adaptation: Genetic Exchange and Recombination

INTRODUCTION

The plasticity of the bacterial genome generated by the mutational processes described in the preceding chapter can account for many aspects of bacterial evolution. A central tenet of evolutionary biology, however, is that the speed of evolution is enormously increased by bringing together mutant alleles of different genes into a single cell for testing in the crucible of natural selection. Two mutant genes of little or no selective advantage individually might offer significant advantage and become established in a population when they are combined in a single cell. Presumably, mixing and reassortment of genes from different individuals provided one of the main forces that led to the evolution of diploidy and sexual reproduction in nonbacterial organisms.

It turns out that bacterial genes can be transferred from one cell to another; so the products of independent mutational events can be combined in one cell and then passed to progeny. In this chapter we examine the processes that enable transfer of DNA between cells and the mechanisms that then permit the transferred DNA to become a stable part of the recipient cell's genome. Together these processes provide an alternative to sexual reproduction for accelerating evolution.

GENETIC EXCHANGE

Three different mechanisms accomplish gene transfer in bacteria. Certain bacteria take soluble DNA from their surroundings by a process called TRANSFORMATION. Certain bacterial viruses occasionally transfer bacterial genes from one cell to another by a process termed TRANSDUC-

TION. Direct transfer of bacterial genes by cell-to-cell contact is brought about by a process termed CONJUGATION. Of the three known mechanisms of bacterial genetic exchange, only transformation appears to have evolved for the function of genetic exchange (having perhaps originally served a nutritional function). The others—transduction and conjugation—may bring about genetic exchange but only as a consequence of errors in phage growth and plasmid transfer, respectively.

Despite the fundamental differences among these three mechanisms, they share certain important common features. With rare exceptions, only a portion of the chromosome is transferred in each case; and the transfer is always polarized, that is, genetic material of the DONOR cell is transferred to the RECIPIENT cell unidirectionally. Because a complete mixing of genomes never occurs, the product of transfer is not a true zygote, so the term MEROZYGOTE is applied to the product to emphasize this partial mixing of genomes. Because the transferred DNA—EXOGENOTE—is only a fragment of the donor chromosome, it usually lacks those features that allow replication; it is not a REPLICON—a DNA molecule capable of self-replication. Thus, unless the exogenote becomes incorporated by crossover into a replicon (the chromosome or a plasmid) of the recipient cell's genome—ENDOGENOTE—it will not be spread among the progeny cells; only a single cell of the clone that develops will contain the unincorporated exogenote. One special case of gene exchange does involve transfer of a replicon—a transmissible plasmid; and integration into the chromosome of the recipient is not necessary for subsequent passage to progeny cells.

Transformation

Transformation holds a special place in the history of molecular genetics and in the biology of procaryotes. It was the first form of genetic exchange to be discovered—by F. Griffith (1928). The finding that the active agent of transformation (the TRANSFORMING PRINCIPLE) is DNA—by O. T. Avery, C. M. Macleod, and M. McCarty (1944)—was the first direct evidence that this macromolecule encodes genetic information.

A wide variety of bacteria undergo transformation (Table 1), and the molecular details of the process are somewhat variable among them. We present two of the best-characterized natural transformation systems and then describe artificial transformation, which can be induced to occur in certain species of bacteria that are naturally impervious to DNA.

Transformation of *Streptococcus* / Griffith discovered transformation in the Gram-positive bacterium *Streptococcus pneumoniae* when he observed that mice died when injected simultaneously with live cells of an avirulent strain and killed cells of a virulent strain. The tissues of the dead mice contained bacterial cells that had genetic characteristics of both injected strains. Later it was established that bits of DNA from the killed bacteria entered cells of the other strain, recombining with the DNA of the avirulent strain. *Streptococcus* has remained a subject for

Table 1. Bacteria known to be capable of transformation

Natural transformation
Gram-positive bacteria
Streptococcus pneumoniae, S. sanguis, Bacillus subtilis, B. cereus, B. stearothermophilus
Gram-negative bacteria
Neisseria gonorrheae, Acinetobacter calcoaceticus, Moraxella osloensis, M. urethalis, Psychrobacter sp., *Azotobacter agilis, Haemophilus influenzae, H. parainfluenzae, Pseudomonas stutzeri*
Artificial transformation
Escherichia coli, Salmonella typhimurium, Pseudomonas aeruginosa

studies on transformation. These organisms and *Bacillus* species, which have also been intensively studied, are paradigms of one brand of transformation (Figure 1).

Cells of a transformable strain of bacteria are not always transformable. When they are, they are said to be COMPETENT. A freshly inoculated culture of *Streptococcus* contains very few competent cells; however, during the later part of the exponential phase of growth, the majority of the cells in the culture rapidly become competent—at a rate much higher than the growth rate. The process of becoming competent appears to be infectious; and, in a sense, it is. Each cell produces a small amount of a low-molecular-weight protein ($\approx$10,000), termed a COMPETENCE FACTOR, that induces cells in the population to synthesize 8 to 10 different proteins that establish the competent state.

The process of release of DNA from donor cells has been little studied. It might occur by the lysis of an occasional cell, or DNA might be extruded, possibly in response to a chemical signal from competent cells. Cultures of *Streptococcus* contain significant amounts of extracellular DNA only when they become competent.

Whatever the mechanism of its release, bits of double-stranded (ds) DNA become reversibly bound to specific proteins on the surface of competent cells. In the case of *S. pneumoniae,* there are 30 to 80 sites. Single-stranded (ss) DNA is not bound; it is completely unable to transform. While still on the surface of the cell—as demonstrated by its sensitivity to DNase—the dsDNA becomes irreversibly bound. Coincident with the binding process, single-strand breaks—or NICKS, are introduced at 6- to 8-kb intervals.

Before the transforming DNA enters the cell, double-strand breaks occur in the DNA, presumably by cleavage opposite the nick introduced during binding. Then one of the strands of DNA is completely hydrolyzed by an envelope-bound exonuclease, and the resulting single strands become associated with molecules of a small (15.5 kDa in the

(A) Development of competence

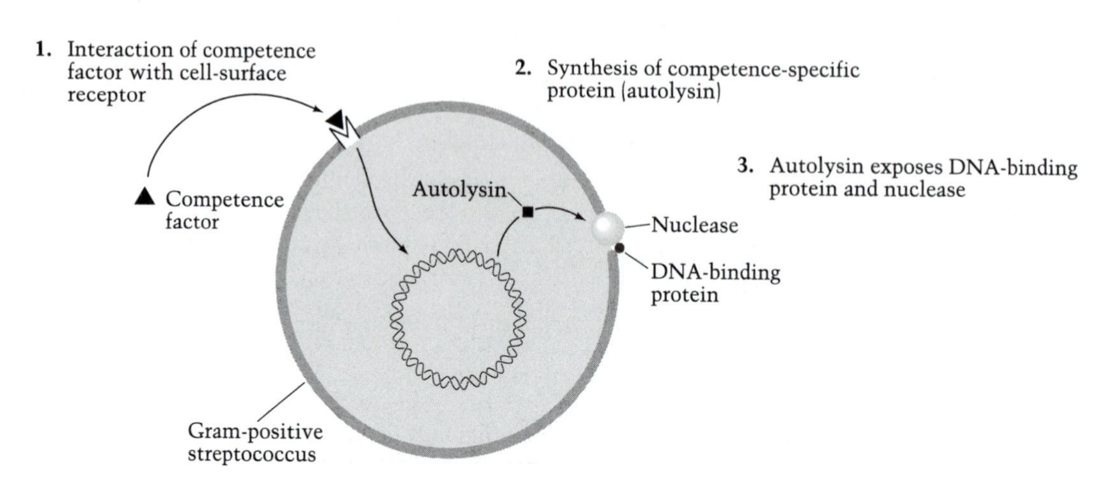

(B) Transformation

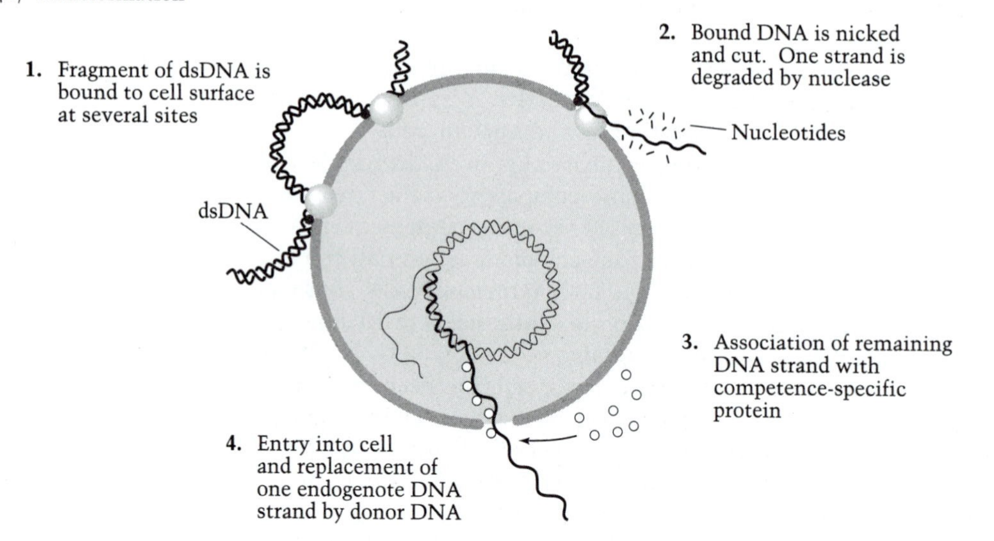

Figure 1

Features of transformation in *Streptococcus*. (After Smith et al., 1981.)

case of *S. sanguis*) polypeptide that coats it. A piece of DNA in this form, termed an ECLIPSE COMPLEX, enters the cell. Still in a single-stranded form, it becomes incorporated into the endogenote. If the transforming DNA differs from the endogenote at specific alleles, the immediate product of incorporation is a HETERODUPLEX because the two strands are not identical. Homogeneous DNA molecules (having perfectly matched complementary strands) are produced either as a consequence of replication of the heteroduplex—a process leading to progeny that are genetically altered—or by a process sometimes termed CORRECTION in which the progeny are not genetically altered. Correction is brought about by DNA repair mechanisms. The imperfectly paired heteroduplex region necessarily forms single-stranded loops. Even if these regions are quite short, the cell's DNA repair mechanism is activated: one strand in the mismatched region is cut, the surrounding regions of that strand are hydrolyzed, and the resulting single-stranded region is refilled by DNA polymerase action using the single strand as a template. If correction occurs, the recipient cell is not genetically altered because, by an unknown mechanism, the donor strand is preferentially used as the template for repair synthesis. The level of correction and hence the efficiency of transformation varies with the genetic markers introduced. Cells are transformed for some markers at high efficiency (1.0) and for others at quite low efficiency (as low as approximately 0.05).

Because this mode of transformation involves the production of a single-stranded DNA intermediate, and because single-stranded DNA cannot cause transformation (because it cannot be bound to surface receptors on the cell), DNA that has just entered a recipient cannot be recovered in a form that is capable of transforming other cells. The entering DNA is said to enter an ECLIPSE PERIOD, apparently disappearing, as judged by a transformation assay, and not reappearing until it integrates into the endogenote and again becomes double stranded.

Transformation of *Haemophilus* / With certain minor exceptions, transformation of other Gram-positive bacteria that can undergo this mechanism of genetic exchange is similar to the process just described. But transformation of Gram-negative bacteria, if one can generalize from the few well-studied examples, appears to be different in a number of important aspects. *Haemophilus influenzae* and *H. parainfluenzae* are by far the best studied of the transformable Gram-negative bacteria.

Soluble competence factors are unknown in Gram-negative bacteria, but cultures can be induced to greater competence by certain growth conditions. Cultures of *Haemophilus* become almost 100% competent when transferred to non–growth-supporting media that permit continued protein synthesis. *Acinetobacter* cells become competent as cultures enter the stationary phase of growth.

Like Gram-positive bacteria, competent Gram-negative bacteria absorb only dsDNA. But Gram-negative bacteria (as has been established for *Haemophilus* and *Neisseria*) only absorb DNA from closely related strains. In the case of *Haemophilus,* this remarkable specificity of absorption has been related to specific sequences in its DNA. These RECOGNITION SITES contain a common 11-bp sequence, 5′-AAGTGCGGTCA-3′, that is essential for uptake. There are approximately 600 such uptake sites on the chromosome of *Haemophilus,* or approximately one site per 4,000 bp; even small segments of DNA have a high probability of containing at least one site and thereby of binding to one of the four to eight receptors on the cell surface. The basis of specificity of DNA recognition by *Neisseria* is not known.

The dsDNA is not degraded to ssDNA as part of the process of entry into the cell. Thus, transforming DNA does not pass through an eclipse period as it does in Gram-positive bacteria; rather, it is thought to enter membrane-bound vesicles, called TRANSFORMASOMES, from which the DNA is somehow able to enter the cells and scan the genome of the recipient cell until a homologous region is found. Only one strand of the transforming DNA is incorporated into the endogenote; the other strand and the displaced strand of the endogenote are degraded, probably simultaneously. No eclipse period occurs, and there are no free, single-stranded intermediates in the process.

Transformation by plasmids / Interestingly, plasmids, many of which can encode functions that guarantee their own transmission from cell to cell, are not usually involved in natural transformation. The reason appears to be the requirement that plasmids remain circular in order to replicate and to avoid degradation. Natural transformation involves cutting DNA at the recipient cell surface. Transformation by plasmids does occur at low frequency, because some plasmids can recircularize after they enter, provided either a single plasmid is dimeric or two plasmids, cut at different sites, enter the same cell. Plasmids can recircularize also if they carry regions that are homologous with the endogenote.

Artificial transformation / The mechanisms of transformation that we have discussed so far are elaborate; they require a number of proteins and, in the case of *Haemophilus*, a specific DNA base sequence. One can hardly doubt that the process evolved in response to the selective advantages to be realized from the exchange of genetic material in nature. But, as far as we know, many species of bacteria, including *E. coli*, have not evolved natural transformation. In fact, the great usefulness of *E. coli* in genetic engineering, which depends on introducing DNA modified chemically in vitro into a living cell for cloning, relies on the success of artificial transformation after permeabilizing the cells to DNA by treatment with Ca^{2+} in the cold. Elaboration of the technique has made it possible to transform approximately 20% of the cells that survive such treatment, or approximately 4% of the cells in the original population. Artificial transformation of *E. coli* is effective only if the DNA used is capable of self-replication without being integrated into the endogenote (i.e., if it is a replicon). Thus, plasmids and viral genomes (transformation by viral genomes is termed TRANFECTION) transform at high efficiency, but bits of chromosomal DNA transform at low efficiency. Apparently the linear bits of chromosomal DNA are hydrolyzed by intracellular nucleases before they can be integrated, because nuclease-deficient strains of *E. coli* (*recBC, sbcB*) can be artificially transformed by chromosomal nonreplicon DNA at significant frequencies.

Calcium treatment renders a number of other bacteria, including both Gram-positive and Gram-negative species, susceptible to tranformation; it appears to be almost universally applicable to other bacteria. Although this technique has already had great impact on biotechnology—it is a cornerstone of the set of procedures termed GENETIC ENGI-

NEERING or RECOMBINANT DNA TECHNOLOGY—the basis of the efficiency and the molecular details of the process have received little attention. However, it is clear that DNA enters and remains in double-stranded, even supercoiled, form. Other artificial means have been developed for introducing replicons into bacteria. A freeze–thaw technique has been used in *Agrobacterium tumefaciens*; protoplast formation followed by polyethylene glycol treatment has been used in *Bacillus subtilis.* More recently, ELECTROPORATION, which involves damaging the cell envelope with pulses of high voltage, has been successfully used to facilitate transformation of a wide number of bacterial species.

Transduction

Natural transformation systems are clearly elaborate mechanisms evolved to mediate genetic exchange between related strains of bacteria. In contrast, transduction, appears to be the consequence of errors that sometimes occur in the development of certain types of bacteriophage. During phage development, an occasional phage particle becomes filled with chromosomal DNA or a mixture of chromosomal and phage DNA rather than with phage DNA alone. Such an aberrant phage is termed a TRANSDUCING PARTICLE. It can attach to a bacterium and introduce bacterial, rather than just phage DNA, into this host cell, thereby transferring bacterial DNA from one cell to another. This process is termed TRANSDUCTION.

Two types of errors in phage development can yield transducing particles: one leads to GENERALIZED TRANSDUCING PARTICLES, which contain DNA from any region of the bacterial chromosome; the other leads to SPECIALIZED TRANSDUCING PARTICLES, which contain phage DNA covalently bound to DNA from one specific region of the chromosome.

Generalized transduction / A number of phages that infect a wide variety of bacteria are known to mediate generalized transduction. Of these, the *Salmonella typhimurium* phage P22, in which the phenomenon was discovered in 1952, is probably the most thoroughly studied. The general scheme of formation of transducing particles by this phage, which is a consequence of the particular way it packages its DNA into the viral capsid, probably applies to all generalized transducing phages.

In infected cells, double-stranded P22 DNA is synthesized in long pieces—CONCATEMERS—of DNA that consist of approximately 10 phage genomes tandemly joined (Figure 2). Certain sites in the phage genome (which tend to be clustered) are susceptible to cleavage by a nuclease. Then, by a so-called HEADFUL MECHANISM, successive cuts at intervals approximately 2% longer than a complete phage genome are made along the concatemer to yield a set of molecules that are packed into capsids to form mature phage particles—VIRIONS. Each virion contains a complete phage genome; each has a different starting and ending point within the genome; and each contains a 2%-long identical region at each end. These phages are said to be TERMINALLY REDUNDANT. By such a packaging mech-

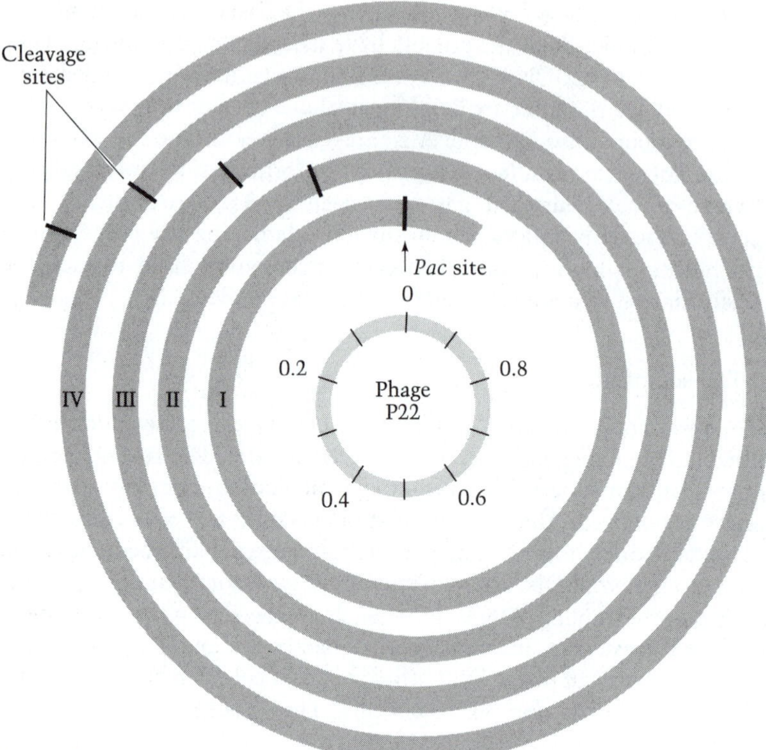

Figure 2

Mechanism of packaging of DNA by phages that mediate generalized transduction. The inner circle represents the genetic map of the generalized transducing phage P22, numbered according to the physical map coordinates. The outer spiral represents a concatemer of phage DNA aligned appropriately with the genetic map. The arrow indicates the *pac* site at which a phage-encoded nuclease cleaves the concatemer. Subsequent cleavages, indicated by the cross lines, are made at distances appropriate to produce a phage-headful amount of DNA. The headful amount of DNA exceeds the amount of DNA in the complete phage genome by approximately 2%. Thus, the ends of each subsequently packaged fragment of the concatemer (I through IV) are different from those of the other fragments. Generalized transducing particles are formed when sites on the host chromosome bearing some resemblance to the *pac* site are cleaved by the phage-encoded endonuclease and sequential headfuls of host DNA are encapsulated. (After Jackson et al., 1982.)

anism, phage particles all contain complete phage genomes, even though they lack specific nuclease-sensitive sites that delineate the genome. There is a low level of specificity for the site of primary cleavage of the concatemer, but subsequent cleavages are made solely on the basis of their distance from the primary cleavage.

If, as sometimes happens, the primary cut is made in the bacterial chromosome rather than in the concatemeric phage DNA, subsequent cuts in the bacterial chromosome by the headful mechanism yield fragments of chromosomal DNA that, when incorporated into the phage capsid, become generalized transducing particles. Presumably all generalized transducing phages package their DNA by a headful mechanism from concatemeric DNA.

The headful mechanism of formation of transducing particles fits with some long-known facts of generalized transduction. Not all chromosomal genes are transduced at the same frequency, presumably because they are at different distances from the sites of primary cleavage. Deletion mutations near, but outside, the region between two closely linked markers affect the frequency with which those two markers appear in the same transducing particle (are COTRANSDUCED), presumably because the deletion changes the register of cuts between the two markers and a nearby probable site for a primary cut (Figure 3).

Some years ago investigators isolated mutants of phage P22 (HT mutants) that were capable of transduction at a greater rate than wild-type P22 was. These mutants have the properties that would be expected if the nuclease that causes the primary cut were altered such that its site specificity was reduced. Wild-type phages incorporate only approximately 1% of chromosomal DNA into phage capsids; HT mutants incorporate up to 50%. Differences among the frequencies with which

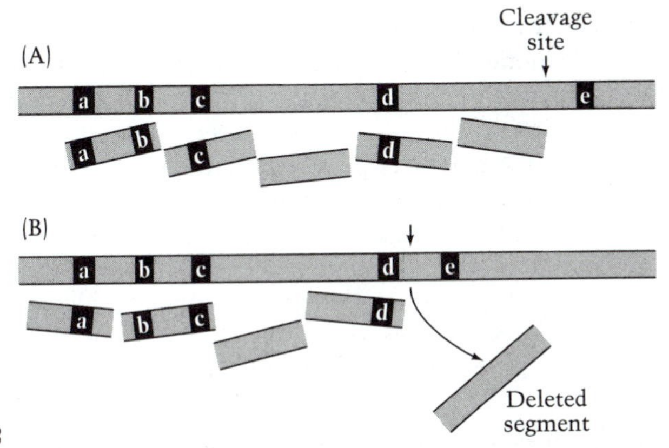

Figure 3

Effect of outside deletions on frequency of cotransduction. (A) Representation of a region of the bacterial chromosome containing a site (↓) susceptible to cleavage by the nuclease encoded on the genome of a generalized transducing phage such as P22. The letters indicate various genetic markers. The sloping bars below represent fragments of chromosomal DNA that would be packaged into transducing particles by the headful mechanism. Note that packaging from this site (packaging from other sites also occurs) produces transducing particles that permit cotransduction of *a* and *b* but none that permit cotransduction of *b* and *c*. (B) A deletion has occurred between markers *d* and *e* (outside the region *a–b–c* in which cotransduction was examined). This deletion changes the relative position of the site of primary nuclease attack (↓) and the markers of interest. Note that now transducing particles that permit cotransduction of *b* and *c* would be produced, but none would cotransduce *a* and *b*. (After Chelala and Margolin, 1974.)

various markers are transduced by wild-type phage almost disappear in HT-mediated transductions. Effects of outside deletions on transduction frequencies are not completely eliminated but vary with each particular HT mutant as though each alters nuclease specificity uniquely.

Once injected into the recipient cell, DNA from transducing particles remains double stranded and becomes incorporated into the endogenote by replacing both strands of the recipient chromosome. However, not all pieces of donor DNA are incorporated; at least 90% fail to recombine. Because the unincorporated pieces of donor DNA are not replicons, they are not replicated; but because they persist within the cytosol and are double stranded, their genes are expressed. Curiously, if incorporation does not occur immediately after injection, it almost never does. Neither the persistence nor the failure to recombine are understood. Cells that contain unincorporated donor DNA are termed ABORTIVE TRANSDUCTANTS. Under conditions that require expression of a donated allele, colonies arising from abortive transductants can be easily distinguished from those arising from COMPLETE TRANDUCTANTS (those in which the donated DNA fragment is recombined into the endogenote). Colonies arising from abortive transductants grow slowly because only a single cell can synthesize the growth-essential gene product. Indeed, although the abortive transductant colony may contain many thousand cells, if it is picked and spread on another selective plate, only a single microcolony will develop. They are approximately 10 times as abundant as complete transductants. Abortive transductants remain stably diploid for the region donated and therefore are quite useful in testing dominance of alleles.

Phage P22 is a TEMPERATE PHAGE, which means that after infection of a sensitive cell the phage DNA can either direct immediate production of phage or enter a repressed state in which one DNA copy, the PROPHAGE, replicates as part of the cell's genome. Later, induction of the prophage leads to phage production. In contrast to temperate phage, VIRULENT PHAGE cannot form a prophage, but rather produces a crop of progeny phage immediately in the original infected cell.

Specialized transduction / Although P22, the example of a generalized transducing phage that was chosen for discussion, happens to be a temperate phage, certain virulent phages also mediate generalized transduction. Specialized transduction, however, can be mediated only by temperate phages, because the developmental aberration that generates specialized transducing particles occurs when the prophage is excised from its chromosomal location. Specialized transduction occurs by ILLEGITIMATE RECOMBINATION, so called because it occurs between nonhomologous sequences within the prophage and the surrounding bacterial genome (see section below on recombination). Because the amount of DNA that can be packaged within the phage capsid is narrowly limited, only genes near the site of integration of the prophage can become part of the specialized transducing particle. In the case of coliphage lambda (λ), which is the most thoroughly studied of the specialized transducing phages, genes encoding the biosynthesis of biotin (*bio*) are on one side of the attachment (integration) site and those encoding the catabolism of galactose (*gal*) are on the other. Although surrounding and intervening genes are also carried, by convention specialized transducing particles of λ are divided into those that carry *bio* and those that carry *gal*. The letter *d* (for defective) is added if they lack sufficient phage genes to form plaques, that is, λ*dgal* or λ*dbio*. If they can form plaques, the letter *p* is sometimes used.

Such aberrant excisions are rare—having a probability of approximately 10^{-6} per excision—so lysates obtained from the induction of prophages are termed LOW FREQUENCY TRANSDUCING (LFT) lysates. However, if such a lysate is used to infect a Gal$^-$ culture and if clones that are able to utilize galactose as a carbon source are selected, one obtains a strain that carries λ*gal* as a prophage. If the λ*gal* is defective and therefore incapable of lysogenization or lytic growth, one selects double lysogens that carry, in addition to λ*dgal*, wild-type λ called HELPER PHAGE. On induction of such a lysogen, the lysate obtained—termed a HIGH FREQUENCY TRANSDUCING (HFT) lysate—contains specialized transducing particles at a frequency of approximately 0.5 of the total virions, because of the presence of the helper phage. The large number of specialized transducing particles results from the faithful excision of both the transducing prophage and the helper prophage.

Specialized transduction has limited value in mapping the location of genes on the bacterial chromosome. It has been useful in the genetic analysis of λ because specialized transducing particles are deletion mutants of the phage. And it has been extremely useful in achieving GENE AMPLIFICATION of segments of the bacterial chromosome. An HFT lysate contains a high concentration of those bacterial genes carried on the transducing particle, and therefore a remarkable gene enrichment is effected. Moreover, during induction of the lysogen, these genes are expressed, and in many instances this expression leads to high levels of the gene products, which are useful for further purifications.

For these reasons, techniques have been developed to isolate specialized transducing phages that carry genes other than those adjacent to the normal attachment site of λ. This arrangement can be accomplished by moving the normal attachment site near the gene of interest, by moving the gene of interest near the phage attachment site, or by deleting the normal phage attachment site and relying on secondary sites of prophage insertion.

Conjugation and F plasmids

Conjugation, the direct transfer of genes from one bacterial cell to another, is mediated by CONJUGATIVE PLASMIDS (Chapter 9). These plasmids are transferred at high frequency, and occasionally chromosomal genes go along for the ride. The first-studied and best-understood of the conjugative plasmids is the F plasmid, which can replicate in *E. coli* and certain closely related enteric bacteria, including *Salmonella typhi-*

murium. Like all plasmids, F is a self-replicating, circular molecule of dsDNA. Like all conjugative plasmids, it carries all the genes necessary to encode its transfer from one cell to another. The 13 transfer genes (*traA* through *traL*) form an operon (Figure 4) that makes up approximately one-third of the 94.5-kbp molecule. The *tra* genes control the formation of a pilus required for DNA transfer—the F-pilus—and also facilitate a special form of DNA replication, called the ROLLING CIRCLE mode. Outside the operon are a cluster of four IS elements, other plasmid genes, and regions that amount to approximately one-fourth of the plasmid and encode no known functions. Apparently F encodes only its own replication and transfer.

If cells containing an F plasmid (designated F^+) are mixed with cells that lack it (designated F^-), conjugal pairs rapidly form by attachment of the end of the F-pilus to the surface of the F^- cells. Attachment then triggers a series of events resulting in transfer of the F plasmid and very little else from the F^+ to the F^- cell. At *oriT* (Figure 5), a nick (a single strand cut) is made in F; the nick serves as the starting point for the replication of the unnicked strand (termed ROLLING CIRCLE REPLICATION) in

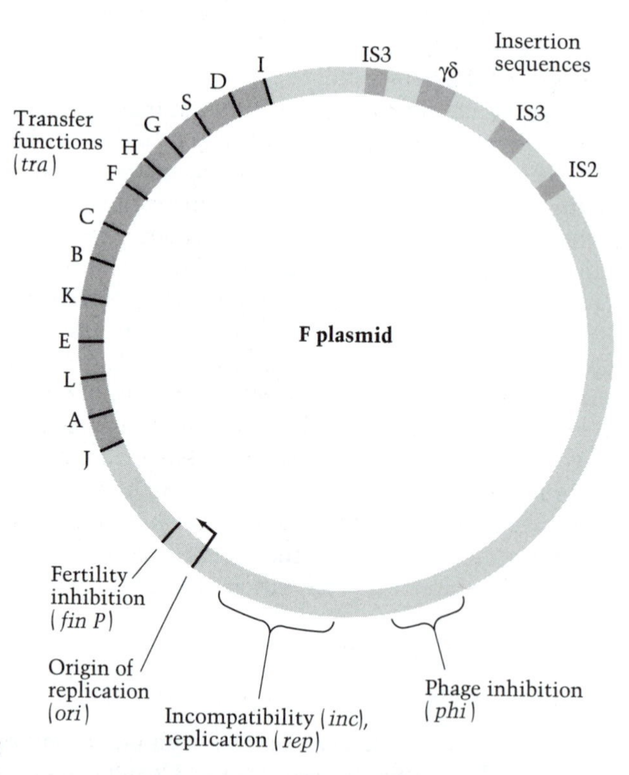

Figure 4

Genetic map of F plasmid. The relative position of genes encoding transfer functions (*tra*), fertility inhibition (*finP*), origin of transfer replication (*ori*), incompatibility (*inc*), replication (*rep*), and phage inhibition (*phi*), along with insertion sequences (IS2, IS3, and γδ) are shown. The length of the plasmid is 94.5 kb. (After Shapiro et al., 1977.)

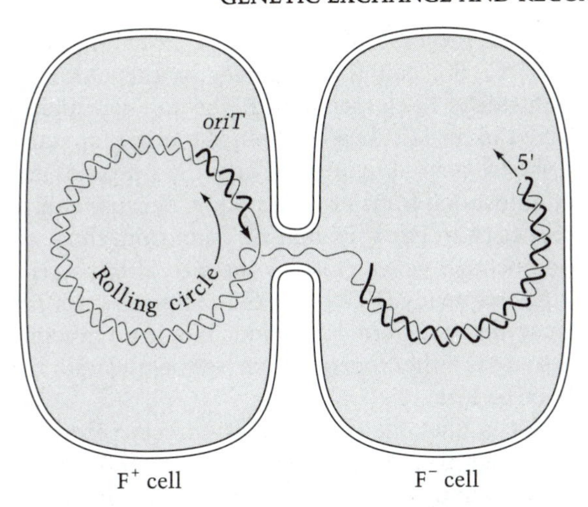

Figure 5

Transfer of F plasmid from an F^+ to an F^- cell. Formation of a mating pair triggers transfer replication of F. By an F-encoded nuclease, one strand is cleaved (nicked) at *oriT*. Then replication (at arrowhead) occurs by a rolling circle mechanism. The newly synthesized DNA displaces a preexisting single strand of F, which enters the F^- cell, where its complementary strand is synthesized.

the 5′ to 3′ direction. The 5′ end of the preexisting nicked strand enters the F^- cell as replication continues. It is not completely clear whether the single strand is conducted by the F-pilus or is passed directly into the F^- cell at some other point where the two cells are in contact, but it is clear that replication and transfer occur simultaneously and that only a single strand is transferred to the recipient. There it is duplicated and recircularized. Recircularization is thought to be a *tra*-encoded function. It will be noted from the form of replication that the donor cell retains a copy of the F plasmid.

Pair formation and plasmid transfer is extremely efficient between F^+ and F^- cells. If two such cultures are mixed, all cells in the culture rapidly become F^+ and acquire all donor (male) properties. At a much lower frequency (10^{-7}), some chromosomal genes are transferred to the F^- (female) cell along with the F plasmid.

How can chromosomal genes be transferred? As stated earlier, the F plasmid contains a cluster of IS elements. The chromosome of *E. coli* also contains IS elements. Homologous recombination (see later) between an IS on the F plasmid and one on the chromosome causes the F plasmid to be inserted into the chromosome. The F genes in such a cell (characterized by its HIGH FREQUENCY OF RECOMBINATION, Hfr) continue to be expressed; the cell remains a donor. But certain of its properties change dramatically. When an Hfr contacts an F^- cell, conjugative transfer of F begins and, because the F plasmid is covalently bound to the chromosome, transfer of one strand of the chromosome follows. The attachment between cells in a mating pair is fragile and breaks frequently, thus interrupting transfer. Transfer is a somewhat lengthy process, requiring, in the case of *E. coli*, approximately 100 minutes to complete at 37°C. So, complete transfer of the chromosome rarely occurs. The cluster of IS elements on F (the site at which the chromosome is inserted in an Hfr) lies between the origin of transfer of the F plasmid and the *tra* genes (Figure 6). Thus, the *tra* genes are transferred only after chromosomal transfer is complete. Because this event is rare, the consequences of an Hfr × F^- mating differ from those of an $F^+ \times F^-$ mating. Chromosomal genes close to the site of F insertion are transferred at a high frequency (10^{-1}), but the recipient cells rarely become donors, because the transferred fragment is not a replicon and cannot circularize. Its fate is either to recombine (see below) with the recipient's chromosome or be lost.

The Hfr trait is unstable; recombination between the homologous ends of the inserted plasmid regenerates an F^+ strain (Figure 6). Occasionally recombination occurs between a site within F and a site on the chromosome or between sites on either side of F. The plasmid product is called F′. Formation of an F′ plasmid generates a chromosomal deletion; the genes carried on the F′ are not duplicated in the chromosome; the cell remains haploid for all genes. Such a cell is termed a PRIMARY F′. But the product of an F′× F^- mating, which usually results in the complete transfer of the F′ plasmid because of its relatively small size, is a partially diploid cell, a MEROZYGOTE, termed a SECONDARY F′.

The F′ merozygotes, incidentally, are useful in studies of dominance of alleles and in mapping the location of genes on the chromosome. They are generally unstable, unless they carry a *recA* mutation (see the section on homologous recombination later in this chapter), because of the considerable homology between the plasmid and the endogenote. Strains of F′ cells are also useful because of the process of HOMOGENOTIZATION. In *recA*$^+$ merozygotes, nonreciprocal crossovers occur frequently between the endogenote and the plasmid. Thus, transfer of a particular allele from one of these genetic elements to the other can be selected by establishing growth conditions that favor growth of the HOMOGENOTE, a cell containing the same allele on both the chromosome and the plasmid. In practice, homogenotization becomes quite valuable in complementation or dominance studies. From a set of bacterial mutants that carry different mutant alleles of a given gene, a set of F′ plasmids carrying these alleles can be easily constructed. Dominance or complementation tests can then be done by transferring the various mutation-bearing F′ plasmids to strains with various mutant alleles in the chromosomally located gene.

A variety of other plasmids are conjugative and capable of mobilizing (facilitating the transfer of) the chromosome as well as any nonconjugative plasmids that might be present in the same cell. The molecular nature of mobilization is unclear, other than that it seems to occur by a process analogous to Hfr formation. Only a portion of chromosome mobilization seen in F^+ strains can be attributed to the Hfr cells that they inevitably contain. The rest remains somewhat mysterious. Possibly transient single-strand connections between plasmid and mobilized DNA account for the phenomenon.

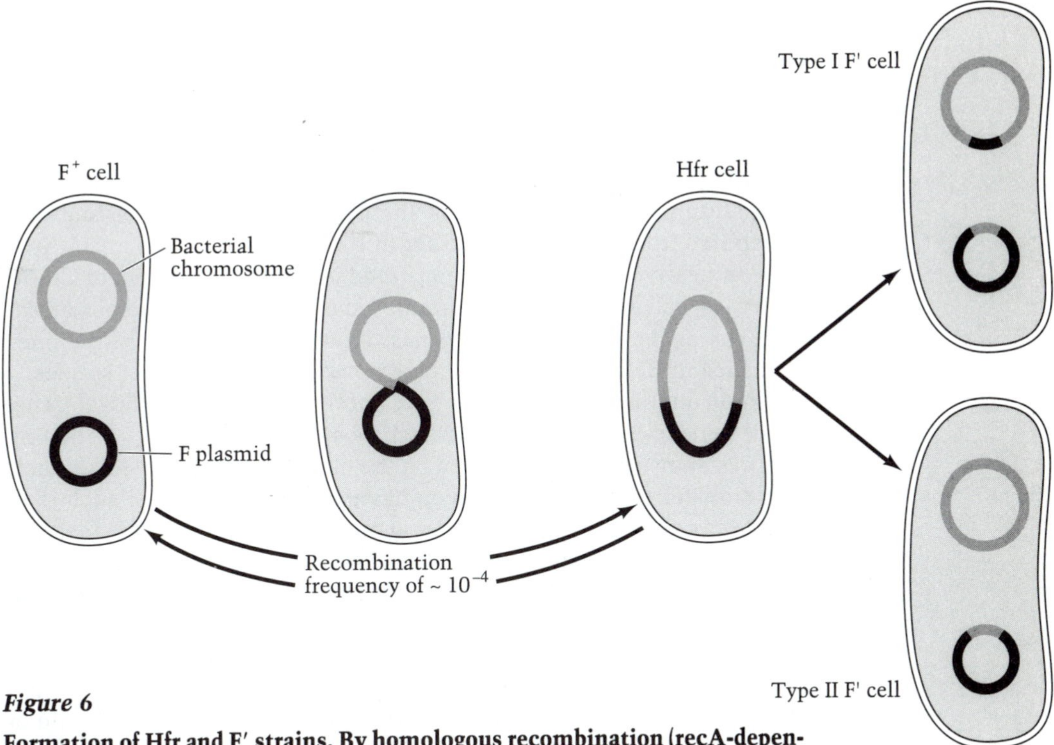

Figure 6

Formation of Hfr and F′ strains. By homologous recombination (recA-dependent) between insertion sequences on the F plasmid and those in the chromosome, Hfr strains are formed from F^+ strains at a frequency of approximately 10^{-4}. At approximately the same frequency, recombination between the homologous ends of the F plasmid DNA regenerates an F^+ strain. Occasionally recombinations between the F plasmid and chromosomal DNA or between regions of chromosomal DNA on either side of the F plasmid occur, thereby generating an F-prime (F′) cell. In the former case, the product (termed a Type I F′) contains an altered F plasmid (an F′ plasmid) that lacks a region of the F plasmid and contains instead certain chromosomal genes adjacent to the site of insertion of the F plasmid in the parental Hfr strain. The chromosome contains a fragment of the F plasmid and lacks the genes carried on the F′ plasmid. In the second case, the F′ plasmid (termed a Type II F′) contains all the DNA of the F plasmid and chromosomal genes adjacent to both ends of the F plasmid DNA in the parental Hfr. The chromosome lacks these genes. Both of these cells (termed primary F′ cells) are haploid for all chromosomal genes; they are carried on the F′ plasmid or on the chromosome. If, by mating, the F′ plasmid from a primary F′ cell is transferred to an F^- cell, the product cell is termed a secondary F′ cell. Secondary F′ cells (both Type I and Type II) are merodiploid; that is, genes carried on the F′ plasmid are also carried on the chromosome.

Conjugation in Gram-positive bacteria

Studies of the conjugation of the Gram-positive bacterium *Enterococcus* (formerly, *Streptococcus*) *faecalis* suggest a mechanism quite different from the better-studied Gram-negative systems. Pili do not play a role in conjugation between strains of this organism; mating signals between the cells are mediated by soluble molecules—PHEROMONES. These protease-sensitive, heat-resistant pheromones are small peptides released by cells that lack a particular plasmid. The pheromones stimulate cells that have the plasmid to produce an AGGREGATION SUBSTANCE on their outer surface. Cells with the aggregation substance form clumps with cells that lack the plasmid (Figure 7). Plasmids are transferred from one cell to another in these clumps.

Conjugative plasmids are being found in a growing number of Gram-positive species in addition to *Enterococcus faecalis,* for example, *Streptococcus agalactiae* and *S. pyogenes.* Pheromones similar to the ones described in *E. faecalis,* however, are not yet known in other species.

Some strains of streptococci have another means for conjugal transfer of genetic material. Certain transposon-like elements containing drug-resistance genes and located on the chromosome can mediate their own transfer to other cells by conjugation. One example of such CONJUGATIVE TRANSPOSONS is Tn*916*, found originally in a strain of *E. faecalis.* This element, approximately 16 kbp in size, contains a gene for tetracycline resistance; therefore, its transfer is easily detected. The element Tn*916*, and similar, closely related elements (Tn*918*, Tn*919*, Tn*920*, Tn*925*, and Tn*1545*) have many properties indistinguishable from transposons, including size, multiple target sites, ability to transfer from a chromosome to a plasmid, and ability to be removed from a plasmid or chromosome by precise excision. But the ability of these elements to mediate their own intercellular transfer places them apart from the more familiar transposons described in Chapter 9. Their conjugational ability is remarkably broad; Tn*916* can transfer among dozens of species of Gram-positive bacteria, including many species of *Streptococcus, Enterococcus, Staphylococcus,* and *Bacillus.* Chromosomal genes are outside these elements and thus are not transferred in the process. However, by inserting them within the borders of these transposons, genes of interest could conceivably be moved readily into species for which other gene transfer processes are unavailable. Although this group of elements cannot transfer themselves into *E. coli,* once introduced by some other vector they can mediate their own transposition between plasmids and chromosome in this organism. Exactly how conjugative transposons bring about their own transfer (or their own transposition) is not known. One idea is that they become excised from the chromosome and circularized; the circle then either inserts at a new site, transfers to another cell, or possibly is lost. Study of the process in *E. coli,* where some late step in transposition seems to be rate-limiting, has permitted the isolation of circular forms that, when introduced into *E. faecalis,* have all the properties of Tn*916*.

Conjugal transfer in *Streptococcus pneumoniae* is mediated also by larger conjugative transposons, some as large as 65 kbp. Several of these carry multiple drug resistance genes (tetracycline, chloramphenicol, and erythromycin) and have the property of transferring always to the same target site in the *S. pneumoniae* chromosome.

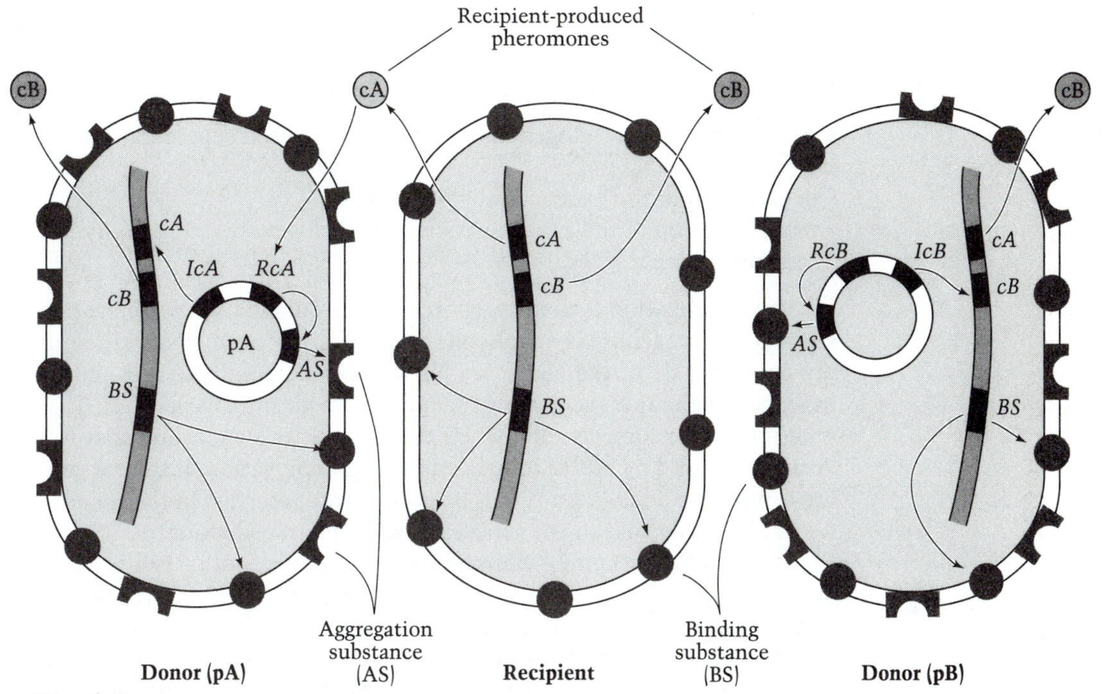

Figure 7

Conjugation of *Enterococcus faecalis*. The model shows the response of two donor cells to two chromosomally encoded pheromones (cA and cB) produced by a recipient. These react with plasmid-encoded genes (*RcA* and *RcB*), which signal another gene (*AS*) to produce aggregation substance (AS). This protein AS causes producing cells to bind to other cells at the binding substance (BS) site. Cells that carry a particular plasmid (e.g., pA) do not produce the corresponding pheromone (cA) because the product of a plasmid gene (IcA) represses the structural gene (cA). (After Clewell, 1981.)

Another Gram-positive bacterium with a much-studied conjugation system is *Streptomyces coelicolor.* This organism is one of a group of closely related *Streptomyces* species, which are all filamentous bacteria of industrial significance—because of their production of antibiotics. Conjugation in *Streptomyces* is brought about by plasmid sex factors (conjugative plasmids) that occur naturally in many strains, often more than one in a given isolate. In *S. coelicolor,* the conjugative plasmid, SCP1, interacts with the bacterial chromosome at different sites, generating different donor strains (in a fashion that seems analogous to the formation of different Hfr strains of *E. coli* K-12). These exhibit high frequencies of conjugation and recombination with strains of *S. coelicolor* lacking this plasmid and are useful in mapping new genes.

FATE OF THE EXOGENOTE

By whatever means an exogenote becomes transferred from donor cell to recipient, little of biological significance has occurred until the fate of the introduced DNA is determined. Basically there are three outcomes of a DNA transfer event:

- stabilization of the exogenote by its circularization separate from the endogenote
- degradation of the exogenote by nucleases
- recombination of the exogenote with the endogenote to form a hybrid chromosome

The stabilization of the exogenote by its circularization can be of two sorts. If the exogenote is a replicon, then the introduced DNA will simply become part of the accessory (plasmid) genome of the cell and be inherited by the rules governing that particular plasmid in that cell. This outcome is the common one when the exogenote is a plasmid that has been introduced by either transformation or conjugation. If the exogenote is not a replicon, then it will be unilinearly inherited in the manner we have already met in the process of abortive transduction.

The second outcome—destruction of the exogenote—is the common one when DNA unrelated to the endogenote has been introduced, and in many instances it is initiated by enzymes called RESTRICTION ENDONUCLEASES. These enzymes, of which several hundred are known, recognize specific nucleotide sequences in DNA and produce double-strand breaks, which lead subsequently to the complete degradation of the DNA (Table 1). Each bacterial strain that possesses a restriction system is able also to disguise these recognition sites in its own DNA by MODIFYING them through methylation of an adenine or cytosine residue within the site. In effect, these RESTRICTION-MODIFICATION SYSTEMS provide surveillance and protection, enabling the cell to tell "self" from "nonself," and then cleaving the "nonself" DNA. The discovery of these systems came about by the observation that λ phage grown on a B strain of *E. coli* could infect other cultures of strain B quite efficiently but infection of K-12 strains occurred at 10,000-fold lower efficiency. Similarly, λ virions grown on K-12 cells infected K-12 cells very efficiently but B cells only poorly. Phage grown on either *E. coli* B or *E. coli* K-12 were each able to infect cells of *E. coli* C with high efficiency. Study revealed that strains B and K-12 each have a different restriction–modification system and that strain C has none. The DNA of phage grown on either B or K-12 cells was marked as "B" or "K-12" by the respective host system. Infection of B or K-12 cells with λ DNA not marked as "B" or "K-12," respectively, is subject to cleavage. In a very large population of infected cells, some phage DNA (approximately 1 in 10,000 molecules) by chance escape the action of the restriction endonuclease, resulting ultimately in the formation of progeny phage DNA appropriately marked as "self."

The many hundreds of procaryotic restriction–modification systems fall into a few broad classes. Type I systems (which include those of *E. coli* B and K-12 strains) combine the modification (methylating) and restriction (cleaving) activities into a single, multisubunit protein that methylates or cleaves in the presence of ATP, Mg^{2+}, and *S*-adenosylmethionine (the methyl source). Type I enzymes cleave unmethylated DNA, not at the recognition site, but at nonrandom sites further along in the DNA molecule. Type II systems consist of separate endonucleases and methylases. Type II endonucleases cleave at the recognition site, producing either blunt ends by a double-strand cut in the center of the site or complementary (cohesive) ends by staggered single-strand nicks. Because of the characteristics of their cleavage specificity, Type II enzymes are valuable in recombinant DNA methodology. Fragments of DNA produced by digesting molecules from different organisms by the same purified restriction enzyme can be readily ligated together (recombined) in vitro because of the complementary ends.

No matter which type of restriction–modification system a cell possesses, a key event is the replication of the cell's DNA. Replication of a segment of DNA containing a methylated restriction site results in two daughter duplexes, each of which has one (old) methylated strand and one (new) unmethylated strand. This pattern occurs because all restriction sites are PALINDROMIC (reading the same backward and forward, see examples in Table 2), so each strand of an unreplicated but modified restriction site has a methylated base. The restriction modification system must correctly recognize the new HEMIMETHYLATED DNA as a site to be methylated rather than cleaved (or ignored). Therefore, successful operation of a restriction–modification system requires that it correctly take one of three courses of action upon encountering its specific recognition sequence: it must *cleave* an unmethylated sequence, *leave intact* a methylated sequence, and *methylate* a hemimethylated sequence.

Although it is undoubtedly correct that restriction–modification evolved in the procaryotes as a system for protection against invading phage and transmissible plasmids, the example we have just described shows that once a viral genome has escaped surveillance and destruction, it and its progeny become marked for protection as host DNA. Furthermore, the trick of restriction–modification has not been lost on the invaders and parasites—some phage, such as P1, and several plasmids encode their own restriction–modification systems.

RECOMBINATION

The third possible fate of an exogenote—recombination with the endogenote to become part of the genome—is the one that first comes to mind when thinking about gene transfer leading to micro- or macroevolution. Recombination is an extremely broad subject. We have already mentioned varieties of recombination when presenting topics such as deletion mutations, the formation of λ prophage, the conversion of F^+ cells into Hfr cells, and the behavior of transposable genetic elements (insertion sequences and transposons). In the following sections we examine the several types of recombination known to occur in bacteria following DNA transfer by transformation, transduction, and conjugation.

Table 2. Action of certain restriction endonucleases

Class	Enzyme	Producing microorganism	Recognized DNA sequence[a]
Six base pairs recognized, complementary single-stranded ends produced	*Eco*RI	*E. coli* (R)[b]	(G↓AATTC CTTAA↑G)
	*Hind*III	*Haemophilus influenzae*	(A↓AGCTT TTCGA↑A)
Six base pairs recognized, blunt ends produced	*Hpa*I	*Haemophilus parainfluenzae*	(GTT↓AAC CAA↑TTG)
	*Hind*II	*Haemophilus influenzae* Rd	C↓A (GT(T)(G)AC CA(A)(T)TG) G↑C
Four base pairs recognized, complementary single-stranded ends produced	*Hha*	*Haemophilus haemolyticus*	(GCG↓C C↑GCG)
	*Mbo*I	*Moraxella bovis*	(↓GATC CTAG↑)
Four base pairs recognized, blunt ends produced	*Hae*III	*Haemophilus aegypticus*	(GG↓CC CC↑GG)

[a] Arrow indicates site of single strand cleavage. Upper sequence of bases is written in the 5′ ⟶ 3′ direction.
[b] Encoded by genes that are plasmid borne.

Homologous recombination

As its name indicates, HOMOLOGOUS RECOMBINATION involves the formation of a recombinant chromosome between two DNA molecules that share reasonably large regions of nucleotide sequence identity or similarity. Less frequently it is referred to as GENERALIZED RECOMBINATION or as *REC*-DEPENDENT RECOMBINATION, terms reflecting two other characteristics of this type of recombination: its nonspecialized nature (with respect to specific genes) and its dependence on a set of enzymes (encoded by genes called *rec*) that must be active in the recipient cell to bring about the covalent substitution of a segment of one DNA molecule (such as an exogenote) for the homologous region of another (such as the endogenote).

The details are far from known, but overall the process of homologous recombination involves breaking one strand of each recombining molecule at a time, pairing it with the unbroken, complementary strand of the other molecule, partially digesting the ends of the broken strands, repairing and joining them so that one rejoined strand is continuous between the recombining chromosomes, forming a bridge or BRANCH. This branch is free to migrate in either direction along the DNA, following the formation of the bridge; at some time later a similar BREAKAGE AND REUNION PROCESS links the second strand of each recombining molecule, and the CROSSOVER is complete. By this pathway a region of heteroduplex DNA is produced near the crossover point (Figure 8). Repeated further along the same molecules, this process results in the insertion of one segment of one molecule into the other.

Although the outcome is readily described, the process of homologous recombination is extremely complex. In fact, it is misleading to refer to it as a single process. It is believed that there may be as many as five partially separate pathways of homologous recombination in *E. coli* alone. More than 30 genes influence homologous recombination in this organism. By studying the properties of mutants altered in these genes, much has been learned of the process, including two helpful generalizations:

- All homologous recombination involves homology matching and strand exchange, and these steps are mediated by the RecA and SSB (single strand-binding) proteins.
- The various pathways of recombination differ in the steps preceding and following homology matching and strand exchange, that is, in the steps in which the molecules are prepared for strand exchange, in which the intermolecular intermediates are processed and resolved into the recombinant products, and in which these recombinant products are replicated.

Not all of the 30 genes that affect homologous recombination (Table 3) have products that are directly responsible for the process, but a large number do. They are responsible for the five pathways discerned in *E. coli*: the RecBCD, RecE, RecF, RecX, and RecY pathways. All pathways require RecA, SSB, PolA, Lig, GyrA, and GyrB for normal functioning (Chapter 3). The multifunctional protein RecA carries out the search for homology and the exchange of strands; its binding to DNA is facilitated by SSB (single-stranded DNA-binding protein). The proteins PolA (DNA polymerase I) and Lig (DNA ligase) are involved in DNA replication and repair. The products GyrA and GyrB are the subunits of DNA gyrase, the enzyme responsible for introducing supercoils into DNA.

The dominant pathway in normal strains of *E. coli* is called RecBCD. It is named for the three gene products that aggregate to form a single, multifunctional enzyme, EXONUCLEASE V. Exonuclease V probably participates in recombination at several steps, including the preparation for homology matching, for which it moves along double-stranded DNA, making single-strand cuts—nicks—near certain nucleotide sequences, called CHI SITES. These nicks probably provide the single strands that are

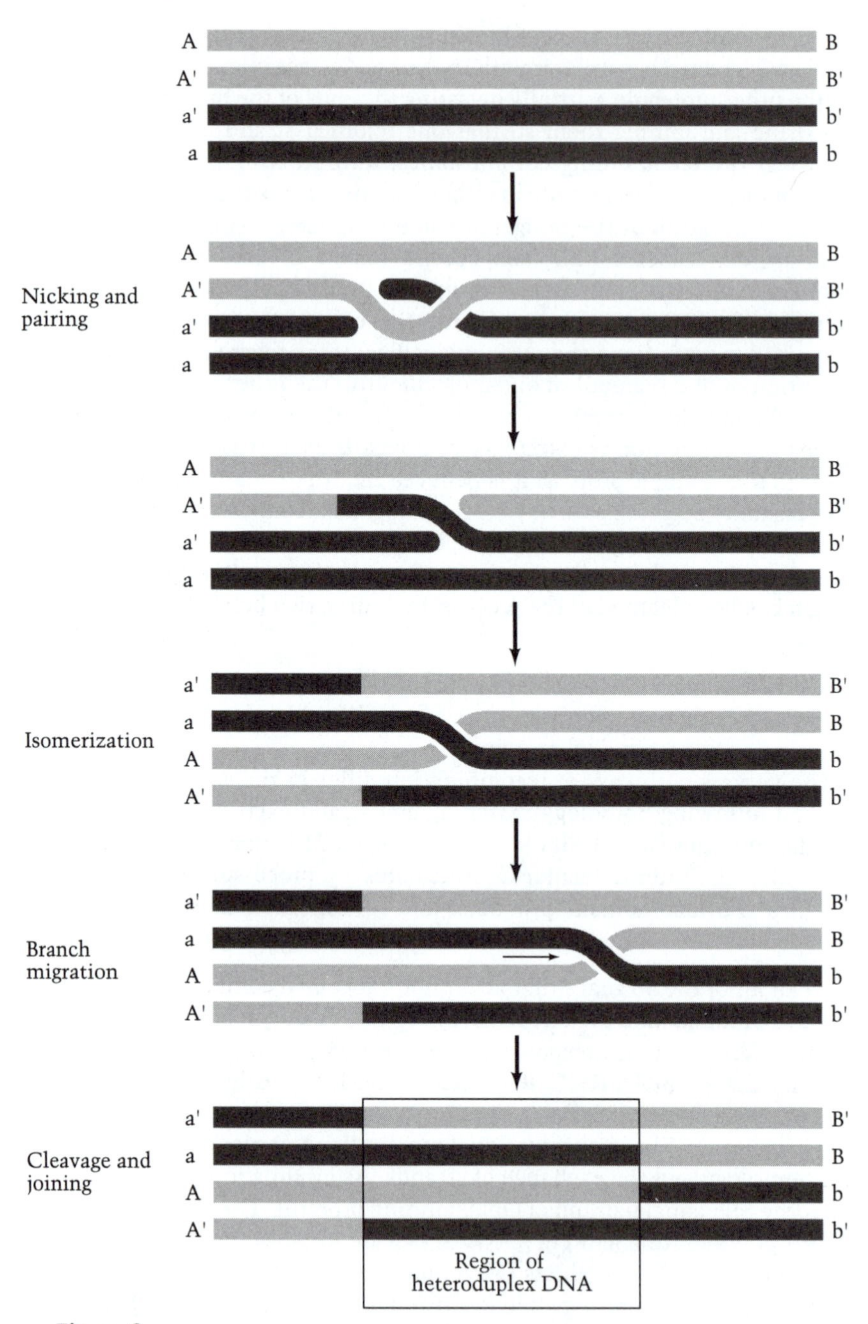

Figure 8

Breakage and reunion to form recombinant molecules of DNA. Homologous regions of two DNA molecules become aligned. One strand of one molecule becomes nicked and its free end invades the duplex of the other molecule in a process aided by RecA. Isomerization of the two linked molecules results in the unbroken strand of each forming a crossover bridge. This form (third from bottom) is referred to as a Holliday structure in Table 3. After branch migration in either direction, cleavage and joining of the bridging strands separates the hybrid molecules, leaving each with a region of heteroduplex DNA.

Table 3. Genes affecting homologous recombinaton in *E. coli*[a]

Gene	Product	Activities
recA	RecA protein	Binds single-stranded DNA (ssDNA), producing helical nucleoprotein filaments that bind dsDNA and generate large DNA networks; aids synapsis of homologous sequences and promotes ATP-dependent assimilation of ssDNA into homologous dsDNA; unwinds dsDNA; promotes pairing and strand exchange between dsDNAs with single-stranded ends to create Holliday structures; anneals complementary ssDNA; cleaves LexA repressor protein in the presence of dsDNA and an adenine nucleotide
recB, C, D	Components of the RecBCD enzyme (ExoV)	Unwinds dsDNA and makes single-strand nicks near Chi sites; has several other activities in vitro
recE	ExoVIII	Degrades dsDNA from the 5′ end, producing long 3′ tails
recF	RecF protein	Unknown
recJ	RecJ protein	Unknown
recN	RecN protein	Unknown
recO	RecO protein	Unknown
recQ	RecQ protein	Unknown
ruv	Ruv protein	Unknown
sbcA	SbcA protein	Unknown
sbcB	Exonuclease I	Attacks ssDNA at free 3′-OH ends
sbcC	SbcC protein	Unknown
ssb	SSB protein	Binds ssDNA; facilitates RecA binding to ssDNA; inhibits nuclease activities of RecBCD enzyme; stimulates DNA polymerases II and III
lexA	LexA protein	Represses *lexA*, *recA*, and other DNA-damage inducible genes
polA	DNA polymerase I	Has 5′ →3′ polymerase activity and 3′ →5′ exonuclease activity on ssDNA
lig	DNA ligase	Seals nicks in dsDNA
gyrA,B	DNA gyrase	Negatively supercoils closed circular dsDNA; forms and resolves knotted and catenated duplexes
topA	DNA topoisomerase I	Nicks and closes duplex DNA, an activity resulting in relaxation of negatively supercoiled DNA
uvrD	Helicase II	Unwinds dsDNA with an adjacent ssDNA
dam	DNA adenine methylase	Methylates adenine residues
dcm	DNA cytosine methylase	Methylates cytosine residues
dut	Deoxyuridine triphosphatase	Converts dUTP to dUMP
rep	Rep protein	Separates strands of dsDNA in the presence of SSB protein
xth	Exonuclease III	Has 3′ →5′ exonuclease activity specific for dsDNA
xseA,B	Exonuclease VII	Degrades linear ssDNA in both directions
mutH	MutH protein	Hemimethylated 5′-GATC-3′ binding protein
mutL	MutL protein	Unknown
mutS	DNA base pair mismatch binding protein	Binds DNA regions containing base-pair mismatches
rdgB	Rdg protein	Unknown

[a] Adapted from Mahajan (1988).

used by RecA in searching for homology, and Chi sites are therefore hot spots where recombination takes place at high frequency.

The other pathways probably come into play under circumstances different from those immediately following transduction or conjugation. The RecF pathway, for example, appears to be responsible for recombination between plasmids or between plasmids and the chromosome.

Homologous recombination is responsible for integrating exogenote fragments introduced by generalized transduction (e.g., by phage P1) and by plasmid-mediated conjugation. Exogenote DNA taken up by natural transformation also recombines with the chromosome by homologous recombination, but in this case only a single strand of the exogenote is exchanged. This process results in a region of HETEROZYGOUS DNA consisting of one strand of the exogenote and one of the endogenote. At the next round of replication, one of the daughter chromosomes will therefore be a HOMOZYGOUS recombinant, the other a homozygous molecule of the original endogenote sequence.

Site-specific recombination

The second major type of recombination is a set of specific mechanisms that

- are independent of RecA
- normally operate at highly preferred sites
- are virtually independent of sequence homology between the interacting DNA molecules

The integration into the *E. coli* chromosome of the specialized transducing phage, λ, that we have studied is a classic example of a site-specific recombination. In this case the site of integration is specific for both the phage DNA and the chromosomal DNA. Another phage, Mu, can integrate at innumerable sites on the *E. coli* chromosome, but the integration (recombination) event is nonetheless site specific within the Mu genome. The replicative transposition of IS elements and of transposons are other examples of site-specific recombination, because, whatever the specificity requirements for the site of integration in the bacterial genome, there is a unique site of the transposable element at which this recombination occurs.

Site-specific recombination has a fourth characteristic feature. The lack of dependence on the RecA protein of the host cell is accompanied by dependence on proteins (TRANSPOSASES and other enzymes) encoded by the phage, IS element, or transposon engaged in the recombination. The RecA-related set of enzymes require extensive base pairing to mediate their task. The enzymes of site-specific recombination work on an entirely different principle: recognition of specific DNA sequences by specific recombination enzymes.

We have already noted (in the section on conjugation) that one example of recombination, the integration of the F plasmid into the *E. coli* chromosome at seven or eight preferred sites, strongly resembles site-specific recombination but, mechanistically, is homologous recombination. The process is RecA-dependent, and the homology is provided by sequences of the IS elements present both in the chromosome and in the F plasmid.

Illegitimate recombination

Some recombinational events, rarer than those we have discussed up to now, defy classification as general or site specific. They require little or no DNA sequence homology, yet do not require special sites. These events occur independently of the proteins that promote homologous or site-specific recombination. As a consequence, the term ILLEGITIMATE RECOMBINATION has been coined as an overall name for events of this sort that occur by unknown mechanisms. The name reflects the frustration of microbiologists attempting to understand recombination and does not signify an underhanded procedure.

SUMMARY

1. The genetic adaptability and evolution of the bacterial genome is greatly enhanced by mechanisms that permit the *inter*cellular transfer of DNA. Gene transfer permits the products of individually rare, independent mutations to be combined within a single cell.
2. Bacterial DNA is transferred in three ways. Transformation is the uptake of DNA by bacteria using fairly complex, highly evolved mechanisms for recognition and transport of DNA. Transduction results essentially from errors in phage growth and transfer. Generalized transduction can introduce into the recipient cell virtually any small region of the donor cell's chromosome. Specialized transduction results in the transfer of only a small region of the chromosome adjacent to the site of integration of a prophage. Conjugation is the transfer of plasmid DNA from donor to recipient cell, sometimes with the attendant transfer of various lengths of the donor cell's genome.
3. Following gene transfer, the fate of the introduced DNA—the exogenote—depends on its chemical nature and its mode of introduction into the recipient cell.
4. DNA fragments of the chromosome of a donor cell of the same species introduced by transformation, generalized transduction, or conjugation are usually integrated into (recombined with) the chromosome of the recipient cell by homologous recombination. Homologous recombination joins segments of separate DNA molecules by a process that involves matching homologous regions of the reactant molecules, breaking, and rejoining strands in such a way as to produce recombinant molecules in which a segment of one has been substituted for the homologous segment of the other. It depends on the bacterial RecA (and SSB) protein and is effective only if extensive regions of sequence similarity (homology) are present in the reactant DNA molecules.

5. Even grossly nonhomologous segments of DNA can undergo recombination. This event occurs by a RecA-independent process involving the action of specific enzymes that recognize special sequences of DNA and catalyze recombination there. This process accounts for the integration of most prophages into the bacterial chromosome and for the transposition of IS elements and transposons.
6. The flexibility of the bacterial chromosome is not confined to the major recombinational mechanisms of homologous and site-specific recombination; other biochemical mechanisms can produce genomic rearrangements and recombinations.

STUDY QUESTIONS

1. Gene transfer between cells is said to enhance genetic evolution. Explain the basis of this enhancement.
2. Cite evidence supporting the view that transformation alone among the mechanisms of natural exchange of genes in bacteria evolved specifically for this function.
3. Consider the merits of the suggestion that transformation arose as a nutritional process rather than as a genetic process, that is, that the ability to take up DNA (competence) made it possible for the cell to use DNA as a source of carbon, nitrogen, and phosphorus.
4. Explain why the frequency of transduction of a gene is greater with an HFT than an LFT lysate.
5. Explain why abortive transduction is a common feature of generalized transduction. Under what circumstances could specialized transduction result in abortive transduction?
6. Explain why recombination of the F plasmid with the *E. coli* chromosome is site specific and yet dependent on the RecA protein.
7. Though bacteria are said to be haploid, give an example of a bacterial cell that is at least partially diploid.
8. Give an example in which the introduction of DNA into a cell leads to a stable, heritable change in the recipient in the absence of genetic recombination.
9. Give an example in which the introduction of DNA into a cell leads to genetic recombination, but without changing the genotype of the recipient cell.
10. The immediate product of a mating of humans is called a zygote, but in bacteria it is called something else. What is it called? Why the different term?
11. Is it more remarkable to you that some bacteria, such as *Streptococcus pneumoniae* are naturally transformable, or that others, such as *Escherichia coli* are not?
12. What structural features might distinguish a competent cell from one that is not competent for transformation?
13. Natural transformation by plasmids does not work, except at very low frequency. Explain why it does not work well, and what accounts for success when it does work.

SUGGESTED READING

Topics in this chapter are covered in the following chapters from *Escherichia coli and Salmonella typhimurium: Cellular and Molecular Biology*, F. C. Neidhardt, J. L. Ingraham, K. B. Low, B. Magasanik, M. Schaechter and H. E. Umbarger (eds.). 1987. American Society for Microbiology, Washington, D.C.

Hanahan, D. Mechanisms of DNA transformation. (Volume 2; Chapter 70)
Holloway, B. and K. B. Low. F-prime and R-prime factors. (Volume 2; Chapter 67)
Margolin, P. Generalized transduction. (Volume 2; Chapter 68)
Weinstock, G. M. General recombination in *Escherichia coli.* (Volume 2; Chapter 60)
Weisberg, R. A. Specialized transduction. (Volume 2; Chapter 69)
Willetts, N. and R. Skurray. Structure and function of the F factor and mechanism of conjugation. (Volume 2; Chapter 64)

General reviews on gene transfer and recombination

Kucherlapati, R. and G. R. Smith (eds.). 1988. *Genetic Recombination.* American Society for Microbiology, Washington, D.C.
Smith, H. O., D. B. Danner and R. A. Deich. 1981. Genetic transformation. Annu. Rev. Biochem. 50:41.

11

Coordination of Metabolic Reactions

INTRODUCTION

In the last two chapters we discussed some of the mechanisms of adaptive genetic change. The next seven chapters deal largely with adaptive mechanisms of a *physiological* nature. These involve changes in the behavior of individual bacteria that are maintained only as long as a given environmental condition persists. The adaptability of individual bacteria to a great variety of environmental conditions, including a wide range of temperature, pressure, osmolarity, and pH, is as much a hallmark of bacteria as is rapid growth. It is all the more remarkable when one considers the challenge of coordinating the several thousand individual chemical reactions to produce a new cell, even in a constant, optimal environment.

Our discussion will deal with how individual bacteria adapt rapidly to their environment—how they coordinate metabolic reactions, how they respond to changes in the availability of nutrients, how they cope with frequent changes in the physical quality of their environment, and how they protect themselves from harmful agents and conditions. Bacterial adaptation has intrigued biologists for over a century, and much has been learned. We begin our discussion of physiological adaptation in this chapter by considering how individual metabolic reactions are coordinated to produce the orderliness and efficiency of bacterial growth.

EVIDENCE FOR COORDINATION OF INDEPENDENT PATHWAYS

Bacterial growth requires coordination of separate metabolic pathways. To give this statement concrete meaning, consider the results of a simple experiment. Prepare a salts medium that will support the growth of a bacterial species such as *E. coli.* Add just enough ^{14}C-labeled glycerol as the sole source of carbon and energy (i.e., as "substrate") to permit growth to a density of 10^8 cells per milliliter (well below what the medium would permit if more glycerol were present). Inoculate the medium with a small number of cells and incubate it aerobically.

Samples taken from this culture during its growth reveal several simple, but informative facts. First, analysis of the culture medium (from which the cells have been removed by filtration or centrifugation) indicates that throughout growth the only radioactive materials in the medium are the unused glycerol and CO_2 formed from its oxidative metabolism. When growth ceases due to the exhaustion of the glycerol, almost all remaining radioactivity is present as CO_2. No more than the smallest traces of labeled amino acids, nucleotides, or other metabolites are found. The growing cells, therefore, spill almost no metabolic intermediates into the medium, so we conclude that the rate of formation of each building block (Chapters 3 and 4) must match its rate of utilization for macromolecule formation, and the flow of carbon through each metabolic pathway must be coordinated with the flow through each of the others.

Repeat this experiment, but now include the amino acid histidine (unlabeled) in the medium. Analysis of the culture at the end of growth reveals a remarkable situation. Hydrolyzing the total protein of the cells to individual amino acid residues shows that the histidine residues, alone among all the amino acid residues, contain almost no traces of radioactivity; they (or more precisely, their carbon atoms) must therefore be derived almost exclusively from the nonradioactive histidine supplied in the medium rather than from the radioactive glycerol. But, as before, the radioactive substance in the medium is almost exclusively CO_2; in particular, there is no radioactive histidine. Thus, almost no histidine was made from glycerol. *The cells not only used the histidine supplied in the medium for protein synthesis but also turned off their own synthesis of histidine.* This experiment can be repeated with virtually any of the building blocks, vitamins, or cofactors with the same result—exclusive use of the substance provided in the medium and shutoff of its endogenous synthesis.

The same outcome is observed if the experiment is performed under different growth conditions—different temperature, pH, O_2 tension, carbon source, mixture of nutrients, and so on. The outcome is independent of the growth rate and the chemical composition of the cells. This result shows that the coordination of one pathway with the others is adjustable, not fixed. The rate of functioning of the histidine pathway relative to that producing leucine, for example, can be altered so that the production of each of these amino acids is exactly appropriate for protein synthesis, even though the production of each amino acid varies over a wide range.

Other equally simple experiments reveal further aspects of metabolic coordination in bacteria. If substrates for two distinct peripheral fueling pathways are simultaneously present in the medium, the substrate supporting the faster growth is utilized preferentially. The very best substrates (glucose in the case of *E. coli,* succinate in the case of *Pseudomonas aeruginosa*) are usually utilized to the exclusion of lesser substrates, until nothing is left of the better substrate. Bacteria are capable of selecting the most nutritionally propitious substrate from a mixture, clearly to their selective advantage.

Yet another demonstration of regulatory versatility emerges from the fact that many bacterial species can grow on a variety of substrates that differ in their molar yield of precursor metabolites, energy, and $NADPH_2$. Because the cellular needs for these resources for biosynthesis, polymerization, and assembly are virtually identical no matter what the carbon and energy source, there must be mechanisms that flexibly adjust the ratios of these resources. Otherwise the cell would run short of either ADP or NADP, or both. Consider, for example, the different situations facing a cell growing in each of two different glucose media, one containing no other organic nutrient and the other (a rich medium) containing a complete array of building blocks and vitamins. In the former case, as we have seen (Chapter 5), the ATP and $NADPH_2$ requirements of the cell are adequately met by the amount of glucose metabolism needed to provide the precursor metabolites for making all the building blocks. In the rich medium, however, all of the major biosynthetic pathways shut off, so the need for precursor metabolites is almost eliminated. Nevertheless, the prodigious demand for ATP to drive polymerization and assembly continues. The demand is met by a metabolism of glucose that loads the cell up with carbon end products and $NAD(P)H_2$—the former requiring excretion and the latter, oxidation.

It should be recalled that the macromolecular composition of a bacterial cell is greatly affected by the nature of the growth medium (Chapter 1). This relationship is not what one might anticipate: that cells growing on a mixture of amino acids would be richer in protein; those growing on fatty acids would be richer in lipid; on nucleosides, richer in nucleic acids; and on a sugar, richer in carbohydrate. Quite the contrary, media of very different chemical natures may produce cells of nearly identical macromolecular composition. As discussed in Chapter 15, the single most important factor relating to the macromolecular makeup of bacterial cells is the rate of growth; medium composition affects cellular composition only to the extent that it affects growth rate.

In summary, all the major fueling, biosynthetic, and polymerization pathways of the cell are subject to powerful, independently adjustable controls that bring order out of the potential chaos of a system composed of thousands of individual working parts.

TWO MODES OF REGULATION: ENZYME ACTIVITY AND ENZYME LEVEL

Considering the formidable job of regulating metabolism, we should not be surprised to learn that there are multiple mechanisms of regulation. One fundamental distinction between regulatory controls is whether they change the activity of a protein already present in the cell or whether they alter the net rate of synthesis and, therefore, the cellular concentration of that protein. In this chapter we discuss how enzyme *activity* is regulated and leave for Chapter 12 a discussion of the regulation of enzyme *synthesis*.

Controls on enzyme activity work in many ways. In some cases a protein is inactivated (or activated) by covalent modification (phosphorylation is common, but addition or removal of adenylyl, acetyl, methyl, and other residues also occurs; see Chapter 3). In other cases, the activity of the protein is modulated by reversible association with another molecule—called a LIGAND if it is a small molecule, a MODULATOR if it is a large one. Controls on enzyme levels are also of several sorts, because the cellular level of a protein can be altered by changing its rate of synthesis or, much more rarely, its rate of degradation.

Both types of controls abound in all cells. As a general rule, however, differentiated cells of higher plants and animals display fewer and smaller changes in enzyme levels than do bacteria. Of course, there are great differences in the levels of many proteins between, say, the hepatic cells and the neurons of an animal, but for each cell type, the hour-to-hour and day-to-day regulation of metabolism involves modest changes in enzyme levels. Bacteria, too, rely heavily on controls of the activity of enzymes to coordinate different reactions, but in addition they display major changes in enzyme levels. A trivial example will illustrate this point: the diet of a human has little effect on the protein profile of hepatic cells, but the protein profile of a bacterium growing slowly in a succinate minimal medium is markedly different from that of one growing rapidly in a rich broth.

Reasons for this different reliance on modulation of enzyme levels are not hard to find. It is uneconomical to make an enzyme and not use it (recall the energy it takes to make a protein; Chapter 3); it is cheaper to adjust an enzyme level by modulating enzyme synthesis. But, in a nongrowing, differentiated eucaryotic cell, controls on synthesis alone cannot reduce the level of an already existing enzyme, except by the slow attrition caused by general protein degradation. The bacterial lifestyle, on the other hand, is admirably suited to upward or downward adjustments in the level of an enzyme merely by changing its rate of synthesis. By turning off the synthesis of a protein, a bacterial cell will halve the level of that protein with each doubling of cell mass. For a cell with a generation time of 20 minutes, the level of a (stable) protein can be reduced eightfold in an hour simply by terminating its synthesis. As we shall see (Chapter 12), the level can also be elevated rapidly by changing the rate of synthesis.

Another characteristic of procaryotes that facilitates this reliance on enzyme synthesis to adjust protein levels is the rapid turnover of their mRNA (Chapters 1 and 3). Every few minutes these blueprints for protein synthesis are renewed, a situation offering the opportunity to control enzyme synthesis rapidly at the transcriptional level (Chapter 12).

Finally, we must emphasize that it is not simply *efficacy* that has given rise to the prominence of controls on synthesis as the mode of regulation in bacteria. In fact, the truth may well lie in the opposite direction—the necessity for bacteria to adjust enzyme synthesis may have led to features that facilitate transcriptional control. The need for environmental *adaptability*, coupled with the selection pressure for rapid growth, forces very small cells to develop intricate control mechanisms that allow gene expression to match demands of a given environment. Bacteria cannot afford to make useless or redundant enzymes, and they cannot survive without the ability to produce essential ones; utility and essentiality are defined by the ever-changing environment.

REGULATION OF ENZYME ACTIVITY

When we discussed the ability of bacteria to grow over a range of temperatures, we mentioned a helpful way to assess the extent to which metabolic flow is coordinated by controlling enzyme activity. The profile of protein spots (i.e., their distribution and abundance) displayed on two-dimensional gels made from extracts of *E. coli* grown at 20°C is almost indistinguishable from that of cells grown at 37°C, despite the fact that enormous adjustments must have been made to accommodate the various effects of temperature on different enzymatic reactions. The adjustments must have been largely accomplished by altering the activities of enzymes—increasing the activity of some, decreasing that of others—while leaving their relative levels untouched.

In the remainder of this chapter we examine the rapidly acting controls that modulate the activity of enzymes and bring about the continuous, second-by-second adjustment of metabolic flow in the bacterial cell.

Allosteric interactions

We opened this chapter by noting that a bacterium will utilize a building block, such as histidine, supplied in the medium, rather than make it. We can press this point further by examining the immediate consequences of adding histidine to a culture already growing in a glucose minimal medium. Using radioactively labeled substrate, investigators have determined that *within seconds of the addition of histidine*, almost all carbon flow through the histidine biosynthetic pathway ceases.

This immediate response is an exaggeration of a process occurring continuously to regulate histidine biosynthesis and is the result of the most pervasive control system in cell biology—the inhibition of an ALLOSTERIC ENZYME by the binding of a molecule of small size, a ligand called an ALLOSTERIC EFFECTOR (allosteric = different shape). In the case at hand, the allosteric enzyme is the first enzyme in the biosynthetic pathway leading specifically to histidine, and the ligand is the histidine molecule itself.

Allosteric enzymes are proteins that have, in addition to their ACTIVE SITE (the site on the protein molecule at which substrates bind and catalysis occurs), another site with its own stereospecific affinity for binding small molecules. Allosteric effectors are thought to work by changing the conformational state of the enzyme. By this model, allosteric enzymes are assumed to exist in at least two conformational states: one in which their active sites have high affinity for their substrates and another in which the active sites have low affinity for substrates. The binding of an effector to the allosteric site changes the properties of the enzyme's active site by favoring one conformation over the other: negative allosteric effectors diminish the enzymatic activity of the enzyme by stabilizing the low-affinity conformation; positive effectors increase activity by favoring the conformation in which the active sites have high affinity for substrates. An example of the sensitivity of one allosteric enzyme (aspartate transcarbamylase) to one of its effectors, cytidine triphosphate (CTP), is shown in Figure 1. In the presence of CTP, the affinity of this enzyme for one of its substrates (aspartate) is markedly decreased (K_m is increased). Thus, enzyme activity decreases with an increase in CTP concentration.

In some cases allosteric effectors alter the V_{max} of their enzyme. One of the enzymes—DAHP (3-deoxy-D-arabino-heptulosonate-7-P) synthetase—in the first chemical reaction leading to the synthesis of the aromatic amino acids in *E. coli* is feedback inhibited by tryptophan, largely as a result of a change in its V_{max}.

Control of biosynthetic pathways by feedback inhibition

It is worth emphasizing that allosteric interactions provide a means for the activity of an enzyme to be modified by substances not even remotely resembling the substrates or products of the enzyme itself.

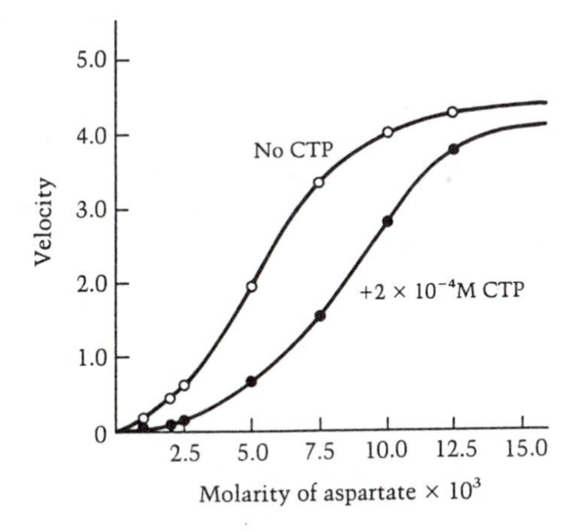

Figure 1

Allosteric effects on aspartate transcarbamylase. There is a sigmoidal relationship between substrate (aspartate) concentration and reaction velocity. Note the effect of the allosteric inhibitor, CTP, on this activity. The topological distinctness of the allosteric (regulatory) and catalytic sites of this enzyme is easily demonstrated, because they are found on two separate subunits of the enzyme. (After Gerhart and Pardee, 1962.)

Thus, allostery can be used for controlling metabolic flows over large chemical distances. The biological significance of the aspartate transcarbamylase example becomes evident when one realizes that CTP is the end product of the biosynthetic pathway (pyrimidine biosynthesis) in which this enzyme participates. The pattern is exactly what we noted in the case of histidine. The ability of a building block to inhibit an early enzyme of its own biosynthetic pathway constitutes a negative feedback loop, and for this reason the inhibition is called FEEDBACK INHIBITION (Figure 2). This phenomenon was discovered in 1956 simultaneously by H. E. Umbarger and by R. A. Yates and A. B. Pardee. Umbarger made the seminal observation that exogenous isoleucine reduces the threonine requirement of a mutant strain of *E. coli* blocked in threonine synthesis. In one of the key experiments of cellular physiology, he followed up this observation by demonstrating that isoleucine had the ability, in vitro, to inhibit threonine deaminase, the first enzyme in the biosynthetic pathway converting threonine into isoleucine. Umbarger recognized that the inhibition could explain the sparing effect he had observed, but he also saw the long-range implication of the inhibition. Feedback inhibition provides the mechanism needed to explain how the addition of an amino acid or any other building block to the culture medium of bacteria can lead to the shutoff of the cognate biosynthetic pathway. In essence, feedback inhibition maintains constant internal concentrations of building blocks in the face of changes in demands or availability.

Note that by this control system, biosynthetic pathways work by "demand feeding"—they produce their building block end products at the rate that these are being consumed by polymerization. It turns out that the situation we have described for CTP and its effect on ornithine transcarbamylase is repeated over and over again throughout the biosynthetic phase of metabolism. The rules are that (1) the first enzyme of a pathway is allosteric and (2) its negative allosteric effector is the end product of the pathway. These rules introduce another characteristic of the 12 precursor metabolites: *they enter biosynthetic pathways via reactions catalyzed by allosterically feedback-inhibited enzymes.* In one fell swoop we have come close to accounting for the coordination of biosynthetic pathways with one another and with the total cellular demand for polymerization.

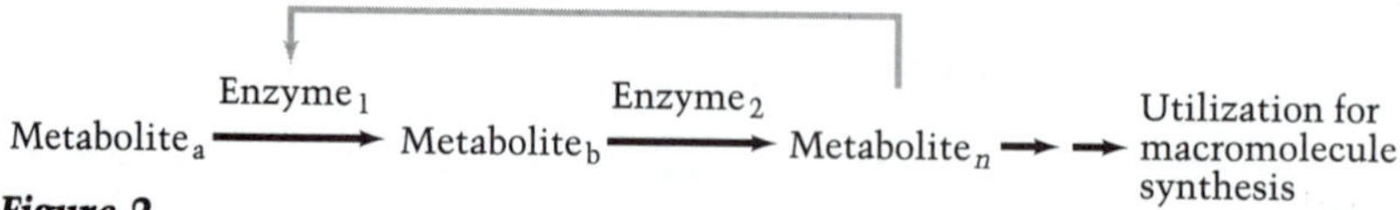

Figure 2

Feedback inhibition. The final product of a series of enzymatic reactions, metabolite$_n$, has the property of binding to the regulatory site of the allosteric protein, enzyme$_1$, and thereby inhibiting it. A scheme of this sort, with appropriate assignment of kinetic parameters, can guarantee that metabolite$_n$ is produced only as rapidly as it is used in some subsequent process.

(A) Isofunctional enzymes

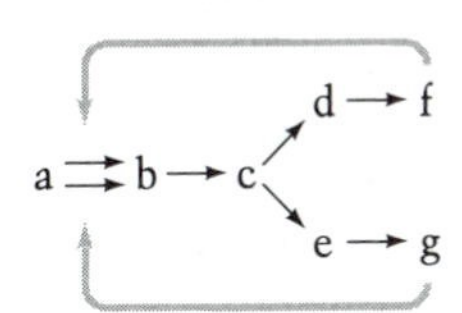

(B) Cumulative feedback inhibition

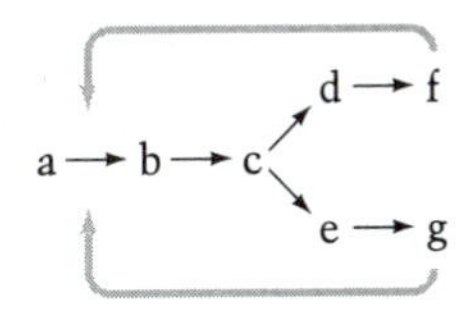

(C) Sequential feedback inhibition

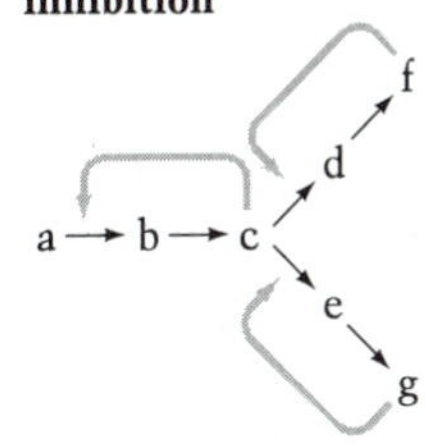

(D) Inhibition plus activation

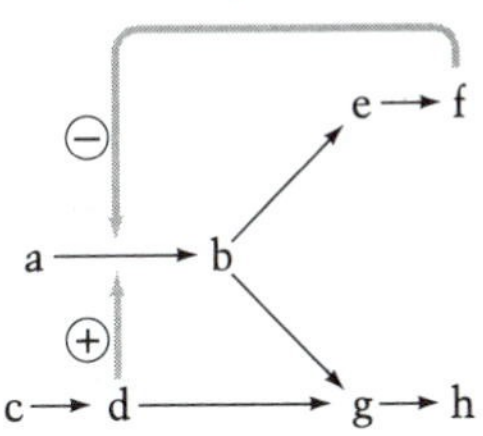

Figure 3

Patterns of feedback inhibition found in bacterial biosynthetic pathways. (A) Isofunctional enzymes for the regulated reaction allow differential feedback effects by the two pathway end products. (B) Cumulative feedback inhibition involves multiple allosteric sites on the regulated enzyme, assuring that there will be some activity unless all of the end products are in excess. (C) In sequential feedback inhibition, different end products operate separately on the various branches of the pathway. (D) Inhibition plus activation uses both positive (⊕) and negative (⊖) allosteric effectors to coordinate complex pathways.

The simple rule that the end product of a biosynthetic pathway is the feedback inhibitor of the first reaction needs clarification for the case of pathways that lead to more than one building block, either sequentially through a linear pathway (e.g., threonine and isoleucine) or divergently through branches from a common stem (e.g., tryptophan, phenylalanine, and tyrosine) (Figure 3). In such cases, feedback inhibition of the first reaction by any single end product could starve the cell for the others. A number of patterns of feedback inhibition have evolved to avoid this metabolic problem.

- ISOFUNCTIONAL ENZYMES Two or more species of an enzyme catalyze the first reaction of the common stem. Each form is sensitive to feedback inhibition by one of the end products of the pathway (Figure 3A).
- CUMULATIVE FEEDBACK INHIBITION A single species of enzyme has distinct allosteric binding sites for each end product of the pathway; the inhibitory consequence of their being bound is cumulative (Figure 3B). (A kinetic variation on this pattern in which singly bound effectors have no effect is sometimes termed CONCERTED FEEDBACK INHIBITION.)
- SEQUENTIAL FEEDBACK INHIBITION Each end product inhibits the first reaction of its own branch and the intermediate at the branch-point inhibits the first reaction of the common stem (Figure 3C).

- INHIBITION PLUS ACTIVATION More complex pathways in which intermediates synthesized by one branch are fed into another are sometimes regulated by a combination of allosteric inhibition and activation; the end product of one pathway is the inhibitor, and the last intermediate product in the other pathway before the pathways join is the activator (Figure 3D). This pattern regulates carbamoylphosphate synthase in *E. coli.*

These patterns of feedback inhibition are highly variable among bacteria. Examples of each can be found to regulate different pathways in a single bacterium, and the same pathway in different bacteria has been found to be regulated in many ways.

Finally, it should be noted that biosynthetic reactions in general rely on a supply of ATP. Obviously, if ATP is in short supply in the cell, biosynthesis suffers. But the decrease in rate of flow of metabolites through biosynthetic pathways is not caused simply by a shortage of an essential reactant. The high-energy compound ATP, and its related products, ADP and AMP, are actively involved as allosteric effectors of key biosynthetic enzymes. To appreciate this point fully, we need first to consider how fueling reactions are controlled.

Control of fueling reactions

Allosteric inhibition and activation play important roles in regulating the flow through fueling pathways. Here the simple device of end product control of the first or an early step in its formation cannot apply. The pathways of formation of the 12 precursor metabolites are far too interrelated (some are cyclic) for each to control its own formation by inhibiting a unique early step—*there are no unique early steps.* Instead, controls work internally in each of the main fueling pathways (Figure 4). For example, phosphoenolpyruvate, an intermediate of glycolysis, inhibits phosphofructokinase, an enzyme catalyzing a reaction three steps back in the same pathway; and α-ketoglutarate, an intermediate of the tricarboxylic acid cycle (TCA), inhibits citrate synthase, which catalyzes the reaction two steps back.

It should be noted that not all the control on central fueling reactions can be accomplished by allosteric interactions. The reversal of flow in the Embden-Meyerhof-Parnas (EMP) pathway to accomplish gluconeogenesis during growth of cells on malate (Figure 7 in Chapter 5) or pyruvate, for example, can occur because all but two of the reactions between pyruvate and hexose phosphate have equilibrium constants not far from 1, and the two that are practically irreversible are bypassed by the synthesis of new enzymes—phosphoenolpyruvate synthetase to bypass pyruvate kinase (reaction C in Figure 7 in Chapter 5), and a specific phosphatase to hydrolyze fructose 1,6-diphosphate (reaction D in Figure 7 in Chapter 5).

Balancing the rates of formation of the 12 precursor metabolites to one another and to the demands of the biosynthetic pathways involves the action of these compounds as allosteric effectors, as illustrated partially in Figure 4. But the need for coordination goes beyond the carbon-containing metabolites; fueling reactions must supply ATP and reduced pyridine nucleotides (both $NADH_2$ and $NADPH_2$). It is not surprising, therefore, to find that these compounds are allosteric effectors at several points in the central fueling pathways (Figure 4).

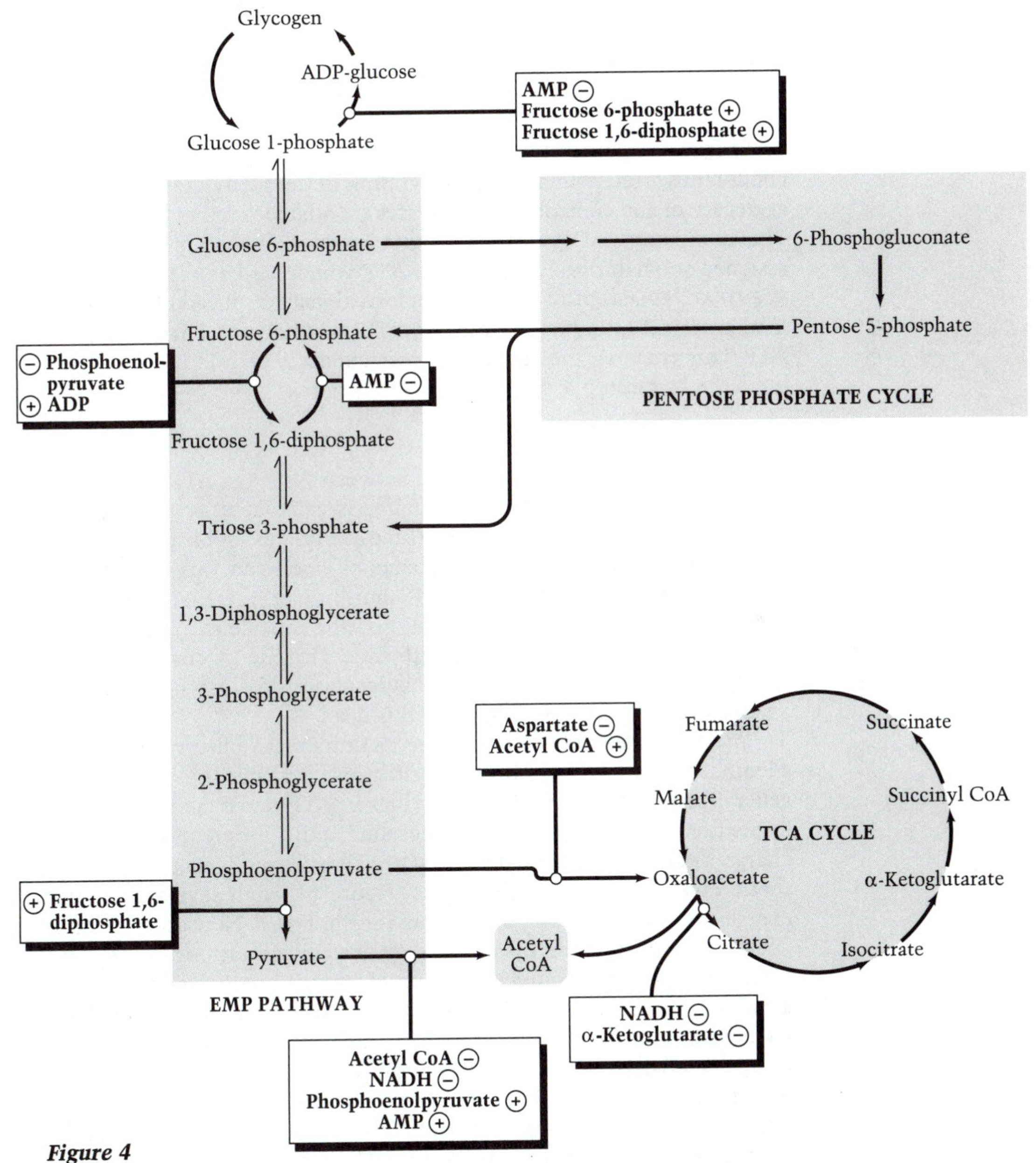

Figure 4

Central pathways of fueling reactions showing some of the allosterically controlled steps. ⊖, negative allosteric effector; ⊕, positive allosteric effector.

Because ATP synthesis and utilization involves a cyclic flow through ADP and AMP, it makes sense to find that all three adenylates play regulatory roles in fueling reactions as well as in biosynthetic pathways. Some enzymes are regulated primarily by the concentration of ATP, ADP, or AMP; others respond to the ratios [ATP]:[ADP] or [ATP]:[AMP]. A summary index of a cell's energy status can be expressed by a convention introduced by D. Atkinson (1968). He defined ENERGY CHARGE as

$$\text{Energy charge} = \frac{[\text{ATP}] + [\text{ADP}]/2}{[\text{ATP}] + [\text{ADP}] + [\text{AMP}]}$$

Energy charge reflects the relative number of high-energy phosphate bonds (anhydride-bound phosphate groups) in the adenylate pool and the aggregate of the controls by adenylates on a large number of enzymes. The compound ADP, with a single anhydride-bound phosphate, is assigned one-half the charge value of ATP, which has two. The activity of ADENYLATE KINASE provides a physiological reason, in addition to the mathemetical one, for assigning to ADP one-half the charge value of ATP. This enzyme, the activity of which is quite high in bacteria, catalyzes the reaction

$$\text{AMP} + \text{ATP} \rightleftharpoons 2\text{ADP}$$

thereby establishing an equilibrium between ADP and ATP with a stoichiometry of 2 to 1. The energy charge is calculated from the actual intracellular concentrations of the three adenylates, but it reflects their relative proportions. The energy charge of a cell can vary mathematically from 0 (all AMP) to 1 (all ATP); but, in fact, the energy charge of bacteria under normal conditions will not lie outside the range 0.87 to 0.95, and it is invariant with growth rate. The energy charge of a cell deprived of a substrate for long periods decreases slowly; when it reaches a value of approximately 0.5, the cell is dead.

The notion of energy charge helps us unify many observations about adenine nucleotides as allosteric effectors. Perhaps the most useful generalization is that regulated enzymes in ATP-replenishing pathways (generally the fueling pathways) respond quite differently to energy charge than do regulated enzymes in ATP-utilizing (generally biosynthetic and polymerizing) pathways (Figure 5). High levels of energy charge inhibit the former and stimulate the latter. Note that it is not energy charge per se that is sensed, but the concentrations of the three adenylate species. Around the value of energy charge maintained in growing cells (approximately 0.9), slight changes in energy charge have large effects on the activity of regulated enzymes. As a result, energy charge is sensitively and precisely poised. Any decrease in it stimulates ATP formation and inhibits its utilization, and an increase in energy charge has the reverse consequence. The critical nature of this regulation can be appreciated by considering that the turnover time for ATP in a growing bacterium is about 0.01 second. Some of the allosteric effects of adenine nucleotides on fueling reactions are shown in Figure 4.

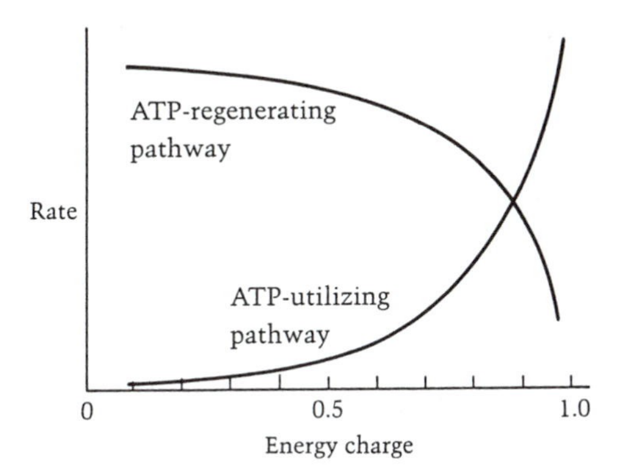

Figure 5

Generalized response to energy charge of the rate of reactions catalyzed by regulated enzymes in ATP-regenerating and ATP-utilizing pathways. Examples of enzymes in ATP-regenerating pathways are phosphofructokinase, pyruvate dehydrogenase, citrate synthase, and isocitrate dehydrogenase; of enzymes in ATP-utilizing pathways are phosphoribosyl-pyrophosphate synthetase, aspartate kinase, and phosphoribosyl-ATP pyrophosphorylase. (After Atkinson, 1968.)

The precise setting of intracellular levels of ATP and the speed with which the controlling mechanisms respond to environmental changes can be illustrated by a direct experiment. If a culture of *E. coli* growing aerobically on glycerol (a nonfermentable substrate) as the sole source of carbon and energy is suddenly deprived of air and sparged with nitrogen, growth and presumably generation of ATP stop abruptly. However, the energy charge changes only to a small extent, although the change is sufficient to essentially freeze polymerization and other ATP-consuming processes.

The opposite problem must also be met. How do cells counteract the tendency to overproduce ATP? Under conditions that stimulate rapid regeneration of ATP, overproduction is dampened by the operation of futile cycles. A FUTILE CYCLE is the operation of two or more reactions that have the net effect of hydrolyzing ATP without changing the concentration of any other metabolite. An example is the concerted action of glutamine synthetase and glutaminase:

$$\text{glutamate} + NH_3 + \text{ATP} \xrightarrow{\text{glutamine synthetase}} \text{glutamine} + \text{ADP} + P_i$$

$$\text{glutamine} + H_2O \xrightarrow{\text{glutaminase}} \text{glutamate} + NH_3$$

$$\text{net reaction: ATP} + H_2O \longrightarrow \text{ADP} + P_i$$

A number of other futile cycles can be postulated from reactions known to exist in procaryotes. Clearly, the operation of futile cycles must be strictly regulated to avoid wasting ATP when it is not in excess.

What about regulation of the pyridine nucleotides (NAD and NADP) that serve as reservoirs and carriers of hydrogen atoms (or reducing power)? Perhaps the first question is, Why does a cell have two pyridine nucleotides as carriers of hydrogen? Recall (from Chapter 5) that NAD functions primarily in oxidoreduction reactions of fueling pathways and NADP in biosynthetic reactions. The necessity for the existence of two species of pyridine nucleotides in cells undergoing aerobic metabolism seems apparent as a consequence of three facts:

- dehydrogenase reactions involving pyridine nucleotides have equilibrium constants near 1
- in fueling reactions, *oxidized* pyridine nucleotides are the reactant
- in biosynthetic reactions, *reduced* pyridine nucleotides are the reactant

Thus, for fueling reactions to proceed, the participating pyridine nucleotide must be largely oxidized; for biosynthetic reactions, it must be largely reduced. Indeed, within the cell, most NAD is in the oxidized form and most NADP in the reduced form. Useful parameters of the oxidoreduction state of pyridine nucleotides are

$$\text{Catabolic reduction charge (CRC)} = \frac{NADH_2}{NADH_2 + NAD}$$

$$\text{Anabolic reduction charge (ARC)} = \frac{NADPH_2}{NADPH_2 + NADP}$$

The CRC in growing cells is maintained at a low level of 0.03 to 0.07, and the ARC is approximately 10-fold higher.

The levels of these charges undoubtedly plays a vital role in coordinating fueling reactions with biosynthesis. The way the levels are set is considerably less clear. A future solution must accommodate two facts. First, as a striking exception to the rule that fueling reactions utilize NAD, two reactions of the pentose phosphate pathway require NADP. These oxidative fueling reactions can proceed despite the unfavorable oxidation–reduction state of NADP, because they are followed by a decarboxylation reaction: the rapid loss of the volatile product, CO_2, shifts the reaction equilibrium toward product formation. Two other fueling reactions, catalyzed by the enzymes isocitrate dehydrogenase and malic enzyme of the TCA cycle, also preferentially utilize NADP; the former is immediately followed by a decarboxylation reaction, and the latter itself involves decarboxylation.

Second, bacteria produce an enzyme that is called TRANSHYDROGENASE and catalyzes the reaction:

$$NADP + NADH_2 \rightleftharpoons NADPH_2 + NAD$$

Transhydrogenase is an integral membrane protein, and the equilibrium of this reaction is believed to be shifted by protonmotive force.

It would be satisfying to conclude that the four fueling reactions in which NADP participates are the primary generators of the reduced form of this coenzyme that is used in biosynthesis, and that transhydrogenase is used to make adjustments in the settings of CRC and ARC. However, other unknown factors must be important also, because mutant strains of *E. coli* that lack the hexose monophosphate shunt, or transhydrogenase, or both, grow well.

Covalent modification of enzymes

Conventional wisdom has it that the use of covalent modification to control enzyme systems is more common among eucaryotes than among procaryotes. In fact, control of enzyme activity by covalent modification (e.g., acetylation/deacetylation, phosphorylation/dephosphorylation, adenylylation/deadenylylation, uridylylation/deuridylylation, methylation/demethylation) occurs in both eucaryotes and procaryotes. Bacterial examples are given in Table 1. An example of methylation has already been discussed in connection with bacterial chemotaxis (Chapter 6).

The regulation of ammonia assimilation is a fascinating example of the utilization of covalent modification of proteins to control their activity. The glutamine synthetase of *E. coli* and of many other bacteria is responsible for ammonia assimilation when the external ammonia concentration is low (1 m*M*), as we have seen in Chapter 5. But, assimilation by this route costs ATP, and therefore it is advantageous to the bacterium to be able to curtail the reaction quickly when it is no longer needed. This response is precisely what is accomplished by a posttranslational modification of glutamine synthetase (Figure 6). A specific enzyme

Table 1. Examples of covalent modification of bacterial enzymes and other proteins

Enzyme	Organism	Modification
Glutamine synthetase	*E. coli* and others	Adenylylation
Isocitrate lyase	*E. coli* and others	Phosphorylation
Isocitrate dehydrogenase	*E. coli* and others	Phosphorylation
Chemotaxis proteins	*E. coli* and others	Methylation
P_{II}[a]	*E. coli* and others	Uridylylation
Ribosomal protein L7	*E. coli* and others	Acetylation
Citrate lyase	*Rhodopseudomonas gelatinosa*	Acetylation
Histidine protein kinases	Many bacteria	Phosphorylation
Phosphorylated response regulators	Many bacteria	Phosphorylation

[a]See Figure 6.

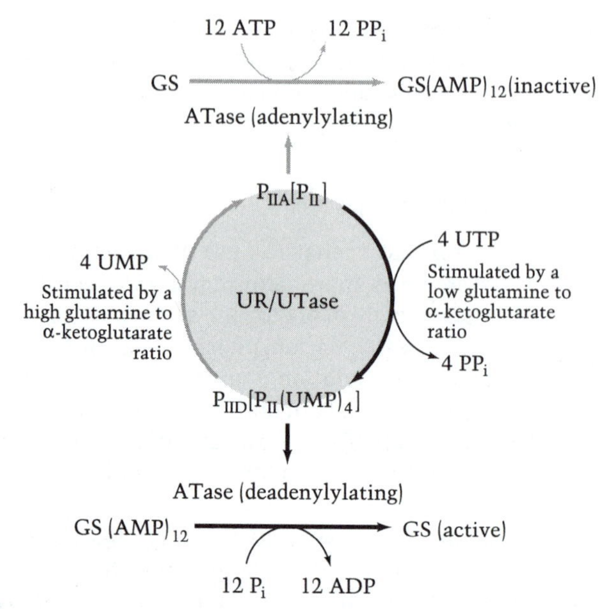

Figure 6

Control of glutamine synthetase (GS) in enteric bacteria by covalent modification. Three proteins regulate GS activity: uridylyltransferase/uridylyl-removing enzyme (UR/UTase), regulatory protein P_{II}, and adenylyltransferase (ATase). Gray arrows show the events leading to adenylylation (inactivation) of GS. High internal levels of glutamine relative to levels of α-ketoglutarate activates UR, thereby causing deuridylylation of the regulatory protein P_{II}. The deuridylylated (unmodified) form of P_{II} is designated P_{IIA} (for P_{II}-adenylylating); it interacts with ATase, enabling it to catalyze the adenylylation of GS. The fully adenylylated GS dodecamer—GS $(AMP)_{12}$—is inactive; as a result of its accumulation, the cell ceases to form glutamine. Black arrows show the events leading to deadenylyation (activation) of GS. High internal levels of α-ketoglutarate relative to glutamine activates UTase, which transfers a UMP group to each of the four subunits of P_{II}, forming $P_{II}(UMP)_4$ or P_{IID} (P_{II}-deadenylylating). P_{IID} interacts with ATase, causing it to deadenylylate (activate) GS, and the cell can once again synthesize glutamine. (After Reitzer and Magasanik, 1987.)

adenylylates it (adding an adenylyl residue to a particular amino acid residue), thereby converting the synthetase into a less active form. Glutamine synthetase is a dodecamer, and each of the 12 subunits can be separately adenylylated. The same modifying enzyme deadenylylates glutamine synthetase, reactivating it. The adenylyltransferase-deadenylylating enzyme switches between its two activities according to the extent to which it itself is modified by uridylylation (the addition of uridylyl groups), and its uridylylation occurs in response to the ratio of the intracellular pools of glutamine and α-ketoglutarate. Albeit complex, this system makes it possible for cells to grow in an environment with limiting nitrogen but to avoid the wasteful operation of this emergency pathway should nitrogen become abundant. When cells find themselves in an environment with a high concentration of ammonia, they assimilate the ammonia through the activity of glutamate dehydrogenase, which has a high K_m for ammonia. This enzyme adds ammonia to α-ketoglutarate, thereby creating glutamate without concomitant cleavage of ATP.

The elaborate system for controlling glutamine synthetase activity by covalent modification has a hidden twist. There is an interplay between *covalent modification* and *feedback inhibition* of this enzyme. Adenylylation not only reduces the activity of glutamine synthetase, it also renders it susceptible to cumulative feedback inhibition by certain products of glutamine metabolism: L-alanine, glycine, histidine, tryptophan, CTP, AMP, carbamoylphosphate and glucosamine 6-phosphate. This situation requires some explanation. When nitrogen is limiting, glutamine synthetase is important for ammonia assimilation; it would be detrimental for it to be inhibited by building blocks made from glutamine, because the synthesis of glutamate and all the metabolites formed using glutamate nitrogen would be restricted; eight times more glutamate than glutamine is used for synthesis of building blocks. This problem is avoided by designing the active, unadenylylated glutamine synthetase to be resistant to the products of glutamine metabolism. It is only when glutamine synthetase is not needed for ammonia assimilation (because glutamate dehydrogenase is active), when it is used just for glutamine formation, that feedback inhibition occurs.

The care with which bacteria control this enzyme makes sense when one recalls that 14% of the dry mass of the cell is nitrogen; the economy of growth is greatly affected by the economy of ammonia assimilation.

It is possible that each instance of regulation of enzyme activity by covalent modification involves allosteric control of the modifying enzyme. In the case just discussed, the uridylylation of the adenylyl transferase-deadenylylating enzyme occurs by an allosteric enzyme sensitive to the intracellular pools of glutamine and α-ketoglutarate.

SUMMARY

1. Bacterial growth involves the intimate coordination of hundreds of separate reactions and pathways. There is much evidence that this coordination is precise and complete. In addition, growth in a variety of environments places special demands on the control of metabolic reactions.
2. Multiple mechanisms regulate metabolism. Some operate by changing the enzymatic activity of a protein, others by altering the cellular amount of that protein.
3. Controls on enzymatic activity can involve either covalent modification of the protein, or, more commonly, reversible association of a ligand that alters the conformation of an allosteric protein. Control by allosteric proteins is particularly significant and widespread, because it offers a means by which the activity of an enzyme can be controlled by a metabolite far removed from its substrates or prod-

ucts. Allosteric effectors can have a positive or negative effect on enyzyme activity.

4. Biosynthetic pathways are commonly controlled by feedback inhibition, by which the end product of the pathway (a building block) negatively influences an early enzyme in the pathway. Because of the complexity of metabolic pathways, several patterns of feedback inhibition have evolved to avoid the situation in which one end product of a pathway might inappropriately turn off synthesis of a second product of the same pathway. Besides the building block end products, adenine nucleotides and pyridine coenzymes also act as allosteric effectors of biosynthetic pathways.
5. Fueling reactions are controlled at key steps by allosteric interactions with the precursor metabolites, with adenine nucleotides, and with pyridine coenzymes.
6. The energy charge of a cell is an index of the extent to which its pool of adenine nucleotides is charged with high-energy phosphate. In general, pathways that consume ATP are inhibited by low energy charge (a measure of the degree to which the adenine nucleotide pool is charged with high-energy phosphate); pathways that produce ATP are inhibited by high energy charge. As a consequence, the energy charge of the cell is strongly buffered in a narrow range of 0.87 to 0.95, even though, in principle, it can vary from 0 to 1.0.
7. The reduction charge (extent of reduction) of each of the two pyridine nucleotide coenzymes, NAD and NADP, can vary mathematically from 0 to 1.0, but each is strongly buffered within a narrow range. The reduction charge of NAD is kept at 0.03 to 0.07, that of NADP at the higher level of 0.4 to 0.5. These respective settings help drive the dehydrogenase reactions in which they participate, because NAD (oxidized form) is required in dehydrogenation reactions in the fueling pathways, and $NADPH_2$ (reduced form) is required in similar reactions in biosynthetic pathways. The mechanism by which these respective reduction charges are maintained is not fully understood.
8. The activity of some bacterial enzymes are modulated by the reversible, covalent attachment of methyl, phosphate, acetyl, or nucleotidyl residues.

STUDY QUESTIONS

1. In what way is feedback control of the first enzyme in a biosynthetic pathway more advantageous than control of a different enzyme, for example the last one?
2. Predict the phenotype of a bacterial mutant that, through a mutation in the structural gene of the first enzyme in an amino acid biosynthetic pathway, has lost feedback control of that pathway.
3. How might one isolate a mutant of the type described in Question 2? For what might it be used?

4. In principle, what might be the advantages of control of the activity of an enzyme by covalent modification, in contrast to allosteric interaction with a ligand?
5. When a bacterial mutant, deficient in a late enzyme in a multistep amino acid biosynthetic pathway, is grown with limiting amounts of the amino acid, remarkable quantities of one or more pathway metabolites prior to the blocked reaction are excreted. The rates of formation of the excreted intermediates per cell far exceed the normal cellular rate of synthesis of that amino acid. Explain.
6. Predict the probable phenotype of a mutant that has lost the ability to make one of two isofunctional allosteric enzymes that catalyze the initial reaction in a pathway and that are differentially sensitive to the two end products of the pathway.
7. Many bacterial isolates from nature prove to be inhibited by single amino acids or small groups of amino acids supplied in laboratory growth media. Provide a working hypothesis to explain how ordinary building blocks, end products of biosynthesis, might be toxic.
8. Tell how you would design an experiment to estimate the extent to which protein degradation (turnover) is significant as a means of altering levels of enzymes in a new, previously unstudied, bacterium.
9. In Chapters 2, 3, 4, 5, 6, and now 11, we have encountered multiple instances in which protein molecules act, not as static catalysts of a chemical reaction, but in a more dynamic, physical way, changing shape and function. Tabulate them, and consider which involve conformational changes of the sort described for feedback inhibition.

SUGGESTED READING

Anderson, K. B. and K. von Meyenburg. 1977. Charges of nicotinamide adenine nucleotides and adenylate energy charge as regulatory parameters of metabolism in *Escherichia coli.* J. Biol. Chem. 252:4151.

Gottesman, S. 1987. Regulation by proteolysis. Volume 2, Chapter 79 in *Escherichia coli and Salmonella typhimurium: Cellular and Molecular Biology.* F. C. Neidhardt, J. L. Ingraham, K. B. Low, B. Magasanik, M. Schaechter and H. E. Umbarger (eds.). American Society for Microbiology, Washington, D.C.

Sanwal, B. D. 1970. Allosteric controls of amphibolic pathways in bacteria. Bacteriol. Rev. 34:20.

Savageau, M. 1976. *Biochemical Systems Analysis: A Study of Function and Design in Molecular Biology.* Addison-Wesley, Reading, Massachusetts.

Umbarger, H. E. 1978. Amino acid biosynthesis and its regulation. Annu. Rev. Biochem. 47:533.

12

Regulation of Gene Expression: Individual Operons

INTRODUCTION

The adaptability of bacteria is nowhere more evident than in the selective use to which they put their genes. Endowed with three to four thousand genes, bacterial cells like *E. coli* use barely half of their repertoire under any given circumstance; the rest is at-the-ready, to be put to use when needed. It is likely that cells of plants and animals in their evolutionary past had the capacity for readily adjusting the expression of their genome and lost much of this flexibility in the secure and constant environment provided by their tissue of residence. To become cooperative and obedient partners in their particular tissue, they had to opt for a more inflexible pattern of differential gene expression.

What is the evidence that bacteria adapt by calling on different sets of genes to produce new proteins appropriate for a new food supply or environmental challenge? Consider the patterns of protein synthesis revealed by two-dimensional gels of extracts of *E. coli* cultures grown under different growth conditions: one culture restricted in the source of nitrogen, the other restricted in the source of carbon and energy (Figure 1). Even the untrained eye can pick out dozens of proteins that are present as significant spots in one gel and missing or barely detectable in the other; automated, quantitative image analysis reveals that there are hundreds of significant differences in the individual proteins of these two cultures. Multiply these differences by the large number of conceivable environments, and one gains a sense of how flexible gene expression is in bacteria.

The cultures represented in Figure 1 were labeled for a relatively brief period. The pattern of radioactive spots would not have been very different if the labeling period had been 2 minutes or 2 hours. This situation

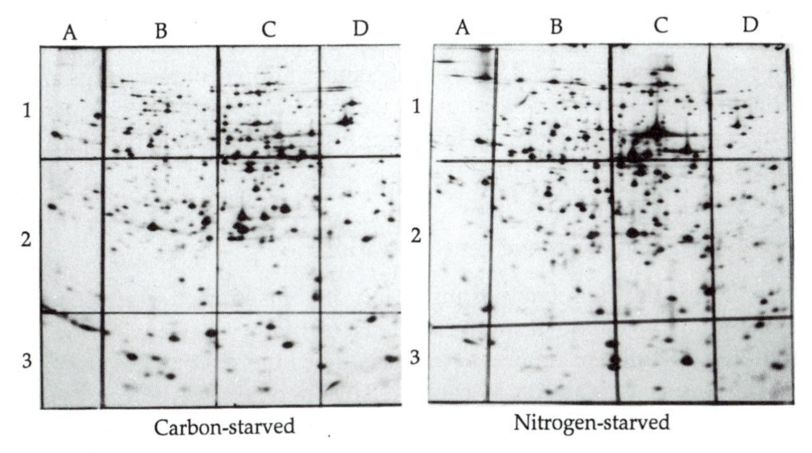

Figure 1

Autoradiograms of two-dimensional polyacrylamide gels of extracts of *E. coli* grown in glucose minimal medium and labeled with $^{35}SO_4$. Left panel: This culture was labeled from 10 to 30 minutes after removal of glucose, the sole source of carbon and energy. Right panel: This culture was labeled from 10 to 30 minutes after starving the culture of ammonia, the sole source of nitrogen. (F. C. Neidhardt, unpublished.)

reflects the fact that the *rate of synthesis* of individual proteins (revealed by short-term labeling) is the main determinant of the relative cellular levels of proteins (measured by long term labeling). There is not much selective degradation of individual proteins; most of the proteins of bacteria are stable, and the loss of any one of them occurs mainly by dilution as the cells grow. To a first approximation, then, *the amount of a protein in a bacterial cell is determined by the overall activity of the gene encoding it.*

Interestingly, there is a subset of bacterial proteins that are either naturally unstable or that can be degraded selectively under special circumstances. Many, if not all, of these "biodegradable" proteins have powerful regulatory roles that affect essential cellular functions, as we shall see later in this chapter and in Chapter 13.

Evidence indicates that flexibility of gene expression is widespread among bacteria. The aim of this chapter is ambitious—to consider the diversity of mechanisms by which bacteria control the functioning of their genome. Our presentation begins with an overview of gene regulation and then moves to a more detailed consideration of selected operons.

CONTROL OF TRANSCRIPTION INITIATION

Throughout the chapters on metabolism, we emphasized the efficiency and economy of bacterial growth. Consistent with this theme, control of the activity of bacterial genes is usually exerted on transcription; and within this process, the controlled step is commonly the very first one—initiation. Control at this point wastes the least amount of energy and other resources, because it avoids the synthesis of useless intermediates (unused mRNA or incomplete proteins). Nevertheless, in other cases to be discussed, control of transcription occurs at steps after initiation, for reasons that are not yet clear in every case. There must be many factors to consider in evaluating the effectiveness and economy of different regulatory mechanisms.

Promoter sequences and gene activation

In *E. coli* the frequency of initiation of transcription varies over a 10,000-fold range as a result of various regulatory devices acting on promoters of different inherent strengths. The inherent or basal strength of a promoter is its ability to capture RNA polymerase molecules and initiate transcription. As we noted in Chapter 3, this process has three steps and can be represented as

$$R + P \underset{K_B}{\overset{\text{Step 1}}{\rightleftharpoons}} RP_c \xrightarrow[k_f]{\text{Step 2}} RP_o \overset{\text{Step 3}}{\rightarrow \rightarrow \ldots \rightarrow} RNA$$

The first step is a reversible binding of polymerase (R) to the promoter (P) to form a CLOSED COMPLEX (RP_c); this reaction is described by an equilibrium constant, K_B. In the second step, a relatively slow isomerization takes place in which an OPEN COMPLEX (RP_o) is formed with a rate constant, k_f. The open complex consists of a TRANSCRIPTION BUBBLE of locally melted DNA duplex; the bubble is about 12 bp in length, and within it polymerase is free to act on the strand that is to be transcribed. The third step is the migration of the polymerase and the transcription bubble (now grown to a length of 18 bp) from the promoter area as the mRNA transcript is formed. The basal or inherent strength of a promoter is determined by its nucleotide sequence, with the −35 region (Chapter 3) affecting both K_B and k_f, and the −10 region affecting primarily k_f. In some promoters, the sequence around the +1 position also seems important.

Adjacent to or overlapping the −35 and −10 regions of promoters are DNA sequences called CONTROL REGIONS. These sequences frequently are binding sites for protein REGULATORS which modulate the basal activity of the adjacent promoter. Control regions may be sequences that direct the formation of secondary structure in their mRNA transcripts, or they even may be promoters for genes transcribed using the opposing strand as template. A control region within which single nucleotide mutations alter the regulation of the adjacent gene was originally called an OPERATOR. The term is less useful now that we know that control regions function in a variety of ways. The term is synonymous with specific sequences to which regulator proteins bind (such as the so-called CAP BOX and the NITROGEN BOX; Chapter 13). Protein regulators that bind to an operator and increase the frequency of transcription of the adjacent promoter are called ACTIVATORS or POSITIVE REGULATORS; those that decrease the frequency are called REPRESSORS or NEGATIVE REGULATORS. Some microbiologists prefer to reserve the term operator for the binding site of a repressor.

Positive regulation

The level of activity of bacterial promoters can be increased by several mechanisms. The best studied of such mechanisms of positive regulation of transcription initiation involve the binding of an activator near the −35 region of the promoter. Promoters that are regulated by activators have a very low basal level of activity. Protein activators increase the rate or extent of open complex formation by affecting either K_B or k_f, or both, but the detailed molecular mechanisms by which they change these parameters are still being elucidated. Some well-studied protein activators are allosteric proteins; their activity changes when small-molecular-weight ligands bind to them. One of the best known activators is the cyclic AMP (cAMP)-binding protein (CAP), which increases K_B at the P1 promoter of the *lac* operon by about 20-fold. The role of CAP in carbon source-related repression is discussed in Chapter 13. The activity of some protein activators is modulated by covalent modification, as in the case of the NR_I protein: when phosphorylated it activates transcription from Ntr (nitrogen-regulated) promoters (Chapter 13). Examples of transcription initiation activators are shown in Table 1.

A second mode of positive regulation of initiation is brought about by the association of various sigma factors (Chapter 3) with the RNA polymerase holoenzyme. This mode is prominently utilized by *Bacillus subtilis* in regulating gene expression during sporulation (Chapter 16). Multiple sigma factors occur in other bacteria as well; at least two sigma factors are involved in the regulation of the heat-shock response (Chapter 13). In general, promoters recognized by a specific sigma factor have

Table 1. Some genes controlled by positive regulators (activators) of transcription initiation

Gene	Organism	Function	Activator
araBAD, E,FG	*E. coli*	L-Arabinose metabolism	AraC
lamB, malPQ, malEFG, malK	*E. coli*	Maltose metabolism	MalT
dsdA	*E. coli*	D-Serine deamination	DsdC
rhaBAD	*E. coli*	Rhamnose metabolism	RhaC
xylABC, xylDEFG	*Pseudomonas putida*	M-Xylene metabolism	XylR
nifLA, nifHDKY	*Klebsiella pneumoniae*	Nitrogen fixation	NifA
lac, mal, gal, ara, deo, tna	*E. coli*	Carbon metabolism	CRP (CAP)
phoA	*E. coli*	Phosphate metabolism	PhoB

a sequence (usually in the −10 region) that distinguishes them from other promoters. Modulation of transcription activity is thought in some cases to be brought about by changes in the cellular levels of the alternate sigma factors (Chapters 13 and 16). The genes involved with nitrogen utilization are recognized by a specific sigma factor, but this sigma factor does not itself modulate the expression of these genes; its role is discussed in Chapter 13.

A third type of positive regulation, less well studied, involves DNA methylation. In most known cases, methylation of DNA in the promoter region has a negative effect on transcription initiation. Because newly replicated DNA is not methylated, a transient region of hemimethylation of DNA—where the old template strand is methylated but the new one is still unmethylated—must follow the replication forks (Chapter 3). While the DNA is in this condition, it can more easily be transcribed. This situation is believed to result in waves of transcription initiations that follows the replication forks around the chromosome.

A fourth mode of regulation, the biological significance of which is still being evaluated, is deduced from the observation that the degree of negative supercoiling of DNA influences the ease with which open complexes are formed. In principle, the expression of some operons could be regulated by the degree of supercoiling of the chromosome. This idea would appear to be an esoteric and purely theoretical possibility were it not for the fact that there is a systematic difference in the degree of supercoiling of the chromosome between anaerobically and aerobically grown cells of *E. coli*. Careful examination of this situation, with the aid of mutants altered in their ability to supercoil DNA, should soon establish the significance of these observations.

Negative regulation

Repressors—negative regulators that bind to operators and decrease promoter activity—are like activators in that they can alter either K_B or k_f for the regulated promoter.

The binding sites of repressors are found in a variety of locations relative to the promoters that they control. Figure 2 shows the control regions of nine *E. coli* genes controlled by a repressor called LexA (Chapter 13). The LexA-binding sites vary in position from upstream of the −35 region to downstream of the −10 region of the promoter. Also, curious operator structures are found. Some *E. coli* operons (*gal, deo,* and *ara*) have two, well-separated, repressor-binding sites. Repressor molecules bound at these sites are thought to interact with each other by looping the intervening DNA segment, thus forming a large repressor–DNA complex.

To be physiologically useful, repressor activity must be capable of being modulated. Like activators, many repressors are allosteric proteins that have one conformation in the absence of their cognate ligand and another when the ligand binds to it. Less frequently, repressors are covalently modified. The NR_I regulator (which acts as a repressor at one site

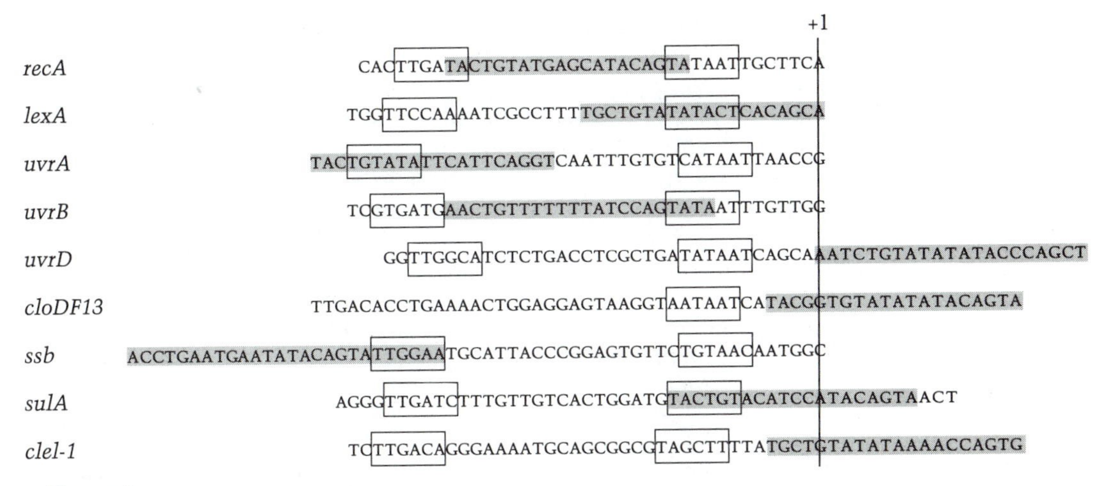

Figure 2

DNA sequences of *lexA*-repressed promoters in *E. coli*. The proposed LexA-binding sites are shaded, and the presumptive −35 and −10 regions of the promoters are boxed. (After Hoopes and McClure, 1987.)

and an activator at a different site) is converted from active to inactive form by reversible phosphorylation. The protein LexA is controlled by an even more drastic modification—an irreversible proteolytic cleavage (Chapter 13).

Autogenous regulation

One common pattern of regulation of transcription initiation is interestingly similar in format to feedback inhibition of enzyme activity (Chapter 11) and has a similar overall outcome. The regulation of genes that encode repressor proteins is ofttimes brought about by their own gene product—the repressor that they produce. This pattern of control is called *autogenous regulation.* At least one example exists of autogenous positive regulation (that brought about by NR_I on the *glnAp2* promoter; see Chapter 13), but most autogenous regulation provides a feedback mechanism by which a repressor can prevent its own overproduction. Examples of autogenous regulation are shown in Table 2.

CONTROL OF TRANSCRIPTION TERMINATION

Control of gene expression by mechanisms that affect the activity of the promoter were thought at one time to account for most, if not all of gene regulation in bacteria, such was the power of the operon model originally proposed by Monod and Jacob (see below in the section on the *lac* operon). It is now known that mechanisms operating after transcription initiation play a major role in the physiology of *E. coli.*

Once initiated, transcription normally proceeds until the polymerase

Table 2. Some autogenously controlled genes in enteric bacteria

Gene	Organism	Function	Mode of control
crp	*E. coli*	Global regulator of carbon metabolism	Negative
araC	*E. coli*	Regulator of operons of L-arabinose metabolism	Negative
trpR	*E. coli*	Repressor of Trp biosynthetic operon	Negative
hutC	*Salmonella typhimurium, Klebsiella pneumoniae*	Repressor of His-degrading operon	Negative
glnG	*E. coli*	Regulator of ammonia assimilation	Positive

encounters one of two possible situations: either (1) a DNA sequence is transcribed that, in the RNA, specifies a GC-rich hairpin structure followed by a series of U residues; or (2) a paused polymerase is exposed to the transcription termination factor, Rho, or some other protein factor (Chapter 3).

Although termination of transcription must occur at the end of an operon, in many instances it can also occur near its beginning. Modulation of the efficiency of termination can serve a regulatory function. The events leading to regulatory termination begin with a pausing of RNA polymerase, which takes place when a hairpin loop forms in the nascent mRNA. The hairpin loop either interferes sterically with the continuation of the polymerase or partially disrupts the integrity of the 12-bp RNA–DNA hybrid structure in the transcription bubble. Whatever the mechanism, the paused polymerase is likely (80–90% of the time) to be ejected from the transcription bubble, thereby terminating elongation. Termination near the beginning of an operon by an "early termination loop" can serve as a regulatory device if there is a physiologically relevant way to modulate it. It turns out that for many different operons early termination indeed aborts most mRNA transcripts, but a special process can override the termination signal when this becomes appropriate for the growth of the cell. This regulated transcription termination is called ATTENUATION.

In most known instances, attenuation operates by a special coupling between translation and transcription. The paradigm for attenuation involving impeded translation of a leader peptide is the tryptophan operon of *E. coli.* As we shall see in detail later in the chapter, if the supply of tryptophan (really tryptophanyl-tRNA) is insufficient to permit rapid translation of a tandem set of tryptophan codons near the beginning of the *trp* transcript, an early termination loop cannot form, and *trp* message is made in its entirety, thereby permitting synthesis of Trp biosynthetic enzymes to relieve the restriction in tryptophan supply. Mechanisms analogous to this control are very common in biosynthetic operons, at least in *E. coli.*

Attenuation need not work by utilizing the sensitivity of coupled translation–transcription to the supply of an aminoacyl-tRNA. Transcription of the *pyrBI* operons of *E. coli* and *S. typhimurium* is controlled by attenuation that is sensitive to the UTP concentration in the cell. Attenuation control of the tryptophan operon in *Bacillus subtilis* is mediated by an allosteric protein that enhances termination when tryptophan is bound to it.

One of the most curious examples of attenuation, called ANTITERMINATION, appears to operate in ribosomal RNA (rRNA) operons in enteric bacteria. These operons are transcribed into a single, untranslated RNA molecule, over 5,000 nucleotides in length. When Rho-dependent termination signals are placed in this sequence, they do not lead to transcription termination. This failure to terminate seems to be a consequence of modification of the transcribing polymerase that occurs as it passes a specific (antitermination) site in the *rrn* leader. The modified RNA polymerase can read through downstream termination signals. An *E. coli* protein, NusA, is thought to participate in this process by binding transiently to RNA polymerase and rendering it susceptible to some unknown factor that makes it immune to Rho-dependent termination. The NusA-associated RNA polymerase, upon encountering a specific nucleotide sequence called BoxA, recognizes a second sequence, the *nut* site, at which certain, as yet undefined cellular proteins convert the transcription complex into the ANTITERMINATION STATE. Although it is not clear that this antitermination state is subject to modulation in response to the cell's physiological state, the possibility is under study that rRNA synthesis is controlled by a termination attenuation mechanism of this sort.

Examples of transcription termination regulatory systems are listed in Table 3.

CONTROL OF TRANSLATION

The expression of some bacterial (and phage) genes is controlled after their mRNA has been made, by what is called POSTTRANSCRIPTIONAL REGULATION. In several well-studied cases, it is the *initiation of translation* that is subject to regulation. Such is the case for several of the operons that encode ribosomal proteins (Chapter 15); in fact, it is the rule rather than the exception in r-protein operons that regulation is by AUTOGENOUS TRANSLATIONAL REPRESSION. How this mechanism works is fascinating. The mRNAs for these operons have been found to contain nucleotide sequences similar to some found in rRNA. In the rRNA, these sequences serve as the sites where certain r-proteins bind during the assembly of the ribosome (Chapter 3). In the mRNA of the r-protein operons, these sequences also serve as binding sites, but in this case the binding of

Table 3. Some genes controlled by transcription termination/antitermination in *E. coli*

Gene (Operon)	Function
hisGDCBHAFIE	Histidine biosynthesis
trpEDCBA	Tryptophan biosynthesis
leuABCD	Leucine biosynthesis
thrABC	Threonine biosynthesis
ilvGMEDA, ilvBN	Isoleucine and valine biosynthesis
pheA	Phenylalanine biosynthesis
pheST	Phenylalanyl-tRNA synthetase
pyrBI	Pyrimidine nucleotide biosynthesis
ampC	β-Lactamase
tna AB	Tryptophan uptake and catabolism
liv, livJ	Branched-chain amino acid uptake

ribosomal protein blocks the initiation of translation of the mRNA. In this way the pool of free (i.e., unassembled into ribosomes) r-protein acts by negative feedback on r-protein synthesis, serving both to match the rate of r-protein synthesis to the rate of assembly of ribosomes, and, in part at least, to coordinate the rates of synthesis of individual r-proteins with each other (Figure 12 in Chapter 15). Not all r-proteins serve as TRANSLATIONAL REPRESSORS; apparently only those that bind directly to rRNA during ribosome assembly have this property. But there appears to be at least one such translational repressor in each of the principal r-protein operons (Figure 11 in Chapter 15).

EVOLUTION OF DIVERSE CONTROL MECHANISMS

Before turning our attention to the details of a few selected operons, we should consider the general principles of regulation of bacterial gene expression that can be inferred from the preceding discussion.

- Bacteria have evolved not one, but multiple ways to modulate gene expression. It seems that every conceivable step in the process that leads from the gene to its finished protein product has been singled out for control in some operon or bacterium (Figure 3).
- Each of the regulated steps is controlled by a plethora of molecular mechanisms in different operons and different bacteria. There are multiple tricks employed to block trancription initiation—or to facilitate it. There are many ways by which regulatory proteins (activators and repressors) are converted from their inactive to their active form—or are elevated in concentration, or are removed by degradation. There are many ways to ensure the synthesis of a full transcript once RNA polymerase has gotten started—or to abort the

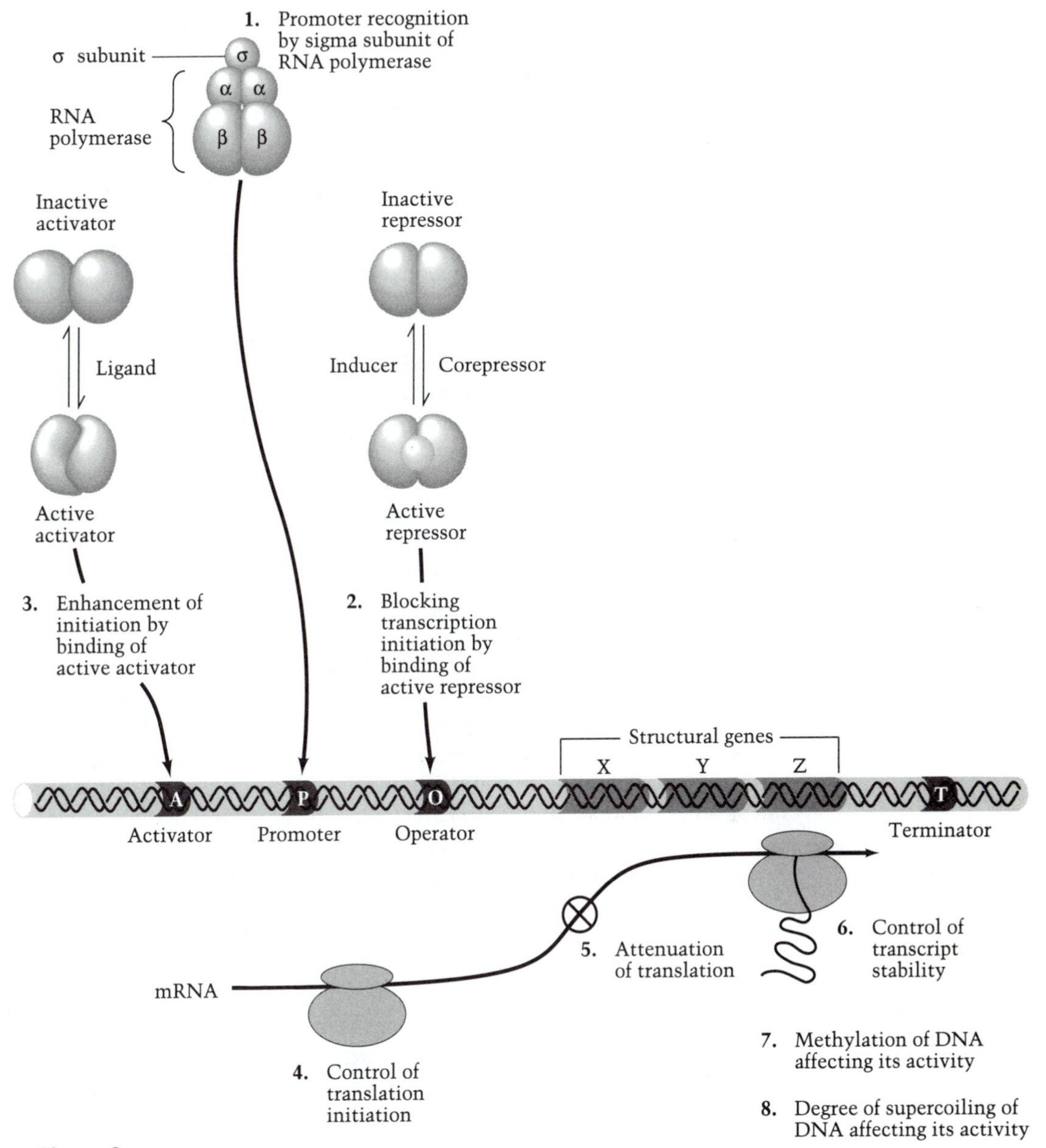

Figure 3

The major sites of control of operon expression in bacteria. 1. Transcription initiation by RNA polymerase (modulated by the availability of polymerase and appropriate sigma factor). 2. Inhibition of transcription initiation by binding of repressor at operator. 3. Enhancement of transcription initiation by binding of activator. 4. Control of ribosome-initiated translation of mRNA. 5. Termination of mRNA synthesis (attenuation). 6. Control of mRNA termination and degradation. 7. Control by methylation of DNA. 8. Control exerted by degree of superhelicity of the DNA.

transcript at an early stage in its synthesis. And there are many ways for the cell to control whether the completed transcript will actually be used to produce protein.
- The same pathway in different species may be subject to different kinds of control.

What is one to make of this diversity? There are (at least) two schools of thought. One holds that the diversity of regulatory mechanisms—their complexity and arcaneness—are not to be understood by studying their properties, because to a large extent they are historical accidents. Natural selection can optimize any of a large number of kinds of control patterns for a given pathway in a given organism. Thus, it is largely chance that determines which mechanism happens to have been optimized. For example, transcription of the genes for the histidine biosynthetic pathway in *E. coli* is controlled by attenuation, the arginine biosynthetic pathway by repression, and the tryptophan pathway by both of these mechanisms.

The opposing view holds that alternative modes of control are not functionally equivalent and that if one fully understood all the relevant metabolic factors, the selection of different regulatory mechanisms during evolution could be explained, even predicted, on the basis of their system behavior. SYSTEM BEHAVIOR refers to integrated system characteristics such as stability, sensitivity, speed of response, and overall economy of function (Savageau, 1976, 1985). The critical test of theories is their predictive value. Some success has been achieved by a method of system analysis that describes the different characteristics of positive, negative, autogenous, and nonautogenous control of transcription in operons encoding peripheral fueling enzymes (catabolic enzymes). From analysis of system behavior, prediction of the choice of transcription control has been made on the basis of the prevalence of the substrate of the pathway in the microbial environment (Savageau, 1974). Besides differences in the resultant behavior of a system, different control modes exact different economic costs from the cell (contrast, for example, control of an enzyme by degrading or modifying it with control by determining whether it is even made). We are far from being able to settle this matter. The question posed by the diversity of regulatory systems is the stimulus for much current investigation, for which system analysis is proving to be an increasingly significant approach.

We turn now to a detailed description of the regulation of two operons that illustrate many of the elements of gene regulation in bacteria and how regulation can be studied. The *lac* operon illustrates negative and positive control of the synthesis of catabolic enzymes; the *trp* operon illustrates the operation of two independent control mechanisms—repression and attenuation—that act on the synthesis of a set of biosynthetic enzymes. These, and the multigene systems to be presented in Chapter 13, illustrate the diversity and the complexity of this subject.

lac OPERON

Background

A vast amount of information has been and continues to be gathered by studying the synthesis of the enzyme β-galactosidase. The reasons for this focus are instructive. ENZYME INDUCTION is defined as an increased differential rate of synthesis of an enzyme (the rate of synthesis of the enzyme expressed relative to the rate of total protein synthesis) when some compound, the INDUCER, is added to the medium. Enzyme induction ("enzymatic adaptation," as it was once called) was a central subject in the early days of bacterial physiology; among the many systems studied was that of lactose fermentation. The strong focus on this system in *E. coli* resulted from the discovery that genetic analysis was possible in this organism and from the introduction of a rapid and sensitive assay for β-galactosidase activity. Both the major discovery and the key technical contribution were by Lederberg (1948).

Work on β-galactosidase induction reached a climax when Jacob and Monod presented the operon model in 1961. This event marked the beginning of a new era, and any modern textbook of biochemistry, genetics, or microbiology offers its own description of the model. An interesting account by Stent and Calendar (1978) emphasizes the logic and the history of the model; the presentation by Schleif (1986) is particularly instructive in illustrating how physical and chemical studies confirmed and extended the molecular aspects of the model suggested by genetic studies; a more recent account by Beckwith (1987) focuses on the pervasive influence the model had on two decades of biology. In our version we stress the questions posed by the model and how they have been answered by the battery of methods available to microbiologists.

When the *lac* operon model was first proposed, strong evidence had just been obtained that growing polypeptide chains were associated with ribosomes and that the increase in β-galactosidase activity when an inducer was added to a culture reflects de novo synthesis of the enzyme (and not, as some had speculated, activation of a hypothetical proenzyme). These experiments, by themselves, suggested that induction somehow released the genetic message contained in the *lac* gene and made it available at the sites where the new enzyme molecules were to be synthesized. Furthermore, the time course of induction had been carefully studied, and it was clear that the then hypothetical message was released quickly by the inducer and that it disappeared quickly when the inducer was removed. It is to be noted that, initially, there was no evidence for the involvement of unstable RNA (mRNA) as a go-between on the way from a gene to its protein product. The proof came somewhat later, when the DNA–RNA hybridization technique was applied to the *lac* system. It was then shown that mRNA appeared and disappeared quickly, as the kinetics of induction demanded and that the quantity of the specific messenger present in the cells paralleled the rate of synthesis of β-galactosidase.

Finally, in several lactose analogue β-galactosides that had been synthesized, the oxygen atom linking the glucose and galactose moieties together had been substituted by sulfur. Some of these thiogalactosides (notably, TMG and IPTG, Figure 4) had been shown to be efficient inducers but not substrates for β-galactosidase. This result tells us two things:

- First, the induction process can be studied in a virtually unchanging state of balanced growth—say, with glycerol as carbon and energy sources and IPTG as inducer. Monod introduced the term GRATUITOUS to describe this condition in which induction does not affect metabolism.
- Second, because no correlation exists between the effectiveness of these compounds as inducers and their affinity for β-galactosidase, the enzyme itself might not be the target of the inducer. Considering the high specificity of induction, it was natural then to think that the target must be a protein—but evidently some protein other than β-galactosidase. This thinking turned out to be correct.

The fact that two other proteins are invariably induced together with β-galactosidase likewise was appreciated at an early time. Galactoside permease (product of *lacY*) is a transmembrane protein involved in actively concentrating β-galactosides within the cell by secondary active transport (Chapter 6); galactoside acetylase (product of *lacA*) transfers acetyl groups to certain galactosides, presumably to reduce their inherent toxicity for *E. coli.*

Such were the physiological and biochemical observations that had to be accounted for by any model claiming to explain the phenomenon of induction. All this is accomplished by the operon model shown in Figure 5.

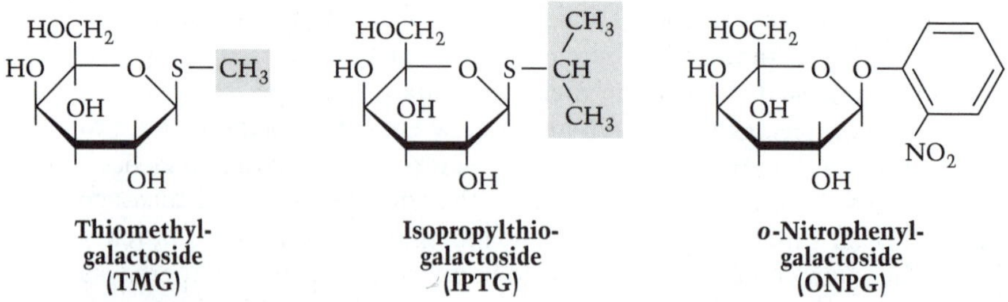

Thiomethyl-galactoside (TMG) **Isopropylthio-galactoside (IPTG)** ***o*-Nitrophenyl-galactoside (ONPG)**

Figure 4

Structure of two nonmetabolized, or gratuitous, inducers of *E. coli* β-galactosidase: thiomethylgalactoside (TMG) and isopropylthiogalactoside (IPTG). Also, structure of the chromogenic enzyme substrate, *o*-nitrophenylgalactoside (ONPG) that is used in the enzyme assay. (From Stent and Calendar, 1978.)

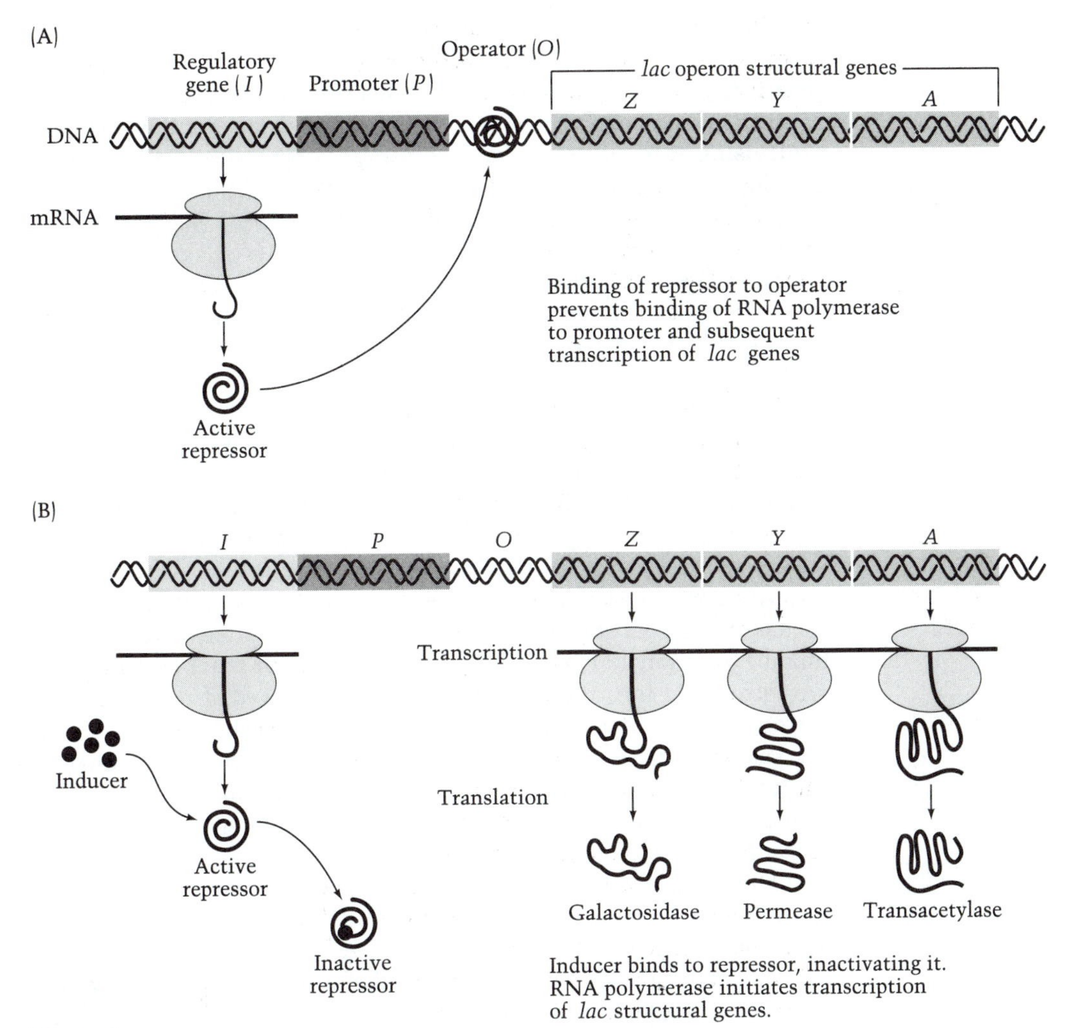

Figure 5

The original operon model of Jacob and Monod, as proposed for the regulation of the *lac* genes of *E. coli* in 1961. (After Jacob and Monod, 1961.)

Genetic analysis

Genetics played a decisive role in carrying the operon model beyond what might otherwise have remained an interesting working hypothesis. The point to be made is at once simple and fundamental. A model involving specific interactions—on the one hand, between proteins and small molecules, and, on the other hand, between proteins and nucleic acids—makes clear predictions about the effects to be expected from changes in specificity of one or another reactant vis-à-vis its target. We have already seen that the small molecule involved—the inducer—can be changed (in this case by organic synthesis) to lose its substrate character while retaining inducer activity. The macromolecular reactants—

proteins as well as DNA target areas—can be changed by mutation. The functioning of a system like the *lac* operon can therefore be analyzed genetically by selecting mutants in which both phenotype and genotype are changed in the ways predicted by the model. The art of this game lies in devising suitable selection procedures. A typical and very simple technique by which *lac* CONSTITUTIVE MUTANTS (that is, mutants producing β-galactosidase in the absence of inducer) have been isolated consists of spreading large numbers of wild-type *E. coli* cells on agar plates with phenyl-β-D-galactoside as the only carbon source. In contrast to IPTG, this galactoside is a substrate for β-galactosidase but not an inducer; in other words it is readily hydrolyzed only by *lac* constitutive mutants, which then grow by using the phenylgalactoside as their carbon source.

Table 4 is reproduced from the original presentation of the operon model and should be read together with Figure 5. The symbols I^s and O^c refer to types of mutants, predicted by the model and isolated by special selection procedures. The mutant line I^s is a cell line producing a "super repressor," that is, an I-product modified in such a way that it is not readily released from the operator site (O). The symbol O^c stands for *operator-constitutive*; this phenotype results from a mutation within the O-site that prevents the normal I-product (the active repressor) from binding to its target. Finally, the genes following the letter F′ in the table are introduced into the cells on various derivatives of the F (*fertility factor*) plasmid (Chapter 9), a process resulting in a cell that is diploid for the set of chromosomal genes carried on the F′ factor. A manual by Miller (1980) contains detailed descriptions of a number of specific selection procedures and also explains how partial diploids, such as those presented in the table, are constructed (Chapter 9).

One example will suffice to indicate how the entries in Table 4 relate to the model in Figure 5: consider line 3, which refers to the cell line with the genetic constitution $lacI^-$, Z^+, A^+ (chromosomal) / F′$lacI^+$, Z^+, A^+ (plasmid). This strain is inducible for β-galactosidase as well as acetylase, just as the haploid wild-type strain is (line 1); this result shows that the chromosomal *lacZ* gene is repressed as effectively as the plasmid *lacZ* gene by the active I-product originating from the plasmid I^+ gene. Thus, as the model indicates, the repressor (the I-product) must be released into the cytoplasm and, by diffusion, must reach the chromosomal O-site as well as the O-site on the plasmid. Therefore, repressor is said to act in the trans as well as in the cis position.

Later, the *lac* genes were introduced onto plasmids that are present in the cell, not as 1 or 2 copies per chromosome as is the F plasmid, but as almost 50 copies. In such cases, it was found that, if the repressor is produced only from the single chromosomal gene, not enough repressor is formed to block all 50 O-sites—the many O-sites "titrate out" the repressor; therefore, the phenotype of the cells is constitutive despite the presence of a normal I^+ gene.

All eight entries in Table 4 should be carefully examined. In particular, the genetic demonstration that the O^c character acts in cis only should be appreciated.

Table 4. Relative concentrations of galactosidase and galactoside transacetylase in *E. coli* variants with various haploid and diploid genomes[a]

Genome[b]	Galactosidase (*lacZ*)		Galactoside transacetylase (*lacA*)	
	Noninduced	Induced	Noninduced	Induced
$lacI^+, Z^+, A^+$	0.1	100	1	100
$lacI, Z^+, A^+$	100	100	90	90
$lacI^-, Z^+, A^+$/F′ $lacI^+, Z^+, A^+$	1	240	1	270
$lacI^s, Z^+, A^+$	0.1	1	1	1
$lacI^s, Z^+, A^+$/F′ $lacI^+, Z^+, A^+$	0.1	2	1	3
$lacO^c, Z^+, A^+$	25	95	15	100
$lacO^+, Z^-, A^+$/F′ $lacO^c, Z^+, A^-$	180	440	1	220
$lacI^s, O^+, Z^+, A^+$/F′ $lacI^+, O^c, Z^+, A^+$	190	219	150	200

[a] From Jacob and Monod (1961).
[b] $lacZ^+$, Inducible for β-galactosidase; $lacZ^-$, noninducible for β-galactosidase; $lacA^+$, inducible for galactoside transacetylase; $lacA^-$, noninducible for galactoside transacetylase; $lacI^+$, *I*-product produced; $lacI^-$, inactive *I*-product produced or no *I*-product produced; $lacI^s$, modified I-product produced, $lacO^c$, operator-constitutive phenotype; $lacO^+$, normal operator; F′, episome carrying *lac* genes.

Isolation of the *lac* repressor

From the beginning to get to the heart of the matter it seemed necessary to isolate the *lacI* gene product and to study its binding properties. Eventually, this task was accomplished, but only after the nucleotide sequences of the operator region that are responsible for the critical and highly specific DNA–protein interactions that make the system work were determined.

As implied in our earlier reference to "titration," the *lac* repressor molecules are normally present in small numbers (about eight per cell). To isolate the repressor, various protein fractions were tested in equilibrium dialysis experiments for their ability to bind, and thus to retain, radioactive IPTG. However, this assay proved insufficiently sensitive at very low concentrations of repressor, and a genetic trick was applied to overcome the difficulty. The procedure is instructive. It was assumed that certain mutations in the *lacI* gene might result in a repressor that bound IPTG more effectively. It was then argued that such tight-binding mutants (with genotype $lacI^t$) would be induced if presented with an inducer concentration too low to be effective in a $lacZ^+$, *Y* cell. The anticipated $lacI^t$ mutants were found; and, given the increased sensitivity of the binding assay with this form of repressor, the way was paved for its isolation. The active form proved to be a tetramer with identical monomers, each with a molecular weight of approximately 25,000, and each of which can bind one molecule of inducer. It appears that cells normally contain 5–10 tetramers, a quantity that corresponds to the yield expected from an average of one or two transcripts from a *lacI* gene per cell cycle.

Analysis of the operator region by DNA sequencing

Once the repressor had been isolated, it was soon demonstrated that the *lac* operator is indeed a small DNA segment. At points along a sequence of approximately 20 DNA base pairs (6–7 nm in length) the repressor makes contacts that result in firm and specific binding. How this and many other protein–nucleic acid interactions are achieved is a subject of intense work in molecular biology.

Sequencing of DNA was the next big advance in our ability to characterize the regulation of the *lac* operon (Figure 6). With efficient methods devised in the 1970s, virtually any genetically defined DNA segment can be recovered from the chromosome, prepared in the requisite amount, characterized by restriction enzymes, and, finally, sequenced.

Sequence information is not the end of the story, but the beginning of a second phase of operon analysis. With the aid of the DNA sequence of a chromosomal segment known to contain the operon of interest, one can scan the DNA segment to locate landmark features that include putative promoters, operators, transcriptional and translational start and stop signals, open reading frames, and possible secondary structures that might form in the mRNA. Catalogs of such landmarks from other operons serve as a valuable resource for this task. Next, one can begin to test the suspected functions of sequences earmarked by the scan. The testing is done with a combination of biochemical, genetic, and physical techniques:

- Structural variants of the sequences of interest—including base substitutions and deletions—can be designed and produced by in vitro chemistry and enzymology or obtained directly from the DNA of appropriate mutants.
- The ability of RNA polymerase and various regulatory proteins to bind to the wild-type and mutant sequences can be tested in vitro by direct binding of these (usually purified) molecules. Binding can be detected by (1) physical measurements, such as changes in the filterability or electrophoretic migration of DNA as a result of protein binding, and (2) chemical techniques, such as protection of the bound DNA segment from nuclease digestion.
- Replacement in vivo of the normal chromosomal DNA segment with engineered variants allows one to obtain important independent verification of the biological function of DNA sequences.

All of these procedures have been employed in analysis of the *lac* operon. A number of the findings are summarized in Figure 6, including the binding sites for repressor and RNA polymerase and the sites of transcription and translation initiation. Also shown are point mutations, marked by arrows at the foot of the figure; these mutations either increase or decrease the strength of the promoter many fold. The entire DNA segment that constitutes the promoter is a unit within which

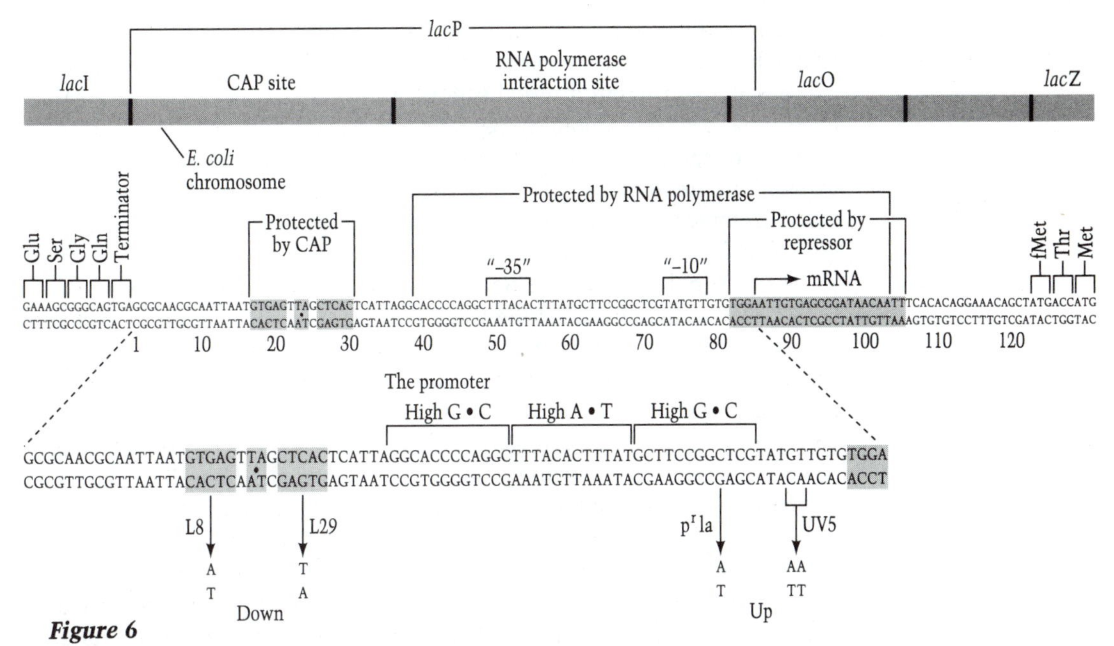

Figure 6

Nucleotide sequence of the regulatory region of the *E. coli lac* operon. The site protected by CAP has a twofold rotational symmetry indicated by a dot at its axis of symmetry. The base-sequence changes associated with promoter mutations are indicated below the vertical arrows. (From Stent and Calendar, 1978.)

many point mutations can change its characteristics dramatically. Although reasonably conserved sequences around the −35 and −10 region can be identified among *E. coli* promoters, there are unexplained observations of the effects of mutations in nonconserved portions of the promoter.

Figure 6 shows one further major feature of the *lac* operon promoter region: just upstream of the RNA polymerase binding site is the binding site for an activator protein called CAP (for, **c**atabolite **a**ctivator **p**rotein). Unless this activator (which is also called CRP, for **c**AMP **r**eceptor **p**rotein) is bound, transcription initiation from the *lac* promoter is weak. This finding correlates with the observation that the promoter structure (−35 = TTTACA; −10 = TATGTT) differs in three significant positions from the optimal, consensus sequence for *E. coli* (−35 = TTGACA; −10 = TATAAT). As a result, normal, full expression of *lac* depends not only on the presence of a specfic galactoside inducer but also on the binding of the activator. The binding of CAP to DNA is itself dependent on the presence of a small nucleotide ligand, cyclic AMP (cAMP), which is required to hold CAP (an allosteric protein) in a suitable conformation for the binding. The requirement for cAMP–CAP for *lac* expression places this operon under an additional control: not only must a galactoside be present, but there must be a high intracellular level of cAMP as well. As a result, the *lac* operon belongs to a larger regulatory unit, the

CATABOLITE REPRESSION NETWORK. Genes within this network are expressed only when two conditions are met: a specific inducer is present, and the cell has an insufficient internal supply of carbon and energy. How cAMP level is related to carbon/energy sufficiency and how it is involved in catabolite repression will be discussed as part of our description of global controls in Chapter 13. The phenomenon of catabolite repression provides, in large measure, the mechanism for achieving the prioritizing of growth substrates that was mentioned in Chapter 11.

Is this, finally, it? Do we now understand at the molecular level the regulation of *lac* operon expression? There are some points, somewhat baroque in nature, that we have not covered. For example, the "natural" inducer of the operon is allolactose, not lactose. This isomer of lactose is produced as a side reaction to the hydrolysis of lactose by β-galactosidase; the small basal level of this enzyme in uninduced cells is sufficient to account for the generation of allolactose internally. Another odd fact is that there is a second *lac* promoter, one that is embedded in the early part of the coding sequence of the *lacZ* gene. The enzyme RNA polymerase binds to this site, but initiates transcription poorly; and the bound polymerase actually interferes with the binding of polymerase at the primary promoter. The binding of cAMP–CAP complex facilitates effective binding of polymerase to the primary site but also *blocks* binding at the secondary site, so it stimulates transcription in two ways.

Finally, it must be admitted that we still do not know everything about the *lac* operon. We have mentioned above that the contributions of the nonconserved parts of the promoter–operator region remain to be learned. Other observations also await explanation. As we shall see in Chapter 13, there are aspects of catabolite repression that are not accounted for by the cAMP–CAP story; another factor is at work in the prioritizing of substrate utilization, and that factor may also affect the *polarity* of the operon—that is, the unequal expression of the three genes of the operon, with the gene products decreasing in molar amounts from the promoter proximal *lacZ* to the distal most, *lacA,* gene. The extent of this polarity can vary under different conditions of *lac* operon expression.

Another area needing clarification is the possible role of termination in regulation. The NusA protein, which we met in our discussion of regulation by transcription termination, is required in vitro for high-level production of β-galactosidase. There is a Rho-dependent transcription termination site (Chapter 3) within the *lacZ* gene, and NusA is required for this site to be ignored by the polymerase. Until the molecular interactions of NusA, RNA polymerase, ribosomes, and termination factors are understood, we cannot claim to have solved *lac* gene expression.

Full understanding of the action of *lac* repressor must also eventually include a clarification of the signficance of two additional weak operators, one within the *lacI* gene and one within the *lacZ* gene. Other unresolved questions concern the possible involvement of other regulatory genes on the *lac* system. One candidate is the *relA* gene, which is involved in the synthesis of the pleiotropic regulatory metabolite, ppGpp (guanosine tetraphosphate); we shall discuss this gene in Chapters 13 and 15. Early in vitro studies indicated that *lac* expression is strongly dependent on ppGpp, but this observation has not been sufficiently extended to settle the issue.

lac as a paradigm

Once the general outline of the regulation of the *lac* operon was solved, there was, understandably, a desire to extend the fruits of this triumph to the study of other genes. And, indeed, many operons do display the pattern of the *lac* system: a promoter–operator complex stands in front of a group of cotranscribed genes encoding proteins of a metabolic pathway or of a closely related set of functions, and a regulatory gene exists somewhere on the chromosome encoding an allosteric repressor protein analogous to the *lacI* gene product. For operons concerned with peripheral fueling pathways, leading into the central pathways, the repressor proteins, like that of *lac,* have reduced affinity for DNA when combined with the inducer, which is usually the initial substrate of the pathway or a close derivative of it. For biosynthetic pathways, on the other hand, the ligand for the repressor is not the initial metabolite of the pathway, but the end product, to which the name COREPRESSOR is given (Figure 3). Frequently the corepressor is the same molecule that is active as a feedback inhibitor of the activity of the pathway (Chapter 11). Combination of the corepressor with the repressor increases greatly the repressor's affinity for binding to the operator. The result is a negative feedback circuit analogous to feedback inhibition, but this mechanism operates at the level of *gene* function rather than at the level of *enzyme* function.

Despite the general applicability of the operon model to many systems, biology proved not to be as simple as all that. After many attempts to force experimental observations into the *lac* control pattern (see the interesting account of this era in Beckwith, 1987), microbiologists eventually realized that the *lac* system represents only one of many varieties of transcriptional regulation used by bacteria and their viruses, and that there are alternatives to transcriptional control itself. Even operons that resemble *lac* in being controlled at transcriptional initiation proved in some cases not to follow the *lac* paradigm. For some operons, control was found to be by positive regulators (activators) rather than by negative ones (repressors); for some operons it turned out that no protein regulator was involved; and for still others it became evident that there were dual modes of control. This last situation is illustrated by the trp operon.

trp OPERON

Background

The *trp* operon of *E. coli* has been analyzed in more detail than any other biosynthetic operon. The five genes of the operon—*trpE, trpD, trpC, trpB,* and *trpA,* in order of transcription—encode the enzymes necessary for the conversion of chorismate, a branch-point intermediate in the aromatic amino acid pathway, to tryptophan. The structure of the operon (Figure 7) contains familiar features: a promoter–operator region precedes the five structural genes of the operon. There are also new structures: notably, a peptide-encoding region within a portion of the operon between the promoter–operator and the first enzyme-encoding gene, and a sequence called an attenuator. This additional complexity reflects the fact that the *trp* operon is under both repression control and the entirely different mode of regulation called attenuation. Not shown in Figure 7 is a gene, *trpR,* located at a distant chromosomal site, which encodes the *trp* repressor protein, TrpR.

Repression by tryptophan

When TrpR is bound by tryptophan, the complex has high binding affinity for the *trp* operator site. Because the *trp* operator lies within the promoter, there is a clear competition between repressor and RNA polymerase, and this competition for binding determines the frequency of transcription initiation. The repressor, free of its ligand, has little binding ability. This property sets up an obvious feedback loop at the level of gene expression. A cell with abundant tryptophan will repress synthesis of the *trp* transcript and be spared synthesis of unnecessary enzymes. A cell with restricting amounts of this amino acid will be DEREPRESSED and will make the transcript necessary for producing the enzymes of the biosynthetic pathway. An additional feature of this system is that the *trp* repressor negatively controls its own structural gene, *trpR,* in addition to the *trp* operon.

During unrestricted growth in minimal medium, tryptophan synthesis is regulated almost exclusively by repression and feedback inhibition. How effective is the repression? This question can be answered by examining what happens when repression is absent. Wild-type cells and a *trp* constitutive mutant ($trpR^-$) grow at practically the same rate in a glucose minimal medium, but the level of the Trp enzymes is approximately five times higher in the cell of the totally derepressed mutant than it is in the normally regulated wild type. The flow of tryptophan needed for protein synthesis is maintained in both cases, and the relatively low enzyme level in the wild-type cells must be the result of internal repression due to the buildup of the appropriate, relatively high concentration of tryptophan. In glucose minimal medium the activity of the *trp* operon is reduced to approximately 20% of its fully derepressed level by sensing the tryptophan produced intracellularly and adjusting operon expression accordingly. The addition of tryptophan to the medium results in a further reduction—to 1 or 2% of the derepressed level—which adds slightly to the economy provided by the regulation.

It is noteworthy that the *trp* operon can satisfy the normal biosynthetic needs of the cell by working at a small fraction of its maximum capacity, because the full capacity for *trp* expression is utilized when cells growing with excess (exogenous) tryptophan suddenly run out. The near instantaneous derepression of the operon can call into play its full

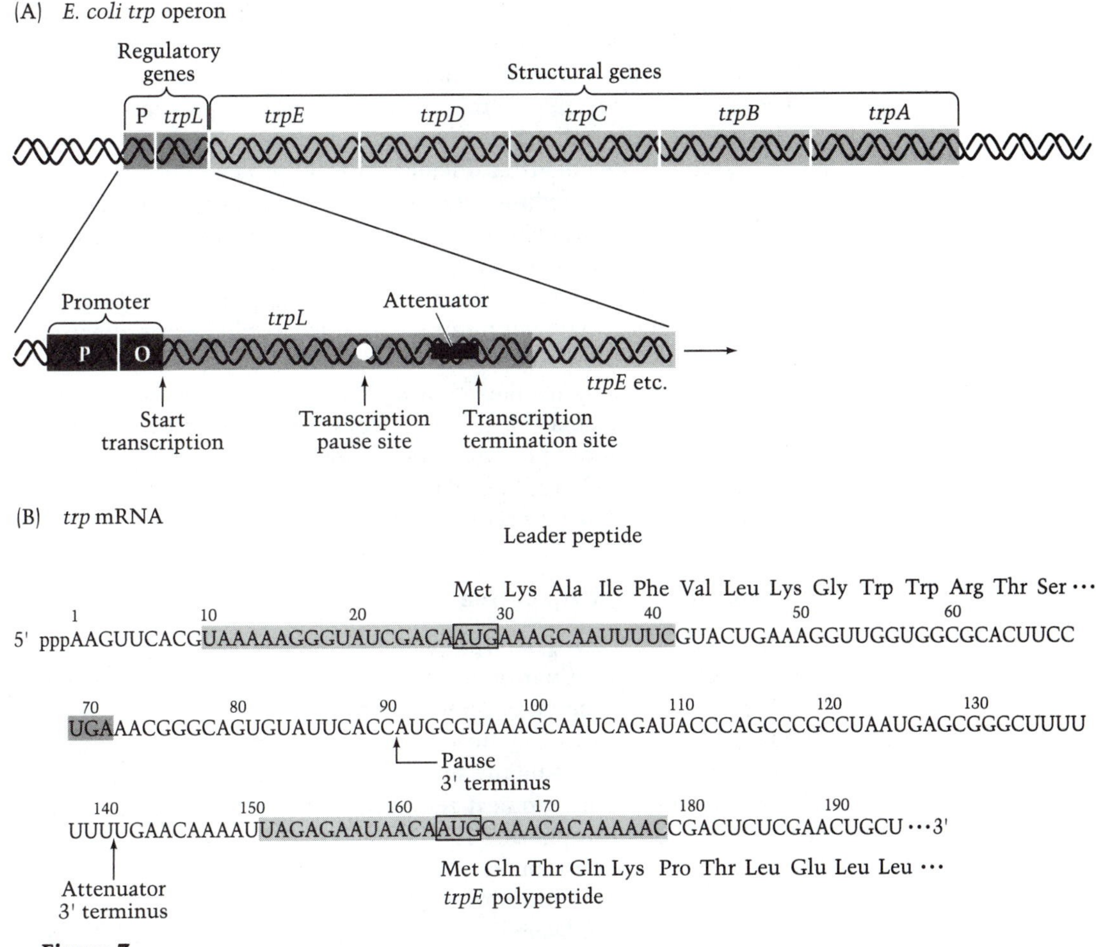

Figure 7

The regulatory and structural regions of the *trp* operon of *E. coli*. (A) Overall view of the structure and major functional features of the operon. Trancription initiation is controlled at a promoter-operator. Transcription termination is regulated at an attenuator in the transcribed 162-base pair leader region, *trpL*. All RNA polymerase molecules transcribing the operon pause at the transcription pause site before proceeding further. (B) The nucleotide sequence of the 5′ end of *trp* messenger RNA. The nonterminated transcript is presented. When transcription is terminated at the attenuator, a 140-nucleotide transcript is produced. Its 3′ terminus is also indicated by an arrow. The 3′ terminus of the pause transcript, at nucleotide 90, is indicated. The two AUG-centered ribosome binding sites in this transcript segment are shaded. The boxed AUG's are where translation starts, and the shaded UGA is where it stops. The predicted amino acid sequence of the *trp* leader peptide is shown. (After Yanofsky, 1981.)

capacity for *trp* mRNA synthesis and translation. The new, higher levels of the Trp biosynthetic enzymes needed in minimal medium can be achieved within a dozen minutes. This capability is a near-universal feature of biosynthetic operons.

The autoregulation of *trpR* could have one (or both) of two functional advantages. The first relates to cells growing in a minimal medium. A cell growing in the presence of added tryptophan maintains greatly reduced levels of the Trp enzymes. The autogenous repression of *trpR* still leaves the cell with approximately 50 molecules of the repressor, and this amount is evidently sufficient for full repression of *trp* in this medium. In minimal medium, the levels of the Trp enzymes are elevated; but, as we have seen, this condition is achieved despite the largely repressed state of the *trp* operon. The level of TrpR is also elevated in this medium, because of the autogenous control circuit. Perhaps the repression of *trp* in cells growing in minimal medium would be less effective without the elevation of TrpR concentration. A second possible advantage of the autogenous regulation of *trpR* is that cells growing in the absence of tryptophan may be, in effect, "looking ahead" to a time when tryptophan is once more present in the medium. The high TrpR levels would then be available to bring about a rapid *over*repression of the *trp* operon to achieve as fast a response as possible.

Attenuation in the *trp* operon

When cells are starved of tryptophan, the *trp* operon opens up in two stages. When the concentration of tryptophan in the cell goes down, the first response of the operon is derepression. As noted earlier, the TrpR-mediated repression mechanism can modulate Trp enzyme synthesis over a 100-fold range. Should the starvation be very severe and the internal tryptophan level drop to near zero, a second process kicks in and is capable of increasing by another 10-fold the rate of *trp* mRNA synthesis. This second process is attenuation, the controlled termination of the *trp* transcript at an early point in its synthesis.

Under normal growth conditions, approximately 9 out of every 10 *trp* transcripts terminate before they reach the structural genes of the *trp* operon. By reducing this high rate of abortion and allowing a near-universal continuation of transcription through the region of termination, the cell can increase its rate of *trp* message formation by a factor of 10. What brings about the termination? What modulates it? How is the process regulated in a manner responsive to the tryptophan supply of the cell? The answers to these three questions are known to a remarkably complete extent.

Components of the attenuation control system

The promoter for the *trp* operon is not adjacent to the first gene, *trpE,* but is located some distance upstream of it (Figure 7); the transcriptional start site is approximately 160 bp from *trpE.* The transcript of the intervening LEADER SEQUENCE (*trpL*) contains the major elements unique to attenuation control. The other components that participate—ribosomes, tRNA, tryptophan, and tryptophanyl-tRNA synthetase—are simply workaday parts of the translation apparatus.

The features of importance to attenuation in this leader sequence are shown in Figures 7 and 8. They are, in order from the transcriptional start site:

- A LEADER PEPTIDE SEQUENCE, which is a sequence of 14 codons, two of them tandem tryptophan codons; the sequence is endowed with a normal AUG start codon and a corresponding ribosome-binding site.

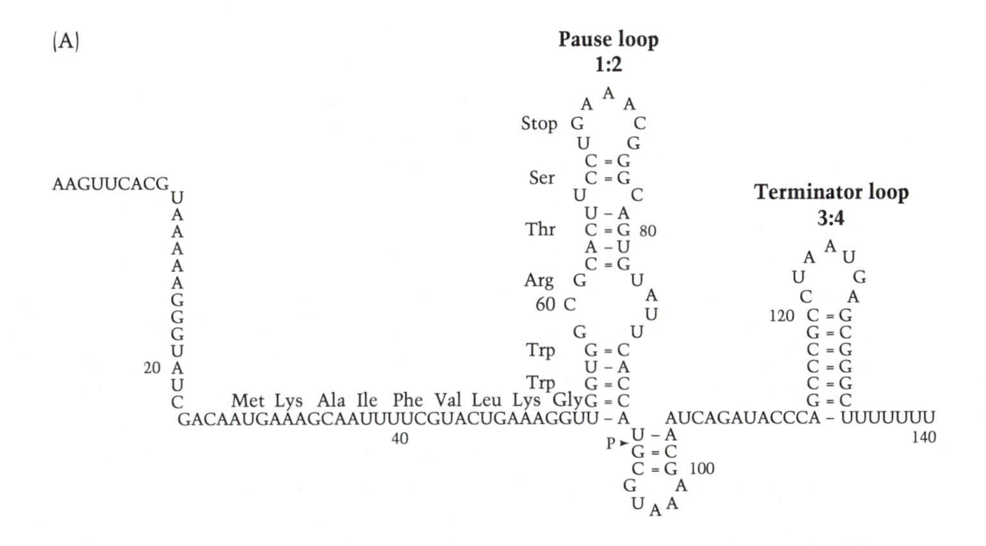

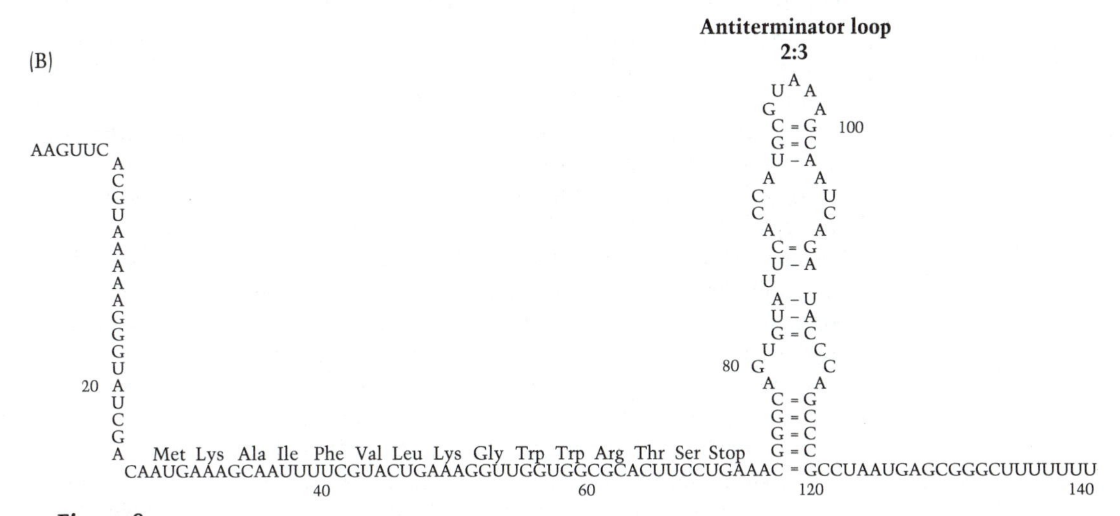

Figure 8

Secondary structures of the *trp* leader region of *E. coli*. Numbers 1, 2, 3, and 4 denote the RNA segments that form the secondary structures pictured. (A) Termination configuration. The arrowhead indicates the site of transcription pausing. (B) Antitermination configuration. (After Landick and Yanofsky, 1987.)

- Four tandem transcript segments, which are designated 1, 2, 3, and 4; by alternative patterns of base pairing, these segments can form either two hairpin structures—called the PAUSE LOOP (1:2) and the TERMINATOR LOOP (3:4)—or a single structure—called the ANTITERMINATOR LOOP (2:3).

With these four features—*leader peptide sequence, pause loop, terminator loop, and antiterminator loop*—the leader transcript is equipped for attenuation control.

Attenuation in action

The story that has emerged from a brilliant series of studies from Yanofsky's laboratory goes as follows. An RNA polymerase molecule initiates transcription of *trp* at the promoter. As transcription proceeds, ribosomes load at the appropriate site to commence translation near the AUG start codon of the leader peptide-coding sequence of the transcript and move in normal fashion closely behind the transcribing RNA polymerase. The first special structure produced by the polymerase is the promoter–proximal hairpin called the pause loop (1:2). As its name indicates, this structure causes the polymerase to cease transcription for a time. The first ribosome translating the transcript is thought to disrupt the pause complex and to restore transcription. What has been accomplished by all this is the assurance that the translating ribosome is positioned on the mRNA close behind the transcribing polymerase. Once synthesis of the leader peptide is complete, the terminator (3:4) forms. This terminator is a stable hairpin structure followed by a run of seven U's, a sequence characteristic of rho-independent transcription terminator signals. Termination is reasonably efficient, and 90% of the *trp* transcripts end at the run of U's, before transcription of *trp* structural genes. So, our first question has been answered: *trp* transcripts terminate at a conventional termination signal.

How is termination modulated, and how is the modulation attuned to tryptophan availability? Here the leader peptide has a decisive role. Two neighboring Trp codons is an unusual situation, considering that in *E. coli* proteins only approximately 1 amino acid residue in 100 is tryptophan. As translation advances and the lead ribosome faces a Trp codon, further progress will depend on the availability of a charged tRNA^{Trp}; in this situation, which calls for two tryptophanyl-tRNAs in tandem, tryptophan limitation can be assumed to cause translation to stall. The location of the stalled ribosome on the transcript prevents the formation of the stem formed by segments 1 and 2 of the transcript, thus favoring the formation of the alternative loop, 2:3. This structure naturally precludes formation of the terminator loop, 3:4; consequently, the RNA polymerase transcribing ahead of the (stalled) ribosome may pass through the "danger zone" without terminating and continue to the end of the operon. Conversely, if the tRNA^{Trp} is more or less fully charged, translation can proceed apace; this event prevents formation of the 2:3 loop and allows the 3:4 terminator loop to persist, a situation resulting in transcription termination 90% of the time.

Our final questions have been answered: termination is relieved by the positioning of a stalled ribosome over the leader transcript in such a way that it prevents formation of the terminator, and this positioning is brought about by an insufficiency of cellular tryptophan. The mechanism is entirely novel; no specific protein is involved, and the DNA–protein interaction that controls access to the promoter according to the classic operon concept is replaced by interactions between the nascent transcript and the common transcription and translation elements of the cell. The entire attenuation mechanism is encoded in the 162 base pairs of the *trp* leader.

How transcription can be stopped by a structure behind the transcribing polymerase has puzzled many. The pictorial view of the process given in Figure 9 may provide some help.

Supporting evidence

Perhaps the most remarkable aspect of attenuation in *trp* is that the story we have just told is buttressed by an immense body of experimental evidence. The obvious corollaries to the model were checked early in the studies; mutations that affect the formation or structure of $tRNA^{Trp}$ affect transcription of the operon. The leader peptide of 14 amino acids has not been isolated, but the proof that translation actually is initiated effectively from the AUG codon of the leader is elegant. Deletion mutants missing a segment between a late codon in the leader sequence and an early codon in the *trpE* gene produce stable fusion polypeptides that begin with the early part of the leader peptide and continue with the TrpE sequence.

Knowing the nucleotide sequence of the leader region was crucial to the development of the hypothesis that the alternative stem-and-loop structures provide the structural basis for attenuation. Computer constructs of transcript structures must be viewed with caution; they are easy to make, but their correspondence to real structures is by no means assured. In the case of the *trp* leader sequence, however, digestion with RNase T1 produces fragments that migrate together in nondenaturing gels and that support the base-pairing pattern illustrated in Figure 8.

A number of mutants with known base substitutions in the leader region have been isolated. Their properties bear out the correctness of the model just described. One mutant harbors a G → A change in the stem of the 2:3 structure, weakening it and stabilizing the 3:4 termination structure; starvation for tryptophan does not relieve attenuation in this mutant. By contrast, approximately 30 mutants in which termination is reduced harbor base changes clustering around the middle of the 3:4 stem. Much additional experimental support exists for the model (Yanofsky and Crawford, 1987).

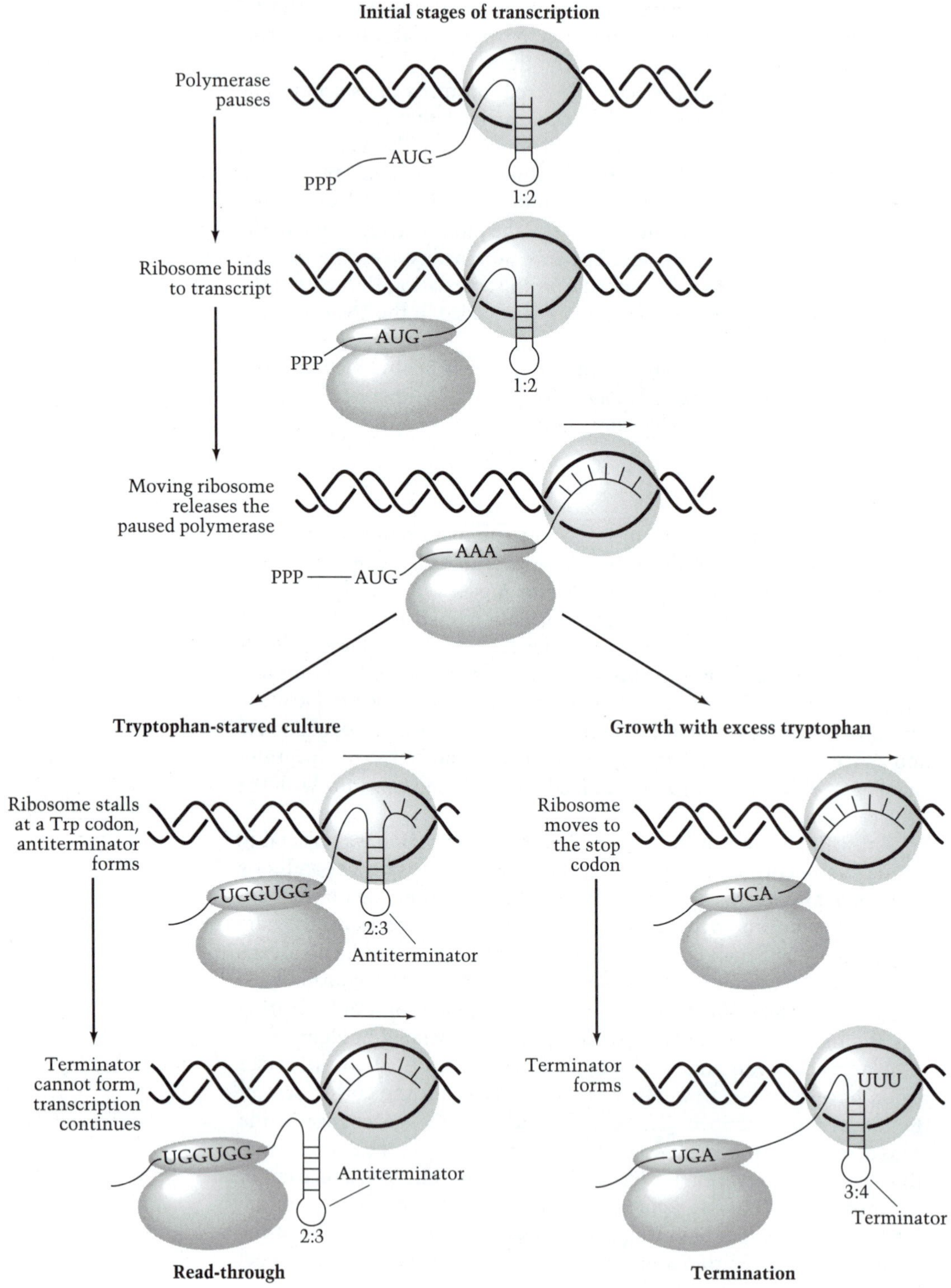

Initial stages of transcription

Polymerase pauses

AUG

PPP

1:2

Ribosome binds to transcript

AUG

PPP

1:2

Moving ribosome releases the paused polymerase

AAA

PPP — AUG

Tryptophan-starved culture

Ribosome stalls at a Trp codon, antiterminator forms

UGGUGG

2:3

Antiterminator

Terminator cannot form, transcription continues

UGGUGG

Antiterminator

2:3

Read-through

Growth with excess tryptophan

Ribosome moves to the stop codon

UGA

Terminator forms

UUU

UGA

3:4

Terminator

Termination

Attenuation and growth physiology

The simultaneous presence in the *trp* operon of two independent control systems invites speculation as to their respective places in evolution and in the growth physiology of present-day *E. coli*. We know that during balanced growth the expression of the operon is regulated by repression and that attenuation is adjusted only when tryptophan is in very short supply. Consider a culture of a wild-type *E. coli* that has grown for a long time with tryptophan supplied in the medium; when this amino acid is exhausted, derepression will, by itself, increase the frequency of transcription from the *trp* promoter approximately 100-fold, and a further 10-fold increase may result from relieving attenuation. The question would seem to be how much a cell gains by being able to increase the potential of the *trp* operon 1,000 rather than 100 times and thus shorten the period necessary to build up an adequate supply of the *trp* enzymes. We do not know the answer, but we must be prepared to accept the possibility that the *trp* attenuator is an evolutionary relic with little or no physiological significance today. If so, this relic has been the source of precious information.

If there can be doubt about the importance of attenuation in regulating *trp*, this uncertainty does not extend to the amino acid biosynthetic operons that appear to be controlled by attenuation alone. For each of the operons listed in Figure 10, the predicted leader peptide is excessively rich in the amino acids thought to be the specific effectors, and an extensive search has failed to turn up repressor-negative mutants. It is therefore natural to assume that attenuation is the only mechanism regulating these five operons.

As noted in our earlier general discussion, attenuation is not restricted to amino acid biosynthetic operons, but attenuation based on the translation of a leader transcript probably applies to these operons exclusively.

SUMMARY

1. Bacteria adapt to environmental change by major adjustments in their pattern of gene expression. Probably fewer than half of their genes are expressed during growth under any given circumstance.
2. Bacterial proteins are for the most part quite stable. Changes in rate of synthesis are responsible for both raising and lowering the cellular level of a given protein. Important exceptions involve proteins that are themselves powerful regulators of essential cell functions.

Figure 9

Pictorial representation of the relationship of RNA polymerase and the first translating ribosome in the region of attentuation control of an operon. (After Landick and Yanofsky, 1987.)

Operon	Leader sequence
pheA	Met–Lys–His–Ile–Pro–Phe–Phe–Phe–Ala–Phe–Phe–Phe–Thr–Phe–Pro
his	Met–Thr–Arg–Val–Gln–Phe–Lys–His–His–His–His–His–His–His–Pro–Asp
leu	Met–Ser–His–Ile–Val–Arg–Phe–Thr–Gly–Leu–Leu–Leu–Leu–Asn–Ala–Phe–Ile–Val–Arg–Gly–Arg–Pro–Val Gly–Gly–Ile–Gln–His
thr	Met–Lys–Arg–Ile–Ser–Thr–Thr–Ile–Thr–Thr–Thr–Ile–Thr–Ile–Thr–Ile–Thr–Thr–Gly–Asn–Gly–Ala–Gly Gly–Ala–Ala–Leu–Gly–Arg–Gly–Lys–Ala
ilv	Met–Thr–Ala–Leu–Leu–Arg–Val–Ile–Ser–Leu–Val–Val–Ile–Ser–Val–Val–Val–Ile–Ile–Ile–Pro–Pro–Cys

Figure 10

Leader sequences in amino acid operons. (After Yanofsky, 1981.)

3. Enzyme synthesis is controlled at a number of different steps in the overall process leading from gene to protein: initiation of transcription, transcription maintenance and termination, translation initiation, and translation efficiency.
4. Transcription initiation can be positively controlled (by protein activators) or negatively (by protein repressors). The activity of different protein regulators of transcription initiation can be modulated by allosterically induced conformational change, by covalent modification, or by selective degradation. Autogenous regulation, by which the protein product of an operon acts as repressor of that operon, is a common pattern in the control of genes with regulatory function.
5. Transcription attenuation provides a means to control the early termination of transcription of an operon. This mechanism is widely used to regulate biosynthetic operons. It is economical genetically, because no regulator protein need be synthesized.
6. Translational repression, by which the binding of a protein to mRNA prevents that transcript from directing protein synthesis, is the most common control device in operons encoding ribosomal proteins.
7. Different control systems have different system properties, but at present it is not possible to predict with confidence the precise mode of gene regulation that will be disclosed for a given operon in a given organism. The diversity and complexity of gene control systems for different operons and different organisms await explanation.
8. The *lac* operon is subject to negative and positive control. A specific repressor protein binds to the operator region, preventing RNA polymerase from initiating transcription at the promoter. Binding of repressor is greatly decreased when a small inducer molecule (allolactose, made from lactose by β-galactosidase) is complexed to the allosteric repressor. High-level expression of the operon requires the intervention of positive control. This control is provided by the binding of an activator protein, CAP, to a site just upstream of the promoter. The binding occurs when the allosteric CAP protein is

complexed with the small nucleotide, cAMP, an internal symbol for carbon–energy source insufficiency. Long thought to be the paradigm for regulation of bacterial operons, the *lac* system actually illustrates only one of the many patterns of transcriptional regulation. Study of this operon continues, because there are still unanswered questions about molecular and physiological aspects of its regulation.

9. The *trp* operon encodes the enzymes of the specific biosynthetic pathway leading from chorismate to the amino acid tryptophan. This operon is under dual control. Repression is brought about by an allosteric protein, the TrpR repressor. This protein, in a complex with trypophan, binds to the operator region, thereby impeding transcription initiation. Repression can provide a 100-fold variation in the level of expression of the operon, depending on the availability of tryptophan. Attenuation is brought about by a mechanism dependent only on the normal components of the cellular translation apparatus and the structure of the leader sequence of 162 nucleotides between the promoter and the first gene of the operon. Normally 90% of the *trp* transcripts terminate within the leader, but this termination is adjustable, a flexibility resulting in a transcription modulation called attenuation. For the *trp* operon, attenuation works on the principle that a deficit of charged $tRNA^{Trp}$ leads to a reduction of termination. Removal of the attenuation control provides a potential 10-fold increase in the production of complete transcripts of the operon. Although attenuation seems to be quantitatively less significant for *trp* than is repression, biosynthetic operons for other amino acids appear to be regulated solely by attenuation.

STUDY QUESTIONS

1. During growth in glucose minimal medium, what is the cost to *E. coli* of attenuation control on the *trp* operon? Express the answer as excess number of high-energy phosphate bonds consumed per complete *trp* transcript made.
2. Compare the answer to Question 1 to the cost of making one molecule of *trp* repressor protein (108 amino acid residues)?
3. Contrast the phenotype of a mutant of *E. coli* that has sustained a knock-out mutation in *lacI* with one in *phoB,* the gene encoding the positive regulator of the gene for alkaline phosphatase, *phoA.*
4. The *his* operon of enteric bacteria appears to be controlled by attenuation. Some mutations, although unlinked to *his,* result in elevated levels of the histidine biosynthetic enzymes in cells growing in minimal medium as well as in the presence of excess histidine. Propose what these genes might encode.
5. Many of the genes encoding regulatory proteins (activators and repressors) are proving to be regulated autogenously. Do any alternative possibilities make sense to you?
6. Using the format and conventions of Table 4, predict the phenotype of *E. coli* variants with the following genotypes:

$lacI^-,O^c,Z^-,A^+$ / F′ $lacI^+,O^+,Z^+,A^-$
$lacI^-,O^+,Z^-,A^+$ / F′ $lacI^+,O^+,Z^+,A^-$
$lacI^-,O^c,Z^-,A^+$ / F′ $lacI^+,O^c,Z^+,A^-$

7. If the *lac* operator were moved slightly downstream, thereby permitting RNA polymerase to initiate transcription and proceed for a few nucleotides until being derailed by encountering the bound *lac* repressor, control of this modified operon would, by definition, be given what name?

SUGGESTED READING

Topics in this chapter are covered in the following chapters from *Escherichia coli and Salmonella typhimurium: Cellular and Molecular Biology*, F. C. Neidhardt, J. L. Ingraham, K. B. Low, B. Magasanik, M. Schaechter and H. E. Umbarger (eds.). 1987. American Society for Microbiology, Washington, D.C.

Beckwith, J. The operon: a historical account. (Volume 2; Chapter 88)
Beckwith, J. The lactose operon. (Volume 2; Chapter 89)
Gold, L. and G. Stormo. Translational initiation. (Volume 2; Chapter 78)
Gottesman, S. Regulation by proteolysis. (Volume 2; Chapter 79)
Hoopes, B. C. and W. R. McClure. Strategies in regulation of transcription initiation. (Volume 2; Chapter 75)
Landick, R. L. and C. Yanofsky. Transcription attenuation. (Volume 2; Chapter 77)
Yager, T. D. and P. H. von Hippel. Transcript elongation and termination in *Escherichia coli.* (Volume 2; Chapter 76)
Yanofsky, C. and I. P. Crawford. The tryptophan operon. (Volume 2; Chapter 90)

Other suggested readings include:

Savageau, M. A. 1985. Coupled circuits of gene regulation. In: *Sequence Specificity in Transcription and Control,* R. Calendar and L. Gold (eds.). Liss, New York.
Schleif, R. 1986. *Genetics and Molecular Biology.* Chapters 12 (Repression and the *lac* operon) and 14 (Attenuation and the *trp* operon). Addison-Wesley, Reading, Massachusetts.

13

Regulation of Gene Expression: Multigene Systems and Global Regulation

INTRODUCTION

Inspired by the model of the *lac* operon presented in 1961, microbial physiologists have spent much of the subsequent 30 years on molecular analysis of operons. The mode of regulation of operon after operon has been solved, revealing the variety of intricate mechanisms by which bacterial cells modulate the expression of their genes. This task is far from complete; many important details and principles remain to be learned by continuing this analysis.

It is now recognized, however, that operon physiology can never be completely understood by the study of individual operons. Many bacterial activities involve gene coordination at levels of organization higher than individual transcriptional units. *Most, if not all, the operons of a bacterial cell belong to higher level regulatory organizations, which we shall call* REGULATORY NETWORKS. In the preceding chapter we examined how operons are regulated; in this chapter and the next three we consider how *networks* of operons are regulated.

COREGULATION OF SEPARATE OPERONS

The operon is a hallmark of the procaryotic cell. This mode of organization—whereby the genes of an entire pathway can be united as a single transcriptional unit—achieves a simple solution to the problem of coregulating genes of related function. Why go further? Is there a need for the bacterial cell to move beyond this obviously successful strategy?

There are at least two answers to these questions. The first is that some bacterial processes involve too many genes to be accommodated in a single workable operon. This situation is illustrated by the translation machinery, a group of at least 150 gene products (ribosomal RNA, ribosomal proteins, initiation, elongation, and termination factors, aminoacyl-tRNA synthetases, and tRNAs) involved directly in making protein from amino acids. They are so numerous that coordinating their synthesis by linking their genes into a single operon would be awkward, if not impossible. Yet, coordinated regulation of these genes is important to the overall efficiency of bacterial growth. As we shall see in Chapter 15, this coordination is accomplished even though these genes are organized into many dozens of unlinked operons.

The second answer is that some bacterial processes involve a number of genes that must be both independently regulated *and* subject to an overriding, coordinating control. This situation is illustrated most clearly by the ensemble of genes encoding catabolic enzymes involved in the utilization of sugars, amino acids, and other compounds for carbon and energy. During growth in an environment containing a mixture of such compounds, economy would demand that only a premium substrate, best able to satisfy the carbon and energy needs of the cell, be metabolized. (For enteric bacteria, glucose is such a substrate; for pseudomonads, succinate is.) The operons encoding enzymes for the metabolism of the secondary, redundant substrates should be repressed. Each operon must, however, be induced individually when its cognate substrate is present in the absence of the premium substrate. This double requirement calls for a level of organization above that of the operon.

OVERVIEW OF MULTIGENE REGULATORY NETWORKS

It is estimated that the bacterial cell has evolved several hundred multigene systems. Their discovery and analysis has only just begun. Some of the better-known examples for the enteric bacterium *Escherichia coli,* are shown in Table 1, along with their mode of regulation, their member genes, and the processes they concern. The entries in Table 1 are arranged by broad categories:

- networks involved in response to limitation of one or another nutrient—carbon and energy sources, ammonia, inorganic phosphate
- networks involved in oxidation–reduction reactions and electron transport
- networks concerned with response to damage by oxidation, radiation, high and low temperature, and extremes of osmotic pressure
- miscellaneous networks, some with subnetworks, that bring about major changes in the physiology and morphology of the cell.

The bacterial cell has evolved diverse ways to weave individual operons into coordinated networks; the various mechanisms are just beginning to be elucidated. In some cases the device of an allosteric protein regulator has simply been borrowed from operon regulation: a protein repressor or activator recognizes a particular sequence common to the controlling regions of the member operons. This device is used in the SOS, oxidation damage, and anaerobic electron transport systems in enteric bacteria. In other systems the network is defined by an alternative sigma factor that reprograms RNA polymerase to recognize the promoters of the member operons (Chapter 3). The heat shock and sporulation systems of various bacterial species illustrate this situation. Other networks involve a combination of protein regulators and sigma factors, as found in the nitrogen utilization system of many bacteria. One of the most pervasive networks, the stringent control system, appears to have no protein modulator at all; the member operons are regulated by the nucleotide guanosine tetraphosphate (ppGpp) in a manner still to be elucidated.

Because many regulatory networks were first recognized as cellular responses to environmental changes, the formal structure of a stimulus–response system can provide a useful organizing function for the mass of information about each network. A STIMULUS–RESPONSE SYSTEM (Figure 1) contains a pathway that begins with a STIMULUS from the environment that affects some cellular target, or SENSOR, which generates a SIGNAL. This stimulus–response system directly—or indirectly when the signal passes through one or more TRANSDUCERS—affects the activity or synthe-

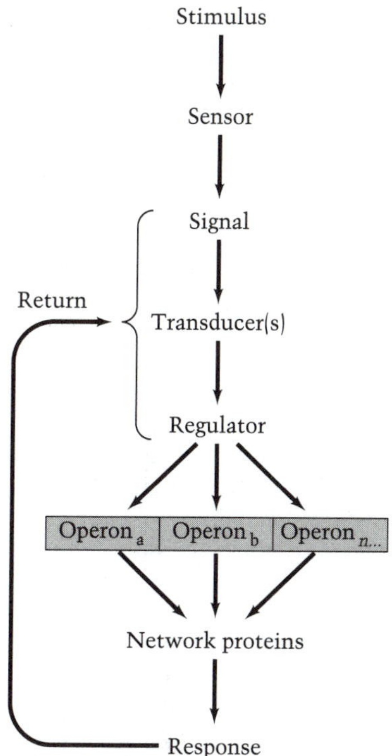

Figure 1

Gene network depicted as a stimulus–response pathway.

Table 1. Multigene systems of bacteria

Stimulus	System	Organism	Regulatory genes	Regulated genes	Type of regulation
Nutrient utilization					
Carbon limitation	Catabolite repression	Enteric bacteria	*crp:* encodes activator *cya:* encodes adenylate cyclase	Catabolic genes, including *lac, mal, gal, ara, tna, dsd, hut,* and many others	Activation by CAP protein complexed with cAMP, as a signal of carbon source limitation
Amino acid or energy limitation	Stringent response	Enteric bacteria and others	*relA* and *spoT:* encode enzymes of (p)ppGpp metabolism	Genes (>200) for ribosomes, other translational proteins, and biosynthetic enzymes	Unknown; see text
Ammonia limitation	Ntr system	Enteric bacteria	*glnB:* encodes protein P_{II} *glnD:* encodes UR/UTase *glnG:* encodes protein NR_I *glnL:* encodes protein NR_{II}	*glnA, hut,* others	Complex; see text and Figure 5
Ammonia limitation	Nif system	*Klebsiella aerogenes,* many others	Multiple genes including those controlling ammonia assimilation	Multiple genes encoding nitrogenase	Complex
Phosphate limitation	Pho system	Enteric bacteria	*phoB:* encodes activator *phoR:* encodes modulator of PhoB activity *phoU:* encodes sensor of phosphate transport	*phoA,* and others	Activation by PhoB after signal from PhoR; see text and Figure 6

Stimulus	System	Organism	Regulatory genes	Regulated genes	Type of regulation
Energy metabolism					
Presence of oxygen	Arc system (Aerobic respiration)	*E. coli*	*arcA:* encodes repressor *arcB:* encodes a modulator of ArcA activity	Many genes for aerobic enzymes; see text	Repression by ArcA after signal by ArcB; see text
Presence of electron acceptors other than oxygen	Anaerobic respiration	*E. coli*	*fnr:* encodes activator	Genes for nitrate reductase and other enzymes of anaerobic respiration	Activation by Fnr protein
Absence of usable electron acceptors	Fermentation	*E. coli* and other facultatives	Unknown	Genes (>20) for enzymes of fermentation pathways	Unknown
Stress responses					
UV and other DNA damagers	SOS response	*E. coli* and others	*lexA:* encodes repressor *recA:* encodes modulator of LexA activity	About 20 genes for repair of UV-damaged DNA	Repression by LexA protein
Alkylation of DNA	Ada system (Alkylation response)	*E. coli* and others	*ada:* encodes activator	Four genes for removal of alkylated bases from DNA	Activation by Ada protein
Presence of H_2O_2 or similar oxidants	Oxidation response	Enteric bacteria	*oxyR:* encodes repressor	About 12 genes for protection from H_2O_2 and similar oxidants	Repression by OxyR protein

Stimulus	System	Organism	Regulatory genes	Regulated genes	Type of regulation
Shift to high temperature	Heat-shock	*E. coli* and all other bacteria	*htpR* (= *hin* = *rpoH*): encodes σ^{32}	About 20 genes for proteins involved in macromolecule synthesis, processing, and degradation	Programming by σ^{32}
Shift to low temperature	Cold-shock	*E. coli*	Unknown	Several (12) genes for macromolecule synthesis	Unknown
High osmolarity	Porin response	*E. coli* and others	*envZ:* encodes sensor *ompR:* encodes DNA-binding protein	Genes for porins	Complex; see text and Figure 7
Miscellaneous global systems					
Growth supporting property of environment	Growth rate control	All bacteria	Unknown; probably several	Hundreds of genes	
Starvation or inhibition	Stationary phase	All bacteria	Unknown	Hundreds of genes	
Starvation	Competence	*Bacillus subtilis*	*comA:* encodes regulator	Seven or more genes for DNA uptake	Unknown
Starvation	Sporulation	*Bacillus subtilis*	*spoOA:* encodes activator *spoOF:* encodes modulator	Many (>100) genes for spore formation	Complex; see Chapter 16
Shift to high temperature	Virulence	*Bordetella pertussis*	*bvgA:* encodes activator *bvgB* and *bvgC:* encode modulators	Many (>10) genes for virulence	Activation by BvgA protein

sis of a REGULATOR, which controls the output, usually some ADAPTIVE RESPONSE to the environmental change. This system often includes a feedback control mechanism that permits a RETURN to the prestimulus condition or to a new equilibrium consonant with the changed environment. In a bacterial gene network, a stimulus from the environment (for example, changed temperature, nutrient status, or toxicity) would be depicted as affecting some sensor which would generate a signal to affect the activity or synthesis of (usually) a protein regulator of the member genes. The products of the member genes would perform some cellular function(s) to promote the growth or survival of the cell. Modulators of some sort would enable a return to the prestimulus state of expression of the member operons or would establish a new level for their expression.

It is clear from the entries in Table 1 that many of the operons of *E. coli* are already assigned to one or more networks. With the help of the framework provided in Figure 1, we shall probe a few of these networks in depth to illustrate principles of multigene physiology and molecular biology. The first two networks to be discussed have been chosen partly because of their quantitative importance in cell physiology; *together they directly or indirectly control probably three-fourths of the protein-synthesizing capacity of the bacterial cell.*

TWO MAJOR MULTIGENE SYSTEMS

Catabolite repression of fueling genes

Most species of bacteria possess an extensive repertoire of genes encoding catabolic enzymes directed against dozens of different compounds of diverse nutritional value. Growth on a single source of carbon and energy—a SUBSTRATE—requires a relatively high cellular level of the enzymes that metabolize the substrate and feed the catabolic products into the central fueling pathways, because all metabolic pathways in the cell flow from the metabolites produced by these catabolic enzymes. Bacteria appear able to sense the appropriateness of each catabolic pathway in a given circumstance and to regulate gene expression accordingly. We have already noted in our discussion of the *lac* operon (Chapter 12), and above, that the program for deciding what enzymes to make and in what quantities must be more sophisticated than one which simply calls for the cell to induce catabolic enzymes for substrates sensed in the environment. Indeed, despite a half-century of intensive effort, the sophistication of this cellular program has exceeded the ability of physiologists to understand its intricacies.

Enteric bacteria employ at least four different processes to assure the priority status of glucose as a carbon and energy source:

- CONSTITUTIVE SYNTHESIS: The enzymes of glucose catabolism, being in large measure a part of the EMP central fueling pathway (Chapter 5), are produced at high levels no matter what substrates are present in the environment.
- INDUCER EXCLUSION: Glucose prevents the entry of certain other inducing substrates by inactivation of their permeases.

- TRANSIENT REPRESSION: Addition of glucose produces a transient, severe inhibition of synthesis of inducible catabolic enzymes for approximately half of a generation despite the presence of their inducer.
- CATABOLITE REPRESSION: Metabolism of glucose produces a continued, but less severe inhibition of induced enzyme synthesis, even in the presence of the appropriate inducers.

None of these processes are fully understood, and catabolite repression is particularly complex. It includes one element, however, about which much has been learned: the CAMP–CAP NETWORK. In *E. coli* the *lac* operon (Chapter 12) is a member of this network. The total number of member operons is not known; but it may be several dozen and includes those concerned with the metabolism of galactose, arabinose, maltose, tryptophan, D-serine, and histidine. Member operons possess a characteristic nucleotide sequence to which the regulatory protein CAP (**c**atabolite gene **a**ctivator **p**rotein, product of the *crp* gene) binds. The binding site (AANTGTGANNTNNNNCA, where N = any nucleotide) occurs at different distances upstream from the transcriptional start site in different operons. Each of the two 22,500-Da subunits of CAP binds a molecule of cAMP, and it is the cAMP–CAP complex that binds to DNA. The binding of cAMP–CAP to the promoters of this multigene network is essential for initiation of transcription. In the case of *lac*, binding of cAMP–CAP increases the promoter's K_B (binding constant) for RNA polymerase, leading to a 10-fold increase in the frequency of *lac* transcription (Chapter 12). From the measured cellular level of cAMP and the binding constant of CAP for this nucleotide, one can tell that in cells growing in glucose there is only a low level of the cAMP–CAP complex—an amount insufficient to provide activation of operons of this network. This limiting level of cAMP–CAP is one of the main factors in glucose-mediated catabolite repression, given that many of the operons that require cAMP–CAP for activation encode enzymes of catabolism of secondary carbon and energy sources.

Growing in the presence of a mixture of glucose and a less-preferred carbon and energy source (e.g., lactose), enteric bacteria, therefore, cannot induce the catabolic enzymes directed at the secondary source until glucose is exhausted from the medium. At this point a dramatic rise in cellular cAMP can be observed, and the bacteria now initiate transcription of those operons (*lac* in this case) that are induced by the secondary substrate.

How is this brought about? That is, how does cAMP become an accurate signaler of carbon and energy insufficiency? Cyclic AMP can be quickly synthesized (by the enzyme ADENYLATE CYCLASE, product of the *cya* gene) and quickly degraded (by a phosphodiesterase). The rate of synthesis is the primary determinant of the cellular level of cAMP. Adenylate cyclase is a membrane-bound enzyme, and its activity appears to be modulated by the state of a second membrane protein, called III^{Glc}. Protein III^{Glc} is a component of the phosphoenolpyruvate:sugar phosphotransferase system for transporting glucose into the cell (Chapter 5), and during its function it becomes reversibly phosphorylated. For adenylate cyclase to be active, III^{Glc} must be in its phosphorylated form, III^{Glc}-P. While glucose is being transported, the cellular level of III^{Glc}-P is low because it is rapidly donating its phosphate residue to glucose molecules. When there are no longer glucose molecules to transport, the level of protein III^{Glc}-P quickly rises, adenylate cyclase becomes more active, and the increased synthesis of cAMP raises the cellular level of the cAMP–CAP complex. In this manner catabolic genes remain in a state of readiness when their inducers (secondary carbon and energy sources) are in the environment and become activated to high-level expression upon the exhaustion of glucose. These events are depicted in Figure 2.

The ability of glucose transport to lower the cellular level of cAMP, and thereby lower transcription from operons of the cAMP–CAP network, can explain some features of catabolite repression, but not others. Perhaps the most satisfying fact is that the addition of cAMP to a medium containing glucose and lactose leads to at least a partial reversal of the repressive effect of glucose on *lac* expression, and analogous results obtain with other catabolic operons. On the other hand, left unexplained are the following observations: (1) No amount of cAMP added to cells growing on glucose with a gratuitous *lac* inducer (Chapter 12) can bring *lac* expression to levels as high as those achieved during growth with a very poor carbon source plus the gratuitous inducer. (2) The protein III^{Glc} mechanism for modulation of adenylate cyclase activity does not explain the fact that many high-priority carbon and energy sources other than glucose can affect cAMP synthesis and exert catabolite repression to at least some extent. These and other puzzling aspects of the cAMP–CAP system require further investigation. Catabolite repression is observed even in (Gram-positive) bacterial species lacking cAMP, and it is possible that studies in these organisms will shed light on some aspects of the phenomenon in enteric bacteria.

What is very clear is that the cAMP–CAP system is extraordinarily important in the physiology of many bacteria. Mutant strains of enteric bacteria that are defective in adenylate cyclase, or in CAP, cannot grow on substrates that are metabolized by enzymes encoded by the cAMP-CAP network. This result is to be expected. But their growth on glucose is also abnormal, with growth rates as low as half that of the wild-type parental strain. This observation implies that some genes of the network must produce products other than optional, inducible catabolic enzymes. Studies using two-dimensional gel electrophoresis to monitor the synthesis of individual protein species in such mutants have helped assess the number of genes in this regulon. The results suggest that this multigene system may be large: many protein species are made in much-reduced amounts. The same gels reveal that some protein species are actually *elevated* in amount in these mutants, as if cAMP–CAP participates in negative regulation as well as activation. This explanation proves to be the case; an example is the gene encoding adenylate cyclase, *cya,* which is *negatively* regulated by cAMP–CAP; the DNA-binding complex is a repressor for the *cya* operon.

The cAMP–CAP network illustrates particularly clearly one general

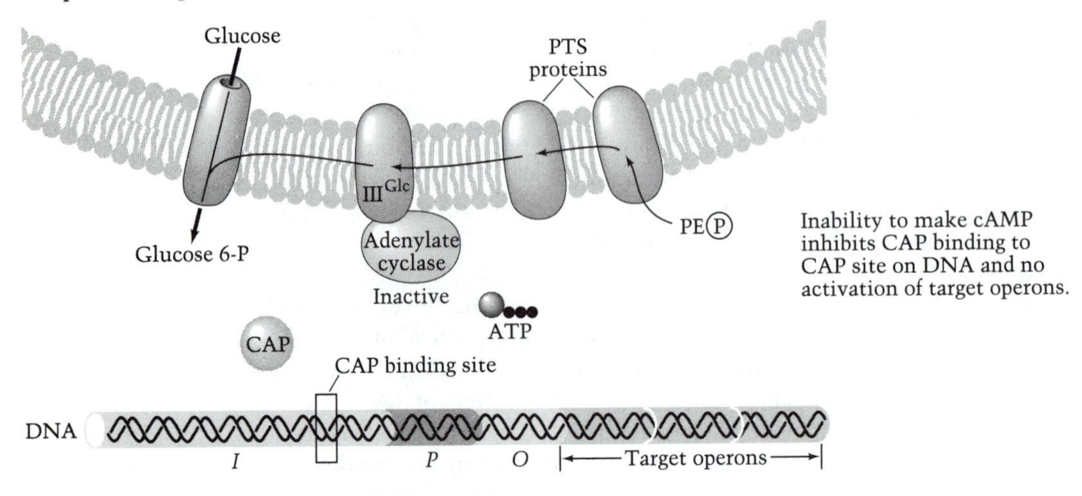

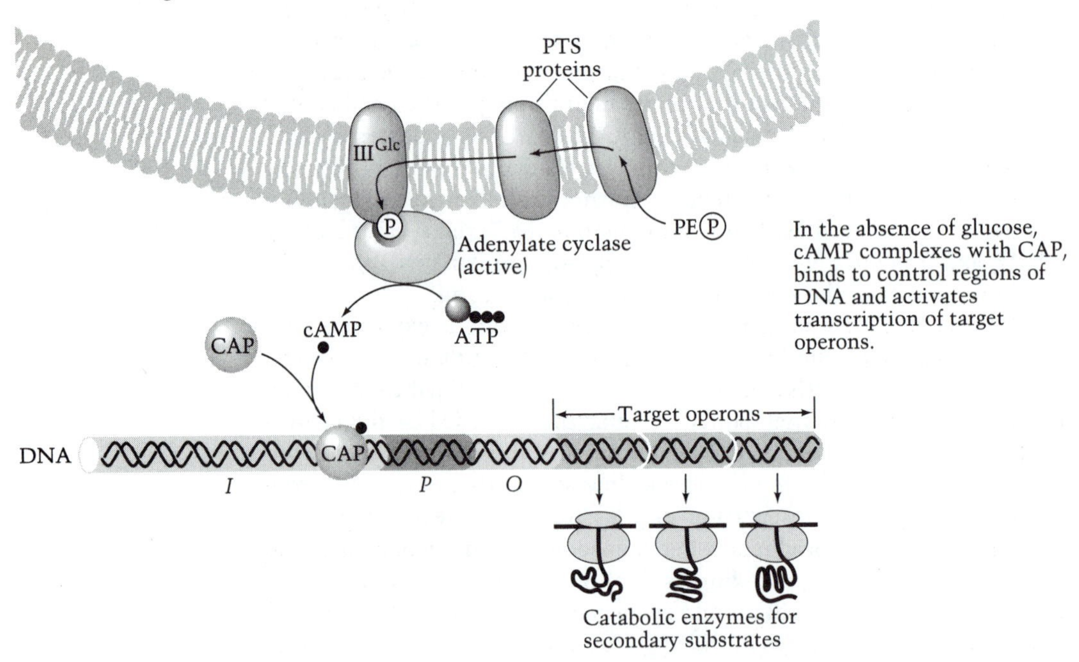

characteristic of the physiology of multigene systems: the individual genes (operons) of a network are in many instances controlled independently of one another by specific regulators. The expression of many of the known operons of the cAMP–CAP system (*lac, gal, ara, mal, tna, dsd, hut*) in enteric bacteria requires the presence of each operon's own specific inducer (for the examples given: lactose, galactose, arabinose,

Figure 2

A model to account for the ability of glucose to interfere with the synthesis of cAMP, an action leading to one component of glucose-mediated catabolite repression. (A) High-energy phosphate from phosphoenolpyruvate (PEP) is transferred through the components of the glucose specific phosphotransferase system (PTS), a reaction resulting in the transport and phosphorylation of a molecule of glucose by the last component, protein III^{Glc}. During this process, the amount of the phosphorylated form of III^{Glc} is very low, because of the rapid phosphorylation of glucose residues; in addition, adenylate cyclase is inhibited by III^{Glc}; the cell's inability to make cAMP leads to the secondary inability to activate catabolite-repressible operons. (B) The pathways used when glucose runs out. Adenylate cyclase becomes active because of the removal of III^{Glc}; cAMP is formed and complexes with CAP protein, which can then bind to the control regions of catabolite-repressible operons, thereby activating their transcription. This model is based on ideas reviewed by Saier (1989).

maltose, tryptophan, D-serine, and histidine, respectively). Abundant intracellular cAMP only provides the *potential* for the cell to express these genes; the inducer of an operon must be present for expression to occur. Other gene networks have alternative motifs. Other networks to be described (SOS and heat-shock) may display independent regulation of individual operons, but coordinate expression of the entire network is the dominant theme. Different networks exhibit as much variety of behavioral properties as do different operons.

By the formalism of Figure 1, the cAMP–CAP network would be described as a pathway in which the stimulus is the change in passage of glucose residues through the membrane; the *sensor* is protein III^{Glc} of the PTS system interacting with adenylate cyclase, the *signal* is the cellular level of cAMP, the *regulator* is the cAMP–CAP complex. The member genes (the catabolic operons with inducers present in the medium) generate their enzyme products, leading to the *response*, which is utilization of the secondary substrate for growth. The *return* circuit (which must exist) is not fully understood, but one component is the repression of CAP synthesis from the *crp* gene by a high intracellular level of cAMP–CAP.

As noted above, the operons of this network are individually capable of higher levels of expression than that of most of the genes of the cell (catabolic enzymes, when made, are more abundant than, for example, biosynthetic ones). By controlling the expression of inducible catabolic operons, the cAMP–CAP network involves a large portion of the cell's protein-synthesizing capacity. Its far reaching consequences are exceeded only by the next network we shall discuss.

Stringent control network

The demand of the ribosomes for aminoacylated tRNA during rapid growth is very great—over half the bacterial cell is protein and that amount of protein must be made every 20 minutes. For decades it has been known that bacteria have some means to sense the sufficiency of their supply of aminoacylated tRNA and to make compensatory adjustments when, for any reason, the supply of one or more amino acids becomes insufficient to maintain the growth rate. The adjustments are truly global in character; few metabolic processes are left unchanged.

Consider, for example, what happens when cells such as *E. coli* or *Salmonella typhimurium* growing rapidly in a rich (amino-acid containing) medium are shifted to a minimal medium. Almost immediately these cells experience a shortage of aminoacyl-tRNA, as the rate of recharging tRNA with amino acid residues fails to keep pace with the rate of consumption by the ribosomes. Not surprisingly, an immediate consequence of this shortage is that protein synthesis is drastically reduced. Not so easily predicted is the fact that the synthesis of all stable RNA species (i.e., tRNA and rRNA), is also curtailed—almost totally halted for at least a temporary period. Replication of DNA continues, but only at replication forks already existing at the time of the amino acid restriction; temporarily, no new replication forks are initiated. The rates of synthesis of phospholipids, carbohydrates, and murein are decreased. In parallel with this general curtailment of macromolecule formation, the biosynthesis of their respective building blocks (nucleotides, sugars and other carbohydrate intermediates, and fatty acids) is reduced. Besides the overall decrease in protein synthesis, the pattern of synthesis of individual proteins changes markedly—many are increased in rate and many decreased, relative to the rate of total protein synthesis. Among those with preferentially increased rates of synthesis are the enzymes of amino acid biosynthetic pathways; among those that are decreased in rate (virtually to zero) are the 50-odd ribosomal proteins. It should be noted that by inhibiting the synthesis of ribosomes (both rRNA and ribosomal proteins) and tRNA, and of certain of the accessory proteins that work with ribosomes (e.g., elongation factors), the cell has shut off processes that had been consuming more than half its amino acid and energy resources. In essence, macromolecules now present in excess of cellular needs are no longer made, thereby conserving the scarce amino acid supply for making precisely those proteins—the amino acid biosynthetic enzymes—that allow the cell to adapt to the new condition and resume growth. Part of this manifold series of events is depicted in Figure 3.

As amino acid supply is increased by endogenous synthesis, the temporary inhibitions are lifted one by one, and growth resumes, albeit at a slower rate than in the nutritionally rich environment. As we shall see (Chapter 15), growth at the new, slower rate is characterized by the synthesis of fewer ribosomes and tRNA per unit cell mass; the old order is replaced by a new leaner one.

The central process in the scenario just described appears to be the transient but near total inhibition of ribosome and tRNA synthesis; almost all of the other events (decrease in synthesis of membrane and wall components and their precursors, decrease in nucleotide synthesis,

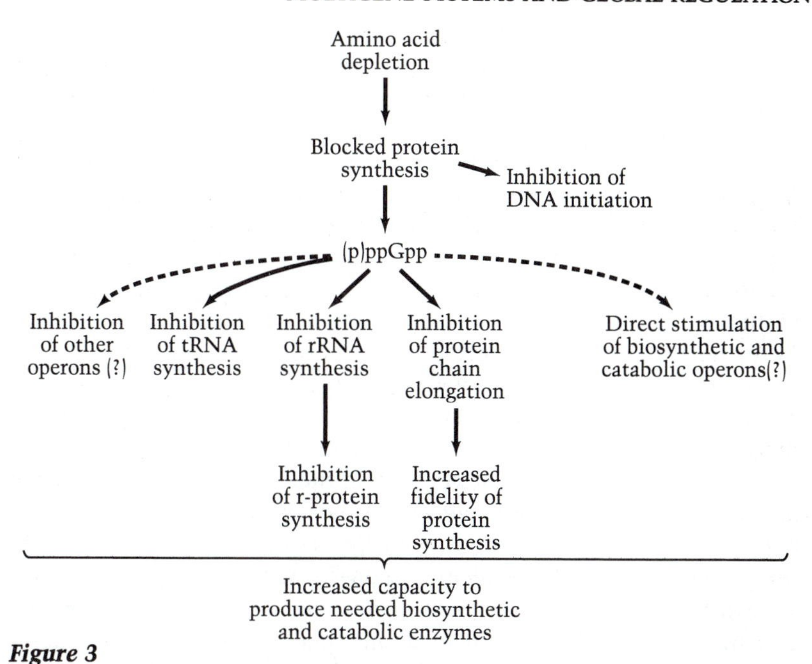

Figure 3

The series of events ensuing after amino acid restriction of bacteria. Solid arrows depict processes known (from reasonably good evidence) to occur. Dashed arrows depict more speculative relations. See text for details of this form of the stringent response network.

and inhibition of initiation of DNA replication) could be secondary consequences of either the restriction in protein synthesis or the block in ribosome and tRNA synthesis. This central inhibition has been called the STRINGENT RESPONSE. With very few exceptions, the stringent response occurs in bacteria whenever their growth rate is decreased by almost any kind of nutritional restriction: a shift from a rich to a minimal medium; starvation of an amino acid auxotroph for its required amino acid; a shift to a poorer carbon and energy source; or the substitution of a poorer source of nitrogen. Likewise, the addition of toxic agents that interfere with energy transduction or with amino acid biosynthesis (directly or indirectly), or the addition of agents that inhibit multiple steps in metabolism (as occurs with oxidants, solvents, and other protein denaturing agents), brings about the stringent response (in addition to whatever *specific* adaptive response might be induced, as we see later in our discussion of responses to radiation, oxidation, and temperature stress).

The stringent response was recognized early in the 1950s, thanks to the discovery of a mutant strain of *E. coli* in which this response is almost absent. In contrast to the classical *stringent* response, the phenotype of the mutant was described as *relaxed* (Rel), and the gene that was mutated in this strain was given the designation *relA*. In contrast to the wild strain, the *relA* mutant failed to halt the synthesis of rRNA and tRNA during amino acid starvation, and the mutant cells displayed poor ability to recover from nutritional down-shifts. Shifts that normal cells can adapt to in minutes produce lags of several hours duration in *relA* mutant cells. This result demonstrates the physiological significance of the stringent response.

The search for biochemical clues to the function of the *relA* gene led to the discovery that the accumulation of two unusual nucleotides correlated with *relA* gene function during the stringent response. These nucleotides, at first given the provocative names magic spots I and II, accumulated to high levels (millimolar concentrations) when environmental conditions led to the stringent response and failed to accumulate in the *relA* mutant during amino acid starvation. The compounds were soon identified as derivatives of GDP and GTP bearing 3′ pyrophosphate residues; the names of these nucleotides are abbreviated as ppGpp and pppGpp, respectively, and as (p)ppGpp, collectively.

The obvious questions are (1) In what manner can the cellular level of (p)ppGpp reflect the sufficiency of aminoacyl-tRNA? and (2) How can (p)ppGpp bring about inhibition of stable RNA synthesis (and the other manifestations of the stringent response)? Only partial answers are known to each of these questions. Let us consider first how the guanosine nucleotides, (p)ppGpp, are made and their levels controlled. What is known of the somewhat complex metabolism of (p)ppGpp in *E. coli* is summarized in Figure 4. Synthesis of (p)ppGpp occurs in two ways. One route occurs when a ribosome is stalled over a codon for which the relevant charged tRNA is in inadequate supply and an uncharged tRNA molecule binds at the ribosomal A site. As a result of the binding of uncharged tRNA, a ribosome-associated protein—the product of the *relA* gene—is activated to produce (p)ppGpp. The RelA protein is known as (p)ppGpp synthetase I; it catalyzes the transfer of a pyrophosphoryl group from ATP to either GDP or GTP, producing ppGpp or pppGpp, respectively. When the concentration of GTP is much higher than that of GDP, as is normally the case in vivo, pppGpp is the product of the RelA-catalyzed, ribosome-dependent reaction. Subsequently, ppGpp is formed in vivo by the hydrolysis of pppGpp through the agency of another enzyme—the product of a gene called *gpp* (Figure 4). The cellular levels of ppGpp and pppGpp are influenced also by their degradation, in a reaction catalyzed by (p)ppGpp 3′-pyrophosphohydrolase—the product of a gene called *spoT.*

The guanosine nucleotide cycle shown in the upper portion of Figure 4 accounts for the accumulation of (p)ppGpp during amino acid restriction; but there is more to the story, because it is known that there is another pathway of (p)ppGpp synthesis. When growth is slowed by the depletion of the primary carbon and energy source, the stringent response is activated by a path that is independent of the *relA* gene product. For years the existence of a second pathway was suspected, because mutants defective in *relA* could mount a stringent response when starved for carbon and energy, but not when starved for amino acids. The biosynthetic enzyme responsible for (p)ppGpp synthesis during carbon and energy source restriction was assigned the name

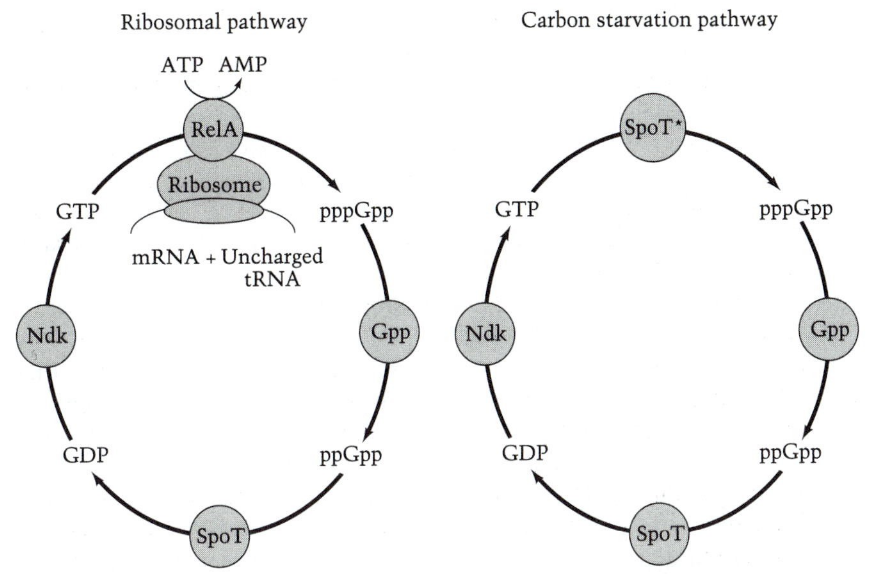

Figure 4

Cellular routes of synthesis of the guanosine nucleotides involved in the stringent response network of enteric bacteria. The left portion depicts the ribosomal pathway of (p)ppGpp synthesis and its subsequent degradation. The RelA protein—(p)ppGpp synthetase I—utilizes GTP and GDP equally well in vitro; and the SpoT protein—(p)ppGpp 3′-pyrophosphohydrolase—degrades ppGpp and pppGpp equally well in vitro. The reactions shown are the likely pathways in vivo based on concentrations of the different nucleotides. The SpoT* catalyzed formation of pppGpp is still unproved, but it is likely to be the long sought-after RelA-independent (p)ppGpp synthetase II reaction. Gpp, pppGpp 5-phosphohydrolase; Ndk, nucleoside diphosphate kinase. (Based on Figure 1, page 1411, in Cashel and Rudd, 1987.)

(p)ppGpp synthetase II, but the search for such an enzyme was unsuccessful until quite recently. Curiously, the long-known SpoT protein appears to be the missing link. Appparently the SpoT protein, which shares much structural similarity to the RelA protein, catalyzes *synthesis* as well as *degradation* of (p)ppGpp. Current suspicion is that the SpoT enzyme can exist in two alternative conformations—one biosynthetic and one degradative—and that the stringent response brought on by energy depletion is produced by modulation of the activities of this protein. Clearly, much remains to be learned about the biochemical details of this second pathway, a general outline of which is presented in the lower portion of Figure 4.

By one or the other of these two pathways, the bacterial cell rapidly elevates its levels of (p)ppGpp in response to nutritional adversity. What happens next? The molecular details are uncertain, and considerable controversy surrounds the question of how the levels of (p)ppGpp bring about the stringent response. The preponderance of evidence indicates that these nucleotides directly inhibit the transcription of tRNA and rRNA genes, probably by a direct action on the β subunit of RNA polymerase. One view is that RNA polymerase can exist in two conformations, only one of which can transcribe stringently controlled operons such as those encoding stable RNA, and that the concentration of (p)ppGpp determines which state will predominate. As a consequence of the inhibition of stable RNA synthesis, the synthesis of ribosomal proteins becomes inhibited by the translational repression mechanism that we have already encountered (Chapter 12) and that is more fully described in Chapter 15.

There is evidence also that (p)ppGpp affects many other processes. A review by Cashel and Rudd (1987) lists nearly a hundred pleiotropic effects of the stringent response. It is far from established, however, that (p)ppGpp participates directly in the selective activation or suppression of the many dozens of genes reported to be affected by the stringent response; many of the effects on gene expression are likely to be the result of secondary effects. There is some evidence that (p)ppGpp stimulates the transcription of certain operons, specifically operons encoding amino acid biosynthetic enzymes and others encoding catabolic enzymes. The evidence is incomplete, however, and comes from in vitro studies on only a few operons.

The recent construction in *E. coli* of a double mutant yielded an important finding. This mutant is defective in both *relA* and *spoT* and therefore is incapable of making (p)ppGpp. It cannot grow in minimal medium because it requires several amino acids for growth. Secondary mutations that restore the ability to grow without added amino acids are easily selected from populations of this mutant. Many of these compensatory mutations map in the gene for the β subunit of RNA polymerase. Together these findings are consistent with the notion that (p)ppGpp is necessary for the transcription of at least some amino acid biosynthetic operons. The observations do not prove a direct involvement. It is possible, for example, that runaway transcription of stable RNA operons occurs in the complete absence of (p)ppGpp and that this diverts RNA polymerase away from biosynthetic operons. Nevertheless, the existence of these mutants, and of others in which the synthesis of (p)ppGpp has been placed under direct experimental control, offers the hope that in the near future one will be able to sort out which effects of these nucleotides are primary and which secondary.

The stringent response, besides conserving resources that would otherwise be consumed in useless synthesis of ribosomes and tRNA, has another beneficial effect. When cells are deprived of a required amino acid, net protein synthesis ceases but residual synthesis goes on as a consequence of the continued slow supply of the amino acid by protein degradation. It turns out that this residual protein synthesis is very prone to error in *relA* mutants but not in cells capable of a normal stringent response. The mechanism of this participation of (p)ppGpp in translational fidelity appears to be the ability of these nucleotides, perhaps as a result of their similarity to GDP, to inhibit the translation process at several GTP-requiring steps. By slowing the action of ribosomes, (p)ppGpp may provide a greater opportunity for the proofreading mechanism (Chapter 3) to correct errors of incorporation at the codons calling for the limiting amino acid.

Viewed as a response network (Figure 1), the stringent response can be considered as a pathway in which the *stimulus* is a decrease in the supply of nutrients supporting protein synthesis; the *sensor* (at least in part) is the RelA-containing ribosome with an uncharged tRNA in its A site; the *signal* is the pair of guanosine nucleotides, pppGpp and ppGpp; the *regulator* is unknown, but may be RNA polymerase itself; and the *response* is the turn-off (directly or indirectly) of genes encoding ribosomes, tRNA, and other parts of the translation machinery, thereby permitting the diversion of precious resources to make proteins of greater use to the cell during starvation. The *return* circuit is provided by the enzyme(s) that degrade (p)ppGpp; it permits rapid relief from stringency when conditions improve.

In one sense the view of the stringent network as a stimulus–response system fails to highlight one potentially unique aspect of its physiology. There is increasing evidence that even the low basal levels of (p)ppGpp in cells are effective in controlling RNA synthesis and that the growth rate of the cell may, in some profound manner, be determined by a control circuit involving the action of these nucleotides on stable RNA synthesis. This topic is resumed in Chapter 15.

FOUR NETWORKS WITH A COMMON THEME

The main reason a cell integrates the expression of sets of operons is to achieve some result that is beyond the power of a single operon. This task almost always requires information processing; the information being processed about the state of the environment, certainly, but also about the physiological state of the cell. We turn our attention now to several gene networks that illustrate one of the newest aspects of information processing to be learned in multigene systems—signal transduction by protein–protein interactions.

The nitrogen network

In Chapter 11 we noted the care with which enteric bacteria use an elaborate system of covalent modification to regulate the activity of glutamine synthetase—the enzyme that (at considerable energetic expense) is capable of assimilating ammonia present in low concentration in the environment. In this section we view another aspect of this system, because the gene encoding glutamine synthetase is regulated as part of a multigene network, the Ntr system, that is called into play when the supply of ammonia becomes limiting. The study of this network is contributing important ideas in the field of multigene regulation.

Glutamine synthetase is not produced in significant quantities by most bacteria when the concentration of ammonia is high in the environment. But whenever the concentration falls below that sufficient to saturate the other assimilatory enzyme—glutamate dehydrogenase, the cells produce glutamine synthetase in large amounts. The control is transcriptional.

In enteric bacteria the gene for glutamine synthetase (*glnA*) is the promoter-proximal gene of a three-cistron operon—called *glnA-ntrB-ntrC* in *S. typhimurium* and *glnA-glnL-glnG* in *E. coli*). This operon has several promoters, but only one of them—called *glnAp2*—is activated to high-level under conditions that call for high output of glutamine synthetase. This promoter, as well as some others that we shall note later, is recognized by RNA polymerase programmed by σ^{54} rather than by the more common σ^{70}. This alternate sigma factor, the product of the *rpoN* gene (formerly called *glnF* and *ntrA*), confers on core RNA polymerase the ability to bind specifically to certain promoters; but, unlike σ^{70}, it does not confer the ability to form open complexes. The bound polymerase requires an additional event to initiate transcription. In the case of the *glnA*-containing operon in *E. coli*, this event is the binding of the *glnG* gene product, NR_I, to special sequences on the DNA upstream from the *glnAP2* promoter. (In *S. typhimurium*, the gene product is called NtrC and is the product of *ntrC*.) Because these binding sequences can be moved far upstream or downstream of the promoter and still be effective, they are called ENHANCERS, to call to mind their similarity to functionally analogous sequences first discovered in eucaryotic cells.

What causes NR_I to bind to these enhancers and initiate high-level transcription of the *glnA* operon when the ammonia concentration falls in the environment? It turns out that NR_I binds only when it is phosphorylated; and it is phosphorylated under just the appropriate condition of limiting ammonia. The phosphorylation of NR_I is accomplished by NR_{II}, the protein product of *glnL*, which is the gene adjacent to *glnG* and *glnA* in the same operon. (In *S. typhimurium*, the NR_{II} equivalent is called NtrB, and its gene, *ntrB*.) How does the phosphorylation of NR_I by NR_{II} accurately reflect the availability of ammonia in the environment? The wondrous fact is that the activity of NR_{II} as a protein kinase is determined by the same cascade that affects the activity of the enzyme that adenylylates glutamine synthetase. The protein UR/UTase, which has dual uridylyl-removing and uridylyltransferase activities, is stimulated by a low glutamine:α-ketoglutarate ratio to uridylylate protein P_{II}, the product of the *glnB* gene. Protein P_{II} ordinarily interacts with NR_{II}, forcing it to be a protein phosphatase; but when P_{II} is removed by uridylylation from this interaction with NR_{II}, then NR_{II} is free to act as a kinase. NR_{II} phosphorylates itself (ATP is the donor) and then transfers the phosphate to NR_I (NtrC), thereby activating the latter as a transcription stimulator, and stimualting synthesis of glutamine synthetase. (This same uridylylated form of P_{II}, it may be recalled (Chapter 11), promotes the activation of glutamine synthetase by stimulating the adenylyltransferase to deadenylylate the enzyme.) All of this is diagrammed in Figure 5.

There are other subtleties to the regulation of the *glnA* operon. For example, the other two promoters of the operon are involved in main-

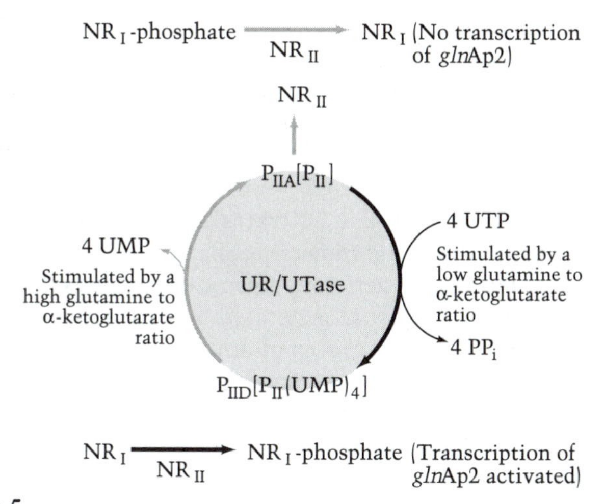

Figure 5

Covalent modification of NR_I, the regulator of the nitrogen utilization gene network. The black arrows indicate the sequence of events when ammonia is limiting. See text for details. (From Reitzer and Magasanik, 1987.)

taining a low basal level of NR_I and NR_{II} during growth in high ammonia; they are recognized by the σ^{70}-containing form of RNA polymerase. But in terms of gene network physiology, the most relevant observations are the following: A large number of operons in enteric bacteria are involved in the synthesis of enzymes that degrade organic compounds to obtain ammonia; the amino acid-degrading enzymes for proline, histidine, and arginine, for example, are part of the Ntr network. Interestingly, whereas in some organisms these operons may be controlled directly by the σ^{54}-NR_I-NR_{II} system, in others it appears that there is still another intermediary at work: a gene called *nac*, discovered in *Klebsiella aerogenes,* is under nitrogen control, and its product, Nac, activates many operons of the Ntr (NR) network.

In organisms that are genetically capable of fixing atmospheric nitrogen (for example, *Klebsiella* spp., *Azotobacter* spp.,) the multiple genes encoding enzymes of this pathway are under control of the same Ntr network.

The lessons learned from analysis of the nitrogen utilization network of enteric bacteria are (1) that the steps between signal and regulator can be multiple and complex, with several signal transducers operating between the generation of a stress signal and the ultimate regulation of gene expression; (2) that a special sigma factor may participate, not as a modulator of gene expression, but as a definer of the member operons of a network; and (3) that a gene network may consist of a cascade in which the products of regulated genes are regulators that determine the expression of other genes.

The nitrogen utilization network is a pathway that begins with the environmental *stimulus* of a restricted supply of ammonia. This stimulus leads to a reduction in the intracellular ratio of glutamine: α-ketoglutarate. The changed ratio is felt by the *sensor*, the UR/UTase protein, which generates a phosphorylation signal in the NR_{II} *transducer* protein, which is passed to the NR_I *regulator*. The activated NR_I regulator stimulates transcription of the central operon of the network, thereby stimualting production of glutamine synthetase and more of the NR_I and NR_{II} proteins. It also activates transcription of the *nac* operon, thereby stimualating the transcriptional activation of the operons involved in extracting nitrogen from organic compounds in the environment. As a result, the *response* generated by this network enables the cell, first, to extract the last traces of ammonia from the environment and, then, to turn to other sources of nitrogen that the cell is genetically capable of using. The *return* circuit is provided by the intracellular accumulation of glutamine, which modifies the activity of the UR/UTase sensor.

The phosphate network

Inorganic phosphate is abundant in some environments and is the major source of phosphate for most bacteria. Still, given the insolubility of many phosphate salts, bacteria have had to evolve efficient systems for utilizing low concentrations of phosphate and for obtaining phosphate from organic compounds. A major enzyme in the degradation of organic phosphate esters is alkaline phosphatase. This enzyme, when phosphate becomes limiting, is synthesized as inactive 43,000-Da monomers that, in Gram-negative bacteria such as *E. coli* and *S. typhimurium*, form active dimers in the periplasm. This enzyme, which is barely detectable during growth in high phosphate, becomes as much as 6% of the protein mass of *E. coli* when phosphate is limiting. The induction of alkaline phosphatase has been intensively studied and is now known to be part of a complex response to phosphate limitation. There are on the order of 100 proteins that become significantly elevated in level in the phosphate starvation-induced (psi) response of enteric bacteria.

The many genes of the phosphate utilization (Pho) network are regulated by a variety of separate and overlapping controls. These include *phoA*, the structural gene for alkaline phosphatase. Like the genes of the Ntr system, the genes of the Pho network are controlled by a positive regulator that binds to their promoters to activate transcription. This regulator—called PhoB, the product of the *phoB* gene—binds to specific sites in DNA only when phosphate is limiting. Modulation of PhoB binding is brought about by its interaction with another regulatory protein, PhoR—the product of the *phoR* gene. It is suspected, but not yet rigorously demonstrated, that PhoR activates PhoB by phosphorylating it. The suspicion is based on the structural similarity of PhoR to NR_{II} (NtrB) and of PhoB to Nr_I (NtrC).

The protein PhoR is believed to be associated with the cytoplasmic membrane; and one might suppose that it senses the extracellular availability of inorganic phosphate and signals the information to PhoB.

This may be the case, but genetic information indicates that PhoR also receives information about phosphate availability by interaction with still another protein, PhoU—the product of the *phoU* gene. The protein PhoU may be functionally analogous to protein P_{II} of the nitrogen network. The protein PhoU apparently can stimulate PhoR to activate PhoB, much as P_{II} can stimulate NtrB to activate NtrC. How does PhoU get its information? On the basis of genetic analysis, its role as a sensor depends on four other proteins—called PstS, PstC, PstA, and PstB—which together function as a high-efficiency transport system (called the PST transport system) for inorganic phosphate. The five genes—*pstS, pstC, pstA, pstB,* and *phoU*—are part of one operon and are transcribed in that order.

A current model for the normal operation of the phosphate utilization network is outlined in Figure 6. For an alternative model that depicts PhoR as a direct sensor of phosphate availability in the periplasm,

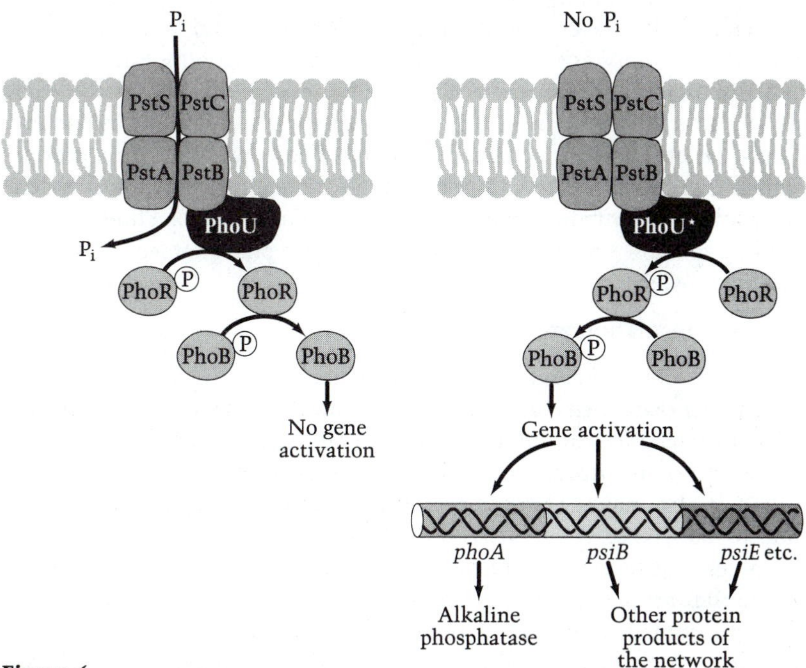

Figure 6

A model for the regulation of the phosphate utilization gene network in enteric bacteria. The model includes some ideas presented by Wanner (1987) and by Stock et al. (1989). P_i, inorganic phosphate; PstA, PstB, PstS, and PstC, components of the high-affinity P_i transport system; PhoU, a sensor of the transport of P_i; PhoU*, form of the sensor when no P_i is entering the cell; PhoR and PhoR-P, native and phosphorylated forms of the signal transducer; PhoB and PhoB-P, native and phosphorylated forms of the activator protein; *phoA, psiB*, and *psiE*, three of the many genes of the Pho network. See text for further explanation.

see Stock et al. (1989). During growth in adequate phosphate, the operation of the PST system in some way places PhoU in a state in which it promotes PhoR to exhibit a phosphatase activity against PhoB. Unphosphorylated, PhoB cannot act as a transcription stimulator of *phoA* or of other *psi* operons it controls. Depletion of phosphate from the medium leads to a changed flux through the PST system, and the interaction of the PST proteins with PhoU enables PhoU to become a stimulator of PhoR's presumed kinase role. The protein PhoR phosphorylates PhoB, which can then stimulate transcription of its target operons, including *phoA*. Alkaline phosphatase and other proteins useful in scavenging phosphate and obtaining it from alternative sources are then made in large amounts.

By this model of the network, the stimulus–response pathway can be analyzed as: *stimulus,* low external concentration of phosphate; *sensor,* the PTS–PhoU complex; *signal,* probably protein phosphorylation; *transducer,* PhoR; *regulator,* PhoB; *output,* proteins made by operons, such as *phoA,* that require PhoB binding for activity; *response,* improvement in supply of inorganic phosphate for biosynthesis. Part of the interesting circuitry governing the response of this system is the control of both the *pst–phoU* operon and the *phoR-phoB* operon by PhoB.

As a final point of interest, it should be noted that, through the study of mutants defective in PhoR, it has been learned that a similar protein—called PhoM—can partially supply PhoR function. This protein, which is presumed to be a kinase normally concerned with another gene network, can provide a low level of phosphorylation of PhoB, thereby leading to a low, constitutive level of alkaline phosphatase.

Osmoregulation and porin expression

One important environmental variable with which bacteria must deal is the osmolarity of their environment. The bacterial cell utilizes several molecular strategies to counteract the effects of deleterious extremes of environmental osmotic pressure (reviewed by Csonka, 1989). It also makes changes that allow it to grow well in dilute media and to withstand toxic environments. One that we shall discuss here involves adjustment of the permeability properties of the outer membrane that are signalled by changes in osmolarity.

The outer membrane is traversed by channels lined by proteins called porins (Chapter 2). In *E. coli* there are two major porins—OmpF and OmpC. They tend to be made as alternatives to each other. The pores made by OmpC are slightly smaller and are made preferentially at higher osmotic pressure and higher temperature. This pattern means that OmpC is the predominant form during growth of enteric bacteria in the intestinal tract. It makes good physiological sense for enteric bacteria to have smaller pores when residing in the intestinal tract and larger ones outside. The smaller pores exclude many of the toxic molecules found in the intestinal tract. The larger pores allow substrates from dilute solution to enter the cell pore readily, because solutes must diffuse through the outer membrane pores and the rate of diffusion is proportional to the cross-sectional area of the pore as well as to the concentration difference across the membrane.

The genes encoding the major porins, as well as certain other genes affecting outer membrane properties, are controlled by two proteins—products of the linked *envZ* (for **env**elope proteins) and *ompR* (for **o**uter **m**embrane **p**rotein **r**egulation) genes. The EnvZ protein is a cytoplasmic membrane protein that is thought to function as an osmosensor. It has strong structural similarity to certain protein kinases; on the basis of this similarity and other evidence, and from the phenotype of missense and other mutations in *envZ*, investigators believe that EnvZ interacts with the second protein, OmpR, affecting its activity by phosphorylating it. Protein OmpR binds specifically to the DNA upstream of promoters that it controls; the binding site is around −50 in the *ompF* promoter, and −90 in the *ompC* promoter. One current model for how alternative expression of *ompF* and *ompC* is accomplished is as follows (Figure 7): high osmotic pressure leads to phosphorylation of EnvZ, and this activated protein in turn phosphorylates OmpR. Phospho-OmpR (OmpR-P) *represses* transcription of *ompF* by binding to a site between −40 and −60 in the *ompF* promoter region; it *activates ompC* transcription by binding to low-affinity sites between −75 and −105 in the *ompC* promoter region. When the osmotic pressure is low, the level of OmpR-P is low, and therefore it is unable to activate *ompC*; but it still is able to activate *ompF* by binding to high-affinity sites in the −60 to −100 region of the *ompF* promoter.

The effects of OmpR-P as a stimulator and an inhibitor of transcription of these two genes is supplemented by another mechanism. The gene *micF* is transcribed in the opposite direction from *ompC* from a promoter near that of *ompC; micF* encodes a 174-base RNA that is complementary to a region near the start of the *ompF* message. The *micF* transcript has been shown to block expression of *ompF*. Because the regulation of *micF* expression is similar to that of *ompC*, the conditions that induce *ompC* will simultaneously induce *micF*, and thereby help repress *ompF* (Figure 7).

There is a network of genes besides those for the two porins that are regulated by the EnvZ/OmpR pair of interacting proteins; some of these genes are induced, others repressed. The overall physiological sense of the choice of genes that are included in this system (Table 1) is still elusive; the importance to us of the EnvZ/OmpR story is that it has provided an important pattern of protein–protein interaction in stimulus–response networks, including many for which the physiological function is crystal clear. The next system to be discussed is a case in point.

Aerobic metabolism

As noted in Chapter 5, major metabolic changes are associated with environmental shifts between aerobic and anaerobic conditions. For facultative organisms (Chapter 5) such as *E. coli* and *S. typhimurium* (and

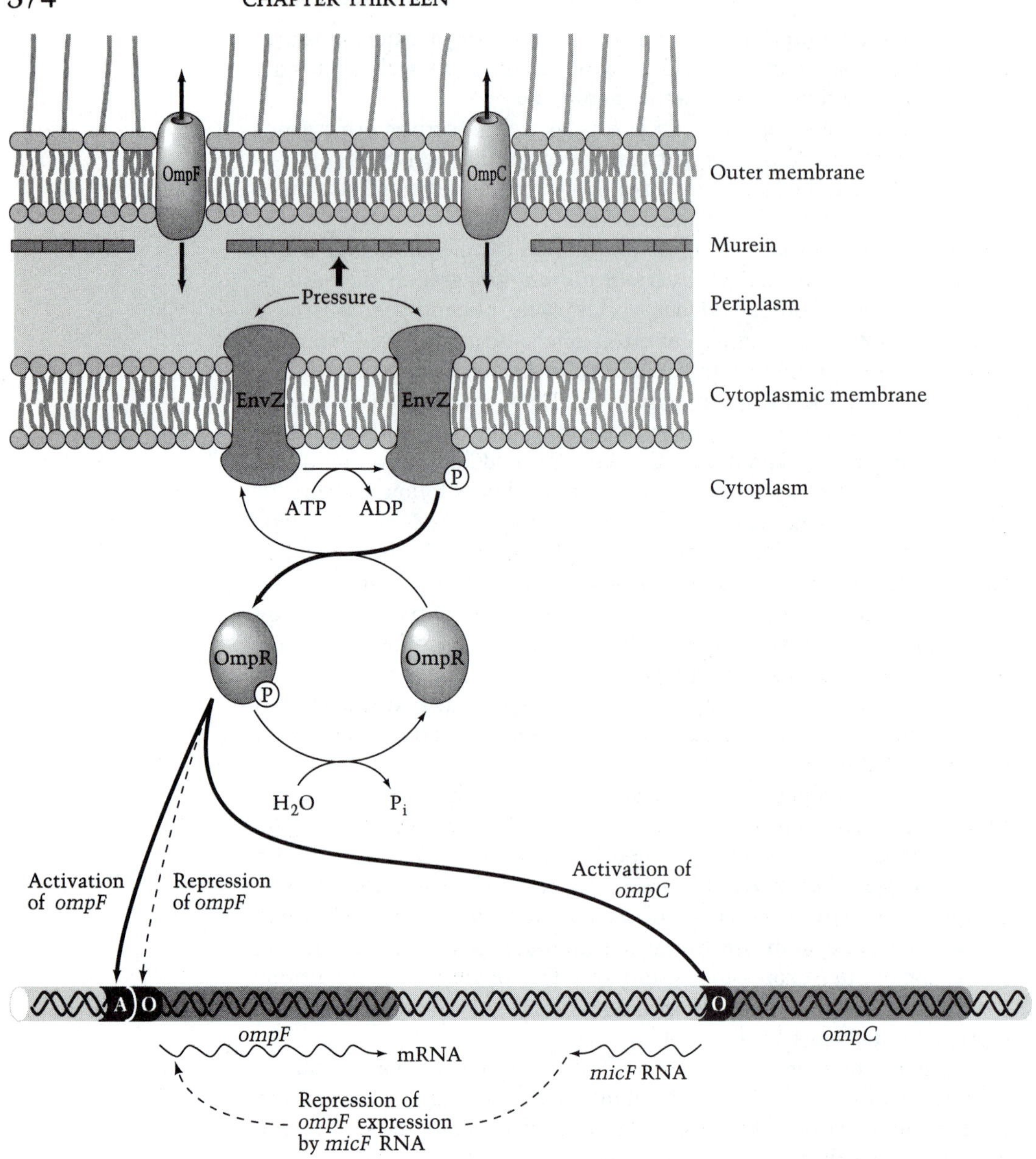

Figure 7

A model for the regulation of major porin synthesis in *E. coli*. OmpF and OmpC, porin proteins; EnvZ and EnvZ-P, native and phosphorylated forms of the signal transducer; OmpR and OmpR-P, native and phosphorylated forms of the regulator protein; *micF*, gene encoding an RNA that blocks expression of *ompF*. See text for further explanation. (From Stock et al., 1989.)

species within many other genera as well), this change in the environment signals a major switch in mode of energy transduction. The cells must adjust from a mode largely dependent on molecular oxygen as an electron acceptor to one in which fermentation and anaerobic respiration dominate. The evidence from two-dimensional gels and from other sources indicated long ago that the regulation of expression of a sizable number of genes is involved in these transitions.

Listed in Table 1 are several gene networks that are important in the adjustment of bacteria to changes in the supply of molecular oxygen—including the *oxyR* system, which provides protection against oxidation damage, and the *fnr* system, which provides the capability of using inorganic electron acceptors such as nitrate and nitrite. Here, for reasons that will become obvious, we wish to highlight a more recently discovered gene network—named *arc* (for **a**erobic **r**espiration **c**ontrol)—that includes as member genes those encoding

- several dehydrogenases of the flavoprotein class
- the cytochrome *o* oxidase complex
- certain enzymes of the TCA cycle, the glyoxylate shunt, and the pathway for fatty acid degradation

All of the proteins encoded by these genes share one function: they participate in pathways, shown in Figure 8, that ultimately donate electrons to molecular oxygen, with the generation of protonmotive force (Chapter 5). These several genes are under different specific controls, but all are repressed anaerobically and induced aerobically. Recently investigators have discovered two genes—*arcA* and *arcB*—that participate in the network regulation of the genes of aerobic metabolism. The gene *arcA* probably encodes a repressor of the *arc* network (mutants defective in *arcA* make high levels of *arc* proteins anaerobically); *arcB* mutants are also derepressed anaerobically for arc-controlled genes, but this phenotype can be overridden by providing a high dose of wild-type *arcA* genes to the *arcB* mutant.

On the basis of sequence analysis of *arcA* and certain other genes that participate in network regulation, researchers have postulated that the *arcA* product—ArcA—is a DNA-binding protein; in one state it represses the *arc* operons and in another state it does not. The product of *arcB*—ArcB—is postulated to be a membrane protein that senses protonmotive force and interacts with ArcA to modulate its activity as a repressor. An analogy is drawn between the *arcA–arcB* pair and the *envZ–ompR* pair of genes that control the osmotic response discussed above.

SIGNAL TRANSDUCTION IN MULTIGENE SYSTEMS

The *arc* network, together with the other three networks just described, share a common theme: information processing by protein–protein interaction. Let us consider this topic in greater depth.

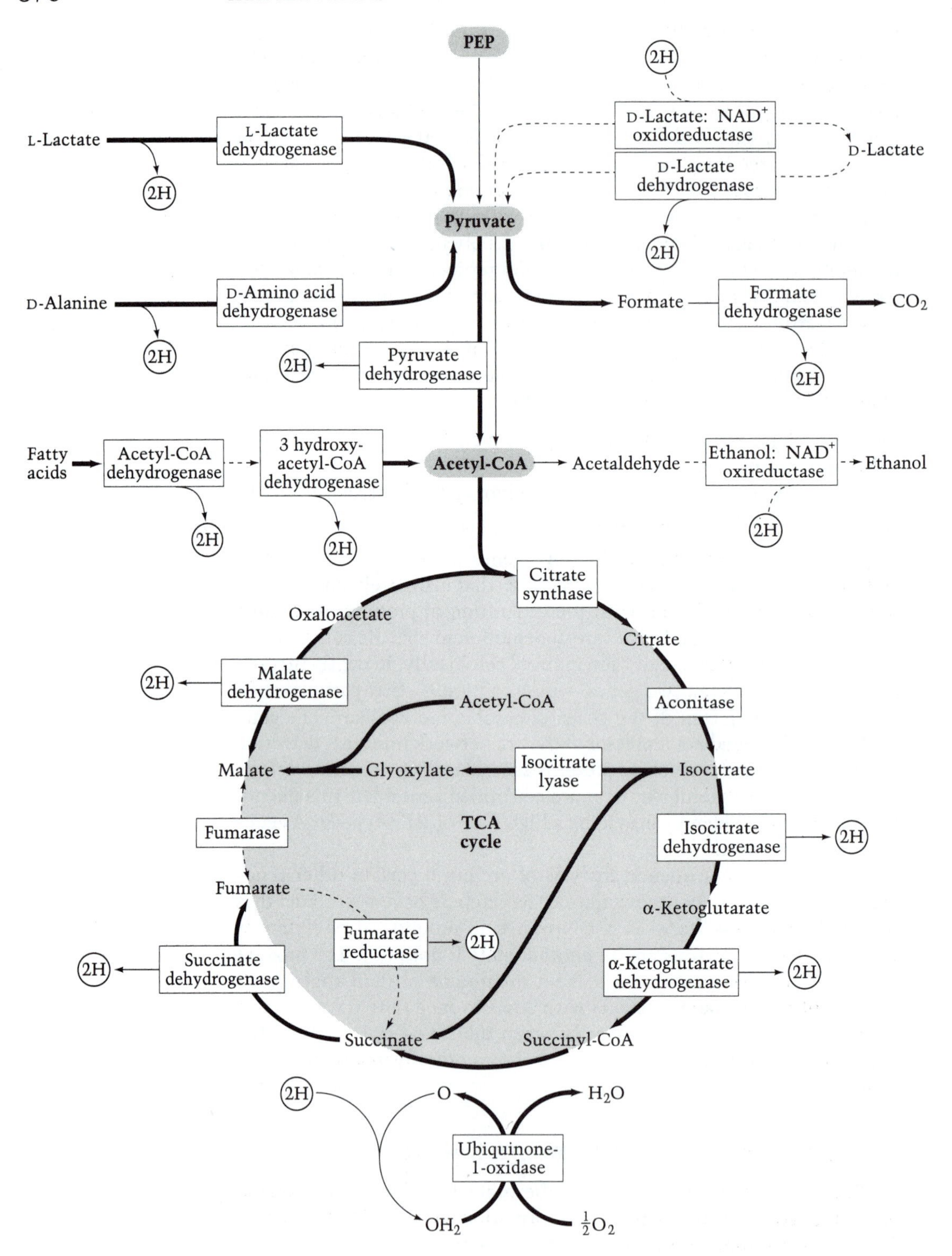

PEP
2H
L-Lactate
L-Lactate dehydrogenase
2H
D-Lactate: NAD+ oxidoreductase
D-Lactate
D-Lactate dehydrogenase
2H
Pyruvate
D-Alanine
D-Amino acid dehydrogenase
2H
Formate
Formate dehydrogenase
CO2
2H
2H
Pyruvate dehydrogenase
Fatty acids
Acetyl-CoA dehydrogenase
3 hydroxy-acetyl-CoA dehydrogenase
Acetyl-CoA
Acetaldehyde
Ethanol: NAD+ oxireductase
Ethanol
2H
2H
2H
Citrate synthase
Oxaloacetate
Citrate
2H
Malate dehydrogenase
Acetyl-CoA
Aconitase
Malate
Glyoxylate
Isocitrate lyase
Isocitrate
TCA cycle
Fumarase
Isocitrate dehydrogenase
2H
Fumarate
α-Ketoglutarate
Fumarate reductase
2H
2H
Succinate dehydrogenase
α-Ketoglutarate dehydrogenase
2H
Succinate
Succinyl-CoA
2H
O
H2O
Ubiquinone-1-oxidase
OH2
½O2

The processing of physiological information is not a new topic for us. We encountered some aspects of information processing when we considered the conformational changes in certain (allosteric) proteins brought about by reversible complexing of specific small molecules. The small molecules—end products of biosynthetic pathways, inducers of catabolic pathways, precursor metabolites, or nucleotides—conveyed *information* (about their cellular concentration) to the allosteric protein. This process accounted for most of the control on enzyme activity in biosynthetic pathways and in fueling reactions, and for the central strategic device of operon physiology—activation and repression of transcription by proteins that in one conformation can bind to DNA and in another, cannot (Chapters 11 and 12).

Although versatile and effective in many circumstances, ligand–protein complexes can be imagined to have some inherent limitations in information processing. There might not always be a convenient metabolite to serve as a unique and sensitive symbol of some metabolic state. How, for example, does one convey information about the rate of passage of a solute through the cell membrane? Or about the turgor pressure on the cell wall? How can different chemicals in the environment be recognized with great specificity, yet lead to a common response by the cell? How can a small signal be quickly amplified to become a large one?

Recently, new insights have been gained about the processing of information in the bacterial cell. Only a few of the answers are in, but some exciting findings have already been made. A few of these are illustrated in the four networks we have just described. What conclusions can we draw from this information?

By the mid 1980s, molecular genetic analysis of a number of different regulatory systems in bacteria had led to a curious finding: many multigene (and multiprotein) systems contained a pair of proteins, one from each of two families of homologous proteins. The amino acid sequence similarities within these families proved to have functional correlates; proteins from one group controlled the activity of their partner from the other group by phosphorylation. The transfer of phosphate is a form of signal transduction. We have already met examples in some of the multigene networks we have examined: NR_{II}, PhoR, and EnvZ (and possibly ArcB) phosphorylate their partners NR_{I}, PhoB, and OmpR (and

Figure 8

Central metabolic reactions and the *arc*-controlled enzymes. Thick arrows indicate reactions catalyzed by enzymes under *arc* control. Reactions catalyzed by enzymes that may or may not be under *arc* control are represented by thin arrows. Dashed arrows indicate reactions catalyzed by enzymes not under *arc* control. The circled 2H symbols indicate reducing equivalent consumed or yielded by the reaction. (After Iuchi et al., 1989).

possibly ArcA), respectively. In these examples, the protein of the first group signals its partner in the second group that, respectively, the supply of glutamine is limiting, the rate of entry of inorganic phosphate into the cell is insufficient to support growth, or the osmotic pressure has increased (and the redox state of the cell has changed).

Members of the first family, which are PROTEIN KINASES (PKs), become phosphorylated by transfer of phosphoryl groups from ATP to histidine residues in the kinase. The phosphoryl groups are then transferred to aspartic acid residues in the members of the second family of proteins, the PHOSPHORYLATED RESPONSE REGULATORS (PRRs); finally the phosphoryl group can be removed from the PRR-phosphoaspartate residues by hydrolysis. These reactions can be summarized as follows:

$$\text{ATP} + \text{PK-His} \rightarrow \text{ADP} + \text{PK-His-P}$$
$$\text{PK-His-P} + \text{PRR-Asp} \rightarrow \text{PK-His} + \text{PRR-Asp-P}$$
$$\text{PRR-Asp-P} \rightarrow \text{PRR-Asp} + \text{P}_i$$

Members of the histidine protein kinase family of proteins have conserved sequences near their carboxyl termini (Figure 9). Many, such as EnvZ, are transmembrane proteins, a location consonant with their postulated role as sensors. These proteins have their divergent amino-terminal regions exposed to the outside of the cell (or the periplasm) and their conserved carboxyl-terminal regions facing the interior of the cell. In this position they are ready to transmit information by phosphorylating their cognate response regulator. The response regulators have a conserved domain of about 100 amino acids near the amino terminus (Figure 10). These proteins are for the most part cytosolic. Some are DNA-binding proteins that regulate transcription of specific operons; others are involved in governing the activity of enzymes; and some—those involved in chemotaxis in *E. coli*—are involved in signal transduction from the methyl-accepting chemotaxis proteins to the proteins that control the direction of rotation of the flagella (Chapter 6).

It has been estimated that there may be as many as 50 of these signal transduction systems employing protein phosphorylation in the single bacterium *E. coli.* Apparently, once a primitive form of this signaling device had evolved the original pair of genes were duplicated over and over, and each time diverged in structure to serve analogous signal transduction functions for diverse processes. Today one finds signaling by phosphorylation throughout the bacterial world, in both Gram-positive and Gram-negative species, where it controls diverse processes such as chemotaxis, regulation of gene expression, virulence, and control of developmental pathways. A number of examples are shown in Table 2.

But the conservation of sequences within the members of the PK and PRR families has more significance than merely serving as a record of common evolutionary origin. In several cases different systems interact. We have noted in the phosphate utilization network that in *phoR* mutants the product of the *phoM* gene, which by sequence analysis appears to be another PK, can substitute partially for the defective PhoR in modifying the response regulator, PhoB. In wild-type cells the PhoM kinase is normally involved with a different response regulator of a

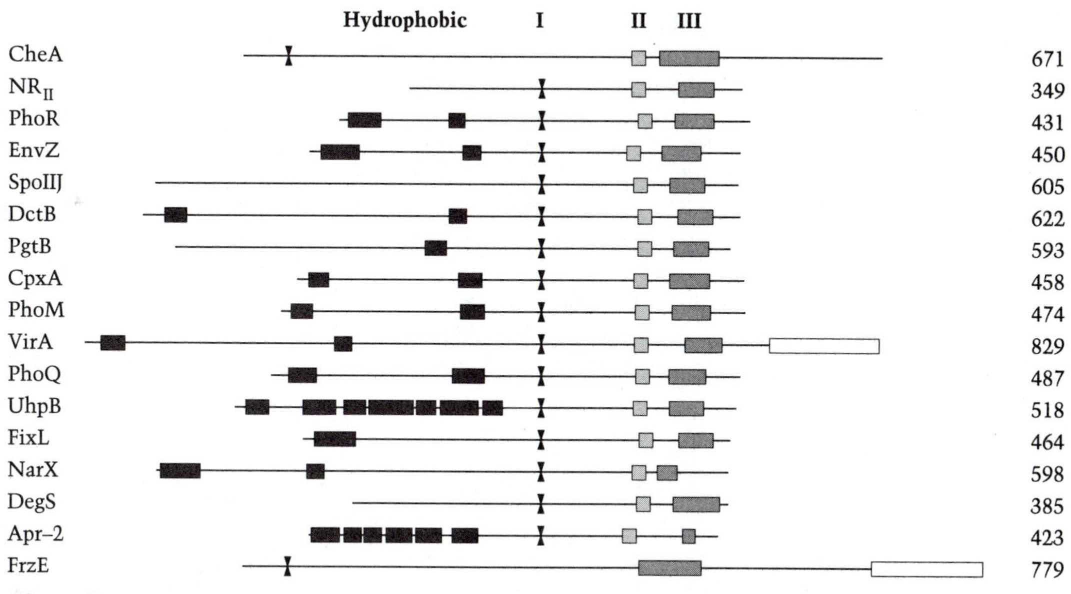

Figure 9

Domain organization of the protein kinases. Hydrophobic, putative membrane-spanning sequences are dark gray boxed regions near the amino termini. Region I contains the conserved histidine residue; it is indicated by thin, indented black bars. Region II, containing the conserved asparagine, is indicated by a light gray box. Region III segments are indicated by medium gray boxes. DegS, NR$_{II}$, SpoIIJ, CheA, and FrzE lack any discernable membrane-spanning sequences. Protein kinase FrzE has no clearly defined region II. Boxed regions at the extreme carboxyl termini of VirA and FrzE are homologous to the phosphorylated response regulator (PRR) domains shown in Figure 10. Numbers to the right designate the total number of amino acid residues in each protein. (From Stock et al., 1989.)

Table 2. Multigene networks controlled by homologous phosphotransfer proteins

System[a]	Protein kinase	Response regulator	Organism
Nitrogen regulation (Ntr)	NR$_{II}$	NR$_{I}$	*E. coli, S. typhimurium, Klebsiella* spp., *Rhizobium* spp.
Phosphate regulation (Pho)	PhoR	PhoB	*E. coli, Bacillus subtilis*
Porin regulation (Omp)	EnvZ	OmpR	*E. coli*
Symbiotic nitrogen fixation (Fix)	FixL	FixJ	*Rhizobium meliloti*
Aerobic respiration (Arc)	ArcB	ArcA	*E. coli*
Virulence (Vir)	VirA	VirA/VirG	*Agrobacterium tumefaciens*
Virulence (Vir)	BvgC	BvgA	*Bordetella pertussis*
Sporulation (Spo)	SpoIIJ	SpoOA/SpoOF	*B. subtilis*

Source: Modified from Stock et al. (1989).
[a]In only a few of the examples given has phosphorylation been demonstrated; for most, the primary information is sequence homology.

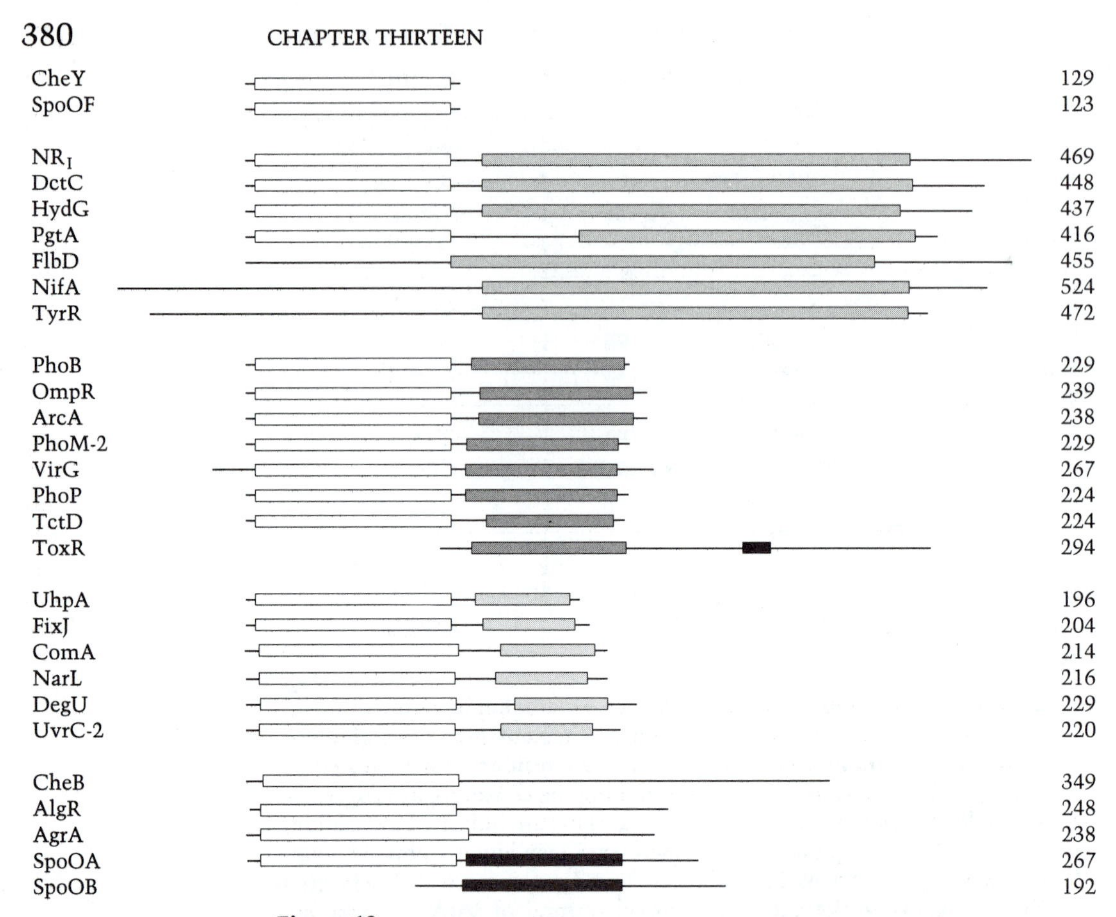

Figure 10

Domain organization of phosphorylated response regulators and related proteins. The response regulator domain at the amino terminus of each protein is indicated by an open box. Other homologous regions are indicated by shaded boxes. Nonconserved regions are indicated by a line. The small black box in ToxR corresponds to the single membrane-spanning sequence within this protein. Numbers to the right designate the total number of amino acid residues in each protein. (From Stock et al., 1989.)

different system. This sort of CROSS TALK occurs because the kinases have considerable cross specificity toward different response regulators. As a result, a cell that has activated one adaptive response system may, by cross phosphorylation, have another response system partially activated, or raised to near activation status.

DIVERSE MEANS OF REGULATING MULTIGENE NETWORKS

Protein–protein interactions are important in regulating the networks described above but are not the sole means of regulating multigene networks. In fact, the molecular means by which groups of different operons are coordinated in their expression are quite diverse. To illustrate this diversity, we consider three networks that resemble one another in one functional aspect. Each is induced to high-level expression as the result of a single, well-defined environmental parameter change, but the mechanism of the induction is very different in each instance.

A shift to high temperature induces a set of approximately 20 genes in most bacterial cells. This induction constitutes the HEAT-SHOCK RESPONSE, which, to an incredible extent, is conserved from archaebacteria and eubacteria up through lower eucaryotes to higher plants and animals. Virtually every microbial cell and every plant and animal cell responds to a shift-up in temperature to a range just above that which is normal for the given organism by inducing the synthesis of its heat-shock proteins.

One of the most fascinating of current studies is that devoted to understanding the function of the heat-shock response, not only because it is so prevalent, but also because it has been genetically conserved in a large number of its features from bacteria to humans. Here there is not space to discuss the entire biology of the heat-shock response, interesting as that is turning out to be. Rather, we point out that the 20-odd genes constituting the heat-shock network of bacteria such as *E. coli* are controlled by the cellular level of an RNA polymerase sigma factor, σ^{32}. Core polymerase programmed by this sigma factor preferentially transcribes the heat-shock genes, because they are governed by promoters that are recognized specifically by this factor. The mechanism of induction of the heat-shock response consists of three processes that, in concert, rapidly elevate the level of σ^{32} upon a shift-up in temperature: decreased degradation of the factor, increased synthesis of its message, and increased translation of the high level of its messsage normally present in the cells growing at any temperature.

Another network—with approximately the same number of genes—is the SOS system (Table 1). This group of genes is induced to high-level expression when the cell experiences damage to its DNA, as by UV irradiation. The protein products of these genes are involved in various processes that repair damage to DNA and that prevent the cells from dividing until such damage has been repaired. This network is induced by a proteolytic cleavage of the repressor of the member operons. The repressor is the product of the *lexA* gene; upon damage to the DNA of a cell, the RecA protein (Chapter 10) is activated to facilitate the proteolytic cleavage of LexA, thereby inducing the LexA-controlled genes, which constitute the SOS system.

Still another mode is illustrated by the *oxyR* network (Table 1). This group of a dozen or so genes is induced by oxidative damage brought on by, for example, hydrogen peroxide. The member genes encode protective enzymes such as catalase-hydroperoxidase I and alkylhydroxyperoxidase. This network is induced by activation of the positive regulator OxyR, product of the *oxyR* gene.

These examples serve to illustrate the fact that the regulation of gene networks is as diverse as is the regulation of individual operons.

NAMING MULTIGENE SYSTEMS: REGULONS, MODULONS AND STIMULONS

Thus far in our discussion of multigene systems in bacteria we have not introduced special terms for them. We have used the phrase *regulatory network* to refer to any and all of these systems, in spite of the vast differences in their size, degree of complexity, and mode of regulation.

This is not to say that the field of hierarchical control of gene expression is lacking in specialized terms. Each of the following has been used to designate one or another regulatory network or class of networks:

- regulon
- modulon
- stimulon
- global control system
- multigene system
- adaptive response system

A REGULON is a relatively simple network of operons that is controlled only by a common regulatory protein and its effector ligand. The name was first given to the set of genes encoding enzymes of the arginine biosynthetic pathway in enteric bacteria. Unlike the genes for most amino acid biosynthetic pathways, the eight genes of the arginine pathway are arranged as multiple, unlinked operons, but all six operons are controlled by the same protein repressor to which arginine binds to produce the DNA-binding conformation. Some microbiologists prefer to limit use of the name *regulon* for an operon network in which the member operons are associated with a single pathway or function and are controlled solely by a common regulatory protein. The *argR* regulon (bearing the name of the gene encoding the repressor protein) serves as the prime example. Networks of operons with more complex controls would not be called regulons. The term has, however, been used in a more general sense to refer even to complex systems involving multiple individual regulators in addition to a pleiotropic regulatory protein.

The term MODULON has been proposed to designate an operon network concerned with multiple pathways, in which the member operons may be governed by individual regulatory proteins and, in addition, are under the control of a common, pleiotropic regulatory protein. The network of operons that are subject to catabolite repression (by the cAMP–CAP system) would be an unambiguous example of a modulon; this network may be called the *crp* modulon (again, naming it after the structural gene of the common pleiotropic regulatory protein).

The term STIMULON refers to all the operons responding together to an environmental stimulus, no matter how many regulons and modulons may be involved. The operons of a regulon or of a modulon are expressed together under particular circumstances—that is the whole point of regulons and modulons—but it is conceivable that some operons that respond together to an environmental change do not share a regulatory protein. Furthermore, even a simple environmental change may induce several regulons and modulons. A different term is needed to describe these sets of independently coinduced or corepressed operons; that term is *stimulon.* Its usefulness stems from the fact that usually the initial observation of a cellular response describes the coinduced or corepressed proteins (or their genes) without shedding light on the mechanism of the response. The terms *regulon* and *modulon,* on the other hand, by their strict definitions cannot be applied until analysis has revealed the molecular details of the regulation of the responding operons. The genes that respond to phosphate limitation belong to the *psi*—or **p**hosphate **s**tarvation-**i**nducible—stimulon, because they consist of dozens of operons controlled as several regulons, the details of which are still being sorted out.

The operon networks of bacteria are complex and diverse. The terms *regulon* and *stimulon* are useful to a point, particularly in describing some of the well-defined and specific responses of a cell to nutrient limitation, cell damage, or other stress brought on by the environment. For other situations, the structure and physiology of a particular network may not be accommodated readily within these concepts, as, for example, the networks regulating carbon and energy metabolism, solute transport, sporulation, chemotaxis, the stringent response, and cell division. For this reason, some general terms with minimal structural and mechanistic implications are useful in this field; global control system, multigene system, and adaptive response system are all terms that have been used to designate a network of bacterial operons with coordinated regulation. GLOBAL CONTROL SYSTEM is a term that emphasizes the multipathway nature of some networks (e.g., the SOS response, the heat-shock response). MULTIGENE SYSTEM is a term emphasizing the large number of operons encompassed by many networks (and, therefore, would be particularly applicable to the cell division and sporulation systems). ADAPTIVE RESPONSE is used for the networks that respond to specific, well-defined, environmental stresses (such as the SOS response and the heat-shock response). The general nature of these terms guarantees that many networks can be described by more than one.

METHODS OF RECOGNIZING MULTIGENE SYSTEMS

The genes and gene products of a stimulon are easily defined operationally—they are the joint responders to some environmental perturbation—and they can be sought out systematically by both biochemical and genetic methods. The biochemical methods establish the protein products of the stimulon, and the task is then to find their genes. Genetic methodology follows the opposite path: genes are found, and then their protein products must be sought.

A useful biochemical approach is the application of two-dimensional gel electrophoresis to monitor changes in the pattern of synthesis of individual proteins. One customarily subjects a culture in balanced growth (Chapter 7) to the environmental change of interest; a portion of the culture is labeled before the shift with a radioactive protein precursor to establish the control pattern of protein synthesis. A similar culture is then labeled shortly after the environmental shift, and the resultant protein pattern is compared with the control pattern. The result is a list of the set of proteins induced (or repressed) by the perturbation. These are the gene products of the stimulon being investigated. (The biochemical approach has been successful also in describing the genes that encode protein components of structures that can be purified from cell extracts, such as the ribosome. Separation of the proteins from the purified structures by column chromatography, electrophoresis, or other means provides the tally of proteins that can then be used to search for their genes.)

The path from an unknown spot on a gel to the cloning of its gene was until recently a daunting journey, but it is now accomplished routinely. One method, referred to as REVERSE GENETICS, involves eluting sufficient protein directly from gels to obtain a partial amino acid sequence. From this information the researchers construct an oligodeoxyribonucleotide that should hybridize with the coding region of the gene of this protein. The oligonucleotide is then labeled radioactively to produce a probe useful in screening one or another GENOMIC LIBRARY for the gene in question. The library usually consists of a collection of a few thousand bacterial clones containing hybrid phage or plasmids that carry small spliced segments of the bacterial chromosome. Once a hybridizing clone is found, its resident phage or plasmid can then be prepared in larger amounts and examined for its ability to produce the protein under study. For this, one employs special strains of bacteria to produce MINICELLS (chromosome-less cells produced by a faulty division of a mutant strain; Chapter 12) or MAXICELLS (cells that have degraded their own chromosomal DNA following UV irradiation) in which the introduced hybrid phage or plasmid can express its genes without a background of expression from the bacterial chromosome. The labeled extracts from the expression cultures are then examined by two-dimensional gel electrophoresis to verify that a protein of the desired mobility has been produced. Subcloning, sequence analysis, and genetic mapping are the next steps in characterizing the gene. This path is diagrammed in Figure 11.

The genetic approach to stimulon analysis is a valuable alternative to and complement of the biochemical approach. In studies with *E. coli*, for

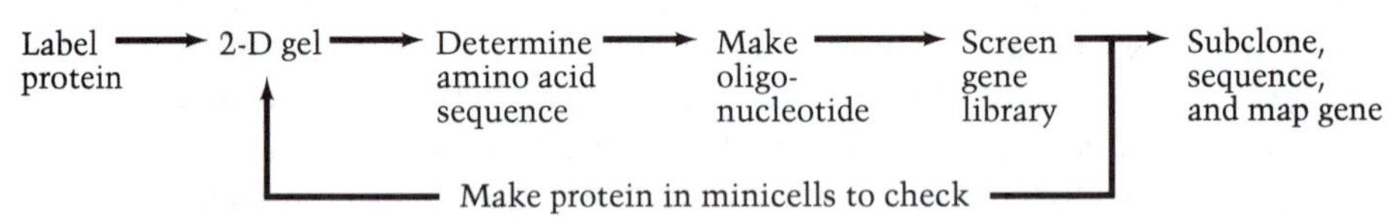

Figure 11

The general pathway of reverse genetics. This process enables the investigator to proceed from the identification of a protein of interest to the isolation, mapping, and sequencing of the gene encoding this protein. It is particularly useful in the analysis of networks of coregulated genes when their products have been identified on two-dimensional gels.

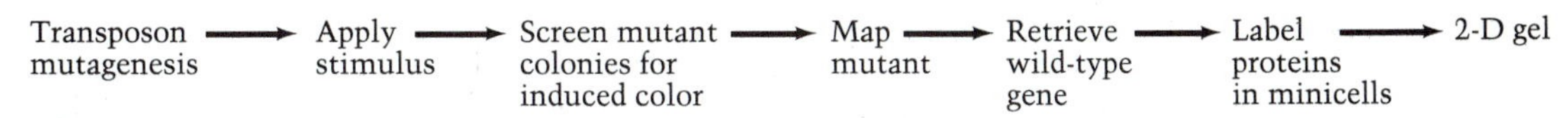

Figure 12

The process of finding genes that share a common pattern of regulation, and then of identifying their protein products.

example, one mutagenizes a population of cells using a transposable element such as the phage Mu (Chapter 15) engineered to contain a promoterless β-galactosidase gene. The mutagenized clones are then grown on solid media containing a chromogenic substrate of β-galactosidase to reveal those that produce the enzyme only under the condition of interest. In such a mutant, the transposable element must have inserted within some operon of the stimulon of interest, so the reporter gene is now governed by this operon's promoter. Genetic methods then permit one to map the location of each such insertion, thereby defining the genes of the stimulon. It should be recognized that this procedure, as described, can identify only those operons that encode products nonessential for cell growth under normal conditions; by using strains that are partially diploid for different regions of the chromosome (Chapter 15), this limitation can be removed. From here, the analytical path is the opposite of that which commenced with two-dimensional gels; the task now is to find the unknown protein products of the genes of the stimulon. This process involves cloning the unmutated allele of the gene that had originally been disrupted by the insertion mutation and then identifying its protein product by maxicell (or minicell) expression followed by two-dimensional gel electrophoresis. This approach is diagrammed in Figure 12.

Do the genes and gene products that are part of a particular *stimulon* belong to one, or to more than one, *regulon* or *modulon*? To assign genes and their products to either a regulon or a modulon requires an additional step: discovery of the common regulator or its gene. Sometimes this can be accomplished by selection of mutants that are unable to adapt to the new environmental condition. Those proteins of the stimulon that fail to be induced in the mutant after the shift are presumed to be part of a *regulon* or a *modulon*.

Finally, genetic approaches have been quite successful in defining the multiple components of complex systems such as motility–chemotaxis, sporulation, and cell division. Mutants deficient in the process of interest are isolated and used to map the mutant gene, followed by its cloning, sequencing, and the identification of its product.

SUMMARY

1. Operons of bacteria are organized into higher order, multigene systems for regulatory and functional purposes. These multigene systems, called *regulatory networks*, display a great variety of structural

and mechanistic diversity. Some regulatory networks involve the coordinate induction of a set of operons controlled by a common regulatory protein (repressor or activator); these have been called *regulons* if their sole control is by the common regulatory protein, *modulons* if the individual operons have separate regulatory proteins in addition to the pleiotropic one that governs all of them. The total group of operons responding to some environmental stimulus - is called a *stimulon*. Many multigene systems contain complex subsystems.

2. Biochemical and genetic approaches that are now available facilitate the recognition of genes and their products that belong to stimulons, modulons, regulons, or other regulatory networks. Two-dimensional gel electrophoresis and transposon mutagenesis combined with classical mutant analysis and protein chemistry are particularly useful.
3. Multigene systems in *E. coli* encompass most of the known operons of this cell. These systems may be grouped as those (i) induced in response to nutrient limitation, (ii) involved in redox reactions and electron transport, (iii) concerned with adaptation to damage or stress caused by oxidation, radiation, and temperature and osmotic pressure extremes, and (iv) involved in major morphological and developmental processes such as adaptation to stationary phase, sporulation, and cell division.
4. The cAMP–CAP network consists of a large number of operons that encode inducible catabolic enzymes. High-level transcription of these operons requires that CAP protein complexed with cAMP be bound to special sites in their promoters to improve the binding of RNA polymerase. Modulation of the system is accomplished in part by inhibition of cAMP synthesis by protein III^{Glc} of the PTS system during glucose transport. The operation of the network helps assure that a premier substrate (glucose in the case of *E. coli*) is used preferentially to poorer substrates.
5. The stringent response network shifts biosynthetic resources from the wasteful synthesis of ribosomes, tRNA, and perhaps other components of the translation machinery when growth is restricted by the supply of one or more amino acids or by the supply of a carbon and energy source. Important elements in the pathway are the guanosine nucleotides, pppGpp and ppGpp, which are made by either of two mechanisms: one involving ribosomes stalled in translation by the unavailability of an aminoacylated tRNA and one that is ribosome independent and sensitive to the carbon and energy supply. The signal nucleotides act directly to prevent transcription of rRNA and tRNA operons and indirectly to affect many others. The primary action appears to be on RNA polymerase.
6. Together, the catabolite repression and stringent control networks control the majority of the protein-synthesizing capacity of the cell.
7. Several gene networks, including those concerned with ammonia assimilation, with phosphate assimilation, with porin synthesis and related responses to osmotic stress, and with aerobic respiration are

controlled by pairs of proteins in which the first member of the pair becomes autophosphorylated as the result of sensing some specific change in the environment, and transduces this signal by phosphorylating its partner. The former proteins appear to be hisitidine protein kinases, with a strong family resemblance to one another, and the latter are phophorylated response regulators, again bearing a strong family resemblance to one another. To some extent cross specificity of the kinases for noncognate response regulators enables cross talk between separate adaptive responses.

8. The heat-shock, SOS, and oxidation stress response networks illustrate the variety of molecular mechanisms employed to govern networks: cellular level of a special sigma factor, a repressor protein that is aided by a second protein to proteolyze itself, and a positive transcription regulator, respectively.

STUDY QUESTIONS

1. Review the definitions of genetic units by defining, with two or three examples, each of the following: cistron, operon, regulon, modulon.
2. Mutants in the *cya* gene of *E. coli* are unable to produce cAMP and grow slowly on glucose; they have lost the ability to grow on substrates such as lactose, maltose, and arabinose. This pleiotropic phenotype occurs because these cells are permanently catabolite repressed. Incubating large numbers of such mutant cells on plates containing only lactose as carbon and energy sources yields several mutant colonies. Further tests reveal that these secondary mutants are of three classes: class A have regained the ability to make cAMP and to grow on the whole range of substrates; similarly, class B can grow on all substrates but cannot make cAMP; class C cannot make cAMP but can grow on lactose, although not on any of the other secondary substrates. Each class maps in a different position on the chromosome. Suggest what gene is likely to be mutated in each class. Tell what experiments would help verify the correctness of your answers.
3. All of the *E. coli* mutants that display the relaxed phenotype are altered in some aspect of (p)ppGpp accumulation. Contrast this with the situation you would predict for the range of mutants defective in catabolite repression, those unable to induce glutamine synthetase, those unable to induce alkaline phosphatase, and those defective in the SOS response.
4. Several global control systems employ protein phosphorylation as a means of signal transduction. What properties of this type of information transfer might have contributed to its widespread use?
5. Three proteins of the approximately 25 that are induced by a shift-up in temperature in *E. coli* are also induced by a concentration of hydrogen peroxide that induces the OxyR response. What questions are raised by these observations? What experimental approaches could address these questions?

6. Some missense mutations in the *phoR* gene of *E. coli* result in the failure of phosphate limitation to induce alkaline phosphatase and several other proteins. Other missense mutations lead to constitutive synthesis of these proteins during phosphate limitation. Propose an explanation based on the suggested mechanism of the phosphate limitation response described in Figure 6.
7. Many circumstances in nature produce multiple changes in the environment of microbes. For example, excretion of *E. coli* cells from a mammalian gut may involve, within a short time period, shift-down in temperature, decrease of nutrient availability, decrease in concentration of toxic substances (bile salts and waste products of microbial metabolism), increase in oxygen concentration, change in pH, increase in redox potential, decrease in osmotic pressure, and exposure to UV radiation. What feature of cellular stress responses might facilitate the adaptation of cells to multiple changes of this sort?
8. Assume that you have isolated a new bacterial species in the genus *Streptococcus* and that you wish to learn whether it has a multigene response to phosphate restriction. How would you go about learning this? If early results were to indicate that such a system exists, outline what plan you would use to characterize its molecular mechanism and function.

SUGGESTED READING

Topics in this chapter are covered in the following chapters from *Escherichia coli and Salmonella typhimurium: Cellular and Molecular Biology*, F. C. Neidhardt, J. L. Ingraham, K. B. Low, B. Magasanik, M. Schaechter and H. E. Umbarger (eds.). 1987. American Society for Microbiology, Washington, D.C.

Cashel, M. and K. E. Rudd. The stringent response. (Volume 2; Chapter 87)
Magasanik, B. and F. C. Neidhardt. Regulation of carbon and nitrogen utilization. (Volume 2; Chapter 81)
Neidhardt, F. C. Multigene systems and regulons. (Volume 2; Chapter 80)
Neidhardt, F. C. and R. A. VanBogelen. Heat shock response. (Volume 2; Chapter 83)
Walker, G. C. The SOS response of *Escherichia coli*. (Volume 2; Chapter 84)
Wanner, B. L. Phosphate regulation of gene expression in *Escherichia coli*. (Volume 2; Chapter 82)

Other suggested readings include:

Csonka, L. N. 1989. Physiological and genetic responses of bacteria to osmotic stress. Microbiol. Rev. 53:121.
Saier, M. H., Jr. 1989. Protein phosphorylation and allosteric control of inducer exclusion and catabolite repression by the bacterial phosphoenolpyruvate:sugar phosphotransferase system. Microbiol. Rev. 53:109.
Stock, J. B., A. J. Ninfa and A. N. Stock. 1989. Protein phosphorylation and the regulation of adaptive responses in bacteria. Microbiol. Rev. 53:450.

14

Cell Cycle

INTRODUCTION

A growing bacterial cell leads an ephemeral existence. Its life span is the interval from its birth by division of the mother cell to its own division into two daughters. This interval is the CELL CYCLE of an individual bacterium. The regulatory system that operates the cell cycle stands at the highest level of organization of any global control system, because the formation of new cells is the ultimate function of all the structure, metabolism, and physiological processes of the cell. In this chapter we examine molecular approaches to the study of the cell cycle and its regulation. For all its fundamental importance, the cell cycle has been studied far less intensively than the growth of the population of which the cell is a part. The principal reasons are that an individual cell is too small to be studied chemically and that, under normal circumstances, bacteria divide asynchronously and a sample of a culture taken at any one time reflects the randomized behavior of cells at different stages of their cell cycles. Thus, the bacterial cell cycle is a specialized topic of investigation.

Many questions arise regarding the cell cycle. What are the key biosynthetic and regulatory events that control this cycle? Are these events arranged as a series of sequential steps or do they take place independently of one another? How are morphogenetic and biochemical events interrelated? As you will see, it is possible to provide some general answers to some of these questions. Two events punctuate the cell cycle—DNA replication and the process of cell division. These two processes result in single structural units—the chromosome and the cell itself. As a result, each of these processes can be considered to constitute a cycle with a discrete beginning and a discrete end. The synthesis of components that are present as greater numbers of individual units (e.g., individual enzymes, ribosomes) takes place in most instances throughout the cell cycle. Thus, it appears that in bacteria, cell growth—

the generation of cellular mass—proceeds without pause throughout the cell cycle, showing no evidence of large changes in the rate of mass synthesis. To understand how these phenomena are studied, it is necessary to have some familiarity with the experimental methods used to study the cell cycle in bacteria.

A single bacterium is obviously too small for chemical measurements; so, to analyze a large sample, cells must be aligned by artificial means. At first glance, it may seem that the study of the bacterial cell cycle can best be carried out by synchronizing the division of all bacteria in a population, lining them up to move together through the cycle. However, much can be learned from the ingenious exploitation of experimental situations that do not require induction of division synchrony. A consideration of these various strategies will help us understand the issues concerning the bacterial cell cycle. We focus first, therefore, on the methodological aspects of this subject.

STRATEGIES FOR STUDYING THE BACTERIAL CELL CYCLE

Following the growth of individual bacteria

A limited amount of information can be obtained by following single bacteria under a microscope. This experiment can be done by taking frequent photomicrographs of cells growing on a small agar block placed on a microscope slide and covered with a cover glass. If the proper temperature is maintained by warming the microscope's stage, bacteria will grow as rapidly as they do in liquid media under unrestricted conditions. Thus, observations can be carried out without perturbing the growth of the cells.

The data of greatest consequence that have been derived from microscopic observations of individual rod-shaped cells concern the kinetics of elongation (Figure 1). Several workers concluded that individual *Escherichia coli* cells elongate exponentially with time. Others have proposed that elongation follows linear kinetics. This point is difficult to establish experimentally because the difference between exponential and linear elongation is very small over one doubling (6% at the most), and microscopic measurements are not that accurate. Nonetheless, this point is important because cell elongation is proportional to growth in volume (the diameter of rod-shaped bacteria does not vary appreciably during the growth of an individual cell). During exponential elongation, growth is proportional to the mass present, a relationship implying that growth is not regulated by a cell cycle-dependent mechanism. In other words, the machinery that carries out the biosynthesis of the major cell components does not distinguish between an old cell and a young cell. A model of linear elongation, on the other hand, must include a special mechanism that senses the mass and thus the age of the cell, because otherwise the rate of mass increase would increase as new synthetic machinery is made. Note that in a population of cells that are dividing asynchronously, growth of the *culture* can be exponential regardless of

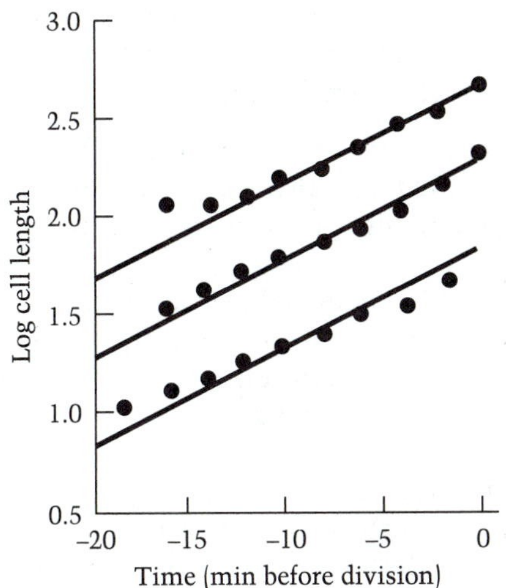

Figure 1

Elongation of three individual *E. coli* cells growing on a thin agar slab on a heated microscope stage at 37°C. The mean doubling time was 20 min. The straight lines represent exponential elongation. For ease of representation, the data for individual cells are displaced along the ordinate. (From Schaechter et al., 1962.)

the growth kinetics of individual cells. Also note how such simple observations bring up global issues of cell growth regulation, such as the law of growth of individual cells.

Studying the cell cycle using undisturbed populations

Certain aspects of the cell cycle can be studied using undisturbed, nonsynchronously dividing populations. To a culture in balanced growth, you can add tritiated thymidine for one-tenth of the culture's doubling time. You then wash the cells of external thymidine and make autoradiograms, which enable you to count the photographic grains that develop from the radioactive decay of tritium in a single bacterium. You will see under the microscope that all cells have grains associated with them, an observation indicating that, roughly speaking, all are equally labeled within a factor of 2. You should wonder whether the radioactive precursor was incorporated solely into DNA or whether an appreciable amount was present in the metabolic pool or in macromolecules other than DNA. We will leave you to think about how you would determine this; but we can tell you that, for practical purposes, the label is overwhelmingly in DNA. You may, then, be justified in concluding that, during the labeling pulse, all bacteria were in the process of synthesizing DNA and that the bacteria in this particular culture, unlike higher cells, do not have appreciable gaps in the synthesis of DNA during the cell cycle (the "G_1" and "G_2" periods of eucaryotic cells). As will be discussed later, this situation is true only for fast-growing bacteria.

Suppose that you wanted to determine quantitative aspects of the pattern of DNA replication during the cell cycle, for example, to learn the kinetics of DNA replication in the life of a single cell. The autoradiographic experiment has several limitations:

- The pulse must be sufficiently long to enable each cell to take up a measurable amount of tritiated thymidine. This requirement introduces significant uncertainty (10% of the cell cycle in the above experiment).
- Counting the photographic grains per cell under the microscope not only is tedious but is meaningful only for cells with small numbers of grains, because when there are many grains per cell, they may overlap and two or more may be counted as one. The range of countability can be extended by using the electron microscope, but even this technique has limitations.
- More fundamentally, it is necessary to establish the relationship between the amount of radioactivity per cell and the age of the cell. How can one establish this relationship? It can be expected that the length (i.e., volume) of the cells in the population is related to their age—older cells are longer than young ones (note that for this purpose it makes little difference whether growth of individual cells is linear or exponential). However, a precise relationship obtains only if cell division always takes place when cells reach the same size. In fact, this condition is not the case for *E. coli* and similar rod-shaped bacteria—here some cells divide before the mean size for division is reached, others after it. The coefficient of variation is 10–20% of the mean value; in other words, the size–age relationship has considerable built-in uncertainty.

These considerations illustrate how experiments with undisturbed cultures may be intrinsically difficult to carry out and to interpret. Can one circumvent these difficulties? In practice, more precise data can be obtained with the use of techniques that are more rapid and sensitive than optical microscopy. As shown in Figure 2, the size distribution of a culture can be estimated with considerable precision with an electronic particle analyzer (see Chapter 7) or under the electron microscope. Both these methods suffer from biases, but with some corrections they yield accurate and useful measurements.

Another technique that is useful in the study the cell cycle is FLOW CYTOMETRY. Flow cytometers employ laser beams to sort cells in a population according to their content of a fluorescent dye, which sometimes is linked to antibodies. These devices are used extensively to measure size distributions and the chemical content of eucaryotic cells. A flow cytometer sufficiently sensitive to measure individual bacteria has recently become commercially available. This method allows one to estimate not only the size distribution of a bacterial culture but also the distribution of DNA per cell (Figure 3).

Another way of studying the cell cycle in an unperturbed population would be to separate its members by age (or, if this task is not possible, by size). Imagine physically dividing a population into classes of young,

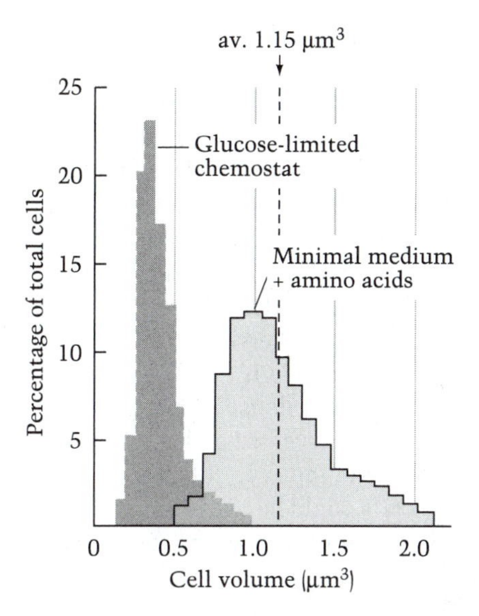

Figure 2

Size distribution of fast- and slow-growing cells of *Salmonella typhimurium*. Dark-shaded area represents cells from a glucose-limited chemostat; growth rate, 0.13 doublings/hr. Light-shaded area shows cells growing in a minimal medium enriched with amino acids; growth rate, 2.0 doublings/hr. (From Ecker and Schaechter, 1963.)

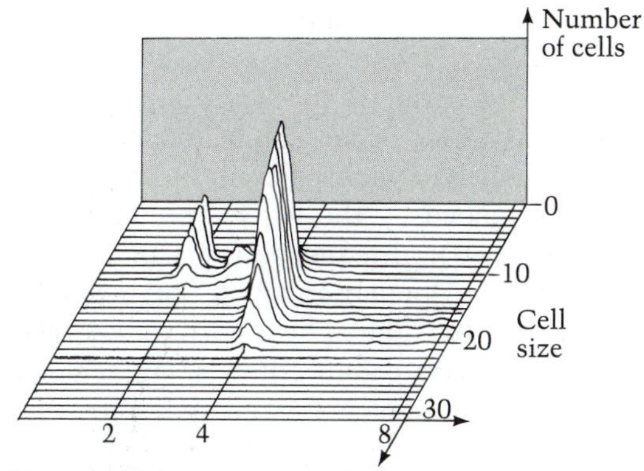

Figure 3

Distribution of cell sizes and DNA/cell measured by flow cytometry. The content of DNA per cell is shown on the *x*-axis, the cell size distribution on the *y*-axis, and the number of cells on the *z*-axis. Cells from an exponential culture were treated with rifampin for 90 minutes before staining with a DNA-specific fluorescent probe. The drug treatment allowed all chromosomes in the process of replication to terminate their cycle but not initiate new rounds of replication. The absence of cells with other than two or four genome equivalents suggests that within each cell chromosomes initiated replication synchronously. (From Skarstad et al., 1986.)

middle-aged, and old cells. It would then be possible to analyze each class and relate the measurement to stages in the cell cycle. How can cells of different size be separated? In principle, a population of cells can be centrifuged through a density gradient, preferably of an osmotically inert material. Because the density of bacteria (weight/volume) does not vary a great deal over the cell cycle, the cells will sediment at a rate that is proportional to their mass. The lighter cells found near the top of the gradient will represent the smallest (thus, the youngest) ones, whereas the oldest ones will be found at the bottom of the band of cells. In practice, polysaccharides such as Percoll have been used to generate a gradient. This technique is limited by the perturbations that must be introduced: to place a sufficiently dense suspension of cells on top of the gradient, the culture must be harvested and concentrated. Although this procedure can be carried out rapidly, subtle changes in the cells are inevitable. In principle, cells could be chemically "fixed" before harvesting by the addition of cross-linking agents, but this approach has not been developed sufficiently.

Synchronizing cell division

If one inoculates a culture with a single bacterium, divisions of its progeny would be synchronous if all cells took the same period of time to divide. In reality, the variation in interdivision time blurs the picture, and sooner or later the cells of the culture will divide asynchronously. Synchronous divisions of a population can be induced by several methods. These include procedures that alter the physiological state of the cell, such as mild temperature shocks or nutritional deprivation. Other methods that have been devised try to avoid such physiological perturbations. Invariably, they deal with the separation of young cells ("babies"), which, if sufficiently alike in age, can be inoculated into fresh medium to divide synchronously. Synchrony will be maintained until the division steps are dampened by the variation in individual interdivision times, usually after two or three generations. In practice, a suspension of the smallest cells of the population can be obtained by gradient centrifugation (as above), by filtration, or by CELL ELUTRIATION (a special centrifugation technique that uses a counterflow of buffer against cells spinning in a special wedged-shaped chamber).

Selection by age: The "baby machine"

The synchronization methods mentioned above have the disadvantages that they induce physiological perturbations and that they select cells by size, not by age. These two drawbacks have been largely circumvented by a fortuitous discovery, which has been exploited by Helmstetter and Cooper. A culture of *E. coli* B/r is filtered through a membrane filter and the filter is inverted over the holder. When fresh medium is slowly poured through, the only cells that detach from the filter are those that just divided (Figure 4). Only one of the two daughter cells will

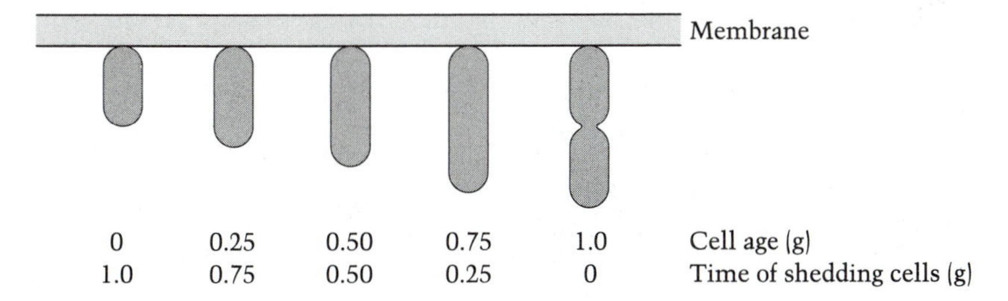

Figure 4

Kinetics of the detachment of cells from a "baby machine." First, the division products of old cells are released, then those from progressively younger cells. (After Helmstetter and Cooper, 1968.)

detach, whereas the other remains on the filter, apparently bound by electrostatic attraction. The resident cell continues to grow and, in time, to divide, eventually to release one of its daughter cells into the medium. This device, which has become known as the "baby machine," sorts cells by age: the first cells to detach come from cells that were oldest at the time the culture was placed on the membrane; the next ones to detach were formed by division of cells that were middle aged; and the last ones to be collected are from the youngest cells in the original culture (Figure 5). The baby machine has some practical limitations: it works better with some strains of *E. coli* than with others, and its capacity is limited (maximally 2×10^6 cells/ml of effluent). Nonetheless, the ingenious use of this device has resulted in some of the most important conclusions regarding biosynthetic activities throughout the cell cycle, especially DNA replication.

The "baby machine" has been used in two ways. One—the "forward" method—is to collect a sample over a short period of time and use it as an inoculum to obtain a synchronous culture. The other—"backward"—method is to use the device to sort out a population by age. It is used to study events that took place before the population was placed on the filter, a point that needs explaining. Imagine, for example, that a culture growing in a flask under balanced growth conditions is pulse-labeled with tritiated thymidine for one-tenth of a generation time. Now place this culture in the baby machine and collect samples of "babies" over one or two generations. As shown in Figure 4, the first cells to detach will be from those that were the oldest at the time of the pulse. The next ones to be collected will be from those that were middle aged. Only after one generation time, will the youngest cells of the original population divide and give off "babies." By measuring the radioactivity per cell in each time sample, one can reconstruct the events that took place during the cell cycle in the original population. This approach has certain major advantages over synchronization methods and, as we discuss below, has been used extensively to study macromolecular syntheses throughout the cell cycle of *E. coli.*

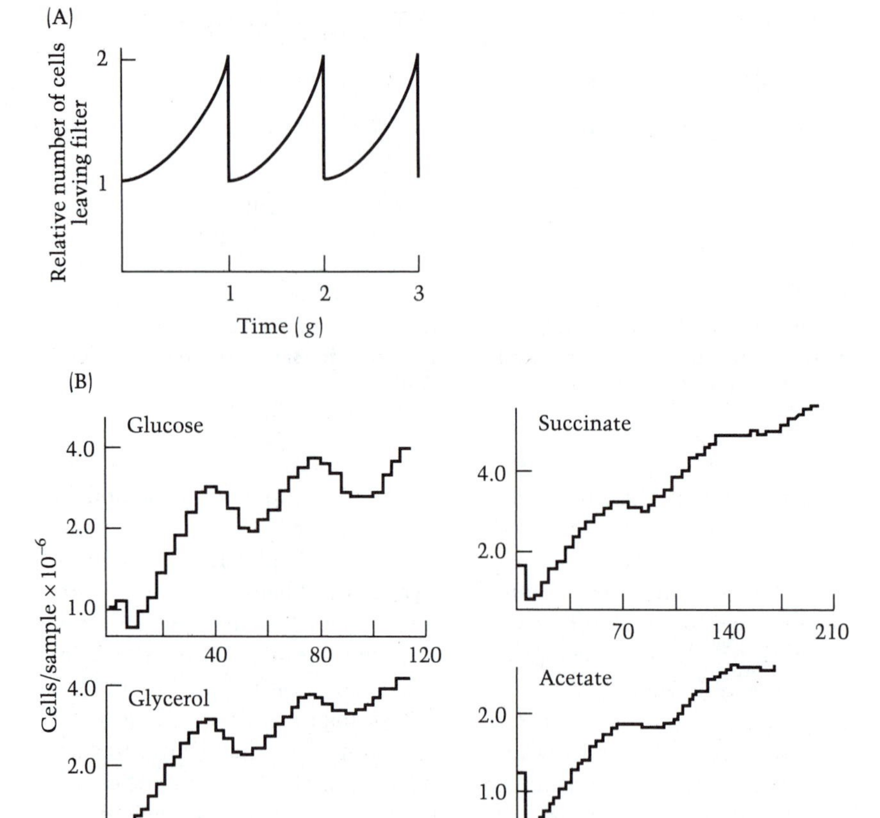

Figure 5

Kinetics of the detachment of cells from a "baby machine" originally seeded with an exponentially growing culture. (A) Idealized relationship. The number of generations, *g*, is shown on the abscissa. (B) Actual data from *E. coli* cultures growing in minimal medium with glucose, glycerol, succinate, or acetate as carbon source. (After Helmstetter and Cooper, 1968.)

DNA REPLICATION DURING THE CELL CYCLE

What has been learned from the studies of the bacterial cell cycle? The emphasis has been on the two unitary events—DNA replication and cell division—both of which we describe in some detail. Some information on the synthesis of surface components is also available. The main studies on DNA replication during the cell cycle were carried out by Helmstetter and Cooper, who used the baby machine to investigate cultures growing at different growth rates. Their experimental findings are presented in detail in Chapter 15, and an example is shown here in Figure 5. Helmstetter and Cooper concluded that DNA replication occu-

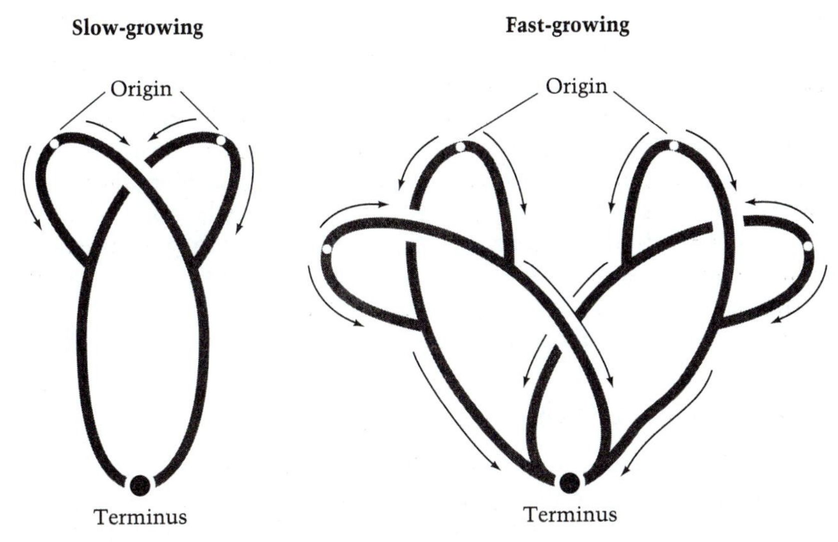

Figure 6

Replication of DNA in slow- and fast-growing *E. coli*. Replication begins at a specific site—the origin—and proceeds in both directions toward the terminus. The process takes about 40 minutes at 37°C. In a culture doubling every 40 minutes or more (left), each chromosome is undergoing a single round of bidirectional replication. In a culture doubling every 20 minutes (right), chromosome replication must be initiated every 20 minutes, that is, before the previous round of replication has terminated. In such a culture, the DNA is undergoing multifork replication. (From Schaechter et al., 1989.)

pies the bulk of the division cycle in cultures growing with doubling times faster than 40 minutes. Only in slower-growing cells did they find a gap in DNA synthesis. They further concluded that, at all doubling times faster than 60 minutes, DNA replication occurs at a steady rate and that it takes 40 minutes at 37°C for a full chromosome to be replicated. In other words, the machinery involved in chromosome replication takes 40 minutes to move from the origin to the terminus. This period—known as the C PERIOD—becomes longer in cells growing at slower rates, but not proportionately longer.

You already know that bacteria such as *E. coli* can grow with doubling times faster than 40 minutes. The question arises, How can a cell grow faster than the time it takes to make its complement of DNA? The answer to this paradox is that in cultures growing with doubling times shorter than 40 minutes, more than one round of replication is going on simultaneously. In other words, the chromosome is being traversed by more than one pair of "replication forks" at any given time (Figure 6). A consequence of this MULTIFORK REPLICATION pattern is that at division each daughter cell receives, not just one completed chromosome, but also the portion of the chromosome that has already replicated. How much of this "extra" DNA is present in newly divided cells depends on the growth rate. In fast-growing cells, whose chromosome is multiforked, the amount of DNA per cell will depend on the distance between their replicating forks and may amount to two or three times the basal level. An important point to note is that the extra DNA always contains the portion of the chromosome that replicates first—that nearer the origin of replication. As a consequence, the ratio of the genes near the origin to those near the terminus will be 1.0 in cells growing at 60-minute doubling times or slower, and will be 2.0 or more for cells growing faster. The resulting frequency distribution has been used to map genes on the chromosome (e.g., in *Staphylococcus aureus*), because, to the extent that it has been examined, the pattern of DNA replication for *E. coli* seems to hold for other bacteria as well.

How is DNA replication regulated?

Clearly, bacteria that grow at high rates synthesize a greater amount of DNA per unit time than do bacteria that grow slowly (Chapter 15). How is this variable output of DNA regulated? For cultures growing slowly (i.e., at doubling times greater than 60 minutes), the rate of elongation of individual DNA molecules (the chain growth rate) varies with the growth rate. The likely reason is that as cultures grow more and more slowly, the substrates for DNA replication become less plentiful. In cultures doubling faster than once every hour, substrates are abundant; and, as demonstrated by the constancy of the C period, the rate of DNA chain elongation does not vary with the growth rate. Under these conditions, the amount of DNA made per cell cycle must then be regulated by the FREQUENCY OF INITIATION of rounds of replication. The faster the cells grow, the more often they must initiate a round of replication. Now the question becomes, How is the frequency of initiation regulated? We do not really know, but an examination of the available facts and ideas is illuminating.

Besides a priori reasons, there are also compelling experimental facts for believing that initiation of DNA replication is a regulated process. One such fact is that in multinucleated cells the replication of sister nucleoids initiates with an astounding degree of synchrony. This fact was demonstrated by flow cytometry measurements, which showed that a population of fast-growing *E. coli* consists of cells with mainly two, four, or eight nucleoids. Very few of the cells had odd numbers of nucleoids, as would be expected from nonsynchronous replication (Skarstad et al., 1986). How is this experiment performed? In order to measure only complete nucleoids and not those in the process of replication, initiation is stopped with rifampin, which inhibits new rounds of replication but does not hinder ongoing ones. Sufficient time for the replication forks to reach the terminus is allowed to pass. At the end of this period, the cells contain only completed nucleoids because no new starts can be made. The DNA is stained with a fluorescent dye, and the distribution of nucleoids per cell is determined in the flow cytometer (Figure 3).

Synchronous replication can also be observed on the population level by studying mutants with defects in DNA metabolism. Certain mutants are thermosensitive in the initiation process but not in the process of replication. At a nonpermissive temperature, the cells continue ongoing rounds of replication until all the replication forks have reached the terminus. At that time all cells contain only completed nucleoids. When such cells are allowed to restart initiation (by lowering the temperature), replication of their nucleoids initiates with amazing precision (Figure 7). Replication forks traverse the chromosomes with such a degree of synchrony that genetic markers along the chromosome are synthesized at practically the same time in all cells of the culture. The synchrony is preserved for at least one doubling time, so the terminus region of the chromosome can be labeled by adding radioactive precursors to the culture at the time when the replication forks are expected to approach the terminus region.

Another conclusion that can be derived from the rifampin experiment described above is that initiation of DNA replication, unlike replication itself, requires the synthesis of new protein. The existence of thermosensitive initiation mutants has allowed the characterization of such an initiation protein—DnaA. A lot of attention has been paid to DnaA, in part because it is the only protein that so far has been irrefutably shown to play a distinct regulatory role in the initiation process. The evidence is derived from both in vitro and in vivo data. An in vitro initiation system was developed by Fuller and Kornberg (1983), who also purified many of its components (Chapter 4). They showed that the addition of DnaA is essential for initiation to proceed in the test tube and that DnaA binds to the replicative origin of *E. coli* at characteristic sequences (Figure 8). In vivo studies revealed not only that initiation does not take place in *dnaA* mutants, but that there is a *quantitative relationship between the amount of DnaA protein and the capacity of the cell to carry out initiation.* Thus, when an excess of DnaA is supplied by a cloned plasmid, the number of initiations increases proportionally. (The extra initiations are aborted after a short time, however. So it appears that the cells have a mechanism for preventing the manufacture of too much DNA).

Having pointed the finger at DnaA as a regulator (albeit not necessarily the only one, just the one we know about), how are its synthesis and function regulated? Definitive evidence on these points is still lacking. We need to explain why DnaA (or any other "initiator protein") acts cyclically and why its synthesis must be renewed at every initiation. The protein DnaA does not appear to be intrinsically unstable, nor does its synthesis fluctuate throughout the cell cycle; thus it must be inactivated or made unavailable. One hint at such a mechanism lies in the observation that DnaA molecules tend to aggregate in solution and also during in vitro binding to the replicative origin. Should such aggregation take place in vivo (for which there is no evidence as yet), it would lead to sequestration of this protein. In addition, sequences to which DnaA binds—the DnaA boxes—are found in locations of the *E. coli* chromosome other than the replicative origin. The importance of titrating the

(A)

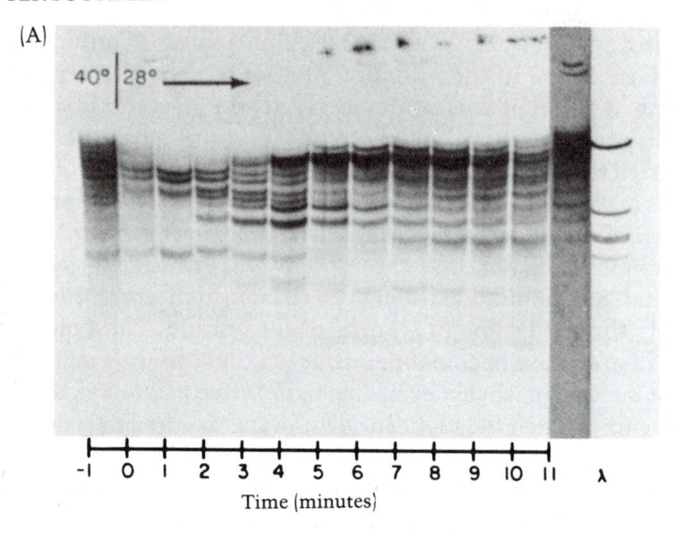

(B)

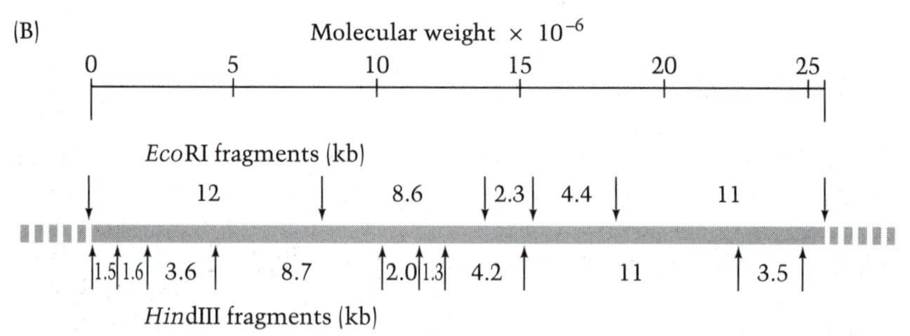

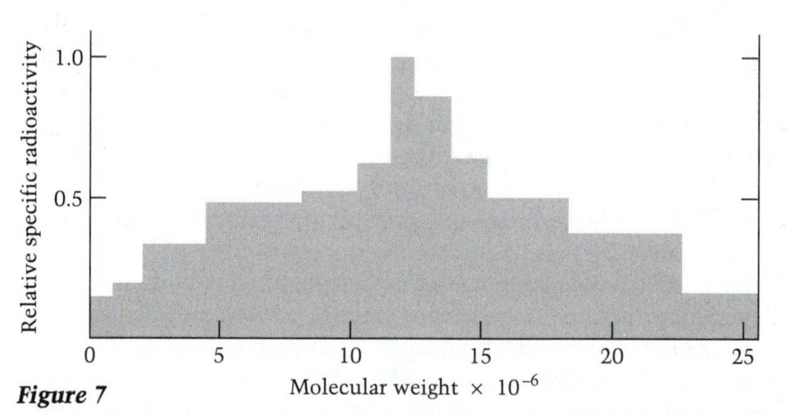

Figure 7

DNA replication starts from the origin. (A) *Escherichia coli* temperature sensitive in initiation were synchronized by being transferred to 40°C for one hour, then shifted to 28°C, the permissive temperature. At different times, the culture was pulse labeled with [^{3}H]thymidine for 1 minute. The DNA was then cut with a restriction enzyme, *Eco*RI, the resulting fragments electrophoresed through an agarose gel, and autoradiograms performed. Note that different fragments become labeled at different times. (B) The order of bidirectional replication was deduced from the quantitation of the radioactivity in the restriction fragment from a sample labeled for 1.5 minutes after the shift from 40°C to 28°C. (From Marsh and Worcel, 1977.)

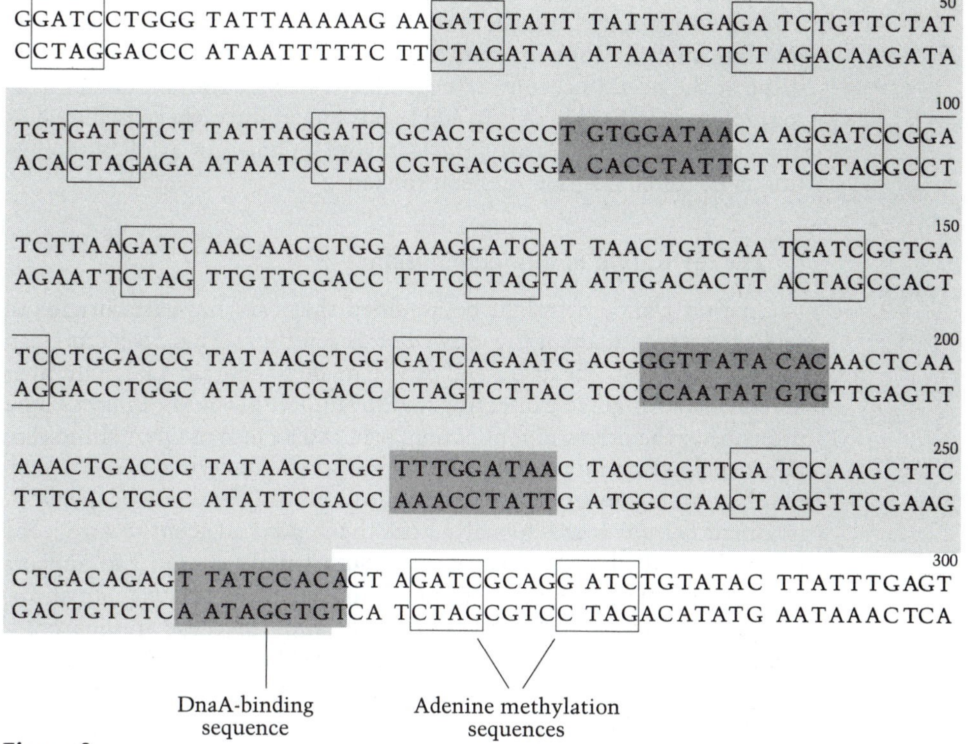

Figure 8

The sequence of *oriC* and some adjacent DNA. Four DnaA-binding sequences are boxed in dark gray and the adenine-methylation sequences (GATC) are in outlined boxes. The minimal origin is indicated in light gray.

number of available DnaA molecules is underscored by the finding that the synthesis of this protein is AUTOREGULATED; that is, it is under the direct control of itself. A molecule of DnaA binds to a DnaA box located between the promoters of the structural gene *dnaA* to inhibit it own synthesis. Thus, when DnaA molecules are used up or sequestered, their effective concentration decreases, thereby allowing their synthesis to proceed. In addition, it has been shown that the synthesis of DnaA is proportional to the growth rate (Chiaramello and Zyskind, 1989). Thus, DnaA is made on demand.

These considerations suggest that initiation may be triggered when DnaA (perhaps with other as yet unidentified factors) reaches a critical effective concentration. How is such a threshold reached during the cell cycle? Donachie (1968) combined the finding that the C time is constant (at doubling times shorter than 40 minutes) with the changes in DNA concentration with growth rate (Chapter 15). He found that *replication initiates at a defined ratio of mass to number of origins of replication.* This relationship is independent of the growth rate, a finding suggesting that initiation is triggered when the cell reaches a critical mass. The inference is that initiation depends on a regulatory substance whose critical concentration is achieved at a certain cell mass. By itself, this finding does not distinguish between two alternative models: the need to dilute an inhibitor of initiation by increasing cell mass past a certain value or the need to accumulate an effective concentration of an initiator protein (such as DnaA). To add to our uncertainty, these two models may both be correct. Clearly, the last word on how initiation of replication is triggered has not yet been spoken.

The replication of minichromosomes

At first glance, it might be assumed that carrying extra origins of replication would affect the growth of a bacterium. This conjecture can be put to the test, because a cell can be made to contain a large number of replicative origins by infecting it with suitable plasmids. In *E. coli,* the presence of the origin of replication, *oriC,* on a piece of DNA suffices to allow replication in vivo. In other words, *oriC* converts any DNA fragment, homologous or heterologous, into a plasmid replicon. If the DNA fragment is from *E. coli* (usually from the regions adjacent to *oriC*), the plasmid is called a MINICHROMOSOME. Surprisingly, *E. coli* can support the presence of as many as 40 minichromosomes per cell without showing any growth perturbation. Is the initiation of these minichromosomes regulated like that of the chromosome? Leonard and Helmstetter (1988) showed that this is indeed the case: in cells growing in the baby machine, minichromosomes initiated at the same time the chromosome did (Figure 9). This result suggests that the cell "measures" the number of target origins in the cell to adjust the concentration of regulators of initiation. If, as we currently suspect, DnaA turns out to be the key regulator, its ability to autoregulate its synthesis may in fact explain how its concentration in the cell is set. When the number of origins increases, more of this protein is sequestered and less becomes available to down-regulate its own synthesis. The result is that the correct ratio of DnaA molecules per origin is maintained, regardless of the actual number of origins. This notion may turn out to be wrong, because DnaA protein readily aggregates, a property that makes it a poor candidate for stoichiometric titration.

CELL DIVISION

General considerations

Most "ordinary" bacteria, such as *E. coli,* divide by binary fission into two equal daughter cells. Others, of more exotic varieties, have more complex division patterns. For example, *Caulobacter* divides into two cells that are unequal with respect to size, shape, flagella, and a stalklike structure (called the prostheca; Chapter 16). Some of these bacteria divide not by fission but by yeastlike budding from one end of the cell. The variation in the details of the division process and in the resulting shape of newly divided cells is quite great, a fact contradicting the superficial notion that all bacteria look pretty much alike.

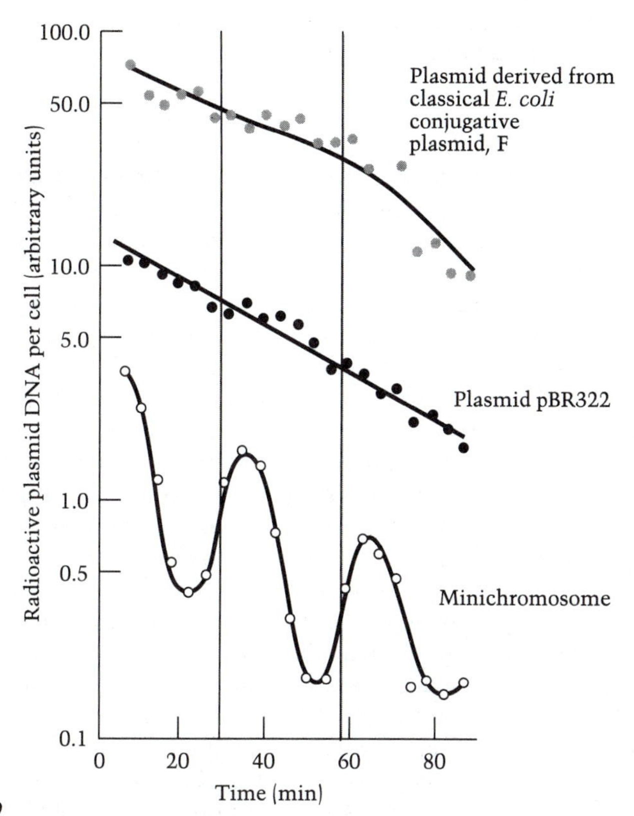

Figure 9

Minichromosome and plasmid replication in the cell cycle of *E. coli*. Cells containing different plasmids or minichromosomes were labeled with [^{3}H]thymidine and eluted from a baby machine. The vertical lines indicate the generations of bacteria eluted from the membrane. The steady decrease in radioactivity in the plasmids indicates that they replicated throughout the cell cycle. The stepwise decrease in radioactive minichromosomes indicates that they replicated at distinct times only. For a more detailed interpretation of the data, see text. (From Leonard and Helmstetter, 1988.)

Even among bacteria that divide by binary fission, there is considerable morphological variation. One extreme is represented by *E. coli* and its relatives, where daughter cells separate from one another upon division and do not make characteristic arrays of progeny cells. Streptococci and some members of the genus *Bacillus,* on the other hand, remain attached after division, thus making chains of individual cells. Other bacteria, such as the purple-sulfur phototroph *Thiopedia,* divide along one plane in one division and at 90° to that in the next division, thus making square sheets of cells. *Sarcina* and certain other bacteria alternate among the three perpendicular division planes, a pattern that leads to the formation of cubical packets of cells. Even more intricate patterns are seen in the procaryotes with complex developmental cycles, such as the myxobacteria, the cyanobacteria, and the actinomycetes (Chapter

16). Some of these higher bacteria resemble fungi and algae in the morphological complexity of their progeny cell arrangements.

Morphological considerations

Cell division requires the partition of the cytoplasmic contents by invagination of the cell membrane, the cell wall, and, in Gram-negative bacteria the outer membrane. Only then can daughter cells separate. Division, therefore, is a multistep process, although several of the steps may take place simultaneously. These steps differ in their details among various species. Generally speaking, most bacteria that have a thin cell wall, such as *E. coli* and other Gram-negative rods, divide by making a midcell constriction of their envelope layers (Figure 10 in Chapter 1). This ring-shaped "furrow" curves inward until two hemispherical caps are formed. At the leading edge of the constriction, there is an infolding of the cell membrane and murein layers. Thus, the daughter cells become separated by a septum before the constriction between them is completed, and the daughter cells remain stuck together for some time after their cytoplasm is physically separated. An analogous pattern is seen with Gram-positive cocci, such as *Enterococcus faecalis,* but in such species the septum is much more evident, in good part because the murein layer of these organisms is very thick (Figure 10). If the splitting of daughter cells is delayed, the chains of cells characteristic of streptococci are formed. Much of what we know about the morphological details of cell division in bacteria comes from electron microscopic studies of ultrathin sections of *E. faecalis* (Higgins et al., 1970). A detailed study of the physics of cell division, especially with regard to the mechanical stresses exerted on the cell wall, has been carried out by Koch (1987).

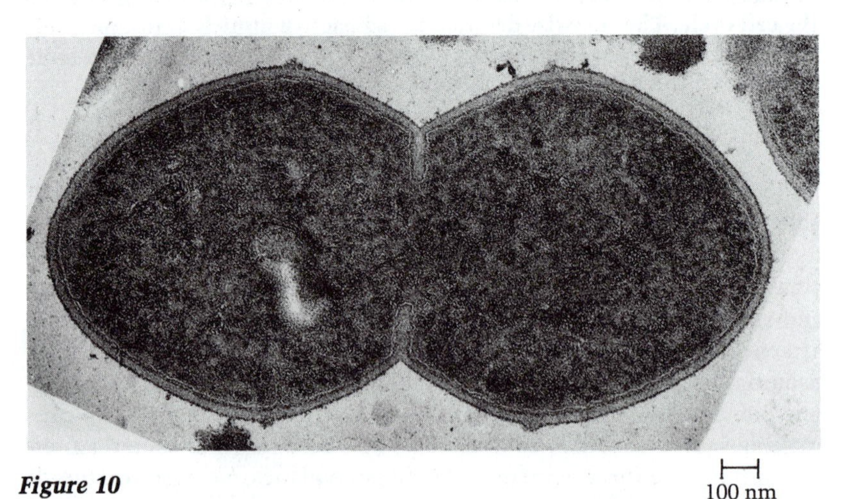

Figure 10

A dividing *Enterococcus faecalis*. (Courtesy of M. Higgins.)

In some Gram-positive cocci and rods, cell division proceeds without apparent constriction of their girth. A septum is formed by the invagination of the envelopes at a 90° angle with the surface, without greatly changing the diameter of the cell (Figure 11 in Chapter 1, showing *Bacillus subtilis*). When the septum is completed, the daughter cells snap apart. In the case of rods, the poles of such cells are sometimes more blunt and less hemispherical than those of rods that divide with constrictions.

Whatever the geometric arrangement, cell division requires that the cell surface component double in thickness at the septum site. These double layers (cell membrane, murein wall, outer membrane in Gram-negative cells) must later separate so that the daughter cells can split apart. Once one cell has laid down its cytoplasmic membrane septum and divided into two distinct cytoplasms, considerable time may pass before its progeny cells physically separate. The offspring may be held together by incompletely separated cell wall murein or outer membrane (in Gram-negative bacteria), or by extracellular capsular material. In *E. coli*, cells split apart shortly after they complete cell division. Certain mutants grow normally but remain attached after cytoplasmic division, to form chains of rods with septa composed of unsplit, double-thick layers of murein (Figure 11). It is likely that such mutants are defective in the action of a murein-splitting amidase required for the separation of wall septa. The outer membrane of these cells is continuous across the

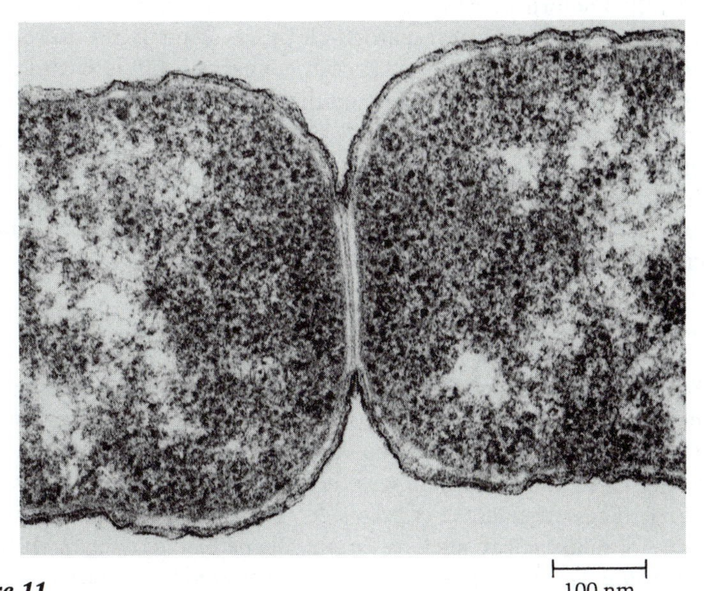

Figure 11

Section of a portion of a chain of rods of an EnvA mutant of *E. coli*. The cells do not readily separate, they have a septum between their poles. (Courtesy of J. T. Park.)

junctions and fully encases the chains of cells. In other bacteria—for example, *B. subtilis* and pneumococci—there is evidence that the action of other hydrolases is required for the last steps of cell separation. Depending on the planes of division, the shapes of the resulting aggregates may be pairs of cocci (*Neisseria gonorrhoeae*), chains of cocci (*Streptococcus pyogenes*), chains of rods (*Bacillus cereus* var. *mycoides*), grapelike bunches (*Staphylococcus aureus*), square sheets (*Thiopedia rosea*), or cubes (*Micrococcus luteus*).

Biochemical considerations

You may notice that so far we have not posed the question, "What is so special about septum formation?" The septal components are physically continuous with the rest of the cell envelope, yet they must be different in some regard. As is often the case, much of our knowledge is derived from the study of mutants.

In *E. coli,* at least 12 genes are known to be specifically involved in septation (Donachie and Robinson, 1987). A great deal is known about the organization of their components and the time in the cell cycle when their products act. However, the relevant biochemical activities are known for the product of only one of these genes; in other words, much remains to be learned. This gene product is a membrane protein called PBP 3. The abbreviation PBP stands for PENICILLIN-BINDING PROTEIN, a class of cytoplasmic membrane proteins that covalently bind β-lactam antibiotics and are involved in murein metabolism (Figure 12). How was the role of PBP 3 determined? A special β-lactam antibiotic—cephalexin—inhibits septation without impairing cell elongation or net murein synthesis. Of the six or so PBPs of *E. coli,* cephalexin binds with greatest affinity to PBP 3, a finding suggesting that this protein is involved in the synthesis of septal murein. Indeed, PBP 3 has been shown to possess enzymatic activities required both in the last steps of murein biosynthesis and in transpeptidation. What is not yet known is how these activities are used for the specific synthesis of septal murein rather than for making the murein of the cylindrical wall.

The site of division

For many rod-shaped bacteria such as *E. coli,* the site of division is the exact center of the cell. At the time of their birth, both daughter cells are the same size. Nothing is known about the mechanisms that measure lengths in such a precise way, although several models have been put forth. The same degree of precision is seen for selection of the division site of many cocci, such as *Enterococcus faecalis* (Figure 10).

Surprisingly, the site of division can be altered by mutation. Certain mutants of *E. coli* divide near the cell poles, apparently with the precision of normal division. The resulting spheres—known as MINICELLS, have the same diameter as the mother cell but contain no DNA (Figure 13). An ingenious proposal about how minicells are formed comes from

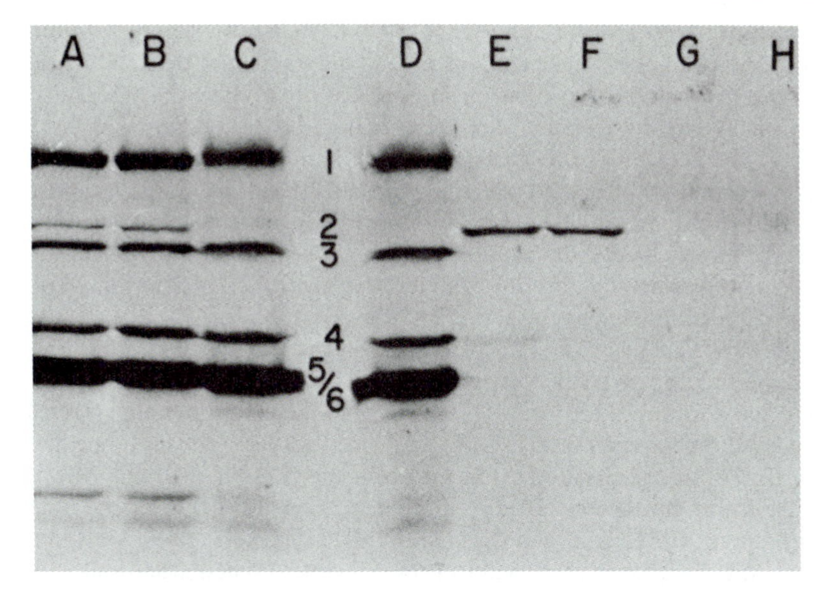

Figure 12

Binding of different penicillins to penicillin-binding proteins (PBPs) of the cytoplasmic membrane of *E. coli*. Radioactive penicillin binds covalently to a number of these proteins. The PBPs are numbered 1 to 6 in order of decreasing molecular weight. ^{14}C-labeled benzylpenicillin (A, B) and ^{14}C-labeled mecillinam (C, D) were added to cells from a wild-type strain (A, B, E, F) and a mecillinam-resistant mutant (C, D, G, H). Note that benzylpenicillin binds to PBPs 1 through 6 in the wild type (A, B) but not to PBP 2 in the mutant (C, D). Mecillinam binds to one protein (PBP 2) in the wild type and not at all in the mutant (G, H). (Modified from Spratt, 1975.)

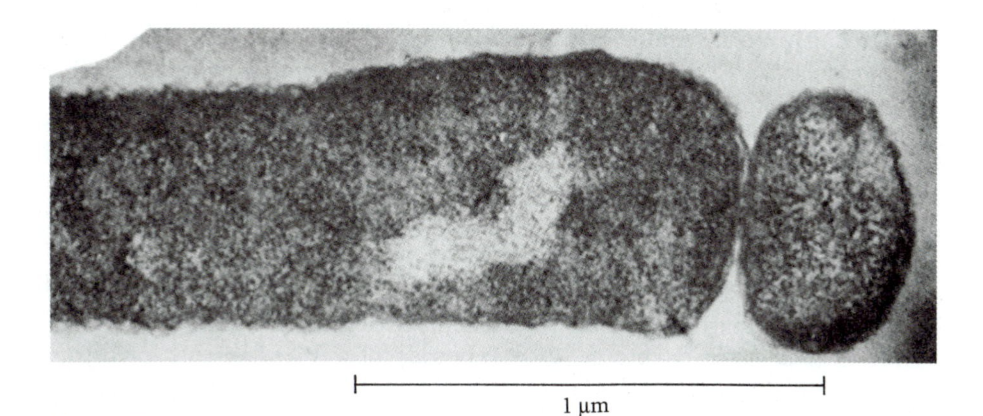

Figure 13

Thin section of *Escherichia coli* showing polar division and production of a minicell. (From Hirota et al., 1968.)

Donachie's laboratory (Teather et al., 1974). The notion recognizes that minicells are not made at random sites along the length of the cell but at one of two alternative division sites: one at the middle, the other near the ends. The subterminal site corresponds to the location of what, before the cell divided, was the periseptal annulus, which is a ring of adhering material between the inner and the outer membrane (Chapter 2). Before the cell divided, the annulus was located near the septum (Figure 11 in Chapter 2). Teather et al. proposed that the subterminal division sites are potentially available for septum formation but are normally inactivated by the action of a specific inhibitor. Minicell-forming mutants are postulated to be deficient in the action of this inhibitor, thus permitting division at the subterminal site. Still unexplained is the fact that division usually takes place either at the cell center or at one of the subterminal sites, not at both. The probabilities either of a normal division or of one that results in minicells are thus about equal. Minicell-forming mutant cells thus skip every other normal division and, accordingly, are longer than those of the wild-type cells.

Minicells have proved useful in several experimental situations

- Certain plasmids (e.g., R plasmids) segregate into minicells, thus allowing one to study the expression of their genes separate from those in the chromosome.
- Half of each minicell is made up of the cell poles, thus allowing fractionation of cell envelope components of this portion of the cell.
- Minicells can be infected with certain phages, thus allowing expression of genes cloned into these phages.

The timing of division

We have already noted that not all the individual bacteria in a population divide after the same time period. Actually, the extent of variation differs even among strains of the same species. When measured by the time of physical separation, some strains of *E. coli* (e.g., B/r) divide with fairly precise timing, whereas others (e.g., B or K12) show greater individual variability (typically, the standard deviation of the interdivision time of these strains is 20% of the mean). This observation is also true for other species. Thus, some bacteria are "faithful partitioners," others are "sloppy splitters." Even among the former strains, the variation in the interdivision time among individual cells is of the order of 10–20% of the mean. This variability may, however, mask underlying processes whose timing may be intrinsically more accurate. As discussed above, the physical separation of progeny cells depends on the completion of the last stage of mechanical detachment. The more meaningful physiological partition into two cytoplasms is probably less variable and better regulated. In multinucleated cells, the initiation of DNA replication is essentially synchronous (Figure 3).

In *E. coli,* division constrictions become visible at about 10 minutes before division, an interval independent of the growth rate. However, the machinery for cell division is set in motion long before that. Thus, a growing cell contains not one potential septal site but at least three: one at the middle and two others located at one-fourth and three-fourths of the cell length, respectively. In *B. subtilis,* the time from inception of a septum to its use is much longer, 138 minutes at 30°C (Figure 14). Interestingly, this time is independent of growth rate. In fast growing cells, septum formation initiates repeatedly before the original cells separate, thus producing cells with three, seven, or more septa. This pattern is reminiscent of the constancy of the C time for DNA replication and the phenomenon of multifork replication in fast-growing cells. Perhaps the reason is the same—both processes require longer than a fast cell's doubling time. Thus, in rapidly growing cells, initiation of both septation and DNA replication must begin in the grandparent. Similar results have been obtained with a relative of *E. coli—Proteus vulgaris* (Gmeiner et al., 1985). One consequence of the existence of more than a single septum in a cell is that the site of cell division must be positioned, not in the middle of the cell, but at sites one-fourth and three-fourths of the distanceee from either end. At fast growth rates, additional intermediate sites will be involved as well.

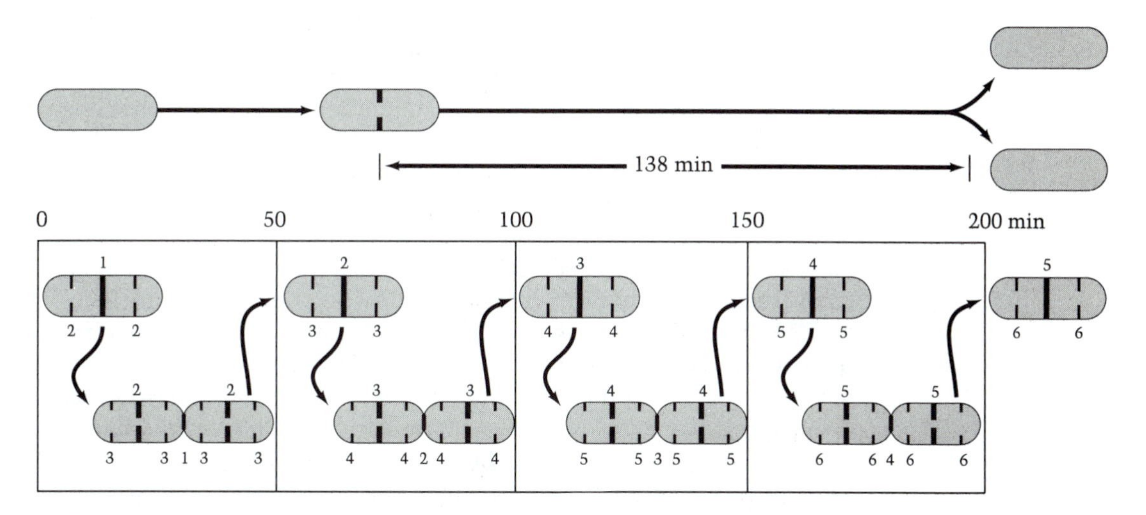

Figure 14

The time from the inception of a septum to its use in cell division is constant and independent of the growth rate. Paulton (1970) measured the number of visible complete and incomplete septa in *B. subtilis* growing at various rates in different media. He visualized the septa by means of a special staining method. The schema he deduced from these measurements is depicted in this figure: Shortly before division, cells growing at a doubling time of 50 minutes at 30°C (lower panel) contain a central septum, two auxiliary nearly complete septa, and four newly formed septa (numbered, respectively, 1, 2, 3 in the dividing cell at lower left). Each newly formed septum is used in a cell division that takes place 138 minutes after its inception. A newly divided cell growing at a doubling time of 200 minutes at 30°C (upper panel) contains no septa at division and makes one after 62 minutes (200 − 138).

The time in the cell cycle when a bacterium becomes committed to division is difficult to define, but it is known that protein synthesis is required over a considerable period. It has been estimated that for *E. coli* growing at 37°C, this period is 40 minutes. The period ends about 20 minutes before division; thus, it has its origin 60 minutes before the cells divide. During the last 20 minutes, cells can divide even if treated with protein synthesis-inhibiting antibiotics such as chloramphenicol. The period of division-required protein synthesis coincides with the C period of DNA synthesis. This finding has led to considerable speculation about the relationship between DNA metabolism and cell division.

Cell division and DNA replication

In principle, a cell ought not to divide unless the chromosome has replicated and each daughter cell can be endowed with its own genome. There are indications that in normal growth such a coupling of cell division and DNA replication indeed exists. This connection can be severed by certain conditional mutations that permit cell division to continue in the absence of DNA replication, a relaxation of normal controls that leads to the formation of anucleated, nonviable cells. Much of what we know about this subject comes from studies of the cell cycle. Experiments carried out with the baby machine showed that cell division takes place at a fixed time after termination of DNA replication, independent of the growth rate (Chapter 15). This interval is called the D PERIOD. For *E. coli* growing at 37°C, the D period is about 22 minutes. As mentioned above, chloramphenicol will inhibit cell division if added before the D period starts. If added during the D period, cell division goes on. This result suggests several possibilities; among them are the proposals that the synthesis of division proteins is coupled to the initiation of DNA replication or, coincidentally, that these proteins reach an effective concentration at the end of the C period. Such division proteins have not been clearly identified and studied. It has also been proposed that cell division is coupled in some way to termination of DNA replication.

One cannot tamper with DNA without stopping cell division. This consequence is seen when DNA replication is impaired, and especially when DNA is damaged. Septum formation is inhibited by a large variety of chemical and physical insults to DNA, such as ultraviolet or ionizing radiation or cross-linking agents such as mitomycin C. Within limits, cells so treated continue to grow, forming long filaments. If the insult is not too massive, these filaments will eventually divide and normal growth will resume. In general, the DNA repair begins quite early, and DNA replication resumes long before septation. In this case, filaments possess nucleoids evenly spaced at normal intervals. In time, when DNA repair has been effected, filaments divide into cells of the proper length.

There is considerable information about the mechanisms that connect the damage to DNA to the inhibition of cell division. The main one of these mechanisms—appropriately known in *E. coli* as the SOS RESPONSE, is one of the global reactions that bacteria exhibit when subjected to chemical and physical insults. The SOS response, which is the main reaction to damage to DNA, is presented in Chapter 13; we discuss it here only with regard to its role in blocking cell division.

Understanding the SOS response requires understanding the action of several gene products. (Because these proteins are known by the name of their gene—a three-letter name—studying the SOS response may be viewed as immersing oneself in an alphabet soup; see Chapter 13.) It will be recalled that when DNA is damaged, RecA becomes activated to cleave the repressor, LexA. Several operons are under the control of this repressor, among them one that contains a gene that encodes an INHIBITOR OF CELL DIVISION—SulA. It is known that SulA protein works by binding to and inactivating a DIVISION ACTIVATOR—FtsZ—which probably operates under normal circumstances. Thus, damage to DNA leads to *inactivation of the repressor* (LexA) *of the division inhibitor* (SulA); *the inhibitor is then synthesized, and division is blocked.* Note that this chain of events is set in motion by nonspecific damage to DNA alone. When DNA repair is completed, the stimulus for the SOS response is gone and cells resume division. This resumption is accelerated by the action of a specific protease—called "Lon"—on the inhibitor of cell division, SulA.

The portion of the SOS mechanism that deals with division inhibition only comes into play when DNA is damaged. Under normal conditions, the SOS mechanism does not appear to have a role in the regulation of division, which proceeds normally in mutants in which the gene encoding SulA has been deleted. Thus, the SOS response has evolved as an anticipatory mechanism, one that is called upon when DNA is damaged. In addition to SOS, there are other, less well studied mechanisms that couple damage to DNA with cell division. Why does *E. coli* go to such considerable trouble to couple damage to its DNA to inhibition of cell division? Several notions seem plausible. Inhibiting cell division ensures that cells have sufficient time to repair their damaged DNA before dividing. In addition, by making filaments, scarce cytoplasmic factors that may be required for recovery from damage to DNA can be shared by several cell-equivalents.

Cell division and plasmid replication

Jealous guard over cell division is not limited to the integrity of the cell's own DNA. Plasmids also contain mechanisms that control the division of their parent cell. For example, in *E. coli,* the fertility plasmid, F, has ways to prevent the production of plasmid-free cells.

How does F ensure its persistence in the population? The mechanism involved has been dissected genetically by Ogura and Hiraga (1983) and shown to be quite intricate. Fertility plasmids carry two genes involved in cell division control: *ccdA* and *ccdB* (*ccd* stands for "**c**ontrol of **c**ell **d**ivision"). The product of *ccdB* inhibits cell division, that of *ccdA* suppresses this inhibition. When the number of plasmids is greater than one, *ccdA* suppresses the inhibitory action of *ccdB*. When the number of plasmids decreases, this suppression is no longer effective (possibly because the *ccdB* gene product is longer lived) and cell division is inhibited. An alternative view is that these gene products play a more drastic role and that they lead to the killing, not just the inhibiting of plasmid-free cells (Jaffé et al., 1985). If so, plasmids have evolved a rather demanding relationship with their host: "Leave me behind and die!".

THE BACTERIAL EQUIVALENT OF MITOSIS

We know a great deal about mitosis in higher cells, the elaborate process that ensures segregation of the two replicated genomes into daughter cells. We know very little about the equivalent process in bacteria. For eucaryotes, we know the region of the chromosome that is directly involved in segregation (the centromere), the mechanical apparatus of chromosome separation (the mitotic spindle), and the mechanism that separates chromosomes (microtubular gliding). In addition, we know a lot about the timetable of mitosis and its placement within the cell cycle. For a few organisms—notably yeast—the genetic analysis of mitosis is in full bloom.

For procaryotes, we still have little insight into equivalent aspects of chromosome segregation. About all we know with certainty is that the bacterial chromosome segregates with fidelity. Anucleated bacteria, which would be expected to arise if segregation were unregulated, are formed very rarely after cell division. As yet, we have not taken full advantage of the ease of genetic manipulation in procaryotes to define the structural and regulatory elements involved. A likely reason for this uncharacteristic lack of progress is that procaryotes, perhaps because of their small size and fewer parts, do not possess a totally separate apparatus for chromosome segregation. Rather, such machinery makes use of cellular and molecular elements that are shared with other functions. The best example is the cell envelope and its constituent membrane and wall, which are widely believed to participate in the procaryotic equivalent of mitosis. Clearly, the mitotic function of the bacterial envelope is difficult to separate from the many other physiological activities of this busy structure.

A notion of long standing—known as the replicon model—is that the bacterial chromosome segregates by attaching to the cell membrane. The original replicon model suggested that the membrane grows outward from the midline of the cell, thereby leading to separation of the daughter chromosomes. Thus, attachment of the chromosome to the membrane would serve two functions that are equivalent to those provided by the eucaryotic kinetochore (attachment) and the spindle fibers (partition). Although the replicon model is popular and reasonable, definitive evidence for some of its aspects is still lacking (see below).

Do bacterial chromosomes have a centromere?

Physical evidence suggests that the origin of replication of *E. coli* and *B. subtilis* has at least one of the functions expected of a bacterial centromere, namely, it binds to the cell membrane. The specificity of this binding is quite remarkable: under defined conditions, only one restriction fragment from the *E. coli* chromosome binds to membranes, the one that contains *oriC,* the replicative origin (Hendrickson et al. 1982). However, neither this physical evidence nor genetic studies are sufficient to evaluate the notion that *oriC* functions as a centromere equivalent.

Cell fractionation of bacteria of several species, mainly *E. coli* and *B. subtilis,* has led to the isolation of membrane–DNA complexes. Despite this suggestive finding, the subject of DNA–membrane interactions has proved difficult to study, because the DNA–membrane complexes obtained are not specific. In addition to attachment at the origin, the bacterial chromosome also is bound to the membrane at random sites along the DNA. It has been estimated that the number of these attachment sites is between 20 and 80 per nucleoid. The function of these nonspecific attachments is not known, but nothing precludes their participation, directly or indirectly, in chromosome segregation.

An alternative to cell fractionation has been the study of DNA–membrane complexes formed by mixing the two constituents in vitro. Using this approach with *E. coli,* investigators have shown that *oriC* DNA binds with great specificity to membrane preparations. The sequences involved are found within two adjacent fragments, 154 bp and 109 bp. The fragment containing *oriC* DNA binds to the membrane only when it is hemimethylated (Ogden et al., 1988). This region of DNA contains an unusually high number of GATC sites—a GATC site is the recognition sequence for the *E. coli* DNA adenine methylase (Figure 8). At 30°C, *oriC* DNA remains in the hemimethylated condition for 8–10 minutes after the start of replication and becomes fully methylated after this time. The *oriC* DNA fragment binds to membranes in vitro only when isolated from cells within 8–10 minutes after initiation. The implication of these findings is that the chromosome attaches to the envelope at the origin only for a short period after initiation.

The viability of *dam* mutants, which do not methylate adenines on the DNA, would seem to contradict the notion that the state of DNA methylation is relevant to chromosome segregation. However, it is not known whether these mutants carry out segregation in a precise manner and whether they replicate exclusively via *oriC.* At present, the role of methylation in segregation remains an open question. Indeed, the state of methylation may be more relevant to initiation of replication, a notion suggested by a finding that fully methylated but not hemimethylated plasmid DNA is capable of in vivo replication (Russell and Zinder, 1987).

At present, we know little about the number, identity, and properties of the origin DNA-binding "membrane receptors." Several *oriC* plasmids normally coexist in a cell—10–20 copies per chromosome. In this case, *oriC* plasmids may act like other medium-to-high copy number plasmids, segregating at random and not requiring a partition system. Alternatively, if these plasmids use the membrane for segregation, could the number of *oriC*-specific membrane binding sites be much greater than three under some circumstances? Are all these sites preexisting, or are they regulated by the number of *oriC* copies—in other words, are they made on demand? We have no answer to these questions yet.

Is the partition device the membrane or the wall?

The original replicon model proposed that chromosomes are partitioned by the outward growth of the membrane from the midline of the cell. However, several lines of experimental evidence favor the idea that the membrane is synthesized at many sites (Chapter 4), an arrangement that would lead to a dispersal of the "old" membrane material over the surface. If this model is correct, the membrane may not be the driving element in chromosome segregation. The membrane may therefore be too fluid to allow segregation without the help of a rigid structure such as the cell wall. Such a rigid structure may be required for carrying out the mechanics of segregation much in the way that spindle fibers function in eucaryotic mitosis. For example, attachment to the murein at the poles could serve to segregate daughter chromosomes.

However, the participation of the cell wall in segregation may have complicated features. At least in the Gram-negative bacteria, membranes and walls are not entirely separate components. In *E. coli*, there is a spatially defined structure—the periseptal annulus, which is a ring-shaped zone of attachment of these two membranes—that flanks the division septum (Chapter 2). There is no direct evidence for the attachment of DNA to these zones of adherence, but their existence allows one to speculate that the chromosome may bind to them.

Is the partition system maintained throughout the cell cycle?

In eucaryotic cells, the partition machinery dissociates after mitosis, although centromeres apparently remain attached to microfibrils throughout the cell cycle. Whereas the bacterial membrane obviously remains intact at all times, elements involved in chromosome segregation may not. Because chromosome segregation occupies the whole of the cell cycle in rapidly growing bacteria, we may ask whether the replicative origin is attached to the membrane over the entire cell cycle or only during a portion of the cycle.

Several investigators have studied the spatial pattern of chromosome segregation. Their basic experiment consists of the following steps. Chromosomes are radioactively labeled with [^{3}H]thymidine. Then the bacteria are grown on microscope slides in unlabeled medium under conditions where they make chains (e.g., by growing on methylcellulose-containing media). The slides are dried, and autoradiograms are made. At the extremes, two results are possible: either the daughter chromosomes segregate randomly with regard to their position in the cell, thus giving "hot spots" of exposed silver grains anywhere along the chains or they are confined to determined sites, such as the original cell poles. Analogous experiments have been carried out with a "baby machine." A random pattern of labeled DNA is consistent with temporary attachment to the membrane, a nonrandom pattern with several mechanisms discussed below.

The conclusion from these experiments, which have been carried out in several laboratories, is that the topological pattern of chromosome distribution falls between random and nonrandom. For *E. coli,* the faster the growth, the more random the distribution becomes. These results have been the basis for several models. Thus, Pierucci and Zuchowski (1973) suggested that one of the two original strands is permanently attached to a pole of the cell, whereas the other segregates randomly. A model based on the alternative view—that both strands behave in the same manner—proposed that nonrandom segregation depends on a measure of "inertia," that is, a probability that either strand moves toward the same pole to which it had previously segregated (review by Leonard and Helmstetter, 1989).

If the attachment of the origin to the envelope were temporary, this result would be explained by the fact that only hemimethylated *oriC* DNA binds to membranes. Adherence to the membrane would take place reversibly for a period of 8–10 minutes (at 30°C) after initiation. This interval might be the time required for the membrane to establish two domains, possibly by the incipient formation of the cell septum. What ensures that each chromosome remains within a given domain and does not cross into the other? *Escherichia coli* DNA binds nonspecifically to membrane at many sites other than *oriC,* and we could imagine that such nonspecific interactions serve to secure each daughter chromosome within its own cell half. Nonrandom segregation may result from attachment of DNA at the nonorigin sites.

Mutant studies

Genetic studies of "bacterial mitosis" are still at an early stage and do not yet allow one to identify specific gene products that act at distinct steps in chromosome segregation. Nonetheless, this approach is highly promising and may well lead to a better formulation of the problem.

A particularly interesting mutation leads to the formation of anucleated cells of apparently normal length. Hiraga et al. (1989) isolated a mutant that forms normal-sized anucleated cells with a frequency of 0.5 to 3% of all cells in a growing population. (This frequency is below that expected from asymmetrical nuclear divisions taking place at every cell cycle, but there are several reasons that explain why the experimental result may be an underestimate of the true frequency.) The gene involved in this phenotype was found to encode for a lesser known membrane protein, a finding suggesting that the product of this gene may well be involved in chromosome segregation.

With these promising beginnings, it seems likely that the genetic approach will help us understand the mechanism of chromosome segregation in bacteria. Further genetic studies, coupled with the judicious use of physiological experiments, may soon propel our understanding of this phenomenon into the depths heretofore reserved for other important aspects of bacterial physiology.

SUMMARY

1. The cell cycle of bacteria can be studied by direct microscopic observation of individual cells or by investigating either normally growing or synchronously dividing populations.
2. An age-selection device—the baby machine—allows one to dissect a bacterial population into cells derived from parents of different age.
3. In *E. coli* growing at fast and medium rates, a round of DNA replication takes 40 minutes at 37°C. Thus, DNA replication is regulated by the frequency of initiation. In fast-growing cells, there are several ongoing rounds of replication, a process described as "multifork replication."
4. The leading candidate for an *E. coli* protein that regulates the initiation of DNA replication is a protein called DnaA, but little is known about how it works.
5. Some bacteria divide by making inward, midline constrictions, others by invagination of their envelopes without constriction.
6. The process of bacterial division takes place in several steps, each leading to the separation of different envelope components. Bacteria that remain attached by their outer layers make characteristic morphological aggregates (e.g., chains, "bunches of grapes," squares, cubes).
7. Certain membrane proteins called PBPs (for "penicillin-binding proteins") play a role in cell wall elongation, others in septum formation.
8. The site of bacterial division can be altered by mutations that apparently impair the inactivation of previously used sites and lead to the formation of small anucleated "minicells."
9. Damage to DNA results in inhibition of cell division via the SOS response.
10. Chromosome segregation appears to involve the attachment of bacterial DNA to cell envelopes. The likely candidate for the partitioning device is the cell wall.

STUDY QUESTIONS

1. A culture of growing *E. coli* was pulse labeled by the addition of [^{3}H]thymidine for one-tenth of a doubling time. Autoradiograms were made and the numbers of grains per cell counted. The mean number of grains per cell was 2.37. Using the equation for the Poisson distribution, $P(f) = m^f e^{-m}/f!$, where f are the classes of cells with 0, 1, 2, . . . grains and m is the mean number of grains per cell, calculate the proportion of cells with 0, 1, 2, 3, 4, and 5 grains per cell. Plot the results. Keep in mind that 0! = 1.
2. In the experiment shown in Figure 3, the distribution of nucleoids per cell was measured after exposing the cells to the antibiotic rifampin for 3 hours. For a normally growing population, what would the distribution of nucleoids per cell be had this step been omitted? Could other antibiotics have been used for the same purpose?

3. Picture a baby machine fed by a minimal medium with glucose in slight excess. Have the outflow of this device run into a chamber whose liquid volume is maintained at a constant level by an overflow syphon. Discuss the differences in age distribution between the population on the filter of the baby machine and that in the liquid vessel. Discuss (preferably mathematically) the conditions under which equilibirium will be reached in the liquid vessel.
4. In cell cycle studies,why is it important to distinguish between the age distribution and the size distribution of a bacterial population? Give examples of experiments that select cells by size and by age and discuss how they differ from one another.
5. What is the basis for the following assertion: "DNA replication in bacteria is regulated largely at the level of initiation rather than elongation"?
6. Discuss the role of DnaA protein in the replication cycle of the *E. coli* chromosome. How is DnaA itself regulated?
7. Discuss the role of penicillin-binding proteins in cell elongation and cell division.
8. What has been learned about bacterial cell division by studying minicells? Propose an experiment that uses minicells to probe into aspects of plasmid replication.
9. How does damage to DNA prevent cell division? Discuss what is known about the genetic and biochemical basis for this phenomenon. What role could this have in evolution?
10. Discuss the evidence for the role of the cell envelope in bacterial chromosome segregation.

SUGGESTED READING

Topics in this chapter are covered in the following chapters from *Escherichia coli and Salmonella typhimurium: Cellular and Molecular Biology*, F. C. Neidhardt, J. L. Ingraham, K. B. Low, B. Magasanik, M. Schaechter and H. E. Umbarger (eds.). 1987. American Society for Microbiology, Washington, D.C.

Donachie, W. D. and A. C. Robinson. Cell division: parameter values and the process. (Volume 2; Chapter 99)

Koch, A. L. The variability and individuality of the bacterium. (Volume 2; Chapter 101)

von Meyenburg, K. and F. G. Hansen. Regulation of chromosome replication. (Volume 2, Chapter 98)

Other suggested readings include:

Koch, A. 1988. Biophysics of bacterial walls viewed as stress-bearing fabric. Microbiol. Rev. 52:337.

Leonard, A.C. and C. Helmstetter. 1989. Replication and segregation control of *Escherichia coli* chromosomes. In: *Chromosomes: Eukaryotic, Prokaryotic, and Viral*, Vol. 2. K. Adolph (ed.). CRC Press, Boca Raton, Florida.

15

Growth Rate as a Variable

INTRODUCTION

All organisms grow faster and do better when provided with a good diet. But the effects on bacteria are especially dramatic: under otherwise constant environmental conditions (pH, temperature, and osmotic strength), the growth rate of a culture of bacteria can vary over more than a 10-fold range, depending on the richness of its growth medium. (The effect of certain media on the growth rate of *Salmonella typhimurium* is shown in Table 1.) In other words, bacteria are particularly well adapted to exploit their nutritional environment and to convert it into their own special form of selective advantage—high growth rate. To accomplish this remarkable feat, the makeup of a bacterial cell changes profoundly with nutrition-imposed growth rate: both macromolecular composition and cell size change with growth rate. The teleologic reasons for some of these changes are obvious. For example, if a bacterium is to grow faster, it needs more protein-synthesizing machinery to accomplish the task. But unused machinery is always a disadvantageous expense. So, for a bacterial cell to grow at the maximum rate that a particular medium will support, it must contain a precisely set optimal amount of protein-synthesizing machinery—more, or less, would decrease growth rate. In contrast, the physiological advantage of some other growth rate-associated changes—for example, DNA content and cell size—are not so immediately apparent; but, as we shall see, they, too, are essential if the bacterial cell is to take full advantage of its nutritional environment. All these changes act coordinately to maximize growth rate of the bacterial cell in the particular environment available to it.

The dependence of bacterial growth rate on nutrition is an observation as old as the science of microbiology itself, but the intermediary role of variable macromolecular composition and cell size was not discovered until 1958. Since then, the growth rate–cell composition relationship—a consequence of the cell's most important global control mechanisms—has been a major theme of research in bacterial physiol-

Table 1. Effect of nutrition on the growth rate of *Salmonella typhimuium*

Medium	Comments	Growth rate (doublings/hour)
Brain-heart infusion	An extract of beef brain and heart	2.80
Nutrient broth	Meat extract + 1% peptone[a]	2.75
20 amino acids	20 natural amino acids + salts[b]	1.88
Amino acid mix #2	amino acids[c] + salts	1.46
Amino acid mix #1	amino acids[d] + salts	1.22
Glucose minimal	glucose[e] + salts	1.20
Succinate minimal	succinate[e] + salts	0.94
Aspartate minimal	aspartate[f] + salts	0.83
Lysine minimal	lysine[g] + salts	0.62

Source: From Schaechter et al. (1958).
[a]A commercially available partial hydrolysate of casein.
[b]$MgSO_4 \cdot 7H_2O$, 0.1g/L citric acid (added as a chelating agent to keep salts in solution), 1.0g/L; $Na_2HPO_4 \cdot 2H_2O$, 5.0 g/L $Na(NH_4)HPO_4 \cdot 4H_2O$, 1.74 g/L; KCl, 0.74 g/L.
[c]Threonine, tyrosine, cysteine, histidine, phenylalanine, isoleucine, hydroxyproline, and arginine.
[d]Glutamate, alanine, serine, valine, glutamine, lysine, methionine, and proline.
[e]0.2%.
[f]0.012%.
[g]0.014%.

ogy. Many important control mechanisms have been elucidated, while others, including some of the most fundamental ones, remain unexplained. Almost all of these studies have been done with *Salmonella typhimurium* and *Escherichia coli.* On the basis of very little experimental data, these principles are presumed to apply, at least in broad outline, to most other species of bacteria as well.

In this chapter we consider the quantitative relationships of cell composition and size with growth rate, the ways that bacteria adjust to sudden changes in growth rate, and the biochemical mechanisms that mediate these changes. It is important at the outset to remind ourselves of the concept introduced in Chapter 7, namely, that these changes are nutrition mediated but not nutrition specific. The growth rate that a particular medium supports, not its specific composition, determines the PHYSIOLOGICAL STATE (cell size and macromolecular composition) of cells growing in it. For example, the physiological state of cells is the same in different media or even in continuous culture if the cultures are all growing at the same rate. As was discussed in Chapter 8, temperature, pH, osmotic strength, and hydrostatic pressure also affect growth rate, but changes in these nonnutritional parameters do not affect overall macromolecular composition and cell size. For example, a change in incubation temperature that would change growth rate several fold will change the intracellular levels of specific proteins (Chapter 8), but it will have an almost imperceptible effect on the fraction of cell mass that is protein or RNA or DNA. The reasons for this fundamental difference between the effect of nutrition-imposed and temperature-imposed changes in growth rate are obvious. A change in temperature affects the rate of all cellular processes by a similar factor, so no modification of the cell's macromolecular composition is needed to adjust to the new environment. In contrast, a change in nutrition differentially affects the rates at which various cellular components must be synthesized. In this chapter we assume, unless otherwise stated, that the cultures we are considering are growing at 37°C in media with values of osmotic strength and pH that are not by themselves rate limiting.

EFFECT OF GROWTH RATE ON PHYSIOLOGICAL STATE

The fundamental relationship between nutrition-imposed growth rate and cell composition was discovered by M. Schaechter, O. Maaløe and N. O. Kjeldgaard in 1958. The essence of these results is shown in Figure 1, where normalized values of the amounts of various cell components per cell are plotted against growth rate. It can be seen that all of them (total cell mass, protein, RNA, and DNA) are exponential functions of growth rate: as the cell grows faster, it becomes bigger and contains more of each of its components. These are major changes: the average cell in a culture of *E. coli* B/r growing at 2.5 doublings per hour contains more than 10 times as much RNA as do cells from a culture growing at 0.6 doubling per hour. But the levels of various components are different exponential functions of growth rate. Ribonucleic acid increases much faster than mass, protein increases somewhat slower, and DNA much slower. So the relative composition of a cell changes

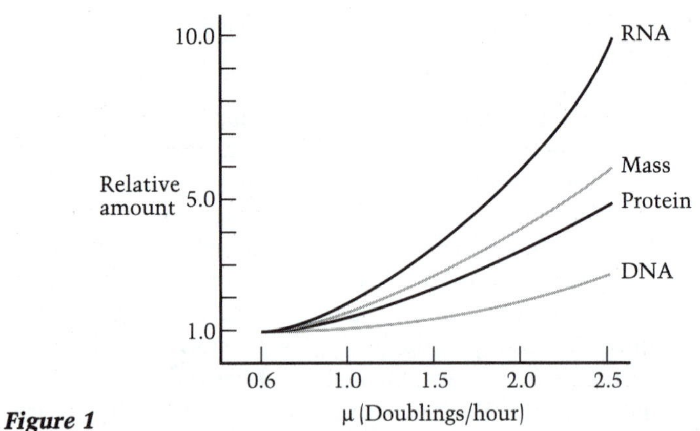

Figure 1

Effect of nutrition-imposed growth rate on the composition of *Escherichia coli* B/r. All values are expressed in amounts per cell normalized to values at $\mu = 0.6$ (mass = 1.48×10^{-13} g; protein = 1.00×10^{-13} g; RNA = 2.0×10^{-14} g; DNA = 6.3×10^{-15} g). (Plotted from data in Bremer and Dennis, 1987.)

predictably with growth rate: cells from rapidly growing cultures are markedly enriched in RNA (because they contain a higher proportion of ribosomes), but they contain a slightly lower proportion of protein and a considerably lower proportion of DNA (Figure 2). At growth rates above about 0.5 doubling per hour, the fraction of the cell that is RNA, protein or DNA is a linear function of growth rate.

The relationship between growth rate and cellular RNA content allows one to make some interesting calculations. Knowing the fraction (0.84–0.85) of stable RNA that is ribosomal and the total molecular weight (1.45×10^6) of the RNA in a ribosome, one can calculate the number of ribosomes a cell contains. Then, using the estimate that 85 to 90% of these ribosomes are actively engaged in synthesizing protein and knowing that a cell must synthesize its protein complement during a doubling time, one can calculate the rate of protein synthesis per ribosome in terms of CHAIN GROWTH RATE (amino acids polymerized

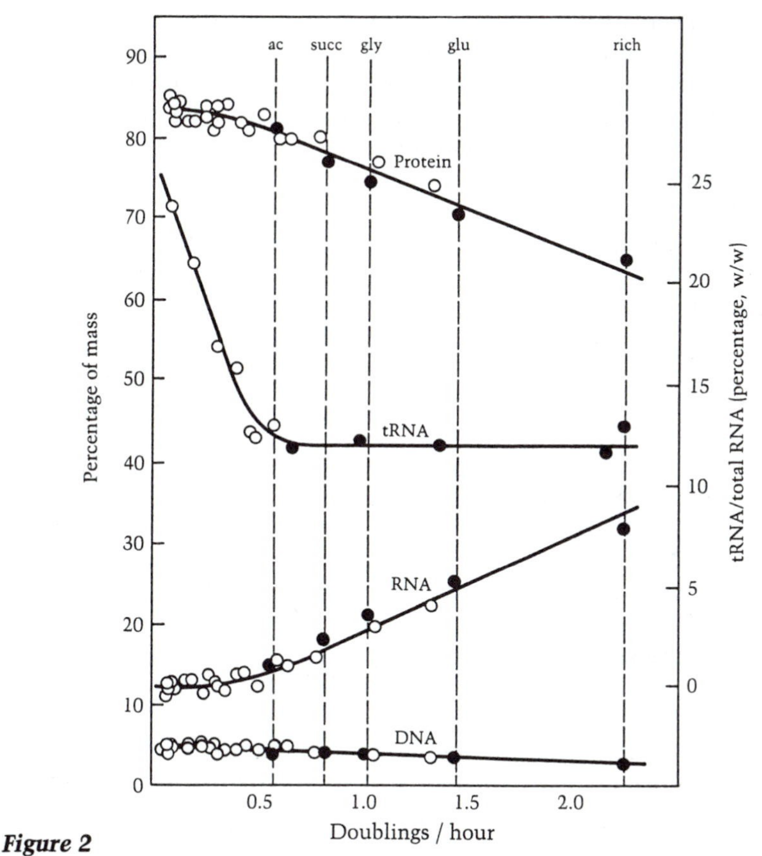

Figure 2

Effect of growth rate on the cellular proportions of protein, RNA, and DNA. Filled circles refer to results from cultures undergoing balanced growth in batch culture in various media; open circles are from cultures growing in a glucose-limited chemostat. (From Jacobsen, 1974.)

per ribosome per second). (The weighted average molecular weight of an amino acid in the protein of an enteric bacterium is 110). Remarkably, to a first approximation, one obtains about the same answer—approximately 16 amino acids are polymerized per ribosome per sec—even when the calculations are made on cultures growing at growth rates as different as 0.5 and 2.5 doublings per hour. This result implies that ribosomes function at a nearly constant rate and that the cell adjusts its ribosome content so that it has just about as many as it needs.

A simpler route to the same conclusion comes from extrapolating the linear portion of the curve of percentage of RNA versus growth rate (Figure 2) to the y-axis, a condition that would describe the hypothetical physiological state of a cell undergoing balanced growth at 0 rate. The value of the extrapolated y intercept is about 6%, not too different from the fraction of cell mass that is mRNA, a result implying that such a theoretical cell would have little or no rRNA or tRNA. It would not need any. Again, this result suggests that the cell adjusts its content of protein-synthesizing machinery so that it can grow at the fastest possible rate in a given medium.

Actually, careful measurements indicated that ribosome function is not absolutely constant but increases throughout the range of 0.5 to 2.5 doublings per hour. Because the increase does not keep pace with the growth rate increase, more ribosomes are needed to maintain protein synthesis.

Below growth rates of about 0.5 doubling per hour, the relationship between the cell's macromolecular composition and growth rate begins to deviate even more markedly from linearity, a finding implying either that the cell makes far more ribosomes than it needs or that chain growth rate is slower at these low growth rates. The latter seems to be the case. The time between the addition of an inducer of β-galactosidase synthesis to the growth medium and the appearance of the first molecule of the completed enzyme can be used to measure chain growth rate, because other processes (such as inducer entry and transcription initiation) are brief compared with the time required to make the β-galactosidase peptide. The time required for the appearance of the first molecule of the completed enzyme can be estimated by extrapolating the linear portion of the rising curve to the basal level of enzyme (Figure 3). Dividing this value by the number of amino acid residues (1,023) in β-galactosidase yields the protein chain growth rate. One sees that the chain growth rate at $\mu = 0.1$ is about one-third less than the chain growth rate at $\mu = 1.5$.

The fundamental observation that fast-growing cells are rich in ribosomes has been extended to many species beyond *E. coli* and *S. typhimurium.* Recently, methods have been developed to exploit the relationship between growth rate of a cell and its RNA content for ecological studies. This procedure makes use of the fact that the ribosomal RNA is highly conserved in evolution, so certain DNA sequences will hybridize to any eubacterial rRNA. By attaching a fluorescent compound to this DNA and hybridizing it to cells in a natural environment, the

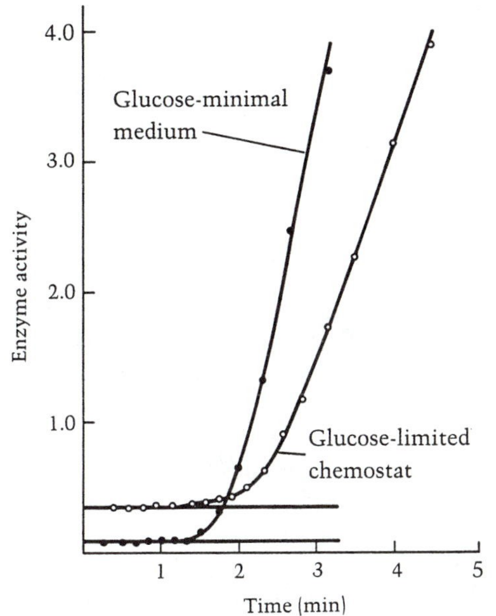

Figure 3

Effect of growth rate of the induction lag for synthesis of β-galactosidase. Isopropyl-thio-β-D-galactoside (IPTG) at $t = 0$ was added at a concentration sufficient to induce synthesis of β-galactosidase at a maximal rate in cultures growing in glucose-minimal medium (•) at $\mu = 1.5$ and in a glucose-limited chemostat (○) at $\mu = 0.1$. (From Jacobsen, 1974.)

amount of fluorescence, measured microscopically, becomes an index of how many ribosomes various individual cells contain and, therefore, of how fast they were growing when sampled.

Transitions between steady states of growth

The kinetics of change of the physiological state of a culture when the medium is abruptly altered can be quite instructive, because they provide a decisive test of any model for bacterial growth. When a culture growing in a steady state is suddenly enriched by adding nutrients to it, the composition of the culture undergoes an orderly and highly reproducible series of changes (Figure 4). In such an experiment—commonly called a SHIFT-UP—the rate of synthesis of new cell mass increases to its new higher value without a perceptible lag. Ribonucleic acid synthesis increases abruptly, almost explosively, for several minutes until the cellular content of this macromolecule is appropriately enriched; then synthesis continues at the new rate. Deoxyribonucleic acid and cell numbers increase at the preshift rate until the appropriate lower content of cellular DNA and larger cell size is established; then they begin to increase at the new rate. The immediate increase in the rate of RNA synthesis following a shift-up suggests that the cell has an unused capac-

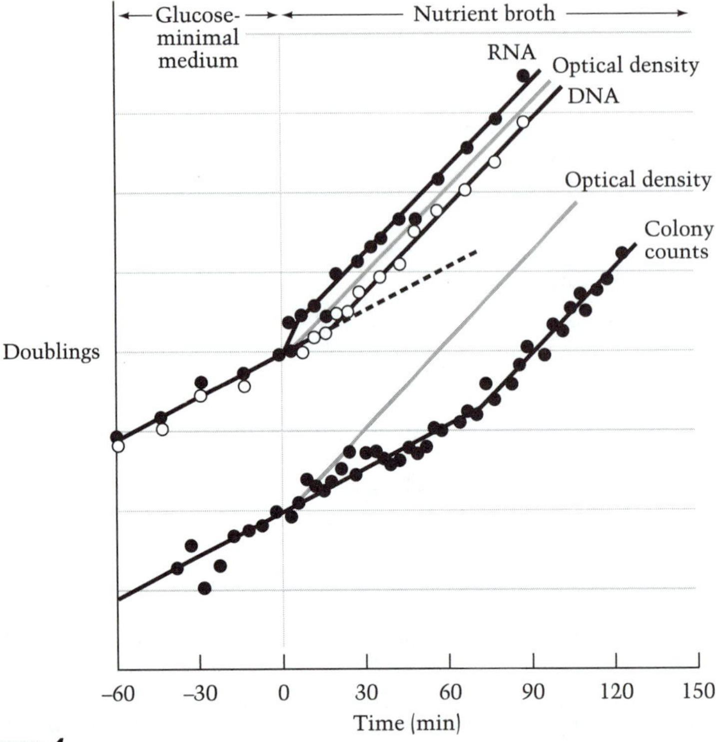

Figure 4

Shift-up of *Salmonella typhimurium*. A culture of *S. typhimurium* growing at 37°C in glucose-minimal medium was shifted to nutrient broth at time 0. The scale of the measured parameters—RNA, DNA, number of cells, and optical density (O.D., a measure of dry weight)—were all adjusted so that they could be described by a single line prior to shift. The distance between horizontal lines corresponds to one doubling. The curve describing colony counts along with a copy of the one describing O.D. are displaced to facilitate interpretation. (After Maaløe and Kjeldgaard, 1966.)

ity to synthesize RNA prior to the shift-up and presumably under most conditions of growth. We shall return to this point later in this chapter.

A shift-down, which can be accomplished in a couple of minutes by filtering cells from a richer medium and resuspending them in a poorer one, brings on a complementary series of changes (Figure 5). After the shift, the number of cells and DNA continue to increase at the preshift rate, while increase in cellular mass and RNA almost ceases. Then, when the macromolecular composition and size of the cells are adjusted appropriately for maximum growth rate in the new medium, synthesis of each component proceeds at the postshift rate.

GROWTH RATE CONTROL

Our knowledge of the mechanisms by which cell constituents are modulated to maximize growth rate in different nutritional environ-

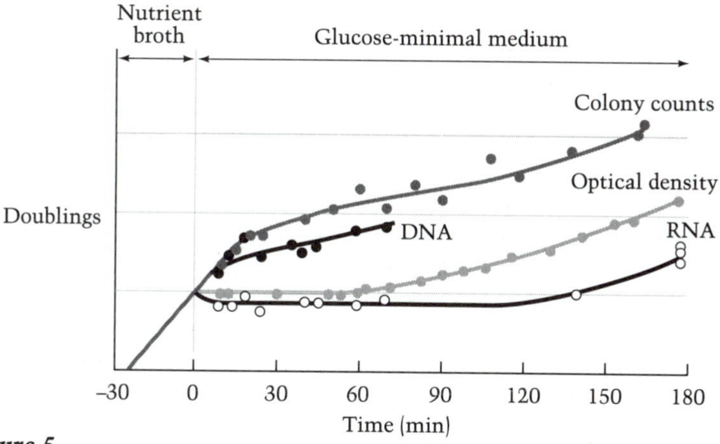

Figure 5

Shift-down of *Salmonella typhimurium*. A culture of *S. typhimurium* growing at 37°C in nutrient broth was shifted to glucose-minimal medium at time 0. The measured parameters are plotted as described in Figure 4. (After Maaløe and Kjeldgaard, 1966.)

ments is in a state of rapid flux, greater than that characteristic of most topics in biology. Some of these regulatory mechanisms—for example those that modulate synthesis of ribosomal proteins—are understood in considerable detail, but others—for example, those modulating synthesis of stable RNA—are poorly understood, and a variety of conflicting models have been proposed. In the following sections we review the present status of growth rate control, as this topic is sometimes called, considering observations that need explanation, hypotheses that have been advanced as well as known regulatory mechanisms.

Modulation of DNA

As we have seen, the amount of DNA a cell contains increases dramatically with growth rate. For example, the average cell in a culture of *E. coli* B/r growing in a medium that supports growth at a rate of 0.6 doublings per hour contains 1.6 genome's worth of DNA, but the same strain in a medium that supports growth at 2.5 doublings per hour contains 3.7 genome's worth. The physiological necessity for this aspect of growth rate control has been known for more than 20 years, but we are just now beginning to understand the control circuits through which it occurs.

The necessity for growth rate control of DNA synthesis stems from the remarkable fact already noted (Chapter 14): *bacteria can grow with doubling times shorter than the time required to replicate a chromosome by a single pair of replication forks.* At 37°C, *E. coli* B/r requires about 40 minutes to replicate its chromosome, but it is able to grow with doubling times of less than 20 minutes. The intuitive explanation of this dilemma is simple and correct:

- In a culture growing with a doubling time shorter than the C PERIOD (the time required to replicate the chromosome), each daughter cell receives a partially replicated chromosome (progressively more replicated as growth rate increases) and, therefore, more than a genome's worth of DNA. At high growth rates, the partially replicated chromosome has more than two replication forks, so the net rate of DNA synthesis is correspondingly increased. For example, in a culture growing at $g = 25$ minutes, each daughter cell receives a chromosome with six replication forks.
- In a culture growing with a doubling time greater than the C period, DNA replication is discontinuous, with progressively longer gaps between periods of DNA synthesis as growth rate decreases.

The consequence of this pattern of DNA synthesis is a positive correlation between DNA per cell and growth rate.

The value of the C period, its relative constancy with growth rate, and the rules governing chromosomal replication (and, therefore, the relationship between cellular DNA content and growth rate) were discovered in a series of remarkably ingenious and insightful experiments by Helmstetter and Cooper in 1968. These experiments remain today so important to our understanding of regulation of DNA synthesis that we consider them here in some detail. (The Helmstetter-Cooper experiments also illustrate that some microbiological experiments can be intellectually challenging even to interpret, let alone to design.)

The Helmstetter-Cooper experiments

Helmstetter and Cooper were able to determine the rate of DNA synthesis during the cell cycle by measuring DNA synthesis in cells separated according to age. They pulse-labeled cultures of *E. coli* B/r with [^{14}C]thymidine during balanced growth, so the amount of radioactivity that each cell incorporated was a measure of its rate of DNA synthesis relative to other cells in the culture. Then they separated the cells by age class using the Helmstetter-Cummings technique of synchronization (Chapter 14). Recall that division products of old cells are the first to appear in the effluent, then the products of progressively younger cells appear (Figure 5 in Chapter 14). Also recall that in any exponentially growing culture, age distribution is an exponential function, with twice as many young cells as old ones (Chapter 1); so an idealized description of the number of cells leaving the filter over time follows the pattern shown in Figure 6 in Chapter 14.

The actual data collected by Helmstetter and Cooper (Figure 6) resembles the ideal except that the numbers do not fall quite so abruptly at the end of a *g* period and the base line drifts upward (probably because one product of each division is not always released into the medium; some cells attach themselves to the membrane, releasing one daughter cell when they divide later). These data showed Helmstetter and Cooper that they could examine individually the division products of cells of

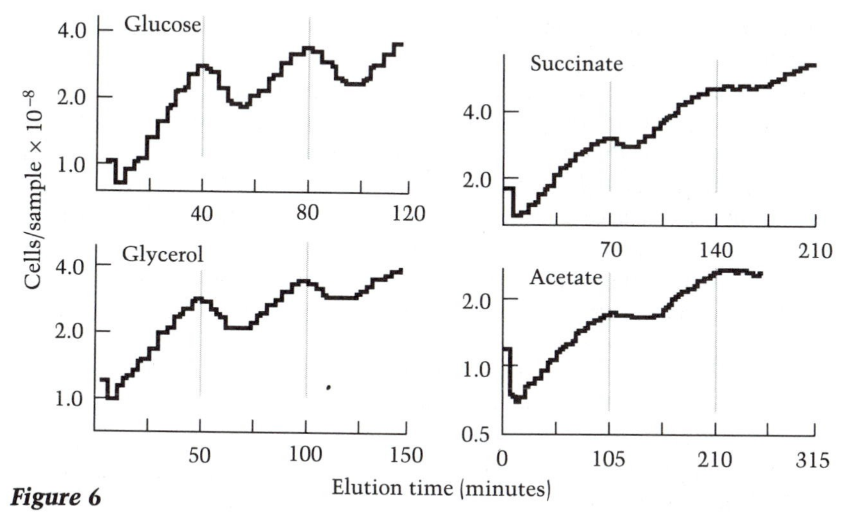

Figure 6

Numbers of cells released by division from a membrane filter covered with a layer of cells from an exponential culture. Data from cultures growing in minimal salts medium with glucose, glycerol, succinate, or acetate as carbon source. (From Helmstetter and Cooper, 1968.)

different ages in the pulse-labeled culture. The radioactivity that each cell contained was an index of its rate of DNA synthesis, so they were able to determine the rate of DNA synthesis during the cell cycle.

We now turn our attention to the kinetics of release of radioactivity from the filter. Because the actual data are complex, it is useful, as they did, to consider first an idealized example. Imagine a culture in which the rate of DNA synthesis doubles (as a consequence of initiating a new round of chromosomal replication) halfway through the cell cycle (Figure 7A). The radioactivity per cell in the effluent of such a pulse-labeled

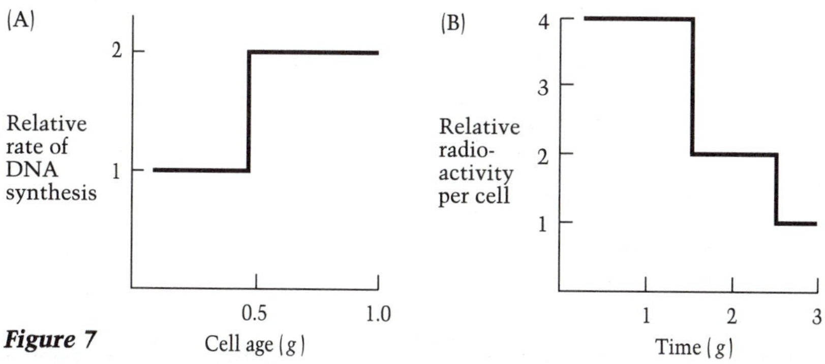

Figure 7

Idealized kinetics of appearance of radioactivity in pulse-labeled cells released by division from a membrane filter. (A) Assumed pattern of rate of DNA synthesis during the cell cycle. (B) Idealized kinetics of appearance of radioactivity in the effluent from the membrane filter. (After Helmstetter and Cooper, 1968.)

membrane-bound culture (Figure 7B) ideally would remain at a constant high value for half a doubling interval (when older, highly labeled cells were released) then decrease to half the previous value at one *g* interval. Conversely, the pattern of shedding radioactive cells shown in Figure 7B establishes the DNA synthesis pattern shown in Figure 7A.

The data they collected (Figure 8) resembles the ideal. In the case of the minimal glucose culture, one can see that the rate of DNA synthesis doubled about halfway through the cell cycle (like the ideal case); this event occurred relatively later in the cell cycle of the minimal glycerol culture, and later still in the cell cycle of the minimal succinate culture. The minimal-succinate and minimal acetate cultures show another distinctive feature: rises in eluted radioactivity with time (indicating times at which DNA synthesis deceased or stopped during the cell cycle). Helmstetter and Cooper summarized the results from many such experiments as shown in Figure 9. From these summaries, they drew the following conclusions:

- The time between initiation and completion of a round of replication (the C PERIOD) is approximately 40 minutes at rapid growth rates but increases somewhat at slow growth rates.
- Cell division occurs about 20 minutes (the D PERIOD; i.e., the interval between the end of DNA replication and cell division) after a round of replication ends.

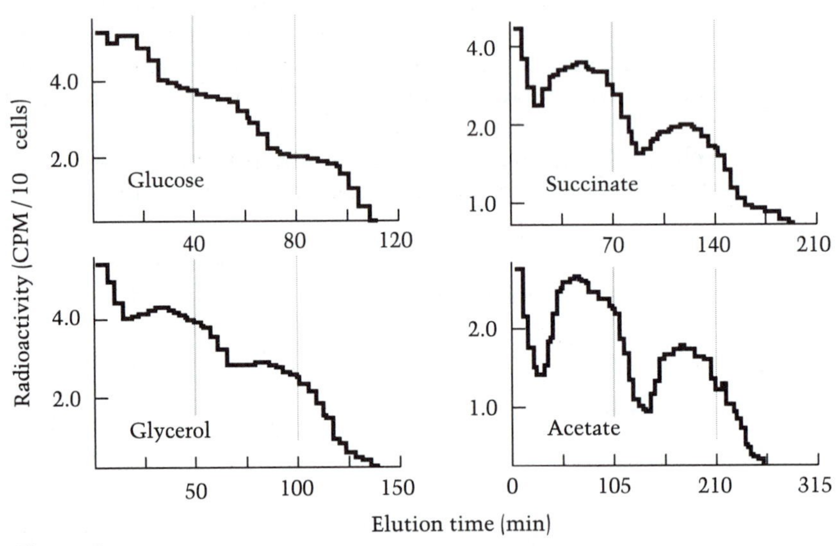

Figure 8

Kinetics of release of radioactivity per cell in the effluent from a membrane-bound culture of *E. coli* B/r which had been pulse-labeled with [^{14}C]thymidine. Results are from glucose, glycerol, succinate, and acetate minimal media. Vertical dotted lines indicate times of cell division. (From Helmstetter and Cooper, 1968.)

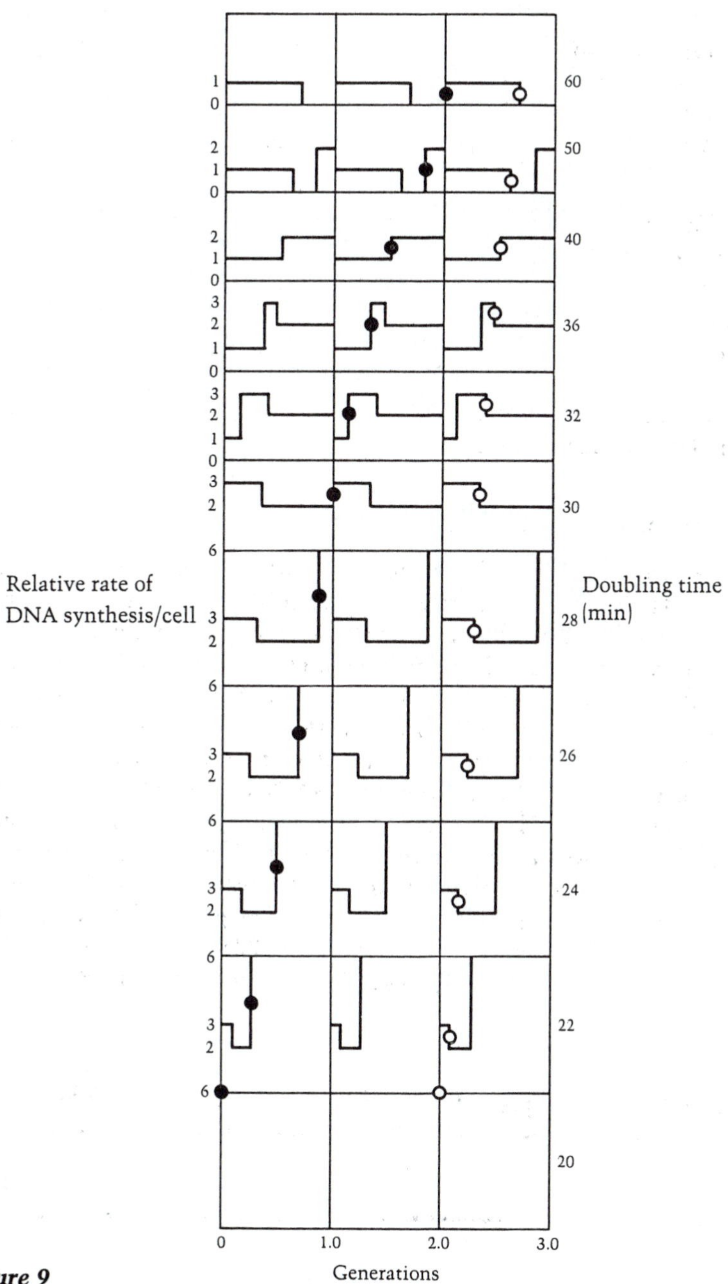

Figure 9

Rate of DNA synthesis during the cell cycle of cultures of *E. coli* growing at different rates. These interpretations are derived from the sort of data shown in Figure 8 in Chapter 12. Filled circles indicate starts, open circles indicate ends of replication. (From Helmstetter and Cooper, 1968.)

- Therefore, chromosomal replication begins about 60 minutes (C + D) before cell division. If the culture is growing with a doubling time of less than 60 minutes, initiation must occur during the previous cell generation. So, at division, daughter cells in cultures growing with doubling times of less than 60 minutes receive partially replicated chromosomes.

We now know that the C and D times are not rigorously invariant with growth rate (Table 2), but the validity of initiation of chromosomal replication and cell division being separated by C + D remains unchallenged. This concept also allows one to calculate the DNA content of the average cell in cultures growing at different rates by determining the state of replication of the chromosome(s) they must contain (Figure 10). We discuss the suspected biochemical signals for chromosomal replication later in this chapter.

Regulation of synthesis of ribosomes

As we have seen, the cellular macromolecule that changes most dramatically with growth rate is RNA. Because almost all of this change is due to stable RNA (tRNA plus rRNA), which participates directly in protein synthesis, the dramatic increases in cellular RNA with growth rate reflect the cell's greater need for protein-synthesizing machinery when it must synthesize its protein complement in less time. The cellular content of the other direct participants in protein synthesis—initiation factors, elongation factors, even RNA polymerase—also increases with growth rate. The regulation of these proteins is discussed later in this chapter. Regulation of the synthesis of ribosomes is central to growth rate control quantitatively as well as mechanistically—a slowly growing *S. typhimurium* cell ($k = 0.2$) has only about 2,000 ribosomes; a rapidly growing one ($k = 2.4$) has about 70,000. At a growth rate of $k = 2.7$, ribosomes make up about 45% of the mass of *E. coli.*

Table 2. Effect of growth rate of *E. coli* B/r on C and D periods

Doubling time (min)	C period (min)	D period (min)
100	67	30
60	50	27
40	45	25
30	43	24
24	42	23

Source: Data from Bremer and Dennis (1987).

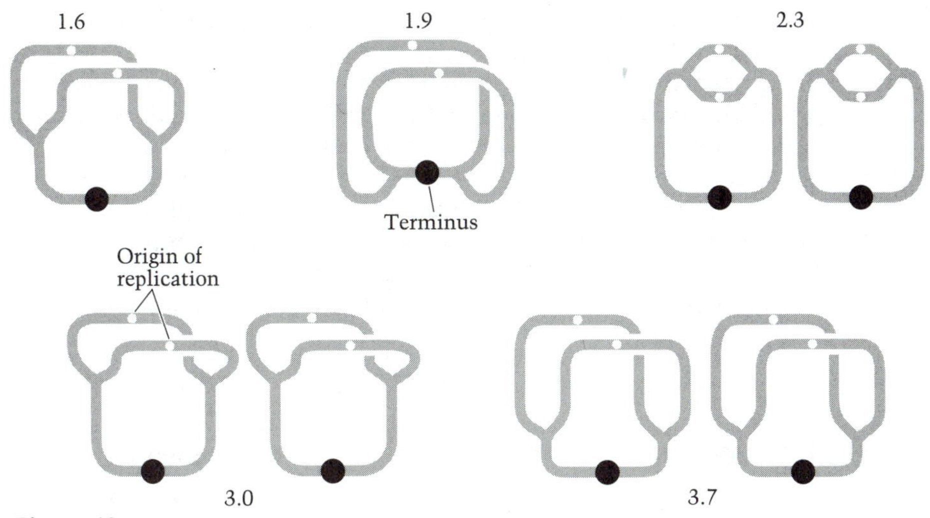

Figure 10

Chromosome structure and equivalent DNA content of the average cell in culture of *E. coli* B/r growing at various rates. The numbers represent doubling time (*g*) in hours. (From Bremer and Dennis, 1987.)

Regulation of synthesis of ribosomal proteins

The regulation of the synthesis of ribosomal proteins (r-proteins) presents a major metabolic challenge:

- Fifty-two distinct proteins (each encoded by a single gene that lies within 1 of at least 20 different operons) are synthesized in the stoichiometric amounts in which they are found in ribosomes (one copy each per ribosome with the exception of r-protein L7/L12, which is present as four copies).
- The rate of synthesis of r-proteins varies more than 500-fold, depending on growth rate; they are not synthesized in excess, because the pool of free r-proteins is quite small.
- The rate of synthesis of r-proteins is precisely coordinated with the rate of synthesis of rRNA.

Regulation of synthesis of r-proteins is an exception to the general rule that expression of most procaryotic genes is regulated at the level of transcription (Chapter 12). Regulation of expression of r-protein genes is, for the most part, an example of autogenous regulation acting at the level of translation.

The discovery that r-protein synthesis is regulated autogenously at the level of translation came from both in vitro and in vivo experiments, most of which were done by M. Nomura and his colleagues. They found that when extra copies of some ribosomal operons were introduced into a cell, the rate of synthesis of their corresponding mRNAs increased in proportion to the gene dosage (number of copies of the genes), but the rate of synthesis of encoded r-proteins was unaffected. These experiments led to an explanatory model: "excess" r-proteins (those not assembled into ribosomes) inhibit by feedback the translation of their corresponding mRNAs. Subsequent in vitro and in vivo experiments showed that one protein in each r-protein operon exerted this feedback control on the entire operon (Figure 11). There is no obvious pattern for the location of the gene of the active protein within the operon—some are first, others are located toward the middle—but there is a clear pattern with the kinetics of incorporation into ribosomes: in all cases

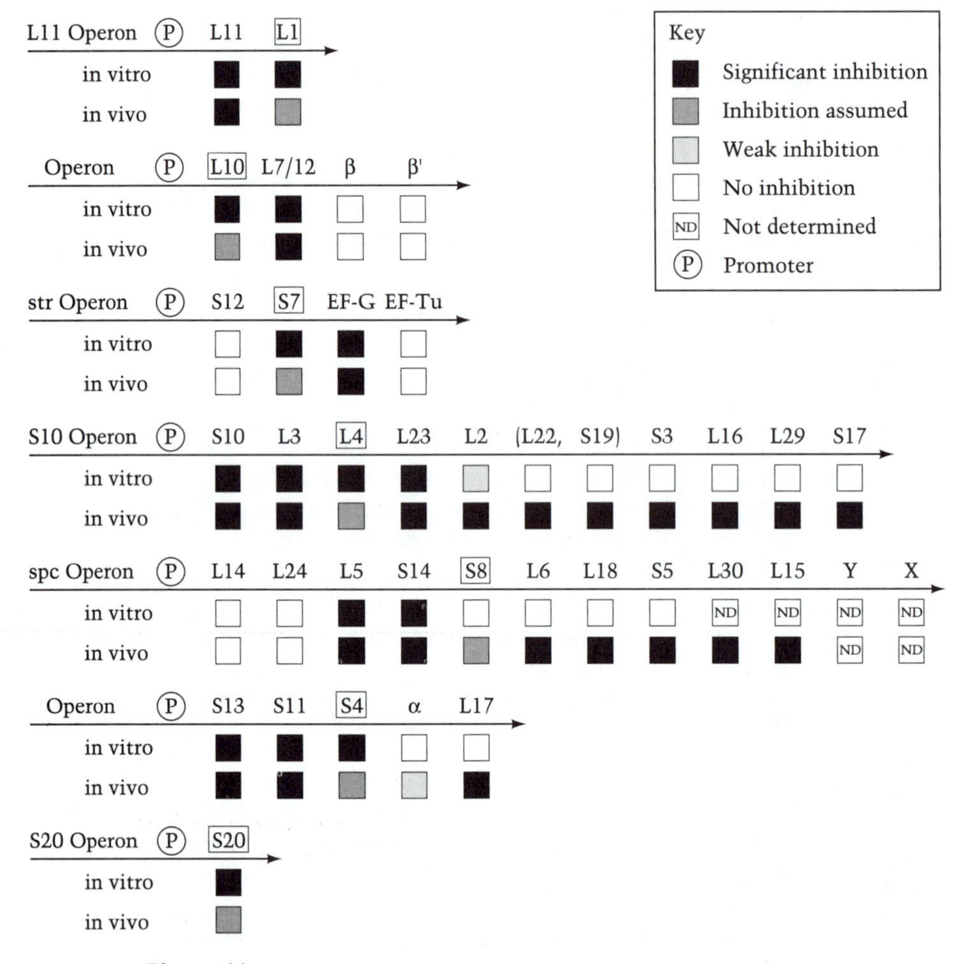

Figure 11

Pattern of feedback inhibition of r-protein operons. The gene products (L [large] indicates proteins from the 50S subunit; S [small] indicates proteins from the 30S subunit) and promoter (P) location are listed above the arrow, which indicates the direction of transcription and translation within the operon. (After Jinks-Robertson and Nomura, 1987.)

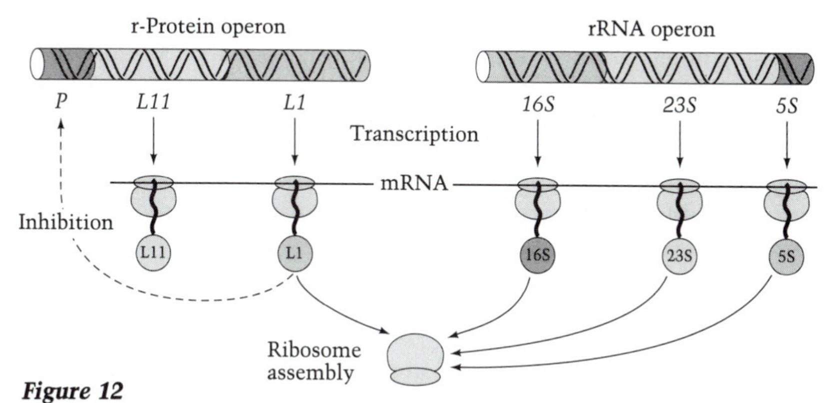

Figure 12

Model for coordination of translational feedback control of synthesis of r-proteins with rate of synthesis of rRNA.

the active protein binds to rRNA at an early stage of ribosomal assembly (Chapter 3). The curious lack of correspondence between in vitro and in vivo control of downstream proteins in certain operons (Figure 11) has been rationalized by differential message stability: in vivo the 3′ (downstream) end of an untranslated message is particularly susceptible to degradation by RNases.

Hints about how these active proteins might exert their feedback effect on translation came from comparing the nucleotide sequences of the leader portion of the mRNAs for r-protein operons with the sequences of the sites on rRNA where the feedback-active protein binds. The striking similarity of these sequences suggested a model for translational regulation (Figure 12): feedback-active r-proteins, termed TRANSLATIONAL REPRESSORS, bind both to rRNA and mRNA. They preferentially bind rRNA and become incorporated into ribosomes, but if the rate of synthesis of these proteins exceeds their rate of incorporation into ribosomes, they bind to their own mRNA near the first Shine-Dalgarno sequence (the critical components of the translational start sequence; Chapter 3), thereby inhibiting translation of the message.

Critical features of the model have been tested and validated: feedback-active proteins do bind preferentially to their rRNA sites, and intact mRNA-binding sites are essential for inhibition of translation. But the model presents a major dilemma: How can blocking the first translational start sequence in a polycistronic message prevent translation of all subsequent genes in the message? After all, with most or all polycistronic messages, each subsequent gene has its own ribosomal start sequence, because translation must reinitiate at the beginning of each protein encoded in the message. Nomura has invoked a hypothetical phenomenon called TRANSLATIONAL COUPLING to resolve the dilemma: the translational starts of downstream cistrons are in some way masked (possibly by secondary structure) in the untranslated message; translation of first cistrons, by an unknown mechanism, unmasks the translational start sequences of all downstream cistrons.

In at least one r-protein operon for certain—the S10 operon—and probably others as well, the translational repressor also controls transcription by a protein-modulated attenuation mechanism similar to the one that controls the tryptophan operon in *Bacillus subtilis* (Chapter 10). But this level of r-protein regulation does not seem to participate in growth rate control. Rather, it probably acts to stimulate synthesis under conditions of exceptionally high demand for r-proteins—for example, during a nutritional shift-up.

Regulation of synthesis of rRNA and tRNA

As we have just seen, regulation of the synthesis of r-proteins is a secondary mechanism with respect to control of ribosome synthesis; it merely coordinates the rate of r-protein synthesis with rRNA synthesis. The rate of rRNA synthesis determines the rate of ribosome synthesis. So, the control of synthesis of rRNA must be the most fundamental aspect of growth rate control.

Synthesis of rRNA is a massive metabolic task. Under certain growth conditions, rRNA transcripts constitute over half the RNA being made by the cell, although the encoding rRNA genes make up less than 0.5% of the *E. coli* genome. As one would expect, the rRNA promoters are by far the strongest ones in the cell. Even so, each rRNA operon has two promoters, an arrangement that allows a potential doubling of the rate of transcription. Owing to the major metabolic cost of making rRNA, precise control of the transcription of rRNA operons is particularly important to the economy of the cell. As the data in Figure 2 suggest, control of synthesis of stable RNA (rRNA and tRNA) is a single topic, because the ratio of tRNA to rRNA remains constant over wide variations of growth rate. (The marked increase in the fraction of tRNA present at low growth rate probably reflects differential rates of rRNA–tRNA breakdown rather than differential rates of synthesis of the two types of stable RNA.)

Although, we do not yet know the precise biochemical mechanism(s) that control the synthesis of stable RNA (and hence ribosomes), we do know quite a bit about its defining parameters. One of these is the lack of effect of gene dosage on synthesis of ribosomes. If the number of rRNA operons in a cell is doubled (by introducing rRNA-containing plasmids into it), the total rate of rRNA synthesis is not increased, rather expression of each rRNA operon is decreased by half. This result implies that a feedback mechanism controls rRNA synthesis—that the amount of gene product, or some derivative of it, sets the rate of gene expression. For such a mechanism to be physiologically relevant, the putative product-inhibitor ought to be an index of excess functional product. An excellent candidate for such a product-inhibitor would be unused ribosomes—"free ribosomes" that are not part of a polysome. This attractive hypothesis was proved to be untrue by a series of incisive experiments, one of which is worth our consideration here.

The gene (*infB*) encoding initiation factor 2 (IF2) was put under the control of the *lac* promoter, thereby making it inducible by IPTG (Chapter 12), so IF2 levels and growth rate of the genetically altered strain were proportional to the concentration of IPTG added to the medium. Under such conditions of IF2-limited growth, large amounts of "free ribosomes" accumulated within the cell in concentrations inversely proportional to growth rate. Were "free ribosomes," indeed, the feedback inhibitor of stable RNA synthesis, one would expect RNA synthesis to increase with growth rate, even more dramatically that it does in wild-type cells. But such was not the case (Figure 13). At lower growth rates, when intracellular concentrations of free ribosomes were high, RNA synthesis was not inhibited; instead, it was stimulated to a level several fold higher than normal. These results establish that "free ribosomes" (and probably any intermediate of ribosome synthesis) cannot be the product-inhibitor or feedback effector; rather, an excess of functioning ribosomes or some product of them must be the negative regulator of RNA synthesis.

The nucleotide ppGpp (Chapter 13) presents itself as an excellent candidate for such a regulatory signal. It is made when concentrations of aminoacyl-tRNAs are rate limiting (an equivalent to excess functioning ribosomes); its intracellular concentration decreases with increasing growth rate; ppGpp is an established inhibitor of transcription under conditions of stringency (Chapter 13). However, for ppGpp to be the primary transducer of growth rate control of rRNA synthesis, it would have to be differentially effective—be a more powerful inhibitor of tran-

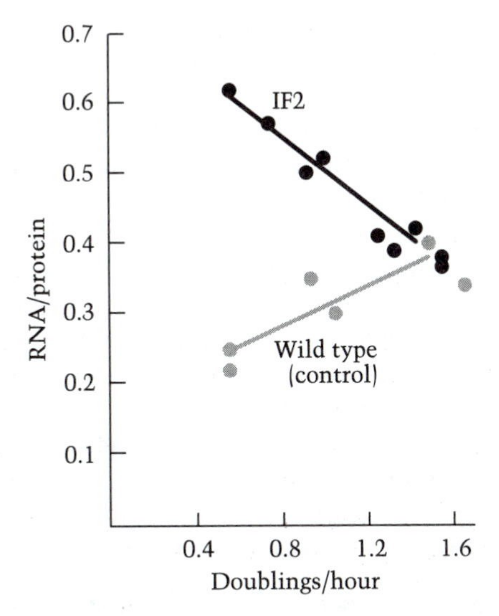

Figure 13

Effect of growth rate on RNA synthesis. In wild-type cells (gray circles), the relative rate of RNA synthesis (as indicated by the RNA:protein ratio) increases with growth rate. In manipulated cells (black circles) in which growth rate is set by intracellular levels of IF2 (see text), relative RNA synthesis decreases with increasing growth rate. (From Cole et al., 1987.)

scription of stable RNA genes than of others. It has been proposed that two forms of RNA polymerase might exist within the cell—one that is ppGpp-modified and one that is not. In vitro experiments to support this hypothesis have produced conflicting results. Some experiments find that ppGpp selectively inhibits transcription of genes encoding stable RNAs; others do not.

An alternate explanation, a form of PASSIVE CONTROL, has recently been proposed by Jensen and Pedersen (1990). This proposal explains a net differential inhibition of stable RNA syntheses by ppGpp even though the inhibitor might not be any more effective in inhibiting transcription of stable RNA genes. The proposal assumes that most genes are not fully saturated by transcribing RNA polymerase molecules and that stable RNA genes (because they have the most active promoters) are the most difficult to saturate. It follows, then, that if ppGpp inhibited equally the rate of transcription of all genes, more RNA polymerase molecules would be engaged in transcription; fewer would be free to bind to promoters, and, stable RNA promoters being the most difficult to saturate, would suffer the greatest decline in transcriptional starts.

In summary, a variety of different experiments suggest that functioning ribosomes exert feedback control on stable RNA, but the biochemical pathway of this control is not yet understood.

Modulation of cell size

As we discussed at the beginning of this chapter, bacterial cells become much bigger as they grow faster; this difference can be noticed in a quick glance at an electron micrograph of a mixture of cells grown at different rates (Figure 14). The mass of the average cell in a culture of *E. coli* grown at 2.5 doublings per hour is almost six times greater than the mass of the average cell of the same strain grown at 0.6 doublings per hour. As Figure 14 shows, growth rate modulates cell diameter. At any particular growth rate, cells elongate during the cell cycle, but their diameter remains constant.

Naively, one might rationalize the larger cell size as being necessary to accommodate more ribosomes and other components of the protein-synthesizing machinery, but this is not true, because a cell needs a high *concentration* of ribosomes, not merely a large *number* of them to grow at a high rate. The relationship between cell size and growth rate seems to depend on the necessity of attaining a particular mass to initiate replication of the chromosome rather than on the need to hold more things. In 1968 Donachie suggested that initiation of chromosomal replication was governed by the following simple rule:

- When a defined mass (M_i) per origin of replication is reached during growth, replication is initiated at every origin (*oriC*) in the cell.

Donachie arrived at this rule by the exercise summarized in Figure 15. He assumed that growth rates of individual cells was exponential; accordingly, he drew the individual cell growth curves on a semiloga-

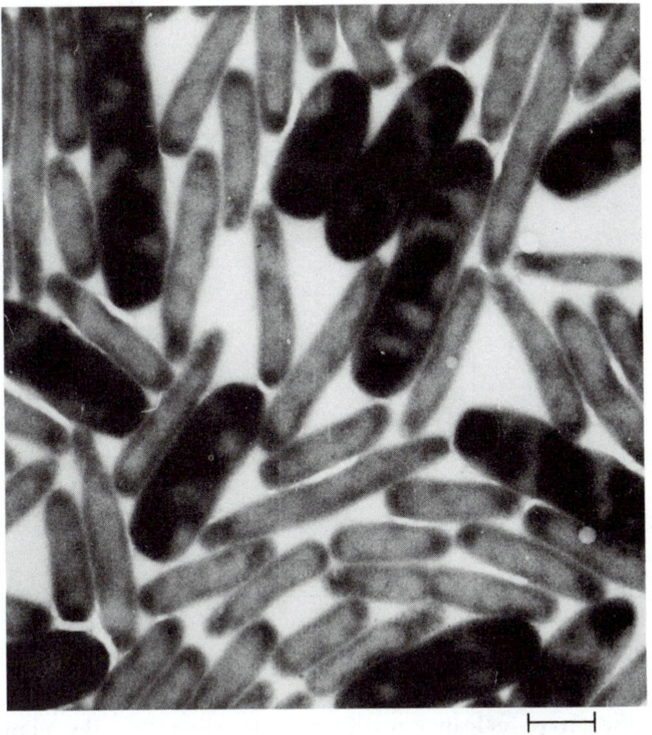

Figure 14

Electron micrograph of a mixture of cells of *E. coli* B/r grown at different rates. The large cells grew with a doubling time of 22 minutes, the small ones with a doubling time of 72 minutes. (From Nanninga and Woldringh, 1985.)

rithmic plot. He then arranged these growth curves on the ordinate according to the relative mass of average cells in cultures growing at the indicated rates. Finally, he located on each growth curve (using the Helmstetter-Cooper rules) the times of initiation of chromosomal replication. The location of these times led to the startling discovery that they all fell at the same mass or multiples of it, causing Donachie to conclude that initiation always occurs at the same cell mass (M_i) per replication origin. Following Donachie's rule, increase of cell size with growth rate is inevitable. Because initiation occurs progressively earlier in the cycle as growth rate increases, the cell must reach M_i sooner; so the average cell has to be bigger.

Of course, we now know that the picture is a little more complicated than the one Donachie presented. The C and D periods are not precisely constant, and neither is M_i. But the essence of Donachie's argument remains sound: if cell mass triggers replication of the chromosome, cell size must increase along with growth rate. One might wonder how a cell can sense its own mass. A reasonable explanation is that a certain amount of a particular protein—possibly the DnaA protein—triggers

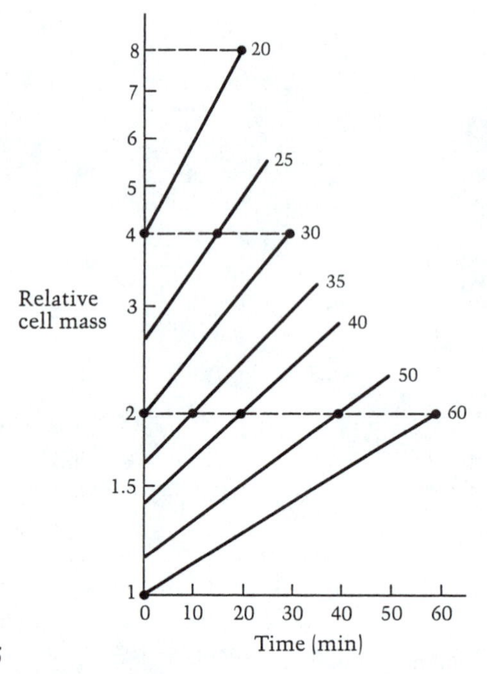

Figure 15

Cell mass and the individual cells growing at different rates. Numbers to the right of the curves indicate growth rate in doubling time; the relative position of the curves at the ordinate is determined by the relative mass of the average cell in cultures growing at the indicated rates. The times during the cell cycle of initiation of chromosome replication is indicated by the black circles. (From Donachie, 1968.)

initiation. If it were synthesized as a constant fraction of cell mass, initiation would always occur at approximately the same cell mass.

The case for DnaA's playing an important role in initiation is a strong one, as discussed in Chapter 14:

- DnaA is essential for initiation at *oriC* in vitro.
- DnaA binds cooperatively to several sites (DnaA boxes) within *oriC.*
- Artificial induction of the *dnaA* gene in vivo stimulates initiation.

Other components of the protein-synthesizing machinery

In our discussion of growth rate-modulated control, we have emphasized ribosomes, because they constitute the major bulk of the cell's protein-synthesizing machinery; and, incomplete as it is, more information is available about this aspect of growth rate control than about any other. But, as one might expect, other components of the cell, including the rest of the cell's protein-synthesizing machinery, are also under growth rate control. We have already considered tRNA; it seems to be subject to the same controls as rRNA. The aminoacyl-tRNA synthetases and the ribosome-associated protein factors (initiation, elongation, and release factors) are under growth rate control, but to a slightly different degree. Thus, the ratio of aminoacyl-tRNA synthetase and ribosome-associated proteins to ribosomes decreases progressively with growth rate. Almost nothing is known about control of synthesis of the ribosome-associated factors, and the same can be said about the growth rate control aspect of synthesis of the aminoacyl-tRNA synthetases, although another aspect (their derepression during amino acid starvation) has been studied is some detail. In this latter respect, the synthetases and enzymes in the pathways of amino acid biosyntheses are controlled similarly.

Passive control

Failure to find a specific mechanism leads one to suspect that a general passive control mechanism might exert growth rate control on expression of aminoacyl-tRNA synthetases and similarly regulated proteins. We have already discussed passive control of rRNA synthesis acting through ppGpp, pointing out that increased ppGpp levels would sequester RNA polymerase molecules if the effect of the inhibitor were to slow all transcription; such sequestration would selectively inhibit transcription of rRNA operons owing to the extremely high activity of their promoters and hence to the difficulty of saturating them with molecules of RNA polymerase. The essence of this and any form of passive control depends on the assumption that intracellular concentrations of RNA polymerase molecules are subsaturating—that is, that the overall rate of transcription is limited by availability of RNA polymerase molecules available to initiate new transcriptional events. Whether or not the intracellular concentration of RNA polymerase molecules is, indeed, subsaturating remains controversial. If RNA polymerase levels are subsaturating, any change of available RNA polymerase (as affected, for example, by changing concentration of ppGpp) or any change in number of competing promoters would shift the pattern of gene expression.

Ole Maaløe suggested that the second possibility—change in number of competing promoters—might contribute to growth rate control. Richer growth media contain more preformed monomers than poorer media, and the monomers cause repression of enzymes in pathways of their own biosynthesis by "turning off" the promoters of the encoding genes. The "turned-off" or "closed" promoters would no longer compete for RNA polymerase molecules; the consequence would be a higher rate of transcription of the remaining "open promoters." The theory of passive control holds that growth rate-controlled genes are simply those with promoters that remain open when the cell is growing in richer media.

SUMMARY

1. More than most organisms, bacteria grow at vastly different rates, depending on their nutrition.
2. Bacteria adjust their macromolecular composition and average cell size with nutrition-imposed growth rate; as growth rate increases, cells become larger, richer in RNA, and slightly poorer in protein and

DNA. These changes allow the cell to grow at the maximum rate that a particular medium is capable of supporting.

3. When their growth medium is suddenly changed by being made either richer or poorer, changes in cellular composition occur that follow a regular, predictable pattern before balanced growth continues at the new rate.
4. Bacteria can grow with doubling times that are shorter than the time required to replicate a chromosome (C period). They overcome this paradox by passing to each daughter cell at division a partially replicated chromosome; sometimes the chromosome has more than two replication forks.
5. Some rules relate chromosome replication to cell division of *E. coli* at 37°C: cell division occurs 20 minutes (the D period) after replication of a chromosome is completed; replication of the chromosome requires 40 minutes (C period). Although we now know that the C and D periods are not constant with changing growth rates, they are close enough for the rules to remain useful.
6. The change in the cell's complement of ribosomes is the largest growth rate-driven cellular change: rate of ribosome synthesis can vary 500-fold with growth rate.
7. The synthesis of ribosomal proteins is coordinated with the rate of synthesis of rRNA through autogenous regulation of translation. One r-protein (the translational repressor) in each r-protein operon can bind to the leader sequence of its own mRNA, thereby preventing translation of the operon. Translational inhibition occurs when there is more translational repressor than is needed for ribosome synthesis; the translational repressor binds more readily to rRNA (resulting in assembly of ribosomes) than to mRNA (resulting in translational repression). How binding to the leader prevents translation of downstream genes remains a mystery; this phenomenon is called translational coupling.
8. Regulation of rRNA synthesis regulates synthesis of ribosomes. Gene dosage experiments indicate that control is exerted by a feedback mechanism probably exerted by an excess of functioning ribosomes. The stringent response effector, ppGpp, might be the transducer for growth rate modulation of rRNA (and hence ribosome) synthesis. Possibly ppGpp selectively inhibits synthesis of stable RNA; or by slowing synthesis of all RNA, it could depress synthesis of stable RNA by decreasing the concentration of free RNA polymerase—a condition that would selectively lower the frequency of initiation of transcription at stable RNA promoters.
9. Increase in cell size with growth rate cannot be explained by a cellular need for more components, because rates of synthesis depend on concentration of reactants, not absolute amounts. Rather, increase in cell size with growth rate appears to be a direct consequence of the fact that initiation of chromosome replication is triggered (probably indirectly) when the cell attains a certain mass (M_i). Because, according to the Helmstetter-Cooper rules, initiation occurs and hence M_i must be attained progressively earlier in the cell cycle as growth rate increases, cell size must increase with growth rate.

10. Growth rate modulation of expression of some genes might be effected through a passive control mechanism. The possibility is not easily tested.

STUDY QUESTIONS

1. What would be the properties of a mutant strain of *E. coli* that had lost the ability to modulate its complement of stable RNA in response to changes in the richness of its growth medium? How would you attempt to enrich for such a mutant?
2. Do you think it would be possible to isolate a mutant strain of *E. coli* that was unable to change its size with growth rate? Explain.
3. Using the Helmstetter-Cooper rules, determine at what point during the interdivision cycle chromosome replication would begin, end, and at what relative rate it would occur in cells of *E. coli* growing at doubling times of 60, 40, and 20 minutes.
4. If you had an in vitro coupled transcription–translation system in which r-proteins were being synthesized from DNA encoding the L11 r-protein operon, what would be the effect of adding the following components: (a) L11 protein; (b) L1 protein; (c) L1 protein + 16S rRNA; (d) L1 protein + 23S rRNA? Explain.
5. Why must translational coupling be invoked to explain regulation of synthesis of r-proteins?
6. Because manipulation of gene dosage by introducing extra copies of rRNA operons into *E. coli* has no effect on the total rate of rRNA synthesis, why do you think that wild-type *E. coli* has seven copies of rRNA operons?
7. What does passive control mean? How does it work?

SUGGESTED READING

Topics in this chapter are covered in the following chapters from *Escherichia coli and Salmonella typhimurium: Cellular and Molecular Biology,* F. C. Neidhardt, J. L. Ingraham, K. B. Low, B. Magasanik, M. Schaechter and H. E. Umbarger (eds.). 1987. American Society for Microbiology, Washington, D.C.

Bremer, H. and P. P. Dennis. Modulation of chemical composition and other parameters of the cell by growth rate. (Volume 2; Chapter 96)

Grunberg-Manago, M. Regulation of expression of aminoacyl-tRNA synthetases and translation factors. (Volume 2; Chapter 86)

Other suggested references include:

Cooper, S. and C. R. Helmstetter. 1968. Chromosome replication and the division cycle of *Escherichia coli* B/r. J. Mol. Biol. 31:519.

Helmstetter, C. R. and S. Cooper. 1968. DNA synthesis during the division cycle of rapidly growing *Escherichia coli* B/r. J. Mol. Biol. 31:507.

Maaløe, O. and N. O. Kjeldgaard. 1966. *Control of Macromolecular Synthesis.* W. A. Benjamin, New York.

Nomura, M., R. Gourse and G. Baughman. 1984. Regulation of synthesis of ribosomes and ribosomal components. Annu. Rev. Biochem. 53:75.

Jensen, K. F. and S. Pedersen. 1990. Metabolic growth rate control in *Escherichia coli* may by a consequence of the subsaturation of the macromolecular biosynthetic appartus with substrates and catalytic components. Microbiol. Rev. (June).

16

Cellular Differentiation

INTRODUCTION

Differentiation has a clear and uncomplicated meaning in common usage—it simply refers to the process of becoming different. But the word has special connotations for most biologists because the mechanisms of cellular differentiation are among the most important, the most complicated, and the least understood biological phenomena. Understanding the pathways of differentiation leading from a fertilized egg to the myriad of morphologically and biochemically distinct cells in the adult plant or animal is the awesome challenge that modern experimental biology has accepted and for which it is seeking feasible routes of attack. One route is the study of the simpler and the more tractable processes of differentiation seen in bacteria. Such studies are tactically appealing because they promise quicker answers to complex questions about the development of bacteria even if they may not necessarily tell us much about differentiation in plants and animals.

In earlier chapters of this book, we have examined many examples of biochemical differentiation: enzyme induction and repression, global control mechanisms, and growth rate control. As we have seen, an *Escherichia coli* cell in a rapidly growing culture differs profoundly from a cell in a slowly growing culture—it is much larger and has a quantitatively different macromolecular composition. It is fair to say that the line delineating global response (Chapter 13) and differentiation is fuzzy; indeed, there is considerable overlap between these terms.

We have also examined some aspects of MORPHOGENESIS (the formation of shape) in connection with our discussions of assembly. We saw that, in some cases, the shape of a cellular organelle was determined by its macromolecular constituents. The intricate shape and macromolecular composition of the ribosome seems to be determined solely by the primary structure of the 54 molecules of protein and 3 molecules of RNA from which it is made. Even more remarkable, self-assembly seems to be the route of formation of the morphologically complex bacterial flagellum. From basal body, through the hook, to the filament, the flagellum seems to form spontaneously, with each set of new proteins assembling on previously assembled components. But each individual kind of protein is morphogenetically inert unless the previously assembled structure is present as a primer. Hook proteins assemble only on a properly assembled basal body; flagellin monomers assemble into a filament of characteristic diameter and wavelength only on a properly assembled hook. Flagellins will spontaneously assemble into filaments, even in vitro, if preformed bits of filaments are present, but initiation of filament assembly only occurs on a hook.

Flagella are the product of a self-assembly line that cannot be shunted or begun any place other than the starting point. The formation of the bacteriophage T4 is considerably more complex; it is the product of three merging self-assembly lines (Figure 1), and T4 assembly introduces an added complication. Assembly of the T4 head is dependent on proteins that are not part of the completed structure. The proteins that appear to function only in assembly are called SCAFFOLDING PROTEINS. The full extent of the role played by scaffolding proteins in the assembly of procaryotic organelles is not yet known, but it must be extensive.

Cell shape is another aspect of procaryotic morphogenesis that we have already touched on. The shape of procaryotic cells is variable among species; some are quite elaborate. Although most are rods, cocci, or comma-shaped forms, some are rectilinear boxes or helices. Some are shaped like a starfish. The shapes of bacterial cells are certainly as complex as those of eucaryotic cells, but the ways in which bacterial cells assume a particular shape are quite different.

The shape of eucaryotic cells is formed from within by the activity of the CYTOSKELETON, which is composed of microtubules and microfilaments. By assembly or disaggregation, the cytoskeleton pushes or pulls the cell into a particular shape—temporarily in the case of animal cells or wall-less protists like amoeba—and this shape is made permanent by the rigid walls of plants and some other protists. Some asymmetric changes in the shape of eucaryotic cells seem to be brought about by the polarizing effects of ion currents, which flow across a cell when ions enter the cell at one point and leave at another.

Because they lack a cytoskeleton, bacterial cells assume their characteristic shapes by completely different mechanisms. As we have seen, the bacterial cell's rigid murein layer maintains cell shape, but it also determines cell shape by the way it is laid down. The driving force of enlargement of a bacterial cell is turgor pressure, which, unlike a cytoskeleton, directs force equally in all directions. So the bacterial cell is shaped by the complex combined activities of autolytic enzymes that break down murein, weakening it in certain regions, and of enzymes (penicillin-binding proteins; Chapter 14) that insert new murein monomers into the cell by cross-linking them to existing polymeric murein already in the wall. There can be no doubt that penicillin-binding proteins contribute to determining the shape of a bacterial cell, because mutant strains of *E. coli* that lack specific penicillin-binding proteins have altered shapes. For example, certain mutant strains of *E. coli* are

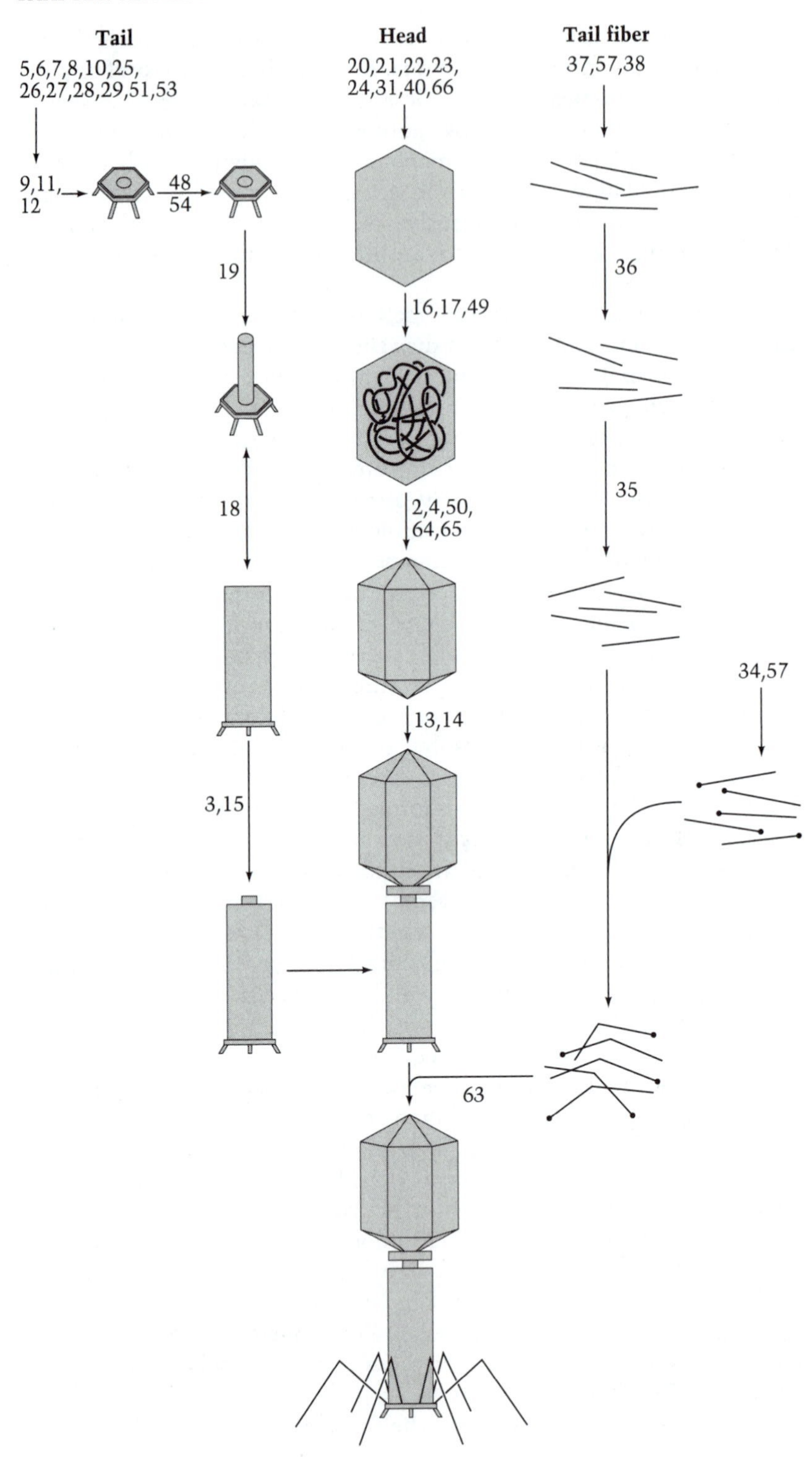

Tail
5,6,7,8,10,25,
26,27,28,29,51,53
9,11,
12
48
54
19
18
3,15
Head
20,21,22,23,
24,31,40,66
16,17,49
2,4,50,
64,65
13,14
63
Tail fiber
37,57,38
36
35
34,57

cocci rather than rods; the penicillin-binding proteins that they lack presumably form the cylindrical midsection of the cell, leaving the cells only the ability to form the two hemispherical ends. But the details of how penicillin-binding proteins in concert with autolytic enzymes determine shape remains a mysterious and challenging question.

Cellular differentiation of bacteria is sometimes dismissed as an oddity that is restricted to a few unusual groups, but that is a limited view. There are many examples of bacterial differentiation, sometimes quite complicated ones—certainly more than we will be able to consider in this chapter. Here we consider CELL DIFFERENTIATION in a strict sense, namely, those morphological and biochemical changes that occur during the organism's life cycle.

A SAMPLING OF EXAMPLES

We mentioned in Chapter 1 that bacteria face three general classes of problems, each of which can be met by differentiation into new forms that are more suitable for the circumstances.

- A life of feast or famine, where periods of nutritional enrichment are interspersed with times of starvation. In some cases, bacteria differentiate to ensure access to nutrients.
- Competition for surface sites. Microorganisms need moist environments for growth. They are often subject to the sweeping action of liquid currents, and some differentiate to ensure residence in their preferred habitat.
- Exposure to noxious agents and damaging conditions. Again, some species differentiate into resistant forms.

We group our examples of procaryotic differentiation according to these three categories, keeping in mind that assignment in some cases is arbitrary.

Strategies to cope with nutritional problems

***Myxococcus xanthus:* A predator** / Myxobacteria are gliding Gram-negative bacteria that undergo, by procaryotic standards, a complex life cycle, the most notable feature of which is the formation of fruiting bodies. These structures vary in complexity (Figure 2). Some, like those of *Stigmatella,* are elaborately sculptured; others, like those of *Myxococcus,* are simple domelike protuberances. Myxobacteria are wide-

Figure 1

Pathway of assembly of the bacteriophage T4. Numbers indicate the genes whose products participate in that step. Most steps of the assembly process occur by self-assembly. (From Wood, 1979.)

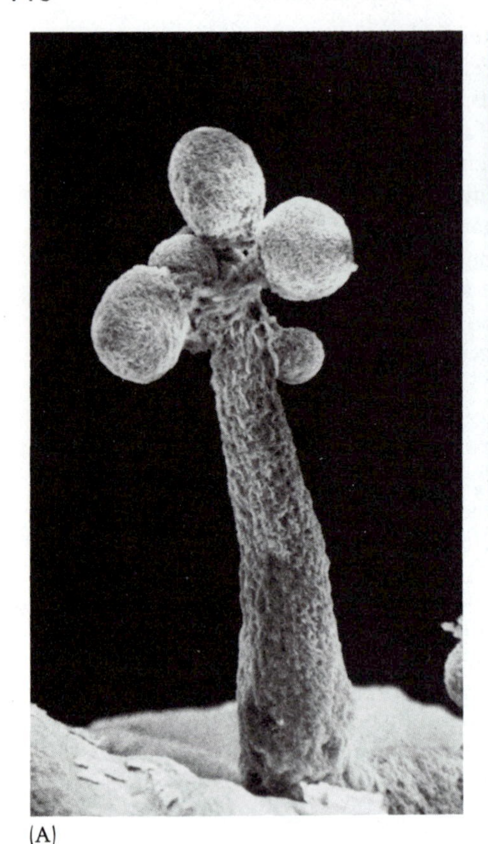

(A)

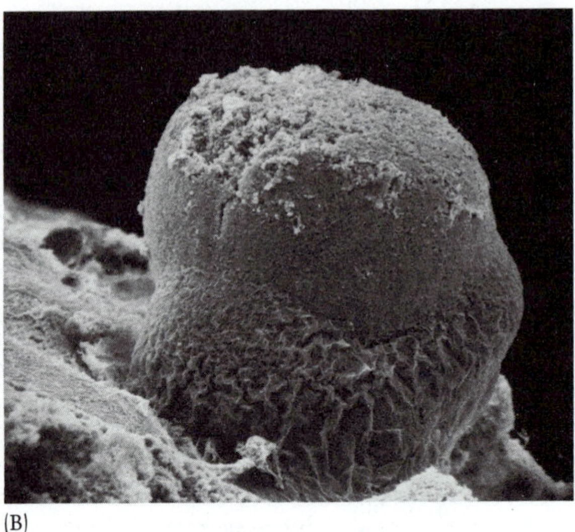

(B)

Figure 2

Fruiting bodies of myxobacteria. (A) *Stigmatella aurantiaca* (× 180). (B) *Myxococcus xanthus* (× 228). (Courtesy of P. Grilione and J. Pangborn.)

spread in nature. If a piece of soil or dung or rotting vegetable material is incubated in a humid environment for a week or so, fruiting bodies of myxobacteria are very likely to appear. They are barely visible to the naked eye but can be seen quite clearly through a dissecting microscope. The ease with which they can be seen accounts for the fact that a myxobacterium—*Polyangium vitellinum*—was the first procaryote to be assigned a scientific name (in 1809).

The relatively simple *Myxococcus xanthus* is the representative upon which most morphogenetic studies have been concentrated because it is so amenable to laboratory investigation. It can be cultivated readily using ordinary microbiological equipment, and genetic transduction between strains can be mediated by P1 phage. The life cycle of *M. xanthus* (Figure 3) alternates between the formation of MYXOSPORES and of vegetative cells. Myxospores are not endospores (see below), rather they are cyst-like structures surrounded by a thick layer of polysaccharide that is resistant to heat and desiccation. When exposed to favorable conditions for growth, myxospores germinate, producing tapered and flexible (probably owing to the fact that their murein layer is discontinuous) vegetative cells. These cells grow and divide, forming swarms of individuals that glide as "wolf packs" over solid surfaces; the cells in the swarm lyse and consume other microbial cells in their path. When conditions for growth becomes unfavorable, about 1,000 cells aggregate in one place to form a fruiting body within which myxospores form. During this process, new proteins are formed at particular times.

During the process of aggregation, formation of fruiting bodies, and conversion into myxospores, myxococcal cells must produce extracellular signals, because certain mutants that cannot form fruiting bodies can do so when mixed with wild-type cells for other classes of mutants. On the basis of this sort of mixing experiment, four classes of mutants have been identified, some of which have defects in genes that encode essential functions.

It is a curious fact that the genome of *M. xanthus* (8.4 × 10^9 Da is about three times larger than *E. coli's* genome. Certainly, the metabolism of *Myxococcus* is no more complicated than that of *E. coli.* One wonders how much of the difference in genome size is attributable to the differentiation that *Myxococcus* undergoes during its life cycle.

***Rhizobium:* Nutritional parasite and nitrogen fixer** / The preceding example of procaryotic differentiation used a free-living organism; but many bacteria form associations, sometimes elaborate ones, with other organisms. One of these is the symbiotic association of bacteria in the genus *Rhizobium* with plants in the legume family. The formation of this symbiosis involves simultaneous differentiation of both partners. The mutualistic benefits of this symbiotic relationship are obvious. The plant offers the bacterium a protected environment, a steady source of nutrients, and a partial pressure of oxygen compatible with fixing nitrogen. In return, the plant receives fixed nitrogen in the form of ammonia from the bacterium.

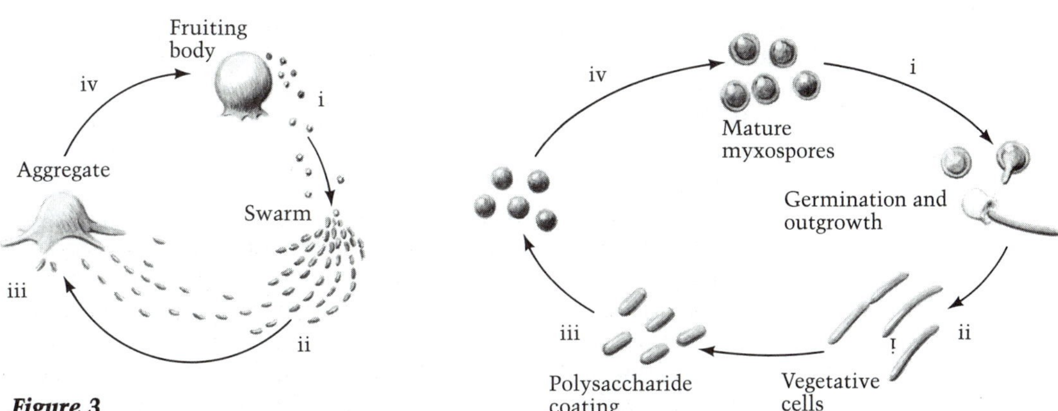

Figure 3

The life cycle of *Myxococcus xanthus*. Myxospores (i) from the fruiting body germinate and grow into vegetative cells (ii), which form swarms. As conditions become less favorable, the vegetative cells aggregate and encyst themselves with polysaccharide (iii), eventually forming the fruiting body. (After Parish, 1979.)

The *Rhizobium*–legume symbiosis is only one of many associations between plants and nitrogen-fixing bacteria. Equally important are the loose associations that cyanobacteria form with bryophytes and ferns. One of the best known of these is between the cyanobacterium *Anabaena* and the miniature aquatic fern *Azolla*. This association has been used for centuries in Southeast Asia to provide fixed nitrogen to rice paddies; *Azolla* is cultured by traditional methods and inoculated into the paddy, a practice resulting in about as much fixed nitrogen as the terrestrial *Rhizobium*–legume association does. This association is now used to grow rice in Africa, in China, and (for organically grown food) in California. The actinomycete *Frankia* forms nitrogen-fixing nodules on a wide variety of nonleguminous plants belonging to 17 genera in 7 orders of plants. *Azospirillum* grows, presumably symbiotically with an exchange of nutrients, in the rhizosphere of plants in the cereal family. But the *Rhizobium*–legume association is by far the most well studied one because of its great importance to Western agriculture.

The *Rhizobium*–legume relationship is a highly evolved and intimate one. Rhizobia are assigned to species on the basis of which legumes they can associate with. Some individual strains of *Rhizobium* (and the closely related genus *Bradyrhizobium*) can form nodules on only a small group of closely related legumes; others have a considerably broader host range. Nitrogen-fixing nodules are easily detectable on the roots of most leguminous plants; they appear as globular or cylindrical swellings protruding from the surface. Pull up any clover or alfalfa plant, brush off the soil, and the nodules are obvious.

Rhizobia are capable of independent growth in soil, but they fix nitrogen only when they are in nodules. Because many legumes are annuals, the rhizobia overwinter in the soil and infect the legume when it germinates in the spring. To begin the process of nodule formation (Figure 4), the *Rhizobium* must find a susceptible plant. This encounter probably occurs by chemotaxis. The chemotactic response of the peritrichous *Rhizobium* follows the run-and-tumble pattern of *E. coli* (Chapter 6). The bacterium is attracted to a glycoprotein produced by the plant. When the bacterium reaches the root surface, it binds specifically to a root hair by a multistep process, one component of which is a reaction between a plant lectin and a polysaccharide on the surface of the *Rhizobium*. The binding triggers a sequence of developmental changes in the plant and the bacterium. The tip of the root hair bends, forming a "shepherd's crook." The bacterium invades the root hair at the pocket of the crook by partial dissolution of the plant cell wall. As one or more bacterial cells proceed into the plant, new plant wall material (pectic substances, cellulose, and hemicellulase) is laid down around them, forming a tube called an INFECTION THREAD that extends into the root tissue. The infection thread elongates and penetrates the root cortex, passing through existing cell walls and branching until it finally connects with a number of plant cells. As the rhizobial cells pass down the infection thread, they multiply, so each branch of the thread contains a number of bacterial cells. Finally these are emptied into the infected cells, where they become surrounded by a plant membrane called the

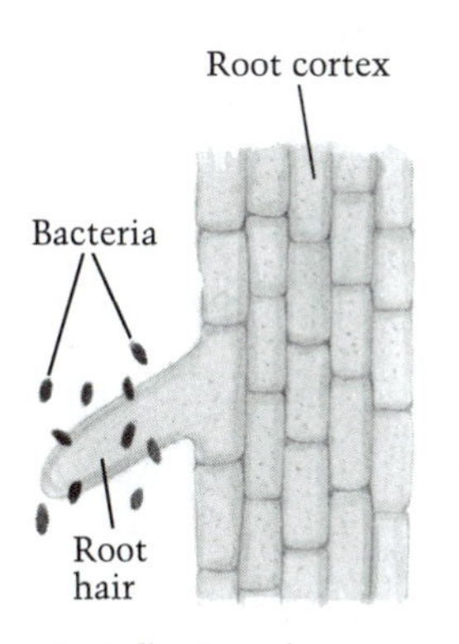

1 Adhesion of *Rhizobium* bacteria to root hair

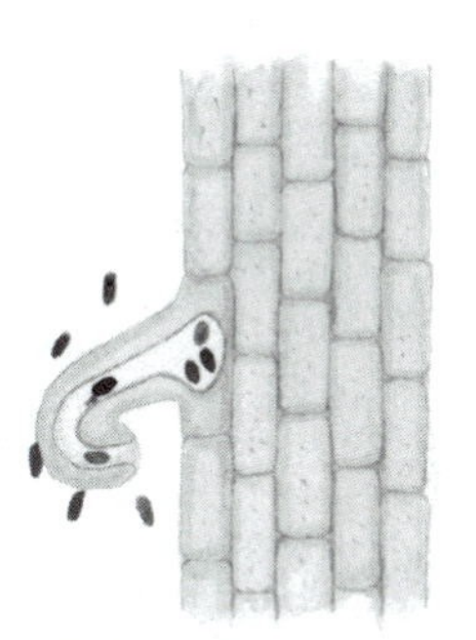

2 Infection: invagination and curling of root hair

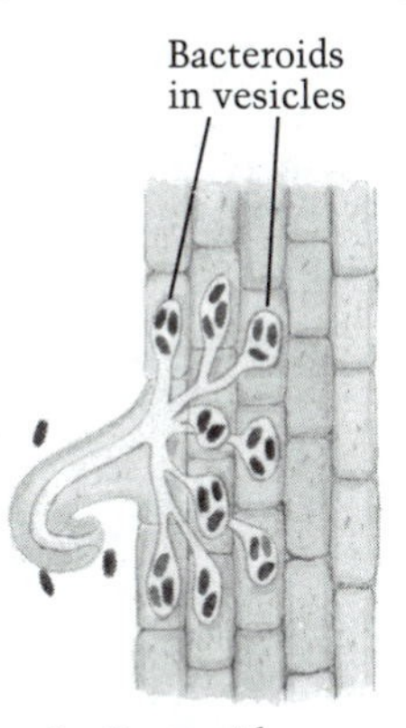

3 Bacteroid development in vesicles in root cortex

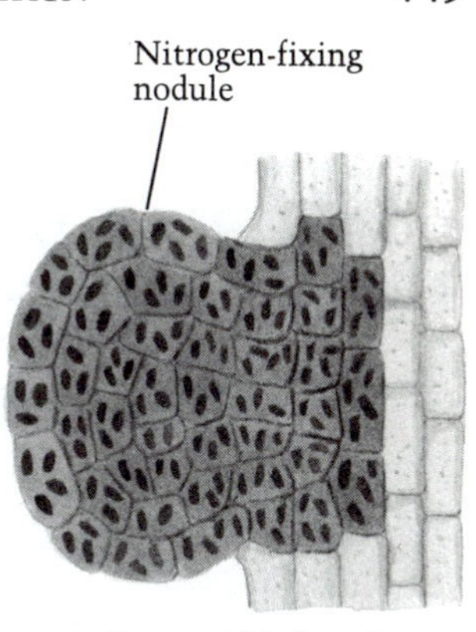

4 Bacteroid-induced cell division of cortex cells forming an nodule

Figure 4

Morphological changes leading to a nitrogen-fixing nodule.

PERIBACTEROID MEMBRANE, which has transport properties appropriate for passing components in both directions—supplying nutrients to the bacterium and fixed nitrogen to the plant. Plant cells in the region are stimulated to grow and divide, thus forming the nodule. The infected plant cells become densely packed with highly differentiated bacteria (Figure 5) that look quite unlike the free-living form. These bacterial cells—called BACTEROIDS—have thinner, less rigid cell walls, altered outer membranes, and a lower content of RNA. They become nitrogen-fixing machines; nitrogenase constitutes 10–12% of the protein in the bacteroid (over 1,000 times more than the amount in free-living rhizobia). When the plant dies, the bacteroids regenerate into normal bacterial cells which survive in the soil until a new plant is available for infection.

The plant creates for the bacteroids optimal conditions for fixing nitrogen. The most important of these are:

- supplying nutrients to meet the heavy energy cost of nitrogen fixation
- removing fixed nitrogen in the form of ammonia so it will not accumulate and inhibit fixation
- maintaining an optimal partial pressure of oxygen (pO_2)—low enough for oxygen-sensitive fixation to occur and high enough for the obligately aerobic rhizobia to generate sufficient ATP to drive fixation at full rate

Optimal pO_2 levels are maintained by LEGHEMOGLOBIN (an oxygen carrier that is structurally and functionally similar to animal hemoglobin), which buffers pO_2 at a constant value. This protein represents 25–30% of the total protein in the infected cell. The production of leghemoglobin emphasizes the intimacy of the legume–*Rhizobium* symbiosis. The protein portion of the molecule—called apoleghemoglobin—is encoded in a plant gene; the heme moiety is synthesized by the bacteroid. The infected cell produces several other new proteins called NODULINS, the functions of which are unknown.

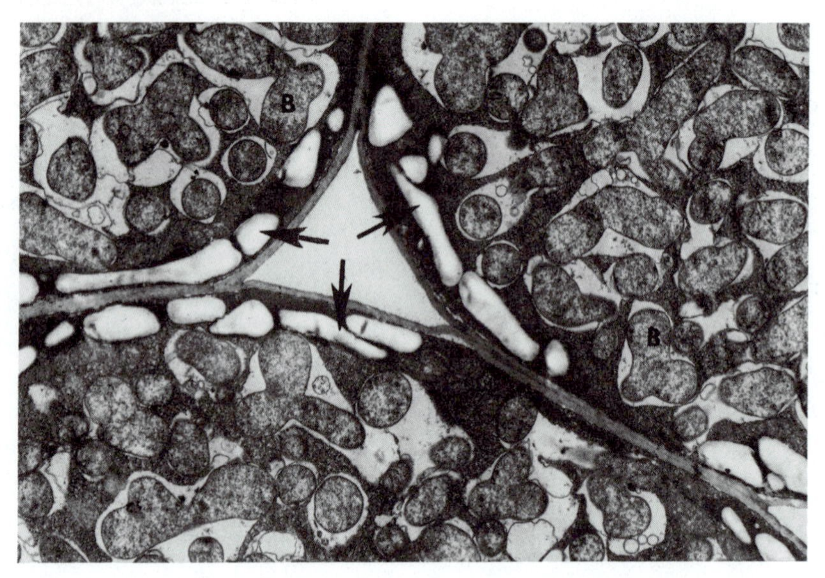

Figure 5

Infected cells in the root nodule of *Pisum sativum* filled with bacteroids (labeled B). Arrows indicate starch granules. (From Phillips et al., 1976.)

Although the sequence of morphological events leading to nodulation is now well understood, the molecular mechanisms by which they are effected remains almost a complete mystery. The promising approach that is now being pursued is the isolation of mutant strains (*nod* mutants) blocked at specific stages of the process. Knowing their biochemical defects should provide important hints about the molecular biology of this process of codifferentiation of the plant and bacterium. Already these studies have produced the surprising result that many *nod* genes are plasmid encoded.

***Cyanobacteria:* Free-living nitrogen fixers** / Many cyanobacteria can fix nitrogen in spite of a seeming paradox: nitrogen fixation is an obligately anaerobic process (nitrogenase is an extremely oxygen-sensitive enzyme) and cyanobacteria are oxygenic (oxygen-producing) phototrophs. Filamentous cyanobacteria solve the problem by producing specialized cells called HETEROCYSTS (Figure 6). Heterocysts seem to be designed for the purpose of fixing nitrogen—they house nitrogenase; they lack photosystem II (thus, they are unable to produce O_2); and they are surrounded by a thickened (presumably protective) cell wall and outer envelope.

The nitrogen-fixing role of heterocysts is further emphasized by the conditions that govern their formation. When ammonium ion is present in the growth medium, the formation of heterocysts is suppressed. (Other forms of fixed nitrogen, including nitrate, also suppress the formation of heterocysts, but not as effectively as ammonium ion, which is the primary product of nitrogen fixation.) Removal of ammonium ion triggers the formation of heterocysts, which in *Anabaena cylindrica* are completed within about 16 hours.

Fulfilling their specialized function of nitrogen fixation, heterocysts receive carbon compounds from adjoining cells and use these compounds as a source of energy to fix nitrogen. In return, heterocysts supply their adjoining cells with fixed nitrogen. Not surprising in view of their physiological role, heterocysts are formed at regular intervals—about every tenth cell—within a cyanobacterial filament.

In contrast to bacteroid formation, heterocyst formation is a TERMINAL DIFFERENTIATION—so called because they never become vegetative

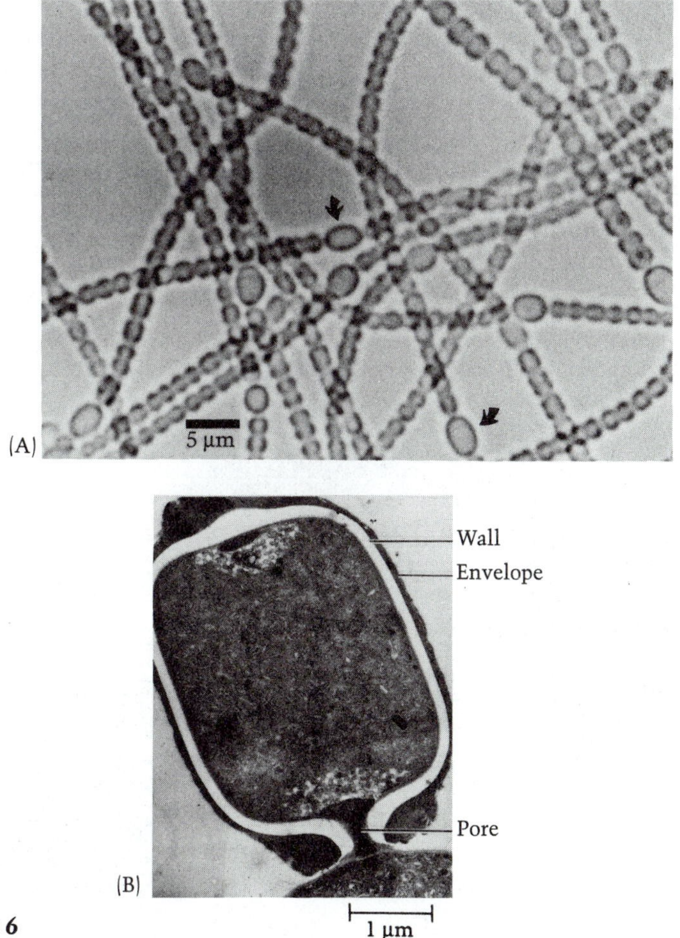

Figure 6

Heterocysts of the cyanobacterium *Anabaena cylindrica*. (A) A filament of *Nostoc* sp. composed of heterocysts (arrows) and vegetative cells. (B) Glutaraldehyde–osmium-fixed thin section of a heterocyst, showing the pore connecting it to adjoining cells, its thickened wall, and its protective outer envelope. (A from Meeks, 1988; B from Fay and Lang, 1971.)

cells again. The differentiation in *Anabaena variabilis* involves an irreversible change in the cell's genome: an 11 kbp segment of the chromosome is excised and becomes a circular molecule.

Site occupancy by *Caulobacter*

A group of Gram-negative bacteria—called PROSTHECATE BACTERIA—found in aqueous environments are characterized by protuberances—called PROSTHECAE—that extend up to a cell length or more from the cell, giving some of these bacteria the appearance of a starfish or a spider (Figure 7). The prostheca is an extension of the cell itself. Like the rest of the cell, the prostheca is surrounded by the typical Gram-negative cell envelope, which is composed of the cytoplasmic membrane, the murein layer, and the outer membrane. The function of prosthecae is probably intimately associated with the aquatic habitat. Prosthecae might be selected because they increase the surface area and therefore the ability of these bacteria to take nutrients from their typically nutrient-poor environment. It has also been postulated that prosthecae are selected because they decrease the rate at which cells settle in a liquid column, thereby keeping them more easily suspended even by quite gentle currents.

The prosthecate bacterium *Caulobacter crescentus* has an elaborate life cycle (Figure 8) that has attracted the interest of scientists who study

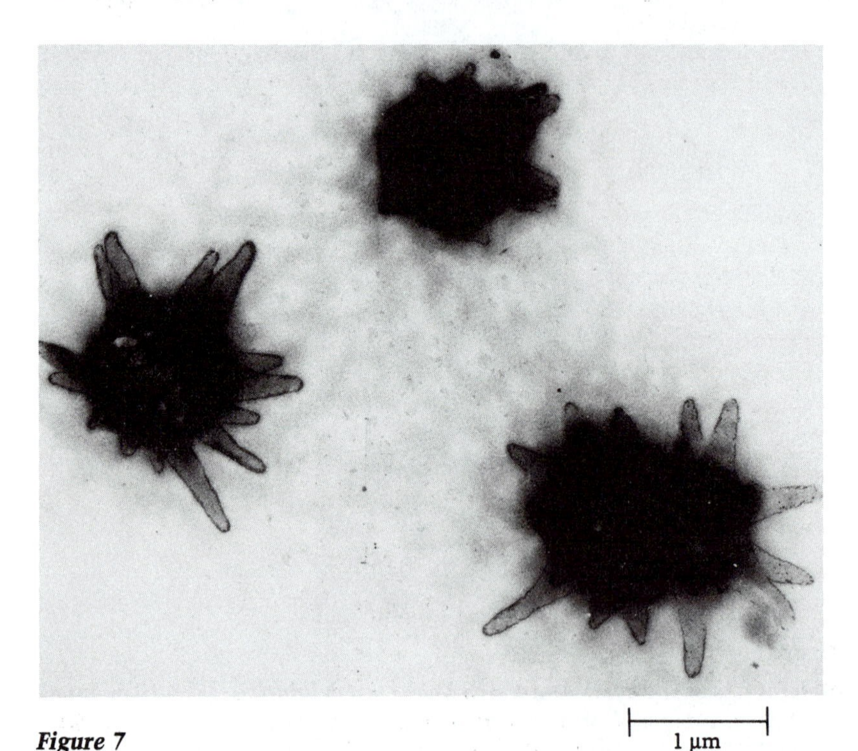

Figure 7

The prosthecate bacterium *Prosthecomicrobium pneumaticum*. The light rod-shaped areas within the cells are gas vacuoles. (From Staley, 1981.)

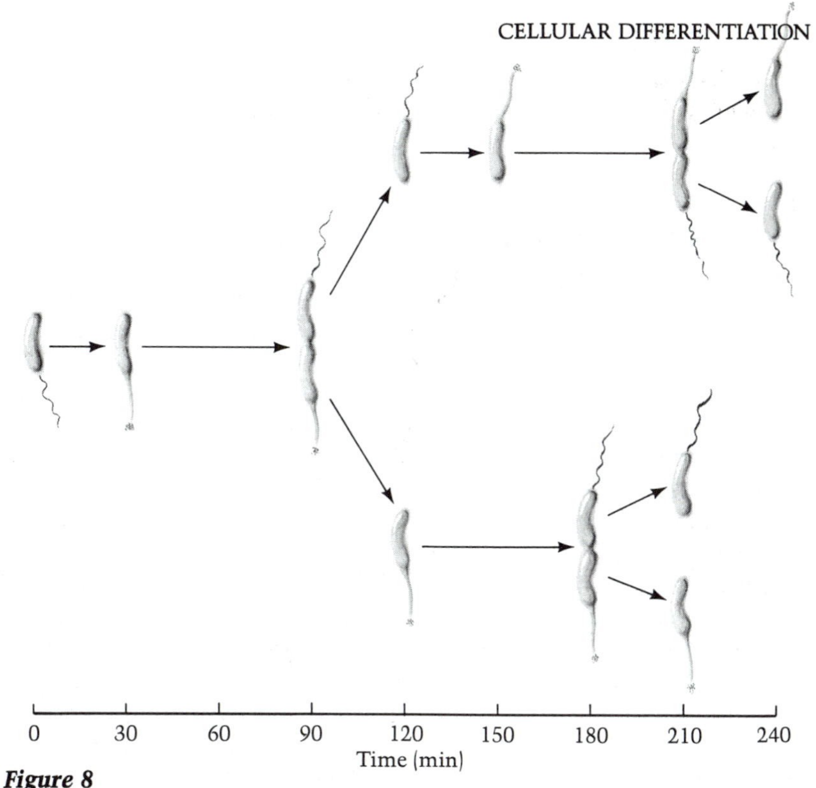

Figure 8

The life cycle of *Caulobacter*. The times at which various events occur in a particular set of conditions are shown on the bottom scale. (After Poindexter, 1962.)

bacterial differentiation. One form of *Caulobacter* is a vibrio-shaped cell with a single prostheca (a stalk) ending in a localized region of sticky material called a HOLDFAST, which attaches to solid surfaces. (In nature, *Caulobacter* seem to attach themselves preferentially to other microorganisms, thereby setting up residence in a nutrient-enriched environment.) The stalked cell elongates and periodically divides to produce a flagellated swarmer cell. Swarmer cells do not divide; rather, each swarmer cell differentiates into a stalked cell that settles in a new location and produces more swarmer cells. The unequal division of a stalked cell to produce a swarmer cell and the eventual development of a swarmer cell into a stalked cell are examples of differentiation that are being intensively studied.

Resistance to environmental insults

Endospores / A few bacteria—notably those belonging to two genera of Gram-positive bacteria, *Bacillus* and *Clostridium*—form ENDOSPORES (so-named because they form inside a vegetative cell). These unique biological structures are the most resistant forms of life we know. They can withstand temperatures well above the boiling point of water and conditions of extreme desiccation. They can remain viable in their metabolically inactive state for at least 300 years (as evidenced by the recovery of viable spores from soil on the roots of specimens stored in a British herbarium) and then germinate in a few minutes when conditions again become favorable for vegetative growth.

Endospores have demanded the attention of microbiologists for very practical reasons. Because they are the most highly resistant life forms known, any procedure to sterilize material—most notably in the preservation of food by canning—must be designed with endospores in mind. If endospores are killed by the procedure, all other microorganisms will surely be dead. Moreover, it is particularly important to eliminate spores from food to be stored for any significant period of time, because some spore-forming bacteria produce lethal toxins. Botulinum toxin, produced by *Clostridium botulinum*, is one of the most deadly substances known. More recently, studies on spore formation have been motivated by a desire to understand more fully this elaborate example of procaryotic differentiation.

Unfavorable conditions—for example, running out of nutrients—trigger the formation of endospores (Figure 9). The nucleoid stretches out, forming an axial filament. It divides, and the cytoplasm undergoes an unequal division. The smaller product of division (the FORESPORE) becomes the endospore. The forespore is engulfed by the remainder of the mother cell, which makes many components for the endospore. Then a heavy wall (the cortex and spore coat) is laid down between the pair of membranes formed around the forespore. When the endospore is mature, it is usually released by lysis of the surrounding cell.

Most endospores cannot germinate immediately after they have formed. But they can after they have rested for several days or after they have been heated briefly (5 minutes at 60°C is optimal for many endospores). Then, if proper nutrients are present, the spore takes in water and swells; the spore coat cracks; and a vegetative cell (germ tube) grows out of it, producing an ordinary looking vegetative cell.

An endospore is quite unlike a vegetative cell. It is smaller and highly refractile. It also differs physically and chemically, having a much lower content of water (which is probably the major contributor to the spore's heat resistance) and containing large amounts of a compound of unknown function (calcium dipicolinate). Vegetative cells contain this compound in miniscule quantities as an intermediate in the synthesis of the amino acid lysine.

Differentiation of a vegetative cell of *Bacillus subtilis* (the most thoroughly studied sporeformer) into an endospore takes about 7 hours under standard conditions in the laboratory. Different morphological and chemical events occur at sequential stages of the process. To distinguish these, investigators have designated the hours after the onset of sporulation T_1 through T_7 and the corresponding stages (Figure 9) I through VII. The initiation of the process is called stage 0; stage I, when the axial filament forms, is followed by stages II through VII, when the mature spore is released from the mother cell.

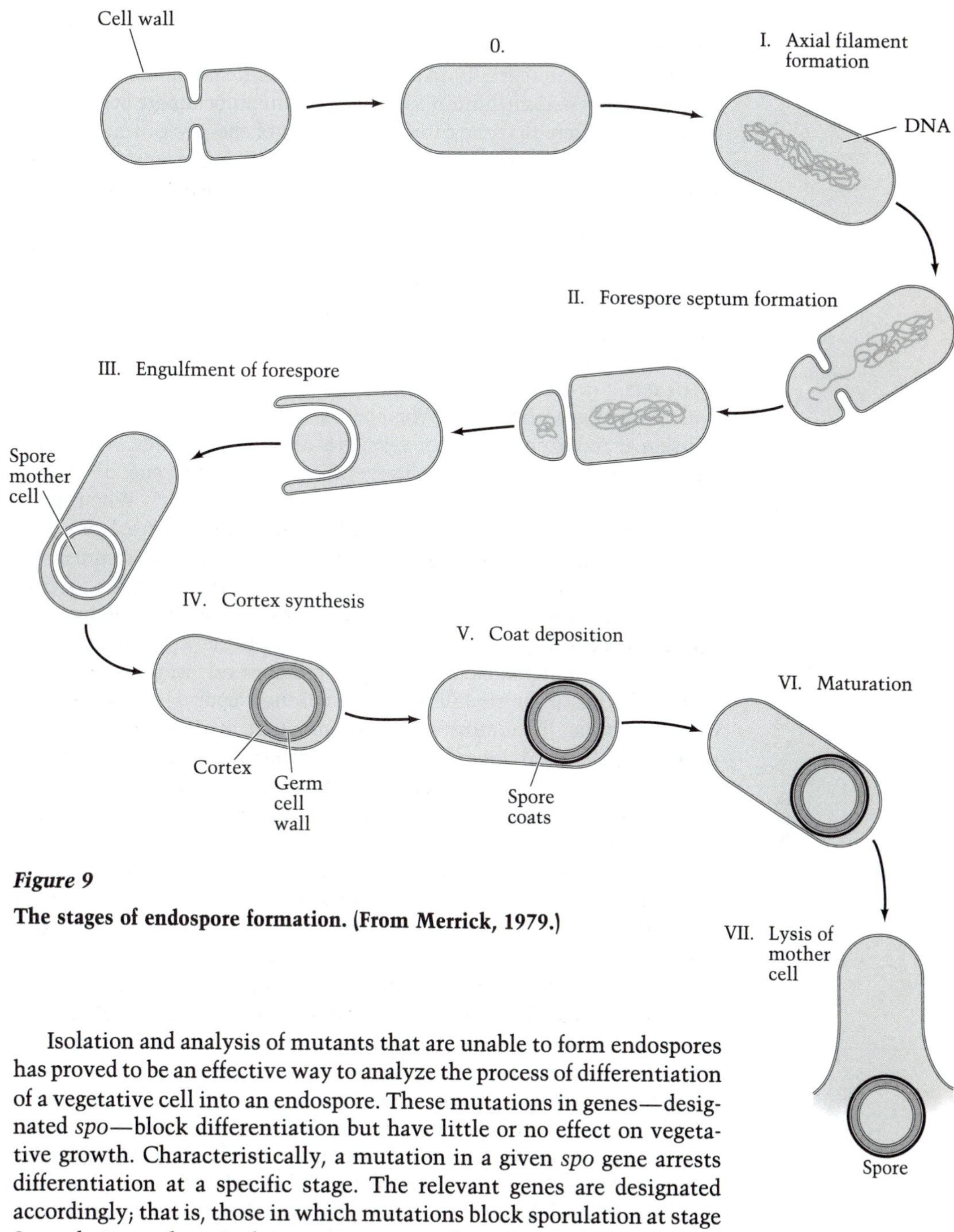

Figure 9

The stages of endospore formation. (From Merrick, 1979.)

Isolation and analysis of mutants that are unable to form endospores has proved to be an effective way to analyze the process of differentiation of a vegetative cell into an endospore. These mutations in genes—designated *spo*—block differentiation but have little or no effect on vegetative growth. Characteristically, a mutation in a given *spo* gene arrests differentiation at a specific stage. The relevant genes are designated accordingly; that is, those in which mutations block sporulation at stage 0 are designated *spo0,* those at stage I are designated *spoI,* and so on. Within each class, individual genes are designated by an additional letter; for example, the majority of mutations that block differentiation before stage I lie in five principal loci, designated *spo0A, spo0B, spo0E, spo0F,* and *spo0H.*

The fact that mutations in *spo* genes arrest differentiation at specific stages implies that the process follows a genetic program in which genes are turned on sequentially to accomplish specific chemical or morphological changes. Some hints about how this might occur are beginning to emerge. It is clear that the turning on and off of specific genes is brought about in large measure by differential expression of sigma (σ) factors. Recall that sigma is the subunit of RNA polymerase that causes it to recognize promoters (Chapter 3). An RNA polymerase molecule with one kind of sigma factor will recognize one set of promoters; with another it will recognize a different set. In *E. coli,* association of RNA polymerase with one kind of sigma factor causes it to recognize promoters of genes controlling nitrogen metabolism; another, heat-shock proteins; a third one, motility and chemotaxis genes; and a fourth one, all others.

In spore formation by *Bacillus* species, different sigma factors control the sequential expression of *spo* genes at appropriate times. Most genes used by *Bacillus subtilis* during vegetative growth are transcribed by an RNA polymerase that contains the major sigma factor, σ^A. When sporulation begins, σ^A loses activity and other sigma factors cause RNA polymerase to transcribe sporulation-specific genes at appropriate times. For example, σ^E-containing RNA polymerase is the principal one functioning at T_2. Others function at other times. Some *spo* genes themselves encode spore-specific sigma factors. Among these are *spo0H,* which encodes σ^H, the sigma factor playing a dominant role in gene expression during the first 2 hours following initiation of sporulation. Other *spo* genes must play regulatory roles. The products of two of them, *spo0A* and *spo0F,* are highly homologous with proteins like the *ntrB* and *ntrC* gene products that comprise the two-component regulatory system controlling nitrogen metabolism in *E. coli* (Chapter 13). Still other *spo* genes must be structural genes that encode specific constituents of spores. These must include enzymes to synthesize assembly proteins and actual structural components. Spores also contain storage material in the form of low-molecular-weight proteins, whose amino acids are used to make vegetative proteins when the spore germinates. These storage proteins, which are made under the direction of a sporulation-specific sigma factor, also bind tightly to DNA and by so doing are thought to contribute to the remarkable radiation resistance of spores. Mutant strains that are unable to make these proteins sporulate normally but suffer decreased resistance to the lethal effects of radiation.

Although the general outlines of differentiation of vegetative cells into spores is beginning to take shape, a thorough understanding of the signals and the orderly cascade of sequential morphogenetic events must be a long way off. Even the primary chemical signal remains unknown. Depriving a culture of a source of carbon, nitrogen, or phosphorus induces sporulation, but what chemical transducers sense these diverse physiological states and respond by initiating sporulation? It has been proposed that a guanine nucleotide might be the chemical signal. It contains the three signal elements, carbon, nitrogen, and phosphorus, and a variety of genetic lesions and inhibitory compounds that reduce intracellular pools of guanine nucleotides induce sporulation. Possibly (p)ppGpp (the nucleotide that mediates the stringent response in *E. coli*) is the actual signal to begin the process of sporulation.

Streptomyces / The actinomycetes are procaryotes with a growth habit that in certain respects resembles that of fungi. The dominant genus of this group—*Streptomyces*—has long been the object of intense genetic and biochemical study because it comprises many antibiotic-producing species. More recently, its developmentally complex life cycle (Figure 10) has caught the interest of microbiologists interested in differentiation. The most highly studied species—*Streptomyces coelicolor*—produces blue-gray spores; they germinate to produce a mat of mycelium, which has only occasional cross-walls, spreads over a solid surface such as an agar medium, and penetrates into it. A few days later, aerial hyphae or filaments develop; these structures coil and, at their ends—following multiple cell divisions and maturation—differentiate into chains of spores. The spores are not particularly resistant to heat or other unfavorable conditions, so they may be primarily organs of dispersal of the species.

The differentiation of *S. coelicolor* colonies into aerial mycelia and spores can be blocked by certain mutations. Mutations in one class are in genes designated *bld* (**b**a**ld**); mutants with *bld* defects are unable to form aerial mycelium. Mutations in another class are in genes designated *whi* (**whi**te). Mutants with *whi* mutations form an aerial mycelium but no spores; so a colony of *whi* mutants is uncolored. About eight *bld* genes and eight *whi* ones have been identified after intensive study, a finding suggesting that the total number of genes dedicated to differentiation of a *S. coelicolor* colony probably does not exceed 20.

The product of one of the *bld* genes, *bldA*, has been identified as a $tRNA^{Leu}$ that recognizes the UUA codon. This codon must be only rarely used in *S. coelicolor.* Because *bldA* mutants are capable of vigorous vegetative growth, the codon must not be used in genes encoding primary metabolism, and indirect evidence suggests that genes encoding the formation and development of aerial mycelia also lack it. But expression of *bldA* is consistent with its role in regulating the development of aerial mycelia: *bldA* transcripts are not detectable in young cultures; they reach maximum levels at about 36 hours. Presumably the few genes in *S. coelicolor* that utilize the UUA codon are regulatory genes that control the development of aerial mycelia.

One of the *whi* genes, *whiG,* is homologous to genes encoding RNA polymerase sigma factors in other procaryotes. Presumably, it acts in differentiation of *S. coelicolor* as the alternate sigma factors in *B. subtilis* do in sporulation.

Stationary phase / Until recently the transition of nonsporulating bacteria into the stationary phase of growth was considered to be a passive event—the automatic consequence of stopping growth. Now it is clear

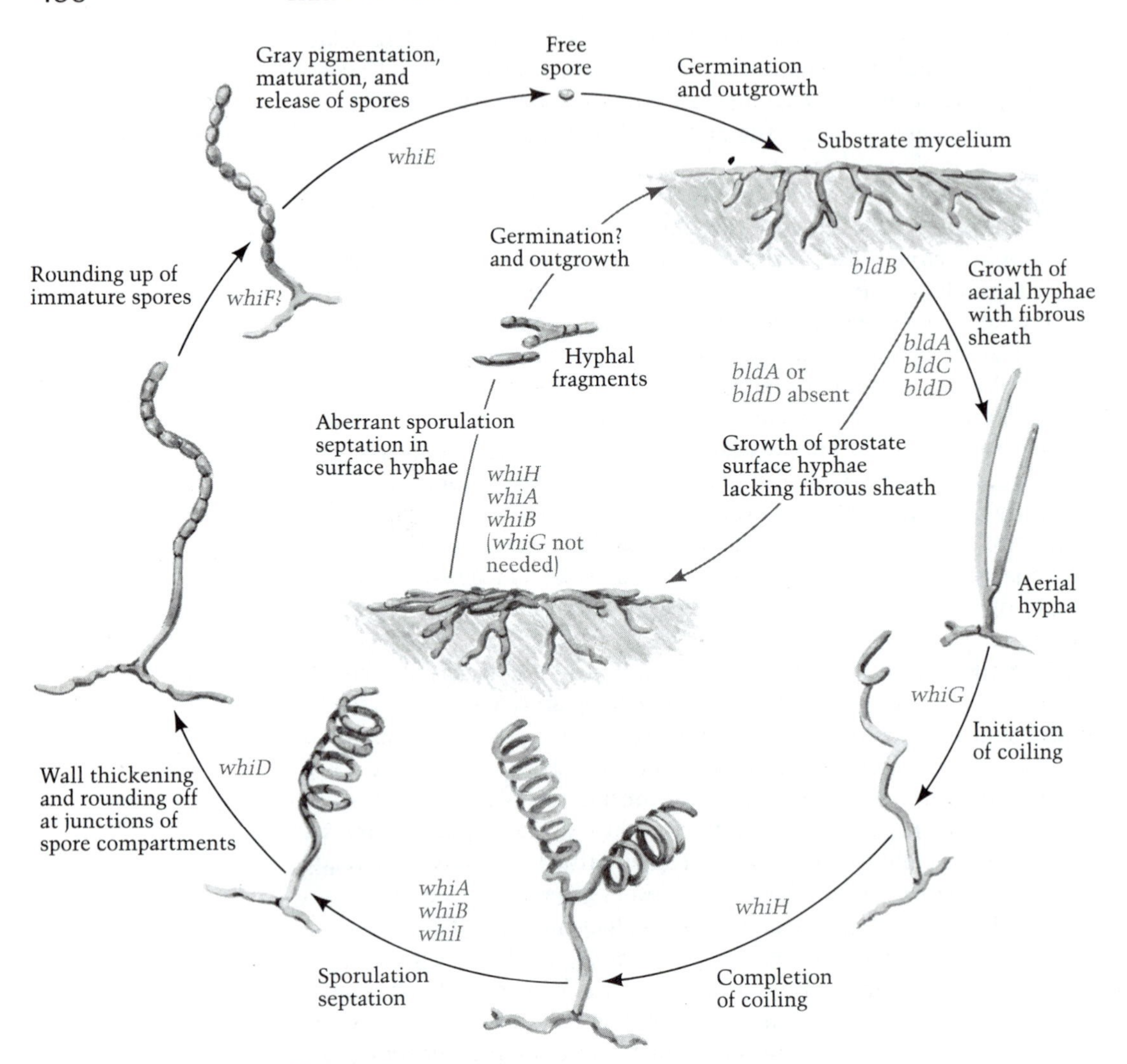

Figure 10

The life cycle of *Streptomyces coelicolor*. The genes known to participate in each stage of differentiation appear inside the arrows. (After Chater and Merrick, 1979.)

that this transition is a form of differentiation that is considerably simpler but functionally analogous to sporulation—a response to starvation that maximizes the cell's chances of survival until conditions again become favorable. It is not surprising that this condition should be finely honed by natural selection because starvation or near starvation is the norm in nature. Owing to the scarcity of nutrients, the mean doubling time of bacteria in deep seawater is 210 days and in deciduous woodland soil, 20 days. Even in the seemingly rich environment of host tissues, pathogenic strains of *E. coli* and *Salmonella typhimurium* may double only every 5 to 20 hours as a result of the restrictions of nutrient limitation.

An obvious difference of stationary-phase cells is their smaller size, but this trait probably is a passive change—reflecting the tendency of cells entering the stationary phase to complete rounds of chromosomal replication, undergo the cell division that this event signals (Chapter 14), but not increase their mass significantly. Other less obvious changes profoundly alter the cell's physiology, rendering it more resistant to a variety of physical and chemical insults, as well as making it better able to withstand prolonged starvation. Stationary-phase cells are more resistant to heat, disinfectants, and osmotic stress.

One of the changes that occurs when *E. coli* enters the stationary phase deals with the protection of its unsaturated fatty acids from oxidation. The rapid addition of methylene groups to the unsaturated fatty acids in its phospholipids forms oxidation-resistant cyclopropane fatty acids. A variety of other changes must also occur, because as soon as growth stops, the turnover of proteins increases about six-fold, thereby making building blocks available for the synthesis of new stationary phase-specific proteins. A surprising number of such proteins are synthesized, the actual number depending on the nutrient that has been depleted. For example, *E. coli* synthesizes at least 30 proteins when deprived of a carbon source, and 26 to 32 when deprived of nitrogen. About 13 of these appear to be general stationary-phase proteins, because they are synthesized following all three types of deprivation.

The pattern of synthesis of stationary-phase proteins varies. Some are synthesized in a burst immediately after nutrient depletion; others are made slowly long into the stationary phase. The early stationary-phase proteins seem to be the most important to the cell's long-term survival during starvation, because, if an inhibitor of protein synthesis (chloramphenicol) is added for a brief period immediately after nutrient depletion, long-term survival is dramatically decreased. Later addition for longer periods has less effect on survival.

The way that certain genes are selectively expressed during starvation remains unanswered, but it is an issue of considerable practical as well as basic significance. Attaching a gene to a promoter of a stationary-phase protein should cause selective expression of the attached gene by nongrowing cells, a condition that could maximize production of commercially useful products by immobilized cells under industrial conditions.

Other paradigms of procaryotic differentiation

Arthrobacter is a genus of soil bacteria belonging to the coryneform group of Gram-positive bacteria. Most of its species undergo simple but easily manipulable morphogenetic changes. Old cultures are composed of nonmotile cocci, but when these cells are inoculated into a rich medium, they elongate and grow as rods; some species become motile during this stage. As the growth of the culture slows, the cells convert themselves again into cocci. These changes do not occur in defined media, a restriction giving the investigator the power to induce the coccus-to-rod shift at will. The cyclic nucleotide cAMP, which is a transducer of so many regulatory signals in eucaryotes as well as procaryotes, seems to play a role in the coccus-to-rod transition: under certain conditions of induction of the rod state, the intracellular concentration of cAMP increases 30-fold.

Vibrio alginolyticus is also an object of investigation, because it undergoes a touch-sensitive differentiation. It is peritrichously flagellated when it is in contact with a solid surface and polarly flagellated when it is not.

SUMMARY

1. Understanding the mechanisms of differentiation is one of the major challenges of modern biology. The complexity of these processes has encouraged biologists to study the more tractable procaryotic differentiations as model systems. But owing to the fundamental differences between procaryotic and eucaryotic cells, processes of differentiation in the two systems might differ fundamentally. For example, procaryotic cells lack a cytoskeleton, so their shape is determined externally by the way the murein layer is laid down rather than internally by the pushing and pulling of microtubules and microfibrils.
2. Although some bacteria—for example, *E. coli*—undergo little or no differentiation during their life cycle, many others do. Examples of procaryotic differentiation that seem to be primarily concerned with nutrition include

   - *Myxococcus xanthus*, a myxobacterium. This species exists at one stage of its life cycle as swarms of individual gliding cells that are highly successful predators on other microorganisms. When nutrients are depleted, these cells aggregate to form a macroscopically visible fruiting body within which resistant myxospores are formed.
   - Establishment of symbiosis between the nitrogen-fixing rhizobia and leguminous plants involves an elaborate and simultaneous differentiation of both partners. This interaction leads to the formation of a nodule on the root of the plant in which differentiated rhizobia are maintained in an optimal nitrogen-fixing environment. The plant host supplies nutrients and a favorable pO_2. The bacterium supplies fixed nitrogen in the form of ammonia.
   - Certain cells of filamentous cyanobacteria differentiate into morphologically and biochemically specialized cells called heterocysts, within which nitrogen fixation takes place.

3. Examples of microbial symbioses that are primarily concerned with finding an optimal location and staying there include *Caulobacter crescentus*, a prosthecate bacterium that alternates during its life cycle between being a sessile, stalked cell and a polarly flagellated, motile one.

4. Microbial differentiations that protect the microorganisms from environmental insults include

   - Spore formation. *Streptomyces* is a soil bacterium that differentiates into substrate and spore-bearing aerial mycelia.
   - Endospore formation. Because of its practical importance and fundamental biological significance, endospore formation by species of *Bacillus* is probably the most thoroughly studied example of bacterial differentiation. The formation of these highly resistant structures is triggered by starvation and seems to proceed by differential gene expression directed by changing mixtures of sigma factors, which control the specificity of RNA polymerase.
   - The stationary phase. Bacteria that are not markedly differentiated, including *E. coli*, undergo a subtle form of differentiation when they are starved and enter the stationary phase. Under these conditons, they produce about 30 new proteins, which help them survive periods of starvation.

5. The selective advantage of some forms of differentiation is not readily apparent. For example,

   - The soil bacterium *Arthrobacter* alternates between coccal and rod forms during its growth cycle.
   - The marine bacterium *Vibrio alginolyticus* is either polarly or peritrichously flagellated, depending on whether or not it is in contact with a solid surface.

STUDY QUESTIONS

1. Which distinctions between procaryotic and eucaryotic cells would lead you to suspect that differentiation would proceed differently in these two groups? Which similarities would suggest comparable patterns of differentiation?
2. Symbiosis between rhizobia and leguminous plants has been proposed to be the evolutionary product of a pathogen that acquired the ability to fix nitrogen and, alternatively, of a nitrogen fixer that acquired the ability to invade plant tissue. Which alternative do you favor? Why?
3. Compare the advantages and disadvantages of the strategies that *Bacillus subtilis* and *E. coli* have evolved to survive periods of starvation.

SUGGESTED READING

Chater, K. F. 1989. Aspects of multicellular differentiation in *Streptomyces coelicolor* A3(2). In *Genetics and Molecular Biology of Industrial Microorganisms*, C. L. Hershberger, S. W. Queener and G. Hegeman (eds.) American Society for Microbiology, Washington, D.C.

Parish, J. H. (ed.). 1979. *Developmental Biology of Prokaryotes*, University of California Press, Berkeley and Los Angeles.

Losick, R. and P. J. Youngman. 1984. Endospore formation in *Bacillus*. In *Microbial Development*, R. Losick and L. K. Shapiro (eds.). Cold Spring Harbor Laboratory, Cold Spring Harbor, New York.

Losick, R., P. J. Youngman and P. J. Piggot. 1986. Genetics of endospore formation in *Bacillus subtilis*. Annu. Rev. Genet. 20:625.

Matin, A., E. A. Auger, P. H. Blum and J. E. Schultz. 1989. Genetic basis of starvation survival in nondifferentiating bacteria. Annu. Rev. Microbiol. 43:293.

Setlow, P., R. Slepecky and I. Smith (eds.) 1989. *Regulation of Prokaryotic Development*. American Society for Microbiology, Washington, D.C.

17

Physiological Ecology

INTRODUCTION

The early development of molecular biology, roughly in the years 1940 through 1970, depended largely on studies of bacteria carried out under laboratory conditions. Recently, it has become apparent that much can be learned from the study of microorganisms in their natural environments. Although the questions are age-old, the use of modern molecular approaches to answer them is in its infancy. At this time, it is not yet possible to provide definitive answers to the major problems of physiological ecology. However, this field has attracted outstanding researchers, and progress can be expected soon. In this chapter, we limit ourselves to examples of the "social life of bacteria."

In most natural environments, bacteria do not live in isolation but share a site with other members of the microbial world, such as fungi, protozoa, and other bacteria. In addition, bacteria often interact with animals and plants in health and disease. The complexities of these interactions are great indeed and do not lend themselves to simple analysis.

As discussed in Chapter 1, bacteria in natural situations face three main classes of problems:

- Starvation or at least gradual depletion of some essential nutrients. This depletion may be due to the metabolic activities of the organisms themselves or to those of other species present in the same environment.
- Competition for surface sites. Many bacteria adhere to natural surfaces, for example, the human gut, plant leaves, and soil particles. Often several species are able to stick to the same kind of surface, thus competing for occupancy of that niche.

- Exposure to noxious chemicals. Again, these may be produced by the organisms themselves (for example, acid from fermentation) or by other species. In addition, bacteria can elaborate specific toxins—termed bacteriocins and microcins—that may kill other species.

It is possible (some microbiologists would say even probable) that a very large part of the genome of bacteria is dedicated to their social life—coding for proteins involved not in growth per se, but in coping with the environment. This conviction stems not only from the growing information about specific survival mechanisms (Chapter 13) but also from the recognition that the biochemistry of bacterial growth may require no more than one-third of the 4,000-protein coding capacity of the chromosome of most bacteria. There is good evidence, for example, that there are no extensive regions of the *E. coli* chromosome that do not encode proteins or RNA products. Under any given laboratory growth condition, however, more than half the genes of this organisms seem dispensable, a conclusion based on the ability of mutants to grow and on the actual fraction of the chromosome that is active under a given growth condition.

The study of how bacteria cope with these challenges inevitably invites teleological interpretations. Many metabolic activities of bacteria appear designed to help the organisms thrive by modifying their environment. Granted that such an interpretation usually reflects our human perception, much can be learned from the examination of a few examples.

INCREASING THE AVAILABILITY OF NUTRIENTS

In the life of a bacterium, any of a number of essential nutrients can and often do become limiting. Many bacterial species do not wait passively for nutrients to become available but take an active role in acquiring them. A general solution is for the organisms to develop the ability to concentrate nutrients when they are present in low concentrations. This tactic helps chiefly in environments where nutrient concentration is low and usually not changing abruptly, such as in lakes and other bodies of water. Bacteria have other strategies for increasing the availability of nutrients. One is to produce extracellular hydrolases, which break down complex molecules in the environment into simpler ones that can be used for nutrition (Chapter 6). The action of these enzymes allows growth in environments that may otherwise appear to be depleted of available nutrients.

Iron

A specific nutritional strategy that deserves special mention is that used by bacteria to take up iron. Iron is essential for the production of enzymes such as peroxidases and cytochromes. Iron, although abundant, is not as readily available as one might think. When oxygen was introduced into the earth's atmosphere, much of the accessible iron became oxidized to an extraordinarily insoluble form—ferric hydroxide (whose solubility constant is less than 10^{-38}!). In addition, in the body of vertebrates, iron is present in small amounts, very little of which is available to microorganisms. Essentially all of the iron in the vertebrate body either is tied up in hemoglobin and cytochromes or is complexed with proteins, such as transferrin and lactoferrin, that have an unusually high affinity for this mineral. Thus, in the blood, bacteria are usually starved for iron. How do they respond to this situation? Some pathogens, for example, the β-hemolytic streptococci, produce toxins (hemolysins) that lyse red blood cells to reach iron in hemoglobin. Many species obtain iron by a more delicate mechanism. These organisms produce and secrete low-molecular-weight iron chelators known as SIDEROPHORES, such as enterochelin and hydroxamates. These compounds have an incredibly high affinity for iron (Figure 1) and scavenge iron molecules in successful competition with the host's iron-binding proteins. How does this characteristic help the bacteria? The outer membrane of Gram-negative cells is, in principle, impervious to these iron chelates. However, many Gram-negative strains have developed in their outer membrane special uptake systems that are specific for their own iron chelates. Because there are many different kinds of siderophores and because they are often species specific, particular iron chelates are usually not taken up by different species. Once taken up, however, siderophores are broken down, thus releasing iron into the cytoplasm.

In pathogenic bacteria a powerful iron-uptake system is often a requirement for virulence, otherwise the organisms could not multiply in tissues. For example, many *Escherichia coli* strains that cause meningitis in young children carry plasmids that code for the manufacture of a potent iron chelator. Iron-uptake systems in general are induced in the absence of the metal and repressed in its presence. In the cases studied,

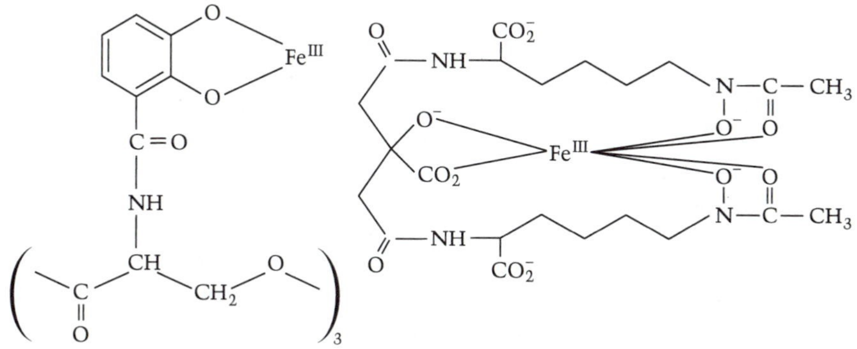

Enterobactin
$K_s = 10^{52}$

Aerobactin
$K_s = 10^{23}$

Figure 1

Structure of two representative iron–siderophore complexes. The thermodynamic stability constants also are shown. (From Bagg and Neilands, 1987.)

regulation is by a repressor that is activated by bound iron. Because these systems are expensive to make, it is not surprising that they are only synthesized when needed. An alternative mechanism that requires less energy has been developed by the meningococci. These bacteria bind transferrin and lactoferrin on their surface. Thus, meningococci can avail themselves of the host's own iron-binding proteins to satisfy their requirement for this metal.

COMPETITION FOR FOOD

The gut

Coping with the shortage of iron is a problem that concerns one nutrient only. What problems do bacteria face in more complex nutritional situations? In many instances, bacteria must compete with many other species and for many kinds of food. Living in the large intestine of humans and other vertebrates illustrates this problem. At least 300 different species of bacteria have been isolated from the feces of a single individual. Most of these organisms are strict anaerobes, and specialized techniques are required for their isolation. These bacteria vary in abundance from about 10^{11} per gram of feces to a barely detectable level (around 10^3 per gram). For all its popularity, *E. coli* is not usually a majority species; it is found at about 10^6 cells per gram of feces. *Escherichia coli* is, however, the most common facultative anaerobic species in feces. The total number of species is probably higher than 300 and appears to be different for each individual human or animal ("to each his own"). These organisms are fed by the periodic influx of partially digested food from the small intestine and from endogenous secretions, including glycoproteins and sloughed cells. The valve that separates the two segments of the gut opens up some 20 times a day in humans, thereby allowing nutrient-rich fluid to squirt into the cecum and the ascending colon.

How can over 300 species of bacteria coexist in the same environment, day after day? Does each species makes exclusive use of one kind of nutrient? This may seem unlikely at first glance because, when tested in vitro, typical intestinal bacteria can utilize a variety of different substances. How relevant is this finding? Throughout most of the large intestine, nutrients are present at low concentrations; thus their utilization at high concentration (as used for in vitro testing) may not be relevant to the situation in the gut. It is possible that some of the intestinal species have evolved to be especially efficient at utilizing one or another nutrient at low concentrations, a trait that would make them highly competitive in the gut. We do not know to what extent this consideration explains why there are so many species of intestinal bacteria or whether the answer has to be sought in other types of competitive advantage. Many other factors must be included, such as the ability to withstand the inhibitors present in the gut contents, for example, H_2S and short-chain fatty acids. Equally important may be the ability of certain species to adhere to the intestinal wall, a subject we consider in more detail later. It seems especially difficult to explain why so many species are present in small but relatively steady numbers. Why are these species not overgrown by the major species? Does part of the answer reside in the occupancy of special sites on the gut wall by different species?

COMPETITION FOR SURFACE SITES

Some species have a great ability to stick to surfaces, such as soil particles or the intestinal wall. Note that, a priori, if an organism is capable of adhering to a surface site, then it does not face the same selective pressure to develop efficient nutrient utilization. In times of starvation, it simply sits it out. Here, occupancy may be more important than affinity for nutrients. This conclusion is particularly true if the surface is invariant or at least if its properties change relatively slowly. In the case of the gut, epithelial cells slough off and regenerate. However, because this process takes days (a relatively slow pace), bacteria may be able to re-attach to new adjacent sites.

When examined under a scanning electron microscope or in thin sections, the surface of a soil particle in a stream or that of the large intestine are indeed covered by layers of bacteria (Figure 2). There is

Figure 2

Scanning electron micrograph (×5,000) of a natural biofilm from a rock in a river near Guelph, Canada. The larger objects are diatoms, the smaller spherical ones, bacteria. (Courtesy of T. J. Beveridge.)

evidence that in the intestine the bacterial mat is a three-dimensional mosaic and that there are microcolonies of different organisms both along the plane of the surface and across the various layers. Some of the bacteria are attached directly to the epithelial cells, others are apparently stuck in the mucus and to each other.

Little is known about the detailed mechanisms involved in adherence of bacteria to the intestinal wall. Obviously, the microbial ecology of the gut is hard to study under natural conditions. It has been possible, however, to reconstruct at least some of the conditions in the intestine by running a chemostat that uses feces as the inoculum. As described in Chapter 7, a chemostat is a continuous-culture device with a constant inflow of nutrients and a corresponding outflow of fluid from a growth vessel. Using anaerobic conditions to simulate the gas phase of the intestine, Freter and colleagues found that if they waited days or weeks, the chemostat walls accumulated a paste of bacteria several layers thick. Interestingly, these bacteria belonged to many species, some of which adhered directly on the glass surface of the vessel, others on top of them. The chemostat-grown bacteria could repopulate germ-free animals to form a fully functional intestinal flora. Thus, the situation mimics that in the gut, but there are obvious differences. The gut surface is not made of glass, and, being semipermeable, it is a considerable source of nutrients for bacteria that attach to it. Nonetheless, this experiment opens the door to the detailed study of the properties of the organisms involved, with regard to both their adherence properties and their metabolic potential.

Pili and other adhesins

How do bacteria adhere to surfaces? Many bacteria have ADHESINS, which are special macromolecules that bind to RECEPTORS on surfaces. Adhesins have been particularly well studied in the case of human and animal pathogens. Some of the best known are the pili of gonococci and a few other bacteria; they permit these organisms to adhere to animal cells and, in some cases, to cause infection. In many species, the most prominent pili—the so-called type 1 pili—are composed of identical structural protein units along their length. Other proteins, often located at the tip of the filament, are directly involved in binding to receptors on animal cells. Unlike the major structural protein, this minor component is conserved among various species of the Enterobacteriaceae.

Adhesion is such an important property that many species do not "put all their eggs in one basket." Rather, a single strain of gonococci may be able to make several kinds of pili. In some instances the difference among pili is not in the major structural protein but in the minor tip protein. This repertoire is varied at little cost and allows the same species to adhere to different types of host cells when circumstances change (Figure 3). We should emphasize that adhesion is far from being the only attribute of pathogenic bacteria; indeed, many factors are involved in the establishment of an infection.

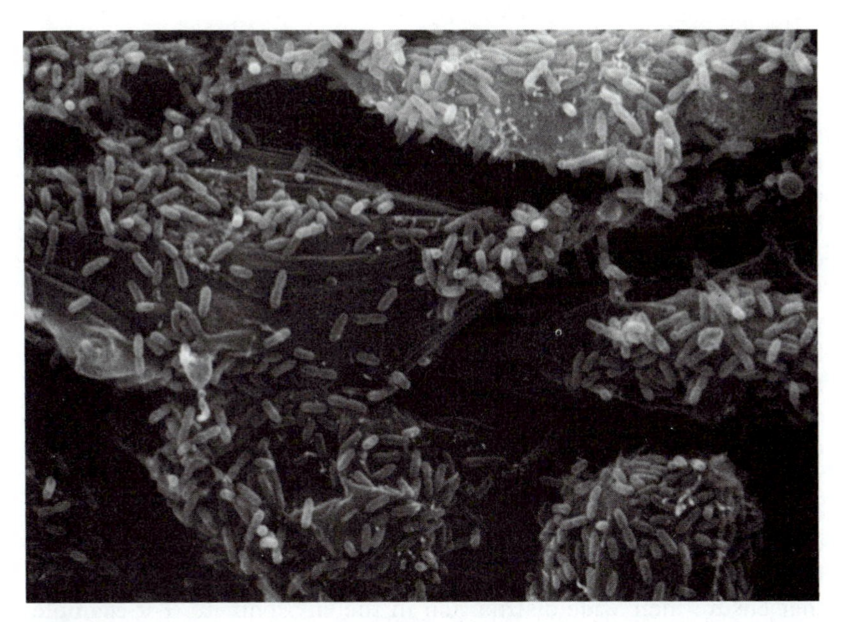

Figure 3

Scanning electron micrograph (×1,900) of *Escherichia coli inv*$^+$ cells adhering to human HeLa cells. The *inv* gene encodes for a protein called invasin, which is responsible for attachment and penetration. Note that no pili are visible on the surface of the bacteria. In this experiment, cells and bacteria were incubated at 0°C, which allowed attachment but not penetration. (Courtesy of R. Isberg.)

Adhesion of bacteria to surfaces may be mediated by structures other than pili. In the case of the intestinal bacteria, it seems that many are mechanically trapped in the thick mucus of the wall, whereas others bind to it chemically. Certain other organisms bind to adaptor molecules made by the host. For example, some streptococci, staphylococci, and the spirochete of syphilis adhere to animal cells indirectly, by binding to a host protein called FIBRONECTIN. This plasma glycoprotein readily sticks to the surface of mucosal cells. In these cases, bacteria do not seek out specific cells for adherence; rather, they rely on fibronectin to select host cells for them. Complexing to fibronectin is mediated in the streptococci by lipoteichoic acid and in the staphylococci by a surface protein.

Adhesion to surfaces may rely on bacterial structures that are less specific and less morphologically distinct than pili. Many species elaborate capsules (extracellular polysaccharides or polypeptides) that are extremely viscous and permit the organisms to stick to particles in the soil, to the surface of teeth, or even to other bacteria that are colonizing a given site. It is the manufacture of polysaccharides from dietary sucrose that leads to the adherence of streptococci and the development of caries. Often, the presence of capsules serves other purposes as well. Some pathogenic species elude host defenses because their capsules greatly hinder phagocytosis by white blood cells.

SPREADING IN THE ENVIRONMENT

Colonization of new sites imparts a clear selective advantage to a given species. Bacteria can spread and be introduced into a new environment in a variety of passive ways. They may be spread through the air in dust particles or aerosols, or they may be swept along by liquid currents. Some species, such as the bacteria that cause tuberculosis, typhoid fever, or brucellosis can be carried inside macrophages to distant sites within the animal hosts. There are even examples in which certain bacteria push others along on surfaces. This poorly studied phenomenon probably takes place when rafts of motile bacteria push along the organisms found in their path. It is thought that this mechanism of spread occurs on the surface of teeth, where the "pushers" are gliding bacteria of the *Cytophaga* group and the passive "pushees," other oral bacteria.

In addition to such passive mechanisms, some bacteria possess active ways of ensuring their own dispersal. In general, motility and chemotaxis are likely to facilitate the dispersion of bacteria to new sites. In some cases, however, bacteria have developed specific mechanisms that ensure their wide distribution in the environment. For example, cholera bacteria and other intestinal pathogens cause copious diarrhea by forming powerful exotoxins. The natural habitat of cholera bacilli is, from all indications, not the human gut, but the surface of estuarine shellfish. Cholera bacilli make a powerful chitinase, which they use to digest the carapace of dead shrimp and other crustaceans. Cholera bacilli are also resistant to salt, like typical marine forms. Teleologically speaking, the diarrhea caused by cholera bacilli serves the spread of these bacteria in waterways and eventually to other host individuals.

Other pathogenic bacteria have developed mechanisms that actively promote their dispersal within the host. Thus, pathogenic group A streptococci (*Streptococcus pyogenes*) elaborate a series of enzymes that facilitate their spread. Among these are hyaluronidase, which degrades hyaluronic acid (the major structural polysaccharide of connective tissue); streptokinase, which facilitates the breakup of fibrin clots; and deoxyribonuclease, which hydrolyzes DNA from lysed white blood cells and thus reduces the viscosity of the environment.

Successful pathogens, by definition, have the ability to spread within their host. However, some are confined to adjacent sites, whereas others are dispersed to distant tissues and organs. In each case, the ability to withstand host defenses, such as phagocytosis, the complement system, or antibodies, is essential if the organisms are to survive. Examples of the manifold repertoire of bacterial antiphagocytic defenses are shown in Table 1.

FINDING THE APPROPRIATE HABITAT

In most instances, bacteria find an appropriate environmental site by a hit-or-miss behavior. As mentioned earlier, they may be swept along by liquid currents and fortuitously find a site that is adequate for their

Table 1. Microbial strategies to evade phagoctyes

Antiphagocyte activity	Mechanism	Examples
Murder	Lysis of phagocyte membrane	Streptococci (streptolysin O); Pseudomonads (exotoxin A); *Staphylococcus aureus* α-toxin
Diversion (to nonproductive use)	Cause aggregation, especially in the lung	Streptococci Gram-negative enterics
Humiliation	Release of adenylate cyclase, which raises cAMP levels and depresses all phagocytic functions	*Bordetella pertussis*
Paralysis	Make cells unresponsive to chemotactic factors; induce inhibitors of phagocyte migration; inactivate chemotoxins that "call up" phagocytic cells	Tubercle bacilli; leprosy bacilli; Pseudomonds
"Playing hard to get"	Slimy capsule on organisms	Pneumococci; meningococci; *Haemophilus influenzae; Bacteroides fragilis;* many others
	M protein	Group A streptococci
	Pili	Gonococci
Prophylaxis	Inhibit phagosome-lysosome fusion	Tubercle bacilli; *Toxoplasma*
Indifference	Resist lysosomal enzymes	*Salmonella typhimurium; Mycobacteria* spp.; *Leishmania* spp.
Escape	Move from phagosomes to cytosol	Rickettsiae; influenza viruses
Inhibit oxidative killing	Inhibit respiratory burst	Virulent salmonellae; *Legionella pneumophila*
	Catalase breaks down H_2O_2	*Staphylococcus aureus*

Source: Modified from a table by M. Klempner in Schaechter et al. (1989).

growth. There are, however, examples where properties of the bacteria contribute with a physiological answer to this ecological problem. We have already discussed how bacteria use chemotaxis to swim up concentration gradients of nutrients (Chapter 6). Here we present the strategy of some bacteria that use the earth's magnetic field to align themselves in the direction of abundant food.

Magnetotaxis

In lakes and ponds, many microorganisms find abundant food at the bottom, where vegetation is being decomposed. How do they reach the bottom? Bacteria are so small that, at sea-level gravity (1 *g*), they fall rather slowly; in addition, they are impeded in their descent by the currents in natural bodies of water. Some species have solved this problem of up and down directionality by developing mechanisms that recognize the earth's magnetic field. North and south of the equator, the magnetic field is not exactly parallel to the surface but has a significant vertical component (for example, 73° from the horizontal in New Hampshire). Thus, by aligning themselves along the magnetic field lines, motile bacteria will tend to swim downward. How do bacteria do this?

Magnetotactic bacteria contain magnetic particles, which are known as MAGNETOSOMES and are visible under the electron microscope (Figure 4). Only a few species have been cultivated, *Aquaspirillum magnetotacticum* being the most studied. Magnetosomes consist of single crystals of magnetic iron oxide—magnetite (Fe_3O_4), also known as lodestone. Magnetosomes can readily be isolated from lysed cells by using a magnet. Each magnetosome is surrounded by a lipid bilayer membrane that contains characteristic proteins. Individual bacterial cells, by virtue of their content of magnetosomes, become biological compasses that, like any compass, recognize magnetic fields.

Magnetotactic bacteria are ubiquitous, but, without exception, those in the northern hemisphere swim predominantly northward and those in the southern hemisphere, southward. The explanation for this curious difference in polarity is that the earth's magnetic field is directed northward over the surface of the globe, but the vertical component is directed downward in the northern hemisphere and upward in the southern hemisphere. The reversed polarity ensures that organisms in both hemispheres swim predominantly downward. The situation is even more interesting at the geomagnetic equator: here magnetotactic bacteria consist of roughly equal numbers of cells of each polarity. This trait does not help them reach the bottom, and these organisms may well have other downward-swimming strategies. However, the reason these bacteria retain their magnetotactic ability has an apparent purpose: by aligning themselves along the magnetic field (which at the equator is horizontal), they avoid upward excursions. The upper layers of bodies of water are not favorable for growth, not only because they may be depleted of nutrients, but also because of their increase oxygen content. Most magnetotactic bacteria are microaerophilic (that is, they will grow at low oxygen pressure) or strictly anaerobic.

Magnetotaxis may be widespread among free-living cells such as unicellular algae and protozoa. In protozoa such as *Tetrahymena,* magnetosomes are apparently derived from ingested bacteria. At present, magnetotaxis has not been studied intensively, but it may prove to be an important tropism toward food-rich areas in aquatic environments.

COMPETITION BY KILLING OTHER MICROORGANISMS

Not infrequently, bacteria kill or are killed by their competitors. Most often they kill by producing toxic compounds to which they themselves are immune. Besides antibiotics, bacteria also elaborate proteins

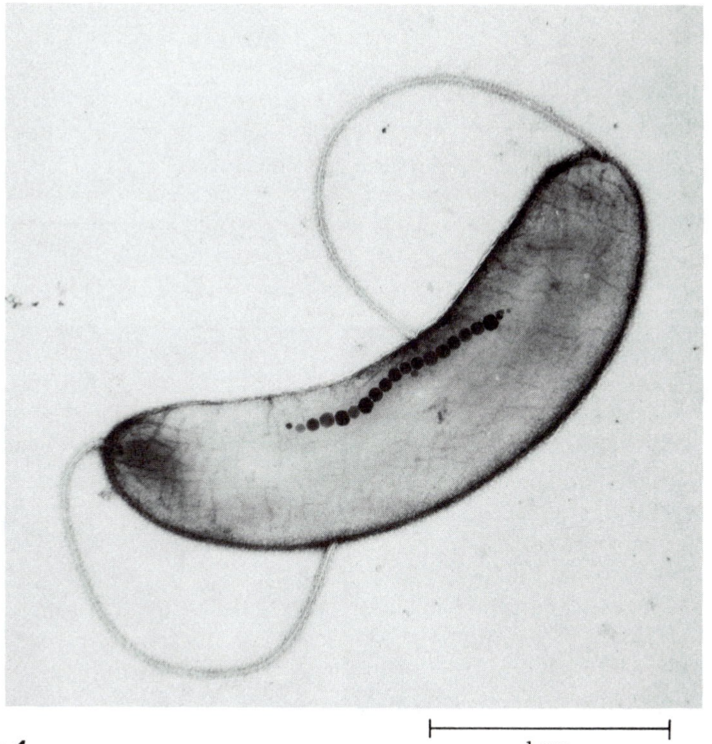

Figure 4

A magnetotactic bacterium found in a freshwater pond in New Hampshire. Note the chain of magnetosomes and the single flagellum at each pole. (Courtesy of R. and N. Blakemore.)

that can damage other species. These proteins are called BACTERIOCINS and MICROCINS. The line between these compounds and antibiotics is not always easy to draw. In addition, a few species of bacteria have mechanisms of predation on other bacteria—by direct attack. *Bdellovibrio bacteriovorans* is such an organism.

Bdellovibrio

The small Gram-negative bacterium *Bdellovibrio* (L. *bdellus*, leech) *bacteriovorans*, is, as the species name implies, a bacterial predator. It has the unusual ability to bore holes through the outer membrane of other, larger Gram-negative bacteria. *Bdellovibrio bacteriovorans* swims extremely rapidly, at about 100 cell lengths per second. When it collides with a susceptible bacterial cell, it causes the bacterium it strikes to recoil noticeably. In fact, this singular property led to its discovery in 1962. Heinz Stolp noticed that a lawn of *Pseudomonas* developed plaques long after the culture had ceased growing. This characteristic is unusual, because bacteriophage plaques do not enlarge after a culture has stopped growing. Stolp made a wet mount of the material within the plaque and, using keen powers of observation, noticed that

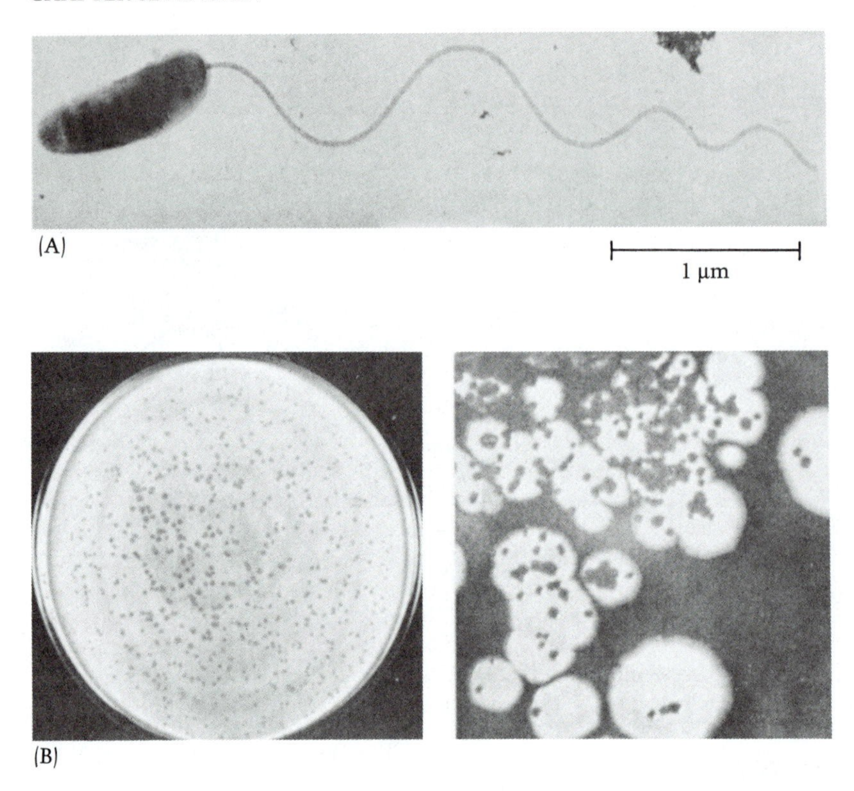

the *Pseudomonas* were being knocked about by something. He was unable to see bdellovibrios because they are so small (0.3 μm × 1.5 μm) and were moving so fast. When he examined the material under the electron microscope, he discovered *B. bacteriovorans.*

When a bdellovibrio attaches to the host bacterium, it spins rapidly (at greater than 6,000 rpm) by rotating a polar flagellum. This action of "boring" through the envelope is facilitated by lytic enzymes made by bdellovibrio. Once a hole is made, the parasite enters the periplasmic space and begins to grow (Figure 5). The prey cell rounds up to form a spheroplast. Bdellovibrios utilize not only energy and building blocks from the host cell but even such preformed materials as fatty acids and purine nucleotides. The organisms grow in the periplasm, increase in length into a cylindrical coil, and eventually, after 3–4 hours, divide. When a sufficient number of bdellovibrios are formed (3–5 per host cell), the host cell lyses and the parasites are released. If the process takes place on an agar surface seeded with a "lawn" of host bacteria, microepidemics of bdellovibrio growth will result in holes that look exactly like plaques caused by bacteriophages.

Bdellovibrios are found in water and can readily be isolated from many natural sources. They are strict intracellular parasites and cannot grow in cell-free medium. However, after mutation they are able to grow on laboratory media. It is not clear why such mutants do not accumulate

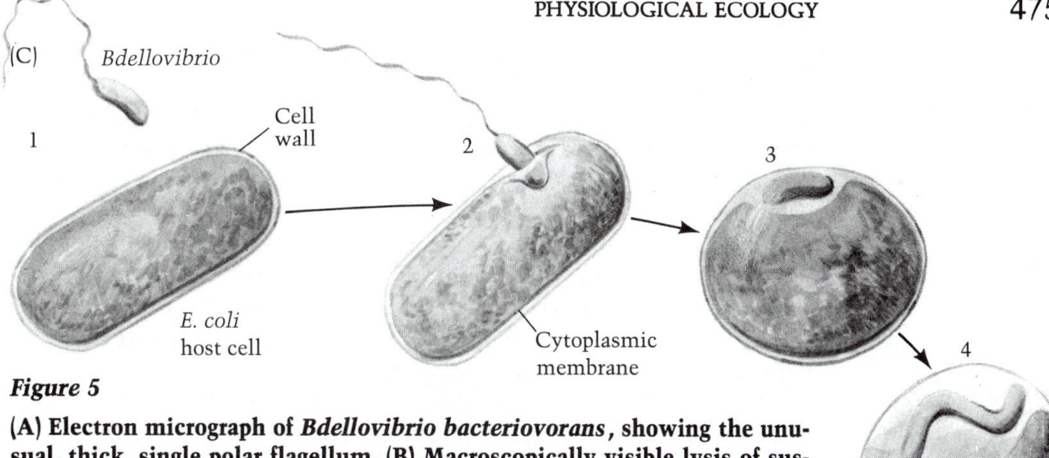

Figure 5

(A) Electron micrograph of *Bdellovibrio bacteriovorans*, showing the unusual, thick, single polar flagellum. (B) Macroscopically visible lysis of susceptible host by *Bdellovibrio*. Left panel: plaque formation in a lawn of *Pseudomonas*. Right panel: Partial lysis of colonies of *Escherichia coli* on an agar plate streaked with a mixture of host and parasite. (C) Life cycle of *Bdellovibrio*. Attachment to the host cell (1) is quickly followed by penetration into the periplasm (2). Growth of the parasite occurs without concomitant cell division, leading to a helical *Bdellovibrio* (3 and 4). Ultimately, the helical *Bdellovibrio* divides into a number of short rods that synthesize flagella (5). The weakened host cell wall (evident from the loss of shape and swelling that occur early in infection) then ruptures and the *Bdellovibrio* progeny cells are released. (A, B from Seidler and Starr, 1968; C after Stanier et al., 1986.)

in nature. Given the existence of these ferocious predators, how can host bacteria survive in nature? The answer seems to be that the developmental cycle of *Bdellovibrio* is sufficiently slow that most host cells will outgrow them and never become infected. This case is a fascinating example of the dynamics involved in predator–host relationships.

Bacteriocins

Many species of bacteria manufacture proteins that kill other strains of the same or closely related species. At present we can only speculate as to their role in nature. Bacteriocins are sometimes called by the name of organisms that makes them (e.g., those made by *E. coli* are known as colicins; those of *Enterobacter cloacae,* cloacins). How do bacteriocins kill sensitive bacteria? The best-studied cases are those of colicins, which exert their killing action in one of three major ways (Table 2):

- They make ion channels in the cytoplasmic membrane, which is depolarized because of loss of ions and membrane potential (see Chapter 6).
- They break down DNA by a nonspecific endonucleolytic activity.
- They damage ribosomes by cleaving the 16S RNA of the smaller ribosomal subunit at a specific site.

Each of these colicins is usually made by a different strain, and their genes are invariably encoded on plasmids. Colicins are rather large proteins (between 5×10^4 and 8×10^4 Da). How do they reach their targets, which are often in the interior of sensitive cells? As the first step in their activity, they attach to specific receptors on the outer membrane. Only a few molecules penetrate further, the rest appear to be destroyed by proteases. We do not know the details of penetration, although the zones of adhesion between the two membranes have been implicated. We know that penetration through the cytoplasmic membrane requires energization of the membrane. This requirement is the same as that for the reverse process—the insertion of proteins from the cytoplasm into the membrane (Chapter 4).

Why are bacteria immune to their own bacteriocins? Each colicinogenic *E. coli* strain produces an IMMUNITY PROTEIN that protects it against the colicin it makes. Immunity is conferred by complexing the colicin with its specific immunity protein within the cytoplasm of the cell that makes them. Immunity proteins are encoded on the same plasmids as colicins are.

It should be emphasized that the role of bacteriocins in nature is unknown. They may play a role in competition between bacterial strains; but, given their high degree of specificity, bacteriocins would only help in competition among related strains. This use may be what is intended, because related strains tend to be found in similar ecological sites.

Killing of other species is not limited to bacteriocins: bacteria can also use viruses for that purpose. Thus, lysogenic bacteria are themselves immune to the phages they carry. Nonlysogenic bacteria of the same or a related strain are sensitive to these phages and, except for a few that are lysogenized, are killed after infection. Like bacteria that manufacture bacteriocins, lysogenic bacteria make an immunity protein that protects them from infecting phages. For the *E. coli* phage λ, the immunity substance is known to be a repressor that acts on viral genes involved in phage replication and maturation. Thus, by making repressor, lysogenic bacteria inhibit both the growth of endogenous phage and of phage introduced by superinfection.

Table 2. Properties of some well-characterized *E. coli* bacteriocins (colicins)

Name	Mode of action
Several (A, E1, K, Ia, Ib)	Membrane depolarization
E2	Breakdown of DNA by nonspecific endonucleolytic activity
E3 and DF13	Ribosome inactivation

GETTING ALONG

Bacteria do not always compete with one another. In fact, they often benefit from mutualistic interactions. Several examples have been studied rather thoroughly.

The rumen

Cows, goats, sheep, and deer cannot get along without bacteria. They are ruminants and obtain most of their energy from the degradation of cellulose and other plant complex polysaccharides in a special organ—the RUMEN. Neither ruminants nor any other animal manufacture enzymes that degrade these polymers. The rumen is indeed rich in bacterial and protozoan species, and their complex interaction is partially understood (Figure 6). The number of bacteria in the rumen is enormous, approximating that of the large intestine or a dense laboratory culture (10^{10} cells per milliliter). In addition, the rumen contains a large amount of protozoa, which account for about half the rumen's biomass.

Polysaccharide degradation in the rumen takes place in various steps, beginning with the breakdown of cellulose by cellulolytic bacteria. These organisms make up about 5% of the rumen flora; their activity serves to feed other species that can ferment breakdown products of cellulose, such as the disaccharide cellobiose. Some of the products of this fermentation are further degraded by yet other species, a process resulting in the production of large amounts of volatile fatty acids (mainly acetic, propionic, and butyric). Ruminants use these metabolic products and the microorganisms themselves instead of sugars as their source of energy.

The concerted biochemical activities of various species of rumen bacteria constitute metabolic FOOD CHAINS in which the products of one organism serve as substrates for other species. In the absence of effective food chains, toxic compounds may accumulate. For example, nonvolatile acids (such as lactic or succinic) formed by fermentation by some species are further utilized by others to produce volatile fatty acids, which are removed from the rumen. Nonvolatile fatty acids are not absorbed by the host; if they were to accumulate, the pH would drop and the rumen would cease to function normally.

Another example of a metabolic chain involves molecular hydrogen, which is formed during certain fermentations. In sufficient amount, H_2 inhibits a number of biochemical reactions, including some involved in its own production, such as the formation of H_2 from reduced pyridine nucleotides ($NADH_2$). In the rumen, H_2 does not accumulate, because H_2 is efficiently utilized by methanogens, such as the archaebacterium *Methanobacterium ruminantium*. The general reaction is

$$4H_2 + HCO_3^- \longrightarrow CH_4 + 2H_2O + OH^-$$

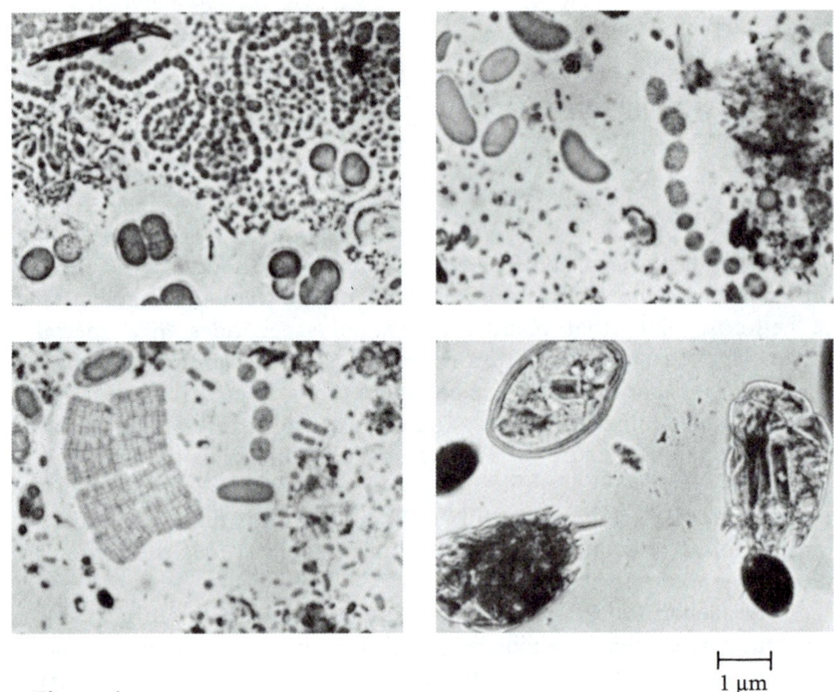

Figure 6

Some bacteria from the rumen of sheep. (From Smiles and Dobson, 1956.)

Thus, the activity of food chains in the rumen results in the orderly and well-coordinated production of metabolic end products.

Sauerkraut

The recipe for making sauerkraut is simple but the underlying microbiology quite complex. As an older relative may be able to tell you, sauerkraut is made by cutting up green cabbage, adding 1 part in 40 of salt by weight, covering the mixture, and leaving it alone for about three weeks in a cool place. The resulting product not only has a characteristic flavor but also extraordinary keeping qualities. What happens during sauerkraut fermentation? Several sequential events can be identified.

First, the addition of salt in the right proportion causes plasmolysis of the cabbage cells, thereby releasing vast amounts of cellular liquid that contains a variety of nutrients. A number of bacterial species present on the cabbage leaves metabolize these compounds and also hydrolyze in part some of the structural macromolecules, polysaccharides, and proteins. This complex fermentation gives way after one or two days to a more specific one, involving chiefly a member of the lactic acid bacteria—*Leuconostoc mesenteroides.* Why this organism takes over is not obvious. *Leuconostoc mesenteroides* carries out heterolactic fermentation (Chapter 5); that is, it does not convert the remaining sugars solely into lactic acid but produces a great number of other metabolic products such as mannitol, acetic acid, ethanol, and CO_2. Of these, CO_2 is visible because the pot begins to bubble vigorously. Some other metabolic products of this fermentation, chiefly esters, give the sauerkraut its characteristic flavor. At this point in the process, the pH has been lowered to about pH 5.5.

Low pH is harmful to *L. mesenteroides* and to many of the remaining bacteria. Other lactic acid bacteria of the genus *Lactobacillus* ferment mannitol and the remaining sugars. These organisms are not only resistant to acid but actually require low pH for growth and metabolism. The resulting homolactic fermentation leads to the accumulation of large amounts of lactic acid and a rapid lowering of the pH to about pH 3.5. The concentration of lactic acid is now about 0.15 *M*. Under these conditions, few microorganisms survive; and the sauerkraut is well preserved if kept under anaerobic conditions.

Sauerkraut fermentation illustrates how a series of microbial populations arises as the result of changes in the chemical and physical properties in a given environment. All the organisms are present from the beginning, but they start to grow and metabolize only when conditions are made suitable by the preceding activities of members of the population.

Mixed bacterial infections

A ruptured appendix invariably results in peritonitis. The reason for this serious condition (nearly always fatal before the use of antibiotics) is the spillage of a mixed bacterial flora from the gut into sterile tissues. Most of the organisms are strict anaerobes, mixed with some facultative anaerobes, a population reflecting the composition of the intestinal flora. At first, the peritoneal cavity is well oxygenated. How can strict anaerobes colonize such a site?

The establishment of peritonitis or related infections (abscesses in various parts of the body) usually requires the interaction of various bacterial species. First, those anaerobes that are sensitive to oxygen are killed off. The remaining ones are tolerant to oxygen, although they will not be able to grow in its presence. Facultative anaerobes make preferential use of their more efficient aerobic metabolism and rapidly deplete the environment of oxygen. Then the strict anaerobes can grow and, when they reach sufficient numbers, elicit the infection.

There are other examples of ways in which pathogenic bacteria use the metabolic activities of other species to colonize tissues. Thus, foot rot of sheep may be caused by two organisms, *Fusiformis necrophorus* and *Actinomyces pyogenes*. *Fusiformis necrophorus* is both invasive and destructive, and makes a toxin that destroys phagocytes. This toxin helps *F. necrophorus* and *A. pyogenes* to avoid phagocytosis. *Actinomyces pyogenes*, in turn, contributes nutritional factors needed for the growth of *F. necrophorus*. This latter organism is a strict anaerobe and is helped by the lowering of oxygen tension and the production of reducing conditions resulting from the metabolism of *Actinomyces*. Neither organism could cause infection by itself, and the coordinated action of both is necessary to cause disease.

SUMMARY

1. Microorganisms often find themselves in competition with others for foodstuffs. They have developed many strategies for optimizing their growth under such conditions, including the ability to utilize nutrients in an efficient manner.
2. Bacteria often develop specific mechanisms to cope with particular circumstances. Thus, to obtain iron, which is not present in most environments in a readily available form, many bacteria secrete highly avid chelators. These organisms possess specific uptake systems to incorporate the iron chelates.
3. Bacteria coexist in extremely complex environments. The large intestine of vertebrates contains hundreds of different bacterial species. Their coexistence requires complex metabolic interactions and also the ability of some species to adhere to the intestinal wall or to other adhering organisms.
4. Adherence to surfaces is sometimes mediated by pili, which are rodlike protein structures extending out from the bacterial surface. In some species, adherence is mediated by the secretion of sticky macromolecules.
5. Bacteria spread to new sites in the environment by being swept passively along by fluid currents. Some also possess active mechanisms to maximize their dispersal, such as enzymes that break down structural tissue constituents.
6. Many species of bacteria swim toward food by a chemotactic response to concentration gradients of nutrient. Others use the earth's magnetic field to swim downward to the nutrient-rich bottom of lakes and ponds.
7. Bacteria sometimes overwhelm their competitors by killing them. A direct example is that of *Bdellovibrio*, which penetrates into the periplasm of host bacteria and eventually kills them. Other bacteria secrete antibiotics and proteins (bacteriocins) lethal for other species and for other strains of their own species. Bacteria that make such compounds, or that are lysogenic for temperate phages, are themselves immune to these noxious agents.
8. In many instances, bacteria have developed a symbiotic coexistence with other species in their environment. Often, a given species produces a metabolite that is used by another species. Example of such food chains can be found in the rumen of cows or in the production of sauerkraut. In other cases, the activities of several species are required to cause disease.
9. A considerable portion of the genome of many bacteria is concerned with their social life rather than with growth per se.

STUDY QUESTIONS

1. Discuss the general ways bacteria cope with starvation for specific nutrients.
2. How do bacteria obtain iron in the face of the reduced availability of this metal?
3. How can large numbers of bacterial species coexist in the intestine? How would you study questions that arise from the consideration of this complex ecological system?
4. Draft a grant application for genetic work on the ecological role of magnetotaxis in bacteria.
5. Suggest an experiment to study the contribution of host cells to the growth and development of *Bdellovibrio.*
6. What is the minimum number of bacterial species required for the efficient function of a cow's rumen? What functions do they carry out?
7. Give three examples not discussed in this chapter that illustrate cooperation between bacterial species.

SUGGESTED READING

Atlas, R. M. and R. Bartha. 1981. *Microbial Ecology: Fundamentals and Applications.* Addison-Wesley, Reading, Massachusetts.

Bagg, A. and J. B. Neilands. 1987. Molecular mechanisms of regulation of siderophore-mediated iron assimilation. Microbiol. Rev. 51:509.

Luria, S. L. and J. L. Suit. 1987. Colicins and Col plasmids. Chapter 102 *in* F. C. Neidhardt, J. L. Ingraham, K. B. Low, B. Magasanik, M. Schaechter and H. E. Umbarger (eds.), *Escherichia coli and Salmonella typhimurium: Cellular and Molecular Biology,* Vol. 2. American Society for Microbiology, Washington, D.C.

Blakemore, R. P. 1982. Magnetotactic bacteria. Annu. Rev. Microbiol. 36:217.

Stanier, R. Y. et al., 1986. *The Microbial World,* 5th Ed. Prentice-Hall, Englewood Cliffs, New Jersey.

Wolin, M. J. 1979. The rumen fermentation: A model for microbial interactions in anaerobic ecosystems. Adv. Microbiol. Ecol. 3:49.

Answers to Study Questions

Answers and some guidance are provided here for all of the numerical problems among the study questions. In addition, guidance is given for a few of the open-ended study questions.

Chapter 1

1. 1,510 (using 110 and 618 as the molecular weights of an average amino acid residue and an average nucleotide base-pair residue, respectively).
2. 3,450 molecules of mRNA per cell ($2.5 \times 1{,}380$).
3. Curve A.
4. Curve B.
5. 14,060 ribosomes per newborn cell.
6. 1.5×10^{-14} g of LPS per oldest cell.

Chapter 2

1. Assuming the cell has the shape of a cylinder, the proportions of the total cell volume are 4.7% for the outer membrane, 5.5% for the periplasmic space, and 4.3% for the cytoplasmic membrane.

Chapter 5

1. From the information in Table 2 (page 142), one can calculate that 16.4% of glucose carbon must flow to the precursor metabolites derived exclusively from the pentose pathway. But owing to the pathway's cyclic nature, 2 molecules of glucose must flow through the pathway in order to accumulate 1 molecule of erythrose 4-phosphate. Taking this into account, the better answer is 21.1%.
2. 6,941 μmol of glucose.
3. Only 5,512 μmol of glucose are required to make precursor metabolites that are synthesized beyond the point that fructose 6-phosphate enters the EMP pathway—i.e., less than the amount that is available.
4. (a) Assuming that the metabolism via the EMP and TCA pathways generate 24 molecules of ATP by oxidative phosphorylation and 4 molecules (net) by substrate-level phosphorylation, 1,325 μmol of additional glucose would have to be consumed. (b) 4,638 μmol.
5. From the information in Table 1 (page 135), calculate the savings in ATP if amino acids need not be synthesized to make protein, lipids, murein, one-carbon units, and polyamines. This answer is 7,461 μmol of ATP. From the information in Table 1 (page 135) and Table 5 (page 155), calculate the ATP that is not generated incidental to forming precursor metabolites. This answer is 3,834 μmol of ATP. The net saving is 3,627 μmol of ATP.

6. The cost to produce a gram of cells [sum of fueling reactions (Table 6, page 156), biosynthesis (Table 13, page 143) and polymerization (Chapters 3 and 4) are given. The yield of ATP per molecule of malate metabolized (about 11) can be deduced from biochemical considerations, and the μmoles of malate metabolized per gram of cells can be calculated. The answer is 5,124 μmol.
7. 8,261 μmol.

Chapter 7

1. First, construct a standard curve relating the dry weight of cells in the suspension to the value of A_{420nm} of the suspension. The sample contains 103.5 mg cells per 100 mL, or 1,035 μg/mL, so the dilutions contain (in the order listed) 1,035, 932, 828, 725, 621, 518, 414, 311, 207, and 104 μg cells (dry weight)/mL. These values can be plotted against the A_{420nm} to produce the standard curve.
   (a) The values of A_{420nm} measured at various times in the growing culture can now be converted to dry weight and log dry weight/mL. (c) A plot of log dry weight/mL versus time on linear graph paper reveals that the rate of increase is a logarithmic function of time between 40 and 180 minutes. Substituting these values into Equation 3 (page 198) and solving for k gives a value of 1.04 hr^{-1}. (d) Using Equation 4 (page 198), one can calculate that g = 0.666 hr, or 40 min. (e) The initial amount of glucose was 2,000 μg/mL, and it was all consumed when growth stopped. Substituting these values and those for dry weight of cells into Equation 10 (page 206), one obtains Y = 0.574 g cells/g glucose. (f) Y_m (glucose) = 0.574 × MW glucose (180) = 103 g cells/mol glucose. (g) Mol ATP/mol glucose = 103 g cells/mol glucose divided by 10.5 g cells/mol ATP = 9.8. (h) Using Equation 10 (page 206), c_0 = 261 μg/mL, or 9.026% (w/v). (i) At 240 min the culture contains 1,205 μg cells/mL, and therefore would contain 1.2×10^9 cells/mL. But since stationary phase cells are about half the size of exponential cells, the number is closer to 2.4×10^9 cells/mL.
2. (a) From the figures given, use Equation 18 (page 213) to calculate c, with the result of 0.18 μg/mL. Then from Equation 21 (page 214), one can calculate c_r = 1,250 μg/mL = 0.125% (w/v). (b) Using Equation 18 (page 213) and a value of $c = 2.0 \times 10^{-7}$ M glucose, one can calculate that D = 0.533 hr^{-1}.
3. (a) To increase the steady state density of cells in a chemostat one could increase C_r, or decrease D; this could be done by either increasing f or decreasing V. (b) To increase k, one must increase D, either by increasing f or decreasing V.
4. At low values of D in a chemostat limited by the source of energy, the requirements for maintenance energy detract significantly from cell yield (page 215). Under these conditions, a strain with a lowered maintenance energy requirement would be expected to displace a strain with a higher requirement. One might expect that the mutation blocked a function that contributed to maintenance energy, an aspect not essential under the particular growth condition.

Chapter 9

4. The synthesis of an adhesin and the synthesis of a protein toxic to human cells are the only ones likely to be encoded on plasmids.

Chapter 11

1. A wasteful synthesis of the biosynthetic intermediates is avoided.
2. More rapid synthesis of the end product would lead to control only by repression. The biosynthetic enzymes might be maintained at a lower level.
3. Mutants resistant to an analog of the particular amino acid are frequently of this sort. They can be used for commercial production of the amino acid.
4. The activity of the enzyme might in this way be made subject to modification by the action of one or more modifying enzymes, the activity of which might be subject to a variety of controls from related pathways.
5. The controlling metabolite ligand is not produced, and as a result the intermediates are made in large quantities by an elevated level of the pathway enzymes, uncontrolled by the end product.
6. The mutant might be unable to grow in the presence of the end product to which it is still sensitive if the other end product is not supplied.
7. These strains might not have the ability to keep open a common biosynthetic pathway in the presence of one of its products.
8. Consider the use of an appropriate radiolabeling protocol followed by two-dimensional gel electrophoresis of total cellular protein.

Chapter 12

1. Nine out of ten transcription events lead to termination, with the product of a 140-nucleotide transcript and a 14-amino acid peptide. Therefore, per complete *trp* transcript formed there must be energy used for 1,260 (9 × 140) phosphodiester bonds in the mRNA, and 126 (9 × 14) peptide bonds, for a total of (2 × 1,260) + (4 × 14) = 2,576 high-energy P bonds (neglecting initiation and formylation costs).
2. A protein of 108 amino acid residues costs (108 × 4) for the peptide and (108 × 0.2) for its mRNA (see Chapter 3), for a total of 454 high-energy P bonds.
6.
   - Constitutive for acetylase, inducible for β-galactosidase.
   - Inducible for acetylase and β-galactosidase.
   - Constitutive for acetylase and β-galactosidase.

Chapter 13

2. Class A: could be mutated in the *cya* gene.
   Class B: could be mutated in the *crp* gene encoding the cAMP receptor protein, enabling it to bind to DNA and activate genes even in the absence of cAMP.
   Class C: could be mutated in the promoter of the *lac* operon, enabling it to be induced in the absence of cAMP-CRP activation.
7. One might consider cross-talk among pathways as being helpful in these situations.

Chapter 14

1. The proportions calculated from the Poisson distribution are: cells with 0 grains, 0.0935; cells with 1 grain, 0.222; cells with 2 grains, 0.263; cells with 3 grains, 0.207; cells with 4 grains, 0.123; cells with 5 grains, 0.0582.

Literature Cited

Andersen, K. B. and K. von Meyenburg. 1980. Are growth rates of *Escherichia coli* in batch cultures limited by respiration? J. Bacteriol. 144:114.

Anderson, R. P. and J. R. Roth. 1977. Tandem genetic duplications in phage and bacteria. Annu. Rev. Microbiol. 31:473.

Apirion, D. and P. Gugenheimer. 1981. Processing of bacterial RNA. FEBS Letters 125:1.

Armitage, J. P. and R. M. Macnab. 1987. Unidirectional, intermittent rotation of the flagellum of *Rhodobacter sphaeroides*. J. Bacteriol. 169:514.

Atkinson, D. E. 1968. The energy charge of the adenylate pool as a regulatory parameter. Interaction with feedback modifiers. Biochemistry 7:430.

Autissier, F., A. Jaffé and A. Képès. 1971. Segregation of galactoside permease, a membrane marker during growth and division of *Escherichia coli*. Mol. Gen. Genet. 112:276.

Avery, O. T., C. M. Macleod and M. McCarty. 1944. Studies on the chemical nature of the substance inducing transformation of pneumococcal types. I. Induction of transformation by a DNA fraction isolated from pneumoccoccus type III. J. Exp. Med. 79:137.

Bachmann, B. J. 1972. Pedigrees of some mutant strains of *Escherichia coli* K-12. Bacteriol. Rev. 36:525.

Bachmann, B. J. 1987. Linkage map of *Escherichia coli* K-12, edition 7. In *Escherichia coli and Salmonella typhimurium: Cellular and Molecular Biology*, Vol. 2, Ch. 53. F. C. Neidhardt, J. L. Ingraham, K. B. Low, B. Magasanik, M. Schaechter and H. E. Umbarger, eds. American Society for Microbiology, Washington, D.C.

Bagg, A. and J. B. Neilands. 1987. Molecular mechanisms of regulation of siderophore-mediated iron assimilation. Microbiol. Rev. 51:509.

Beckwith, J. 1987. The operon: An historical account. In *Escherichia coli and Salmonella typhimurium: Cellular and Molecular Biology*, Vol. 2, Ch. 88. F. C. Neidhardt, J. L. Ingraham, K. B. Low, B. Magasanik, M. Schaechter and H. E. Umbarger, eds. American Society for Microbiology, Washington, D.C.

Bloch, P. L., T. A. Phillips and F. C. Neidhardt. 1980. Protein identifications on O'Farrell two-dimensional gels: Locations of 81 *Escherichia coli* proteins. J. Bacteriol. 141:1409.

Bremer, H. and P. P. Dennis. 1987. Modulation of chemical composition and other parameters of the cell by growth rate. In *Escherichia coli and Salmonella typhimurium: Cellular and Molecular Biology*, Vol. 2, Ch. 96. F. C. Neidhardt, J. L. Ingraham, K. B. Low, B. Magasanik, M. Schaechter and H. E. Umbarger, eds. American Society for Microbiology, Washington, D.C.

Brewer, B. 1988. When polymerases collide: Replication and the transcriptional organization of the *E. coli* chromosome. Cell 53:679.

Buckmire, F. L. A. and R. G. E. Murray. 1970. Studies on the cell wall of *Spirillum serpens*. Can. J. Microbiol. 16:1011.

Bukhari, A. I., J. A. Shapiro and S. L. Adhya, eds. 1977. *DNA: Insertion Elements, Plasmids, and Episomes*. Cold Spring Harbor Laboratory, Cold Spring Harbor, New York.

Burman, L. G. and J. T. Park. 1984. Molecular model for elongation of the murein sacculus of *Escherichia coli*. Proc. Natl. Acad. Sci. USA 81:1844.

Cashel, M. and K. E. Rudd. 1987. The stringent response. In *Escherichia coli and Salmonella typhimurium: Cellular and Molecular Biology*, Vol. 2, Ch. 87. F. C. Neidhardt, J. L. Ingraham, K. B. Low, B. Magasanik, M. Schaechter and H. E. Umbarger, eds. American Society for Microbiology, Washington, D.C.

Chater, K. F. and M. J. Merrick, 1979. *Streptomyces*. In *Developmental Biology of Prokaryotes*, J. H. Parish, ed. University of California Press, Berkeley.

Chelala, C. A. and P. Margolin. 1974. Effects of deletions on cotransduction linkage in *Salmonella typhimurium*: Evidence that bacterial chromosome deletions affect the formation of transducing DNA fragments. Mol. Gen. Genet. 131:97.

Chiaramello, A. E. and J. W. Zyskind. 1989. Expression of *Escherichia coli dnaA* and *mioC* genes as a function of the growth rate. J. Bacteriol. 171: 4272.

Clewell, D. B. 1981. Plasmids, drug resistance, and gene transfer in the genus *Streptococcus*. Microbiol. Rev. 45:409.

Cole, J. R., C. L. Olsson, J. W. B. Hershey, M. Grunberg-Manago and M. Nomura. 1987. Feedback regulation of rRNA synthesis in *Escherichia coli*: Requirement for initiation factor IF2. J. Mol. Biol. 198:383.

Cole, R. M. and J. J. Hahn. 1962. Cell wall replication in *Streptococcus pyogenes*. Science 135: 722.

Crawford, J. P. 1975. Gene arrangement in the evolution of the tryptophan synthetic pathway. Bacteriol. Rev. 39:87.

Csonka, L. N. 1989. Physiological and genetic responses of bacteria to osmotic stress. Microbiol. Rev. 53:121.

Davis, B. D., R. Dulbecco, H. N. Eisen and H. S. Ginsberg. 1980. *Microbiology*, 3rd Edition. Harper & Row, New York.

de Boer, W. E., C. Golten and W. A. Scheffers. 1975. Effects of some physical factors on flagellation and swarming of *Vibrio alginolyticus*. Nether. Sea Res. 9:197.

Dennis, P. P. and H. Bremer. 1974. Macromolecular composition during steady-state growth of *Escherichia coli* B/r. J. Bacteriol. 119:270.

DePamphilis, M. L. and J. Adler. 1971. Attachment of flagellar basal bodies to the cell envelope: Specific attachment to the outer, lipopolysaccharide membrane and the cytoplasmic membrane. J. Bacteriol. 105:396.

Dills, S. S., A. Aperson, M. R. Schmidt and H. M. Saier. 1980. Carbohydrate transport in bacteria. Microbiol. Rev. 44:385.

DiRienzo, J. M., K. Nakamura and M. Inouye. 1978. The outer membrane proteins of Gram-negative bacteria: Biosynthesis, assembly and functions. Annu. Rev. Biochem. 47:481.

Donachie, W. D. 1968. Relationship between cell size and time of initiation of DNA replication. Nature 219:1077.

Drlica, K. 1987. The nucleoid. In *Escherichia coli and Salmonella typhimurium: Cellular and Molecular Biology*, Vol. 1, Ch. 9. F. C. Neidhardt, J. L. Ingraham, K. B. Low, B. Magasanik, M. Schaechter and H. E. Umbarger, eds. American Society for Microbiology, Washington, D.C.

Ecker, R. E. and M. Schaechter. 1963. Bacterial growth under conditions of limited nutrition. Ann. N.Y. Acad. Sci. 102:549.

Epstein, W. and S. G. Schultz. 1965. Cation transport in *Escherichia coli*. V. Regulation of cation content. J. Gen. Physiol. 49:221.

Fay, P. and N. J. Lang. 1971. The heterocysts of blue-green algae. I. Ultrastructural integrity after isolation. Proc. R. Soc. London Ser. B 209: 810.

Fraenkel, D. G. 1962. The control of ribonucleic acid synthesis of *Aerobacter aerogenes*. Ph.D. Dissertation, Harvard University, Cambridge, Massachusetts.

Fuller, R. S. and A. Kornberg. 1983. Purified dnaA protein in initiation of replication at the *Escherichia coli* chromosomal origin of replication. Proc. Natl. Acad. Sci. USA 81:4275.

Gerhart, J. C. and A. B. Pardee. 1962. The enzymology of control by feedback inhibition. J. Biol. Chem. 237:891.

Ghuysen, J. N. 1968. Bacteriolytic enzymes in the determination of wall structure. Bacteriol. Rev. 32:425.

Gmeiner, J. E., E. Sarnow and K. Milde. 1985. Cell cycle parameters of *Proteus mirabilis*: Interdependence of the biosynthetic cell cycle and the interdivision cycle. J. Bacteriol. 164:741.

Gottschalk, G. 1986. *Bacterial Metabolism*, 2nd Edition. Springer-Verlag, New York.

Green, E. W. and M. Schaechter. 1972. The mode of segregation of the bacterial cell membrane. Proc. Natl. Acad. Sci. USA 69:2312.

Griffith, F. 1928. Significance of pneumococcal types. J. Hyg. 27:113.

Helmstetter, C. E. and S. Cooper. 1968. DNA synthesis during the division cycle of rapidly growing *Escherichia coli* B/r. J. Mol. Biol. 31:507.

Hendrickson, W. G., T. Kusano, H. Yamaki, R. Balakrishnan, M. King, J. Murchie and M. Schaechter. 1982. Binding of the origin of replication of *Escherichia coli* to the outer membrane. Cell 30:915.

Hengge, R. and W. Boos. 1983. Maltose and lactose transport in *Escherichia coli*: Examples of two different types of concentrative transport systems. Biochim. Biophys. Acta 783:443.

Henner, D. J. and J. A. Hoch. 1980. The *Bacillus subtilis* chromosome. Microbiol. Rev. 44:57.

Herenden, S. L., R. A. VanGogelen and F. C. Neidhardt. 1979. Levels of major proteins of *Escherichia coli* during growth at different temperatures. J. Bacteriol. 139:185.

Hershey, J. W. B. 1987. Protein synthesis. In *Escherichia coli and Salmonella typhimurium: Cellular and Molecular Biology*, Vol. 1, Ch. 40. F. C. Neidhardt, J. L. Ingraham, K. B. Low, B. Magasanik, M. Schaechter and H. E. Umbarger, eds. American Society for Microbiology, Washington, D.C.

Higgins, M. L. and G. D. Shockman. 1970. Model for cell wall growth of *Streptococcus faecalis*. J. Bacteriol. 101:643.

Hill, C. W., J. Foulds, L. Soll and P. Berg. 1969. Instability of a missense suppressor resulting from a duplication of genetic material. J. Mol. Biol. 39: 562.

Hill, C. W., R. H. Gafstrom and B. S. Hillman. 1977. Chromosomal rearrangements resulting from recombination between ribosomal RNA genes. In *DNA: Insertion Elements, Plasmids, and Episomes*. A. I. Bukhari, J. A. Shapiro and S. L.

Adhya, eds. Cold Spring Harbor Laboratory, Cold Spring Harbor, New York.
Hinkle, P. C. and R. E. McCarthy. 1978. How cells make ATP. Sci. Am. 238:104.
Hirota, Y., A. Ryter and F. Jacob. 1968. Thermosensitive mutants of *E. coli* affected in the processes of DNA synthesis and cellular division. Cold Spring Harbor Symp. Quant. Biol. 33:677.
Holt, J. G. and N. R. Krieg, eds. 1984, 1986. *Bergey's Manual of Systematic Bacteriology*, Vols. 1 and 2. Williams & Wilkins, Baltimore.
Holt, S. C. 1978. Anatomy and chemistry of spirochetes. Microbiol. Rev. 42:114.
Hoopes, B. C. and W. R. McClure. 1987. Strategies in regulation of transcription initiation. In *Escherichia coli and Salmonella typhimurium: Cellular and Molecular Biology*, Vol. 2, Ch. 75. F. C. Neidhardt, J. L. Ingraham, K. B. Low, B. Magasanik, M. Schaechter and H. E. Umbarger, eds. American Society for Microbiology, Washington, D.C.
Inouye, M. 1979. *Bacterial Outer Membranes: Biogenesis and Function*. Wiley, New York.
Inouye, M. and S. Halegoua. 1980. Secretion and membrane localization of proteins in *Escherichia coli*. CRC Critical Reviews in Biochemistry.
Iuchi, S., D. C. Cameron and E. C. C. Lin. 1989. A second global regulator gene (*arcB*) mediating repression of enzymes in aerobic pathways of *Escherichia coli*. J. Bacteriol. 171:868.
Jackson, E. N., F. Laski and C. Andres. 1982. Bacteriophage P22 mutants that alter the specificity of DNA packaging. J. Mol. Biol. 154:551.
Jacob, F. and J. Monod. 1961. Genetic regulatory mechanisms in the synthesis of proteins. J. Mol. Biol. 3:318.
Jacobs, A. E., J. A. Shapiro, L. Yamamoto, P. I. Smith, S. N. Cohen and D. Berg. 1977. In *DNA: Insertion Elements, Plasmids and Episomes*. A. I. Bukhari, J. A. Shapiro and S. L. Adhya, eds. Cold Spring Harbor Laboratory, Cold Spring Harbor, New York.
Jacobsen, H. 1974. Doctoral dissertation, Copenhagen University.
Jaffé, A., T. Ogura and S. Hiraga. 1985. Effects of *ccd* function of the F plasmid on bacterial growth. J. Bacteriol. 163:841.
Jensen, K. F. and S. Pedersen. 1990. Metabolic growth rate control in *Escherichia coli* may be a consequence of subsaturation of the macromolecular biosynthetic apparatus with substrates and catalytic components. Microbiol. Rev. June 1990.
Jinks-Robertson, S. and M. Nomura. 1978. Ribosomes and tRNA. In *Escherichia coli and Salmonella typhimurium: Cellular and Molecular Biology*, Vol. 2, Ch. 85. F. C. Neidhardt, J. L. Ingraham, K. B. Low, B. Magasanik, M. Schaechter and H. E. Umbarger, eds. American Society for Microbiology, Washington, D.C.
Kahn, S., R. M. Macnab, A. L. DeFranco and D. E. Koshland, Jr. 1978. Inversion of a behavioral response in bacterial chemotaxis: Explanation at the molecular level. Proc. Natl. Acad. Sci. USA 75:4150.
Kavenoff, R. and O. Ryder. 1976. Electron microscopy of membrane-associated folded chromosomes of *Escherichia coli*. Chromosoma 55:13.
Képès, A. and F. Autissier. 1972. Topology of membrane growth in bacteria. Biochim. Biophys. Acta 265:443.
Koch, A. L. 1987. The variability and individuality of the bacterium. In *Escherichia coli and Salmonella typhimurium: Cellular and Molecular Biology*, Vol. 2, Ch. 101. F. C. Neidhardt, J. L. Ingraham, K. B. Low, B. Magasanik, M. Schaechter and H. E. Umbarger, eds. American Society for Microbiology, Washington, D.C.
Koppes, L. J. H., C. L. Woldringh and N. Nanninga. 1978. Size variation and correlation of different cell cycle events in slow-growing *Escherichia coli*. J. Bacteriol. 134:423
Koval, S. F. and R. G. E. Murray. 1986. The superficial protein arrays on bacteria. Microbiol. Sci. 3:357.
Lake, J. A. 1988. Origin of the eukaryotic nucleus determined by rate-invariant analysis of rRNA sequences. Nature 331:184.
Landick, R. and C. Yanofsky. 1987. Transcription attenuation. In *Escherichia coli and Salmonella typhimurium: Cellular and Molecular Biology*, Vol. 2, Ch. 77. F. C. Neidhardt, J. L. Ingraham, K. B. Low, B. Magasanik, M. Schaechter and H. E. Umbarger, eds. American Society for Microbiology, Washington, D.C.
Langley, K. E. and E. P. Kennedy. 1979. Energetics of rapid transmembrane movement and of compositional symmetry of phosphatidylethanolamine in membranes of *Bacillus megaterium*. Proc. Natl. Acad. Sci. USA 76:6245.
Langridge, J. 1968. Thermal responses to mutant enzymes and temperature limits to growth. Mol. Gen. Genetics 108:116.
Lapidus, I. R. and H. C. Berg. 1982. Gliding motility of *Cytophaga* sp. strain U67. J. Bacteriol. 151:384.
Laughrea, M. and F. B. Moore. 1978. Ribosomal components required for binding protein S1 to 30S subunit of *Escherichia coli*. J. Mol. Biol. 122:109.
Lederberg, J. 1948. Gene control of β-galactosidase in *E. coli*. Genetics 33:716.
Leonard, A. C. and C. E. Helmstetter. 1988. Replication pattern of multiple plasmids coexisting in *Escherichia coli*. J. Bacteriol. 170:1380.

Maaløe, O. 1979. Regulation of the protein-synthesizing machinery—ribosomes, tRNA, factors, and so on. In *Biological Regulation and Development*, Vol. 1, *Gene Expression*. R. F. Goldberger, ed. Plenum, New York.

Maaløe, O. and N. O. Kjeldgaard. 1966. *Control of Macromolecular Synthesis*. W. A. Benjamin, New York.

MacAlester, T. J., B. MacDonald and L. I. Rothfield. 1983. The periseptal annulus: An organelle associated with cell division in Gram-negative bacteria. J. Bacteriol. 137:574.

Macnab, R. M. 1987. Flagella. In *Escherichia coli and Salmonella typhimurium: Cellular and Molecular Biology*, Vol. 1, Ch. 7. F. C. Neidhardt, J. L. Ingraham, K. B. Low, B. Magasanik, M. Schaechter and H. E. Umbarger, eds. American Society for Microbiology, Washington, D.C.

Macnab, R. M. and M. K. Ornston. 1977. Normal-to-curly flagellar transitions and their role in bacterial tumbling. Stabilization of alternative quaternary structure by mechanical force. J. Mol. Biol. 112:1.

Mahajan, S. K. 1988. Pathways of homologous recombination in *Escherichia coli*. In *Genetic Recombination*. R. Kucherlapati and G. R. Smith, eds. American Society for Microbiology, Washington, D.C.

Mandelstam, J. and K. McQuillen, K. 1973. *Biochemistry of Bacterial Growth*, 2nd Edition. Blackwell Scientific, Oxford.

Marr, A. G. and J. L. Ingraham. 1962. Effect of temperature on the composition of fatty acids in *Escherichia coli*. J. Bacteriol. 84:1260.

Marsh, R. C. and A. Worcel. 1977. A DNA fragment containing the origin of replication of the *Escherichia coli* chromosome. Proc. Natl. Acad. Sci. USA 74:2720.

McMacken, R., L. Silver and C. Georgopoulos. 1987. DNA replication. In *Escherichia coli and Salmonella typhimurium: Cellular and Molecular Biology*, Vol. 1, Ch. 39. F. C. Neidhardt, J. L. Ingraham, K. B. Low, B. Magasanik, M. Schaechter and H. E. Umbarger, eds. American Society for Microbiology, Washington, D.C.

Meeks, J. C. 1988. Cyanobacterial–bryophyte associations. In A. N. Rai, ed. *CRC Handbook of Symbiotic Cyanobacteria*. CRC Press, Boca Raton, Florida.

Merrick, M. J. 1979. *Streptomyces*. In *Developmental Biology of Procaryotes*, J. H. Parish, ed. University of California Press, Berkeley.

Miller, J. H. 1980. *Experiments in Molecular Genetics*. Cold Spring Harbor Laboratory, Cold Spring Harbor, New York.

Miller, O. L., Jr., B. A. Hamkalo and C. A. Thomas. 1980. Visualization of bacterial genes in action. Science 169:392.

Monod, J. 1942. *Recherches sur la Croissance des Cultures Bactériennes*. Hermann et Cie, Paris.

Nanninga, N., ed. 1985. *Molecular Cytology of Escherichia coli*. Academic Press, San Diego.

Neidhardt, F. C., J. L. Ingraham, K. B. Low, B. Magasanik, M. Schaechter and H. E. Umbarger, eds. 1987. *Escherichia coli and Salmonella typhimurium: Cellular and Molecular Biology*. American Society for Microbiology, Washington, D.C.

Nelson, D. C., B. B. Jørgensen and N. P. Revsbech. 1986. Growth pattern and yield of a chemoautotrophic *Beggiatoa* sp. in oxygen–sulfide microgradients. Appl. Environ. Microbiol. 52:225.

Nierhaus, K. H. 1982. Structure, assembly and function of ribosomes. In *Current Topics in Microbiology and Immunology*, Vol. 97. W. Henle, P. H. Hofschnieder, H. Kodprowski, F. Melcherrs, R. Rott, H. G. Schwider and P. K. Vogt, eds. Springer-Verlag, New York.

Nikaido, H. 1973. Biosynthesis and assembly of lipopolysaccharide. In *Bacterial Membranes and Walls*. L. Leive, ed. Marcel Dekker, New York.

Nikaido, H. and T. Nakae. 1979. The outer membrane of Gram-negative bacteria. Adv. Microb. Physiol. 20:163.

Novick, A. and L. Szilard. 1950. Experiments with the chemostat on spontaneous mutations of bacteria. Proc. Natl. Acad. Sci. USA 36:708.

Ogden, G., M. J. Pratt and M. Schaechter. 1988. The replicative origin of the *Escherichia coli* chromosome binds to cell membranes only when hemimethylated. Cell 54:127.

Ogura, T. and S. Hiraga. 1983. Mini-F genes that couple host cell division to plasmid proliferation. Proc. Natl. Acad. Sci. USA 80:4784.

Osborn, M. J. and H. C. P. Wu. 1980. Proteins of the outer membrane of Gram-negative bacteria. Annu. Rev. Biochem. 34:369.

Parish, J. H. 1979. Myxobacteria. In *Developmental Biology of Prokaryotes*. J. H. Parish, ed. University of California Press, Berkeley.

Park, J. T. 1987a. The murein sacculus. In *Escherichia coli and Salmonella typhimurium: Cellular and Molecular Biology*, Vol. 1, Ch. 4. F. C. Neidhardt, J. L. Ingraham, K. B. Low, B. Magasanik, M. Schaechter and H. E. Umbarger, eds. American Society for Microbiology, Washington, D.C.

Park, J. T. 1987b. Murein synthesis. In *Escherichia coli and Salmonella typhimurium: Cellular and Molecular Biology*, Vol. 1, Ch. 42. F. C. Neidhardt, J. L. Ingraham, K. B. Low, B. Magasanik, M. Schaechter and H. E. Umbarger, eds. American Society for Microbiology, Washington, D.C.

Pato, M. L. and K. von Meyenburg. 1970. Residual RNA synthesis in *Escherichia coli* after inhibition of initiation of transcription by rifampicin. Cold Spring Harbor Symp. Quant. Biol. 35:497.

Pato, M. L., P. M. Bennett and K. von Meyenburg. 1973. Messenger ribonucleic acid synthesis and degradation in *Escherichia coli* during inhibition of translation. J. Bacteriol. 116:710.

Paulton, R. J. L. 1970. Analysis of the multiseptate potential of *Bacillus subtilis*. J. Bacteriol. 104: 762.

Phillips, D. A., K. D. Newell, S. A. Hassell and C. E. Felling. 1976. The effect of CO_2 enrichment on root nodule development in *Pisum sativum*. Am. J. Bot. 63:356.

Phillips, T. A., V. Vaughn, P. L. Bloch and F. C. Neidhardt. 1987. Gene–protein index of *Escherichia coli* K-12, Edition 2. In *Escherichia coli and Salmonella typhimurium: Cellular and Molecular Biology*, Vol. 2, Ch. 55. F. C. Neidhardt, J. L. Ingraham, K. B. Low, B. Magasanik, M. Schaechter and H. E. Umbarger, eds. American Society for Microbiology, Washington, D.C.

Pierucci, O. and C. Zuchowski. 1973. Non-random segregation of DNA strands in *Escherichia coli* B/r. J. Mol. Biol. 80:477.

Pirt, L. J. 1965. The maintenance energy of bacteria in growing cultures. Proc. R. Soc. London Ser. B. 163:224.

Poindexter, J. L. S. 1962. Biological properties and classification of the *Caulobacter* group. Bacteriol. Rev. 28:231.

Postma, P. W. and J. W. Lengeler. 1985. Phosphoenolpyruvate-carbohydrate phosphotransferase system of bacteria. Microbiol. Rev. 49:232.

Powell, E. O. 1956. Growth rate and generation time of bacteria, with special reference to continuous culture. J. Gen. Microbiol. 15:492.

Prescott, D. M. and P. L. Kuempel. 1972. Bidirectional replication of the chromosome in *Escherichia coli*. Proc. Natl. Acad. Sci. USA 69:2842.

Pribnow, D. 1979. Genetic control signals in DNA. In *Biological Regulation and Development*. R. F. Goldberger, ed. Plenum, New York.

Prince, J. B., R. R. Gutell and R. A. Garrett. 1983. A consensus model of the *Escherichia coli* ribosomes. Trends Biochem. Sci. 8:359.

Randall, L. L. and S. J. S. Hardy. 1989. Unity in function in the absence of consensus in sequence: Role of leader peptides in export. Science 243:1156.

Randall, L. L., S. J. S. Hardy and J. B. Thom. 1987. Export of protein: A biochemical view. Annu. Rev. Microbiol. 41:507.

Ratkowski, D. A., J. Olley, T. A. Meekin and A. Ball. 1982. Relationship between temperature and growth rate of bacterial cultures. J. Bacteriol. 149:1.

Reitzer, L. J. and B. Magasanik. 1987. Ammonia assimilation and the biosynthesis of glutamine, glutamate, aspartate, asparagine, L-alanine, and D-alanine. In *Escherichia coli and Salmonella typhimurium: Cellular and Molecular Biology*, Vol. 1, Ch. 20. F. C. Neidhardt, J. L. Ingraham, K. B. Low, B. Magasanik, M. Schaechter and H. E. Umbarger, eds. American Society for Microbiology, Washington, D.C.

Riley, M. and A. Anilionis. 1978. Evolution of the bacterial genome. Annu. Rev. Microbiol. 32:519.

Roberts, R. B., R. H. Abelson, D. B. Cowie, E. T. Bolton and R. J. Britten. 1955. *Studies of Biosynthesis of Escherichia coli*. Carnegie Inst. Washington Publ. 607.

Rolfe, B. G., P. M. Gressholf and J. Shine. 1980. Plant Science Lett. 19:277.

Rosen, B. P. and E. R. Kashet. 1978. Energetics of bacterial transport. In *Bacterial Transport*. B. P. Rosen, ed. Marcel Dekker, New York.

Royle, P. L., H. Matsumoto and B. W. Holloway. 1981. Genetic circularity of the *Pseudomonas aeruginosa* PAO chromosome. J. Bacteriol. 145:145.

Russell, D. W. and N. D. Zinder. 1987. Hemimethylation prevents DNA replication in *E. coli*. Cell 50:1071.

Saier, M. H., Jr. 1989. Protein phosphorylation and allosteric control of inducer exclusion and catabolite repression by the bacterial phosphoenolpyruvate:sugar phosphotransferase system. Microbiol. Rev. 53:109.

Savageau, M. A. 1974. Genetic regulatory mechanisms and the ecological niche of *Escherichia coli*. Proc. Natl. Acad. Sci. USA 71:2453.

Savageau, M. A. 1976. *Biochemical Systems Analysis: A Study of Function and Design in Molecular Biology*. Addison-Wesley, Reading, Massachusetts.

Savageau, M. A. 1985. Coupled circuits of gene regulation. In *Sequence Specificity in Transcription and Control*. R. Calendar and L. Gold, eds. Alan R. Liss, New York.

Schaechter, M., G. Medoff and D. Schlessinger. 1989. *Mechanisms of Microbial Disease*. Williams & Wilkins, Baltimore.

Schaechter, M., J. P. Williamson, F. R. Hood, Jr. and A. L. Koch. 1962. Growth, cell and nuclear division in some bacteria. J. Gen. Microbiol. 29:421.

Schaechter, M., O. Maaløe and N. O. Kjeldgaard. 1958. Dependency on medium and temperature of cell size and chemical composition during balanced growth of *Salmonella typhimurium*. J. Gen. Microbiol. 19:592.

Schleif, R. 1986. *Genetics and Molecular Biology*, Ch. 12. Addison-Wesley, Reading, Massachusetts.

Seidler, R. J. and M. Starr. 1968. Structure of the flagellum of *Bdellovibrio bacteriovorans*. J. Bacteriol. 95:1952.

Shapiro, J. A. 1977. F, the *E. coli* sex factor. In *DNA Insertion Elements, Plasmids, and Episomes*.

A. I. Bukhari, J. A. Shapiro and S. L. Adhya, eds. Cold Spring Harbor Laboratory, Cold Spring Harbor, New York.

Shehata, T. E. and A. G. Marr. 1970. Synchronous growth of enteric bacteria. J. Bacteriol. 103:789.

Silhavy, T. J., S. A. Benson and S. D. Emr. 1983. Mechanisms of protein localization. Microbiol. Rev. 47:313.

Sinden, R., J. Carlson and D. Pettijohn. 1980. Torsional tension in the DNA measured with trimethyl-psoralen in living *Escherichia coli* cells. Cell 21:773.

Skarstad, K., E. Boye and H. B. Steen. 1986. Timing of initiation of chromosome replication in individual *Escherichia coli* cells. EMBO J. 5:1711.

Smiles, J. and M. J. Dobson. 1956. Direct ultraviolet and ultraviolet negative phase-contrast micrography of bacteria from the rumen of the sheep. J. R. Microsc. Soc. 75:244.

Smith, H. O., D. B. Danner and R. A. Deich. 1981. Genetic transformation. Annu. Rev. Biochem. 50:41.

Spratt, B. 1975. Distinct penicillin binding proteins involved in the division, elongation and shape of *Escherichia coli* K-12. Proc. Natl. Acad. Sci. USA 72:2999.

Staley, J. T. 1981. The genera *Prosthecomicrobium* and *Ancalomicrobium.* In *The Procaryotes,* M. P. Starr, H. Stolp, H. G. Truper, A. Balows and H. G. Schlegel, eds. Springer-Verlag, New York.

Stanier, R. Y. 1954. Some singular features of bacteria as dynamic systems. In *Cellular Metabolism and Infection*. E. Racker, ed. Academic Press, New York.

Stanier, R. Y., J. L. Ingraham, M. L. Wheelis and P. R. Painter. 1986. *The Microbial World*, 5th Edition. Prentice-Hall, Englewood Cliffs, New Jersey.

Steitz, J. 1979. Genetic signals and nucleotide sequences in messenger RNA. In *Biological Regulation and Development*. R. F. Goldberger, ed. Plenum, New York.

Stent, G. S. and R. Calendar. 1978. *Molecular Genetics*. W. H. Freeman, San Francisco.

Stock, J. B., A. J. Ninfa and A. M. Stock. 1989. Protein phosphorylation and regulation of adaptive responses in bacteria. Microbiol. Rev. 53:450.

Stouthammer, A. H. 1973. A theoretical study of the amount of ATP required for synthesis of microbial cell material. Antonie van Leeuwenhoek 39:545.

Taketa, K. and M. Pogell. 1965. Allosteric inhibition of rat liver fructose-1,6-diphosphatase by adenosine-5′-monophosphate. J. Biol. Chem. 240:651.

Taylor, B. L. 1983. Role of proton motive force in sensory transduction in bacteria. Annu. Rev. Microbiol. 37:551.

Teather, R. M., J. F. Collins and W. D. Donachie. 1974. Quantal behavior of a diffusible factor which initiates septum formation at potential division sites in *Escherichia coli*. J. Bacteriol. 118:407.

Tsang, N., R. Macnab and D. E. Koshland, Jr. 1973. Common mechanism for repellants and attractants in bacterial chemotaxis. Science 181:60.

Umbarger, H. E. 1956. Evidence for a negative-feedback mechanism in the biosynthesis of isoleucine. Science 123:848.

Umbarger, H. E. 1977. A one-semester project for the immersion of graduate students in metabolic pathways. Biochem. Educ. 5:67.

Wanner, B. L. 1987. Phosphate regulation of gene expression in *Escherichia coli*. In *Escherichia coli and Salmonella typhimurium: Cellular and Molecular Biology*, Vol. 2, Ch. 82. F. C. Neidhardt, J. L. Ingraham, K. B. Low, B. Magasanik, M. Schaechter and H. E. Umbarger, eds. American Society for Microbiology, Washington, D.C.

Weiner, J. H., B. D. Lemire, M. L. Elmes, R. D. Bradley and D. G. Scraba. 1984. Overproduction of fumarate reductase in *Escherichia coli* induces a novel intracellular lipid-protein organelle. J. Bacteriol. 158:590.

Woese, C. R. 1987. Bacterial evolution. Microbiol. Rev. 51:221.

Wood, W. B. 1979. Bacteriophage T4 assembly and the morphogenesis of subcellular structure. Harvey Lectures 73:203.

Wu, H. C. 1986. Proteolytic processing of signal peptides. In *Protein Compartmentalization*. A. W. Strauss, I. Boime and G. Kreil, eds. Springer-Verlag, New York.

Yamada, H. and S. Mizushima. 1978. Reconstitution of an ordered structure from major outer membrane constituents and the lipoprotein-bearing peptidoglycan sacculus of *Escherichia coli*. J. Bacteriol. 135:1024.

Yanofsky, C. 1981. Attenuation in the control of expression of bacterial operons. Nature 289:751.

Yanofsky, C. and I. R. Crawford. 1987. The tryptophan operon. In *Escherichia coli and Salmonella typhimurium: Cellular and Molecular Biology*, Vol. 2, Ch. 90. F. C. Neidhardt, J. L. Ingraham, K. B. Low, B. Magasanik, M. Schaechter and H. E. Umbarger, eds. American Society for Microbiology, Washington, D.C.

Yates, R. A. and A. B. Pardee. 1956. Control of pyrimidine biosynthesis in *Escherichia coli* by a feedback mechanism. J. Biol. Chem. 221:757.

Zeikus, J. G. and V. G. Bowen. 1975. Fine structure of *Methanospirillium hungatii*. J. Bacteriol. 121:373.

Index

M. Schaechter F. C. Neidhardt J. L. Ingraham

THE AUTHORS

Frederick C. Neidhardt is F. G. Novy Distinguished Professor of Microbiology and Immunology at the University of Michigan. He received his Ph.D. in 1956 from Harvard University, where he worked with Boris Magasanik. For his research in bacterial physiology, Dr. Neidhardt has received the Eli Lilly Award in Bacteriology and Immunology (1966) and the Alexander von Humboldt Senior U.S. Scientist Award (1979) from the Federal Republic of Germany. He served as President of the American Society of Microbiology in 1981–1982. His research has concerned the regulation of metabolism and gene expression as a function of growth rate, and mechanisms by which multigene networks are regulated in bacteria. Dr. Neidhardt has published six textbooks and research treatises in bacterial physiology.

John L. Ingraham is Professor of Microbiology at the University of California, Davis. He received his Ph.D. in 1951 from the University of California, Berkeley, under the joint direction of Roger Y. Stanier and Ralph Emerson. After employment with the DuPont Company and the U.S. Department of Agriculture, he joined the staff of the University of California, Davis, where he has been ever since. Dr. Ingraham has studied the physiology and genetics of the malo-lactic fermentation, the biosynthesis of fusel oil, the factors precluding growth of bacteria at low temperature, pyrimidine nucleotide metabolism, and, currently, denitrification. He is the editor of *Microbiological Reviews* and has coauthored three textbooks, *The Microbial World, Introduction to the Microbial World,* and *Growth of the Bacterial Cell.*

Moselio Schaechter is Distinguished Professor and Chair of the Department of Molecular Biology and Microbiology at Tufts University. He received his Ph.D. in 1954 from the University of Pennsylvania and carried out postdoctoral research with Ole Maaløe in Copenhagen. Dr. Schaechter has been the recipient of an NIH Research Career Development Award (1961–1968) and is a member of the NIH Recombinant DNA Advisory Committee. He served as President of the American Society of Microbiology in 1985–1986. Dr. Schaechter's research concerns chromosome segregation in bacteria and the regulation of bacterial growth. He has edited or coauthored three previous books, including the introductory medical microbiology textbook *Mechanisms of Microbial Disease* (1989).

ABOUT THE BOOK

This book was typeset at Camden Type 'n Graphics, Camden, Maine on Linotron 202 using Trump Mediaeval and Zapf Book typefaces. Joseph Vesely designed the book and coordinated production. The J/B Woolsey studio provided art direction and produced the art. Jodi Simpson was the book's copy editor. Craig Malone designed the cover, which was then printed at New England Book Components, Inc. The book was manufactured at Hamilton Printing Company.